INTRODUCTION TO MODERN PHYSICS

Some Physical Constants and Their Values

Quantity	*Symbol*	*Value*
Angstrom	Å	10^{-10} m
Atomic mass unit	u	1.66×10^{-27} kg
Avogadro number	N_0	6.023×10^{26} particles/kcal
Bohr magneton	$\mu_B \left(=\frac{e\hbar}{2m}\right)$	9.3×10^{-24} J/weber-m^{-2}
Bohr radius	$a_0 \left(=\frac{4\pi\epsilon_0 \hbar^2}{me^2}\right)$	5.3×10^{-11} m
Boltzmann constant	k	1.38×10^{-23} J/K
Electron charge	e	1.6×10^{-19} C
Electron Compton wavelength	$\lambda_c \left(=\frac{h}{mc}\right)$	2.4×10^{-12} m
Electron mass	m	9.1×10^{-31} kg
Electron rest mass energy	mc^2	0.51 MeV
Electron volt	eV	1.6×10^{-19} J
Fermi	F	10^{-15} m
Fine structure constant	$\alpha \left(=\frac{1}{4\pi\epsilon_0}\frac{e^2}{c\hbar}\right)$	$\frac{1}{137}$
Ionization energy of hydrogen		13.6 eV
Neutron rest mass energy	$m_n c^2$	939.6 MeV
Planck's constant	h	6.6×10^{-34} J-sec
Proton rest mass energy	$m_p c^2$	938.3 MeV
Rydberg constant	R	1.097×10^7 m^{-1}
Speed of light	c	2.998×10^8 m/sec

Introduction to MODERN PHYSICS

H. S. Mani
Professor of Physics
Indian Institute of Technology
Kanpur

G. K. Mehta
Director
Nuclear Science Centre
New Delhi

AFFILIATED EAST-WEST PRESS PVT LTD
NEW DELHI

TO OUR PARENTS

2022

Published by Affiliated East-West Press Private Limited
G-1/16, Ansari Road, Draya Ganj, New Delhi 110002
e-mail: aewp.pub@gmail.com

ISBN 978-81-8509-573-8

Contents

Preface

Our experience has shown that the area of modern physics, being an exciting one, generates a great deal of interest among the undergraduate science and engineering students. We therefore felt that a book emphasizing the fundamental principles, so essential for understanding the modern developments, and the applications of these principles would be welcomed by such students.

The fundamental principles are enunciated in the chapters on quantum mechanics [second and (part of) third], equilibrium statistical mechanics (sixth), and special theory of relativity (eighth). The chapters on atomic physics (fourth) and molecular physics (fifth) use the concepts developed in the chapters on quantum mechanics. The chapter on solid state physics (seventh) utilizes the ideas of equilibrium statistical mechanics, in addition to those of quantum mechanics.

The chapters on nuclear physics (ninth and tenth) require only the famous mass-energy relationship of Einstein's theory of relativity (Chapter 8). Since, presumably, the student would have been exposed to this relationship in high school itself, he can omit Chapter 8 and yet conveniently follow Chapters 9 and 10.

The chapters on particle physics (eleventh) and cosmology (twelfth) bring into play all the concepts elaborated in Chapters 2–10, and give a glimpse of the two most exciting fields of research today. We have included the two topics in the hope that these would become a part of the undergraduate curriculum where it is not so.

This study contains a large number of worked examples; they are an integral part of the text. The importance of estimating the order of magnitude of a physical quantity cannot be overemphasized and several examples are of this nature. Other examples acquaint the student with the techniques required for tackling a problem and illustrate the basic principles involved. The problems at the end of each chapter and questions, interspersed with the text, help the student in a self-assessment of his understanding.

The book has evolved from the lectures delivered for over a decade to the undergraduate students at the Indian Institute of Technology, Kanpur. It is intended as a two-semester text for a student majoring in a science discipline and as a one-semester text for the one majoring in an engineering discipline. For the latter category, a suggested coverage would be Chapters 2, 3, 4, 7, 9, and 10.

We could complete the manuscript mainly because of the encouragement we received from the Educational Development Centre and the Quality Improvement Programme at the Indian Institute of Technology, Kanpur. The financial assistance granted by the Centre enabled us to replicate our notes for distribution to our students over the years. The comments of and our numerous dialogues with these students helped us a great deal in improving the presentation. We are indebted to them, to Professor T. V. Ramakrishnan for his initial participation in drafting Chapter 7, and to Drs. K. Banerjee, D. Chowdhury, A. Ghosh, A. Joshipura, Janaki Krishnamoorthy, A. Mukherjee, R. K. Popli, R. Ramachandran, R. Shankar,

M. M. Sharma, Mitra Sharma, R. Singh, V. A. Singh, A. K. Sinha, and Amrita Tripathi for reading parts of the manuscript. We are thankful to Dr. M. Noman for checking the solutions to most problems. We should also like to warmly thank Mr. L. S. Bajpai, Mr. U. S. Mishra, and Mr. R. N. Srivastava for typing the manuscript, Mr. B. L. Arora, Mr. B. K. Jain, and Mr. B. N. Srivastava for drawing the illustrations, and Mr. H. K. Panda and Mr. L. S. Rathaur for assistance in preparing the notes. Finally, the encouragement provided and the patience shown by our families throughout our writing are gratefully acknowledged.

H. S. MANI
G. K. MEHTA

1 Introduction

The basic philosophy of physics is to observe physical phenomena, conduct experiments accurately to learn as much as is possible from the measurements thus carried out, critically analyze the results so obtained, and, finally, formulate *laws* in terms of which the results can be understood quantitatively. The objective of such an exercise is to devise concepts and principles that can help us understand all the natural phenomena. The physical laws are constructions of the human mind and are based on the available knowledge. These laws establish definite relations among the various quantities observed. Thus, several different phenomena, when analyzed critically, often lead to the same law. The more general the applicability of a law, the more basic the law. The ultimate test for a physical law to be good is the accuracy with which it can predict the results of different experimental observations.

In the later part of the nineteenth century, it seemed that the construction of the basic structure of physics had more or less been completed. Thus, for example, the laws of classical mechanics (i.e., motion) and gravitation as developed by Isaac Newton and Galileo Galilei had provided a solid foundation for dynamics. Also, the complicated motions of the solar system could be predicted, using Newton's laws of motion, with remarkable accuracy in terms of the gravitational attraction between the sun and the planets. These laws later enabled the scientists to discover many new planets and formulate physical mechanisms for studying the possible origins of stars and galaxies. Besides its astronomical applications, classical mechanics has played a crucial role in the dynamical description of moving objects. For example, all the mechanical machines are based on the principles of classical mechanics.

The work of C. A. Coulomb, A. M. Ampère, and M. Faraday led to the laws of electrical and magnetic phenomena. These laws were coherently gathered together and modified to give a unified picture of the laws in the form of electromagnetic theory by J. C. Maxwell. This theory offered a complete account of all the optical phenomena resulting from the description it gave of light as an electromagnetic wave.

The laws of thermodynamics as developed by R. J. E. Clausius and W. T. B. Kelvin gave a satisfactory account of transfer of heat between two bodies. A deeper understanding of the concept of heat and temperature was possible only after the formulation of the concepts of atoms and molecules. The kinetic theory of gases and its success confirmed the atomic hypothesis. Subsequently, the foregoing developments were formalized by L. Boltzmann and J. W. Gibbs in statistical mechanics. This area of physics now forms the basis for understanding the many-particle systems.

The years 1895 to 1905 saw the discovery of a whole series of new phenomena mostly concerned with the physics of atomic particles. The atomic dimension (10^{-10} m) is much too small for direct visual observation even when the most powerful microscope is used. However, it is still possible to study the atomic particles if carefully designed experiments are performed and the results thus obtained are analyzed. It was found that Newton's laws are not adequate to understand such experimental results. The attempts to explain these results led to the development of the laws of quantum mechanics. These laws give a very good account of what goes on inside the atoms and nuclei. For simple atoms, the predictions based on the principles of quantum mechanics agree with the experimental results. When applied to the macroscopic phenomena, the principles of quantum mechanics give results which turn out to be the same as those obtained using classical mechanics. Thus, the laws of quantum mechanics provide the correct picture of all microscopic as well as large objects, but the laws of classical mechanics are valid only in the macroscopic limit.

It was also discovered that Newton's laws are inadequate when applied for studying the particles moving with a velocity comparable to that of light. Instead, the laws of relativity as developed by A. Einstein lead to the correct way to deal with particles of all velocities; for particles with a low velocity, these laws yield the same results as those arrived at by using Newton's laws.

The development of quantum mechanics and relativity took place in this century. The fields of study it covered are atoms, molecules, all forms of matter, nuclei, and subnuclear particles. At about the same time, the laws of quantum statistical mechanics, which provide the method to understand the many-particle systems, were developed.

In what follows, we shall introduce the basic ideas of quantum mechanics, equilibrium statistical mechanics, and special theory of relativity, and illustrate their application to the various areas of modern physics.

We have seen that the laws of physics deal with phenomena involving large objects, such as planets, and small objects, e.g., atoms (see Table 1.1). Similarly, they deal with a tremendous range of time scale, for example, the time light from stars takes (millions of years) to reach us and the time for which some of the particles exist (10^{-24} yr).

TABLE 1.1 Physical dimensions of some objects

Object	Dimension
Macroscopic	
Universe	10^{26} m
Star	10^{10} m
Living being	1 m
Microscopic	
Basic cell	10^{-5} m
Atom	10^{-10} m
Nucleus	10^{-14} m
Elementary particle	10^{-15} m or less

The laws of quantum mechanics become relevant in the microscopic world. Let us get some historical perspective of the way the investigations of physical phenomena entered the microscopic world. The first scientific theory, based on experimental grounds, that everything is made of tiny, indivisible atoms was proposed by J. Dalton at the beginning of the nineteenth century. After considerable arguments among the then leading physicists whether atoms really exist, it was established that the atom is the basic constituent of matter. Towards the end of the nineteenth century, the electron was discovered and shown to be a constituent of atoms and the carrier of electrical current in conductors. The electron is a particle with a mass of about 9.1×10^{-31} kg and a charge equal to -1.6×10^{-19} C. Then, in 1911, E. Rutherford demonstrated that an atom consists of a nucleus surrounded by a cloud of electrons. This was a major discovery in the microscopic world. The tiny, indivisible atom which was supposed to be the basic constituent of matter was observed to have a structure. The nucleus was found to be extremely small ($\sim 10^{-14}$ m) in comparison to the atom which has a dimension 10^{-10} m. However, almost the entire mass of the atom was noted to be concentrated in the nucleus. We can appreciate the smallness of the nucleus by considering the fact that if we could pack just the nuclei of all the atoms of the earth they would be accommodated in a hockey field.

The nucleus of the lightest atom (hydrogen) is called the proton. It has a mass $\sim 1.67 \times 10^{-27}$ kg and a charge equal in magnitude but opposite in sign to that of an electron. At first, it was believed that all nuclei consist of protons and electrons, but this model yielded serious inconsistencies and led to violations of the well established conservation laws. A new particle, namely, neutron, was discovered by J. Chadwick in 1932. The neutron has a mass almost equal to that of a proton, but has no electrical charge. The discovery of the neutron finally solved the riddle about the constituents of the nucleus, and it was shown that neutrons exist in the nucleus of every element except hydrogen, providing the nucleus about half or more of its mass, the rest of its mass and its entire charge being supplied by protons. At this stage, it was believed that the proton and neutron along with the electron formed a set of fundamental particles of nature. The physicists then began to probe the interior of the nucleus and found that the nuclear properties could not be understood only in terms of its basic constituents, i.e., neutrons and protons. This is because when a nucleus is bombarded with high energy particles, a variety of short-lived particles, which do not exist within the atoms of ordinary matter, are created. With the development of large atom smashers (i.e., particle accelerators), a number of new particles have been discovered, e.g., various kinds of mesons, neutrinos, and strange particles and their antiparticles. The basic question, namely, what really are the elementary particles, is still wide open and is currently occupying the attention of the physicists.

An important classification of the various phenomena in physics is based on the nature of the interactions causing them. The motion of planets around the sun is due to the gravitational attraction between the sun and the planets. The forces in atoms and molecules are electrical and magnetic in nature. These two, i.e., gravitational and electromagnetic, were the only kinds of interactions known in the last century. However, after the discovery of radioactivity, when the nucleus was explored for the first time, two new types of interactions were discovered, namely, (1) the strong interaction, which keeps the constituents of the nucleus together and (2) the weak interaction, which causes beta-decay of the nucleus. Thus, the following four types of fundamental interactions are known at present:

(1) *Gravitational interaction* It is the familiar interaction between two bodies due to their masses. The form of this force is given by Newton's law of gravitation.

(2) *Electromagnetic interaction* The force

(a) between the nucleus and the electrons surrounding it,

(b) between the atoms in a molecule, and

(c) holding a solid together

is electromagnetic in nature. It is due to the charges on the particles and the latter's motion. The general law of this interaction is quantitatively expressed in terms of the Lorentz force formula.

(3) *Strong interaction* It occurs, for example, between a neutron and a proton, a proton and a proton, and a neutron and a neutron. This force is responsible for the binding of a nucleus. Its exact nature is not yet known.

(4) *Weak interaction* This occurs, for instance, between electrons, protons, and neutrons. It is weaker than the electromagnetic and strong interactions, and is responsible for the spontaneous decay of a large number of nuclei and particles.

In any physical phenomenon, all these interactions are, in general, present. However, not all of them play an equal role in a given situation because of the differences in their nature. For example, between the planets and the sun the interaction is gravitational and all the other interactions are negligible. Further, in an atom, the interaction due to the charge is dominant. In order to get a picture of the strength and the range of these interactions, we compare them

TABLE 1.2 Strength and range of various interactions

Interaction	Strength	Range
Strong	~ 10	10^{-14} m
Electromagnetic	$\sim 10^{-2}$	∞
Weak	$\sim 10^{-12}$	10^{-18} m
Gravitational	$\sim 10^{-40}$	∞

in Table 1.2. As can be noted, the gravitational interaction is the weakest, but has a long range. Also, both the strong and weak interactions have an extremely short range, and it is for this reason that we do not feel their presence in our daily experience.

In the atomic, molecular, and biological phenomena, the electromagnetic forces play a crucial role. In a nucleus, on the other hand, it is the strong interaction which is vital.

A natural question that arises is: Are the four interactions related to one another or completely independent? This question poses one of the most outstanding problems in physics today, and many ideas to unify these forces are being developed. Before we can understand these exciting ideas, it is necessary that we know of the advances in quantum mechanics and relativity.

2

Quantum Mechanics

2.1 INTRODUCTION

The most important development in physics in the twentieth century is that of quantum mechanics. Most advances that have since taken place in solid state, atomic, and nuclear physics are based on the principles of quantum mechanics. This chapter is devoted to a study of these principles.

By the end of the last century, many physical phenomena had been understood on the basis of some fundamental laws. These laws, now called the laws of classical physics, are those of mechanics and gravitation (as stated by Galileo and Newton), electromagnetism (as formulated by Coulomb, Ampère, Faraday, and Maxwell), and thermodynamics and statistical mechanics (as proposed by Clausius, Kelvin, Boltzmann, Maxwell, and Gibbs).

The motion of planets around the sun, for example, is completely expressed in terms of Newton's laws of mechanics and gravitation. Again, the phenomenon of interference of light rays is understood using the laws of electromagnetism. In fact, at one stage it was believed that all the physical phenomena could be explained in terms of the laws of mechanics, gravitation, and electromagnetism. However, these laws could not explain many experimental observations made towards the end of the last century. As noted in Chapter 1, the attempts to understand these observations led to the laws of quantum mechanics.

In our approach to the study of quantum mechanics, we shall not follow the historical sequence of events. Instead, we shall discuss some key phenomena which will lead us to the laws of quantum mechanics. To begin with, we will consider one such phenomenon, namely, the photoelectric effect. This is totally inexplicable by classical physics.

2.2 PHOTOELECTRIC EFFECT

THE EXPERIMENT

When light falls on the surface of a substance, electrons are emitted. This phenomenon is called the photoelectric effect.

The photoelectric effect can be demonstrated by shining light on a surface of a plate in an evacuated chamber (see Fig. 2.1). This chamber also has another plate. When the two plates are connected by a wire through an ammeter, the flow of current can be measured. This current stops flowing if the light supply is cut off. The current flow here implies the flow of charged particles from one plate to the other. The study of these charged particles has established that they are electrons, called photoelectrons.

> **QUESTION 2.1 What is the direction of the electric current between the two plates?**

QUESTION 2.2 Why is it necessary to have an evacuated chamber?

When conducting the experiment, a potential difference can be applied between the two plates through a voltage source. The current can then be measured for various values of the potential difference, both positive (i.e., the collector being at a potential higher than that of the emitter on which light falls) and negative. The typical curves between V and I are shown in Fig. 2.2. It

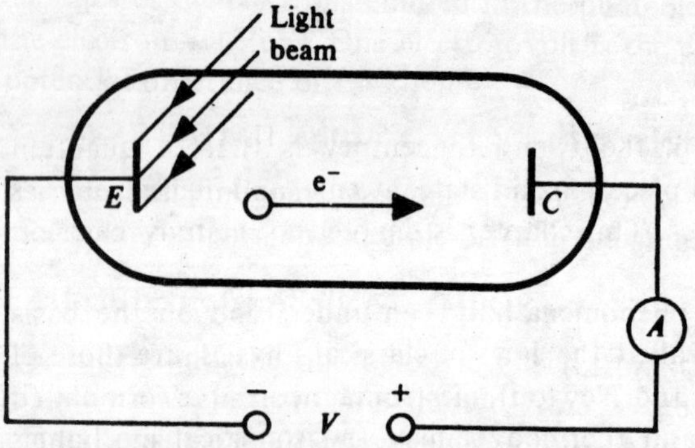

Fig. 2.1 Schematic diagram of apparatus used for studying photoelectric effect. E, emitter; C, collector; V, potential difference; A, ammeter.

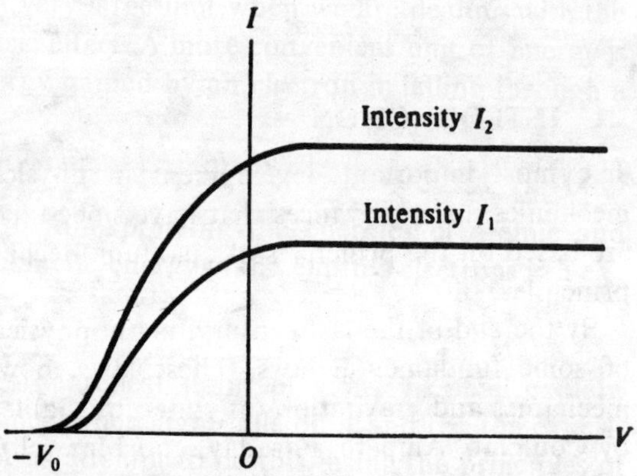

Fig. 2.2 Photoelectric current I as function of potential difference V when light frequency is fixed. The two curves are for intensities I_1 and I_2 ($I_2 > I_1$). Stopping potential V_0 is positive.

should be noted that the current would continue to flow even when the potential difference is negative, that is, when the electric field is directed from the emitter to the collector. In such a situation, the electrons are repelled by the collector and many of these return to the emitter. If this negative potential difference is increased in magnitude, the electric field would increase, and so most of the electrons emitted would go back to the emitter. As such, very few electrons would reach the collector. Therefore, the effect of increasing the magnitude of the negative potential difference is that the flow of current is reduced. If we continue to increase the negative potential difference, then, at some stage, none of the electrons, not even the fastest ones, would reach the collector. As a result, the current will stop flowing. The magnitude of the potential difference at which the current just becomes zero is called the *stopping potential.* A further increase in the magnitude of the potential difference would, of course, give zero current. For most substances, the stopping potential is of the order of a few volts.

The detailed experimental study of the photoelectric effect has revealed several facts. Of these, the important ones are as follows:

(1) The photoelectric current starts flowing as soon as the light supply is turned on. In other words, there is virtually no time lag between the shining of light and the emission of electrons. More precise measurements indicate that the time lag is less than 3×10^{-9} sec.

(2) If the light falling on the emitter has a fixed frequency, then the stopping potential V_0 does not depend on the intensity of the light.

(3) The stopping potential V_0, on the other hand, does depend on the frequency (and hence on the wavelength) of the light used. The relation between V_0 and the frequency ν is given by

$$V_0 = \alpha\nu + \beta, \tag{2.1}$$

where α and β are constants. The constant β depends on the photoelectric substance, but the constant α does not. The linear relationship between the stopping potential and the frequency is illustrated in Fig. 2.3 for two different substances.

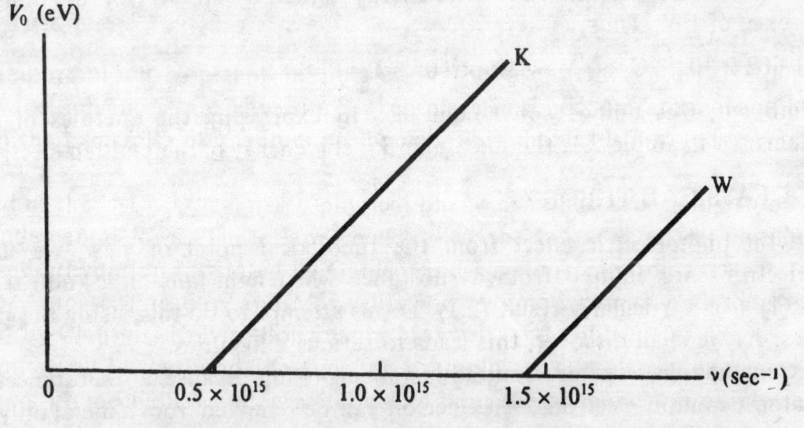

Fig. 2.3 Dependence of stopping potential V_0 on frequency ν of light source for potassium and tungsten.

(4) When the frequency of the light used is held fixed, the current is directly proportional to the intensity of the light.

Physical Significance of Stopping Potential

The electrons that are emitted from the emitter surface do not all have the same energy. This is primarily because the electrons in a solid do not have the same energy (see Section 7.2). The other reason of course is that the electrons, since they are emitted from varying depths of the emitter, lose different amounts of energy before emanating from the emitter surface. When the potential between the emitter and the collector is equal to the stopping potential, then even the most energetic electron would just fail to reach the collector. If v_m is the maximum velocity of the emitted electrons at the emitter, then their kinetic energy at the emitter is $(1/2)mv_m^2$, where m is the mass of an electron. As an electron travels from the emitter to the collector, its kinetic energy decreases and its potential energy increases such that its total energy (i.e., kinetic + potential) is conserved. For those electrons that just manage to reach the collector, the velocity, and hence the kinetic energy at the collector, becomes zero. Thus,

$$\tfrac{1}{2}mv_m^2 = eV_0, \tag{2.2}$$

where e is the magnitude of the electronic charge.

EXAMPLE 2.1 What would be the maximum kinetic energy of the emitted electrons when no potential is applied between the two plates and the stopping potential is 3 V?

The relation between the maximum kinetic energy and the stopping potential is given by Eq. (2.2). Thus, the maximum kinetic energy of the electrons emitted would be

$$\tfrac{1}{2}mv_{\rm m}^2 = eV_0 = 1.602 \times 10^{-19}\ {\rm C} \times 3\ {\rm V} \sim 4.8 \times 10^{-19}\ {\rm J}.$$

Electron Volt

From Example 2.1, it is clear that the joule is a very large unit when we are dealing with the energies of the electrons emitted in the photoelectric effect. A more convenient unit of energy is the electron volt (eV). One electron volt is the energy gained by an electron in falling through a potential difference of 1 V. Thus,

$$1\ {\rm eV} = 1.602 \times 10^{-19}\ {\rm C} \times 1\ {\rm V} \sim 1.6 \times 10^{-19}\ {\rm J}. \tag{2.3}$$

As we shall find out, this unit is convenient also in expressing the energies of atomic and molecular systems. In Example 2.1, the maximum kinetic energy of the emitted electrons is 3 eV.

FAILURE OF CLASSICAL THEORY

To understand the photoelectric effect from the theoretical point of view, we shall explain (a) why the electrons are emitted from a substance when light falls on it and (b) the experimental results (1)–(4), particularly result (2.1). Let us attempt to do this, using the principles of classical physics. As we shall discover, this leads to serious difficulties.

The observation that electrons are emitted is not surprising because all substances are made of atoms and atoms contain electrons. An electron can be removed from a metal only if a certain amount of energy is supplied to it. A metal, being electrically neutral, has both negative and positive charges. Thus, when an electron has to be removed from a metal, it needs to be pulled away from the positive charge and this requires work. Therefore, unless this amount of energy is supplied, the electron will not leave the metal. In the photoelectric effect, the external source of energy is the light beam.

According to classical physics, the transfer of energy from the electromagnetic field to the electron should take a long time, of the order of seconds or more (see Example 2.2). This means that the electron will take some time to be emitted, but this directly contradicts the experimental observation.

EXAMPLE 2.2 Consider a 100-W bulb emitting light in all directions. Suppose that a metallic sodium surface is kept at a distance of 1 m from the bulb. Estimate the time needed by an electron in a Na atom to receive an energy of 1 eV. Assume that all the energy is absorbed by the top layer of the surface and all the energy given to the Na atom is taken by one electron.

Let us assume that the light from the bulb is spreading uniformly in all directions. The energy crossing per second a spherical surface of radius r, with the bulb at the centre, is then 100 J. For $r = 1$ m, the light intensity I will be

$$I = 100/(4\pi) \sim 8\ {\rm J/m^2\text{-}sec}.$$

The number of sodium atoms N per unit area can be estimated approximately because the radius of an atom is about 10^{-10} m. Thus,

$$N = \frac{1}{\pi \times (\text{radius of Na atom})^2} \sim \frac{1}{\pi \times (10^{-10})^2} \sim 3 \times 10^{19}\ \text{atoms/m}^2.$$

The power absorbed by one atom is then approximately

$$\frac{8\ \text{J/m}^2\text{-sec}}{3 \times 10^{19}\ \text{atoms/m}^2} \sim 2.6 \times 10^{-19}\ \text{J/sec} \sim 1.6\ \text{eV/sec}.$$

The time needed by the electron to gain an energy of 1 eV will approximately be

$$\frac{1\ \text{eV}}{1.6\ \text{eV/sec}} \sim 0.6\ \text{sec}.$$

The assumption we have made is not realistic because the energy absorbed by an atom cannot all get transferred to a single electron. As such, the actual energy acquired by an electron will be much less. This means that the actual time required will be much more than that we have estimated.

The maximum energy of an emitted electron (eV_0) is observed to be independent of the light intensity as long as the wavelength of the light beam is kept constant. This means that the maximum energy of the electron emitted cannot be increased by just increasing the intensity. If the classical ideas of energy transfer were correct, then a higher intensity light beam should contribute additional energy, and so a more energetic electron should be emitted.

Result (2.1) clearly cannot be understood in terms of the classical theory because this theory predicts that the maximum energy of the emitted electron, and hence the stopping potential, depends only on the light intensity. In other words, the frequency of the electromagnetic wave finds no significance in the classical explanation.

QUESTION 2.3 An increase in the intensity of a light beam does not change the maximum energy of an electron. What happens then to the extra energy given out by the beam?

The experimental result (4) is obvious because, with the increase in light intensity, the energy absorbed by the emitter surface is more, leading to an increase in the emission of the number of electrons. This in turn increases the photoelectric current.

EINSTEIN'S QUANTUM THEORY

The photoelectric effect was explained by Einstein in 1905. He suggested that a light beam should not be considered as a wave supplying energy to a substance in a continuous manner. Instead, it should be thought of as consisting of particles of light, called *photons*, each photon having a definite energy and a velocity equal to that of light. He made a far-reaching assumption relating the energy E of the photon and the frequency ν of the light beam as

$$E = h\nu, \tag{2.4}$$

where h is known as Planck's constant. The value of Planck's constant is 6.626×10^{-34} J-sec. M. Planck had introduced this constant earlier in his theory of black-body radiation. We shall discuss this in Chapter 6. A monochromatic light beam contains a number of photons, each with an energy $h\nu$. It should be noted that this is a drastically different view of the light beam. Let us compare this view with the classical concept of light. In the latter, the energy of a light beam can vary continuously. In the former, a consequence of Einstein's assumption is that the energy of the light beam having n photons is $nh\nu$. As n can assume only the integral values

1, 2, . . . , n, the energy of the beam can have only the corresponding values $h\nu$, $2h\nu$, . . . , $nh\nu$. The question that now arises is: If a light beam can have only discrete values of energy, i.e., $h\nu$, $2h\nu$, . . . , $nh\nu$, then how are we able to continuously vary its energy? The answer to this will be clear from the example that follows.

EXAMPLE 2.3 If one photon is added per second to a light beam of intensity 10^{-7} J/sec, what will be the fractional change in the light intensity? Assume $\lambda = 6000$ Å.

Following Einstein's theory, the energy of a photon, using Eq. (2.4), is

$$E = h\nu = \frac{hc}{\lambda} = \frac{6.6 \times 10^{-34} \text{ J-sec} \times 3 \times 10^{8} \text{ m/sec}}{6 \times 10^{-7} \text{ m}}$$

$$\sim 3.3 \times 10^{-19} \text{ J}.$$

The number of photons N per second in the light beam is

$$N = \frac{10^{-7}}{3.3 \times 10^{-19}} \sim 0.3 \times 10^{12}.$$

Thus, if we add one photon per second, the fractional change in the number of photons or the fractional change in the intensity will approximately be 3.3×10^{-12} which is a very small number. It should be noted that this is so because Planck's constant has a small value. For all practical purposes, such a change can be considered continuous.

QUESTION 2.4 If Planck's constant were 100 J-sec, what effect will you observe when you see a 100-W bulb?

If the intensity of the light beam is increased (without changing its frequency), another consequence of Einstein's assumption is that the number of photons increases, but the energy of each photon remains unchanged. The energy of a single photon is always given by Eq. (2.4).

We now give Einstein's explanation of the photoelectric effect. Assume that an electron in an atom of a metal is emitted when a photon collides with it. The electron absorbs an energy $h\nu$ from the photon and can leave the metal. The collision is instantaneous so that there is no time lag between the emission of the electron and its collision with the photon. Not all the electrons hit by photons leave the metal. Those that are not emitted may just move further inside the metal. So far, we have only shown how an electron acquires energy but have not discussed the dynamics of electron emission. Let us now see how this picture leads to the experimental results (1)–(4).

The experimental fact (1)—negligible time lag—has essentially been assumed in this picture. That is, when two particles collide, the energy transfer takes place in a very short time.

The experimental observation (2) means that the photoelectrons have a maximum energy eV_0 independent of the intensity. This follows from Einstein's quantum theory because each electron absorbs an energy $h\nu$ from a single photon. If the intensity increases, the number of photons also increases, and hence the probability of collision of photons with electrons goes up. However, in each collision, an electron acquires only an energy $h\nu$. We might wonder why one electron cannot absorb an energy equal to $2h\nu$ by simultaneously colliding with two photons. It can be shown that the probability of such an occurrence is very small. Not all the energy $h\nu$ supplied to the electron appears as kinetic energy. This is because the electron spends some energy for getting itself released from the metal. When an electron moves away from the metal, a positive charge is left in the metal because the metal was initially neutral. This positive charge attracts the electron and, unless enough energy is supplied to the electron to overcome this attraction, the electron will remain in the metal. The minimum energy required to remove an electron from a substance is referred to as the *work function* of the substance. If ϕ is the work function, then the maximum energy E_m of the electron outside the substance is

$$E_m = \tfrac{1}{2}mv_m^2 = h\nu - \phi. \tag{2.5}$$

Combining Eqs. (2.2) and (2.5), we get

$$V_0 = h\nu/e - \phi/e. \tag{2.6}$$

On comparing this equation with Eq. (2.1), we find that the experimental result (3) gets explained with $\alpha = h/e$ and $\beta = -\phi/e$. We see that α depends only on two fundamental constants, namely, Planck's constant and the charge of the electron, and not on the nature of the emitter. On the other hand, β contains the work function which varies from one substance to another. The experimental value of α is in good agreement with the experimental value of h/e.

Result (4), i.e., the photoelectric current is proportional to the intensity of light, can be understood by noting that an increase in the intensity leads to an increase in the number of photons in direct proportion. With more photons, the number of collisions rises and the number of electrons emitted correspondingly goes up. The current, being directly proportional to the number of electrons emitted, is hence proportional to the intensity. Thus, Einstein's quantum theory gives a quantitative explanation of the photoelectric effect.

EXAMPLE 2.4 A light beam of wavelength $\lambda = 4000$ Å falls on a metallic surface used in an experiment to study the photoelectric effect. If the stopping voltage is 1.5 V, calculate (a) the work function of the surface and (b) the maximum wavelength of light that will cause the photoelectric emission.

(a) The maximum energy of an electron is $\tfrac{1}{2}mv_m^2 = 1.5$ eV. The energy of a photon is

$$\frac{hc}{\lambda} = \frac{6.6 \times 10^{-34} \times 3 \times 10^8}{4 \times 10^{-7}}\ \text{J} = \frac{6.6 \times 10^{-34} \times 3 \times 10^8}{4 \times 10^{-7} \times 1.6 \times 10^{-19}}\ \text{eV} \sim 3.1\ \text{eV}.$$

Then, from Eq. (2.5), we have 1.5 eV = 3.1 − ϕ or the work function $\phi = 1.6$ eV.

(b) The energy of a photon has to be more than 1.6 eV in order that the electron absorbing the photon can overcome the force binding it to the metallic surface. The wavelength λ of a photon of energy 1.6 eV is

$$\lambda = \frac{hc}{E} = \frac{6.6 \times 10^{-34} \times 3 \times 10^8}{1.6 \times 1.6 \times 10^{-19}} \sim 7734\ \text{Å}.$$

A light beam of a wavelength larger (i.e., lower frequency) than this will have photons with energy less than 1.6 eV. Therefore, a light beam of a wavelength higher than 7734 Å cannot make the surface emit photoelectrons, irrespective of its intensity. So, this is the maximum wavelength that will cause the photoelectric emission.

QUESTION 2.5 The light from a firefly (*jugnu*) affects a photographic plate, but a powerful radiowave does not! Explain.

2.3 THE de BROGLIE RELATION

Einstein's quantum theory of light is drastically different from the wave theory. We recall that the wave theory of light was developed to understand the phenomena of interference and diffraction. If we completely abandon the wave theory in favour of Einstein's idea that light consists of quantas (i.e., particles of definite energy), we will have difficulties in explaining the interference and diffraction phenomena. It appears that we require both the physical pictures to explain these phenomena. It is paradoxical that two completely different theories of light are needed to understand the results of experiments involving light. In other words, light seems to behave like a particle in certain situations and like a wave in certain others.

L. de Broglie conjectured that it is not just light but all matter that has a wave- as well as a particle-like property. He asserted that the particle-wave nature is a general property of all objects. In other words, we need both the wave-like and particle-like pictures to interpret the results of the experiments involving not only light but also all matter. In particular, he suggested that the electron, which is normally considered as a particle, also exhibits wave-like properties. He postulated that if an electron has a momentum p, then the wavelength λ associated with it is

$$\lambda = h/p, \tag{2.7}$$

where h is Planck's constant. Equation (2.7), called the de Broglie relation, forms the basis for the mathematical development of quantum mechanics. In the case of light, Einstein's equation $E = h\nu$ can be shown to be the same as the de Broglie relation. To do this, we recall that the momentum p of a light wave is related to its energy E [see Eq. (A1.22) in Appendix 1] by

$$p = E/c. \tag{2.8}$$

Then, from Einstein's relation, i.e., Eq. (2.4), we get $p = h\nu/c = h/\lambda$. The de Broglie relation is more general. It is true not only for light waves but for all matter.

EXAMPLE 2.5 Calculate the de Broglie wavelength associated with an electron having a kinetic energy of 10 eV.

To calculate the de Broglie wavelength, we need to know the momentum p of the electron. Since $E = p^2/(2m)$, we have $p = (2mE)^{1/2}$. The de Broglie wavelength λ then is

$$\lambda = \frac{h}{p} = \frac{h}{(2mE)^{1/2}} = \frac{6.6 \times 10^{-34}}{(2 \times 9.1 \times 10^{-31} \times 10 \times 1.6 \times 10^{-19})^{1/2}} \sim 3.8 \times 10^{-10} \text{ m.}$$

The wavelength associated with the electron is 3.8 Å.

In order to see the effects of wave-like nature of light, an experiment for observing the interference or diffraction phenomenon has to be performed. These effects appear when the slit widths or grating spacing used for studying diffraction are comparable to the wavelength. If the width of each slit is much larger than the wavelength, the diffraction effects become small and are difficult to observe. In Example 2.5, we found that the wavelength associated with an electron of energy 10 eV is about 4 Å. Therefore, to observe the wave-like nature of an electron, we would need slits having a width of a few angstroms. But it is not possible to make such small slits by mechanical means. The solution to this problem was provided by the technique developed for studying X-ray diffraction.

X-RAY DIFFRACTION

The X-rays are electromagnetic waves with a wavelength of the order of a few angstroms. The diffraction effects of X-rays were studied by using crystals. A crystal is a solid where the atoms are arranged in a regular pattern. The atoms in a simple cubic crystal are located on the corners of the cube as shown in Fig. 2.4. In a cubic structure, the distance between any two neighbouring atoms is the same and is of the order of a few angstroms.

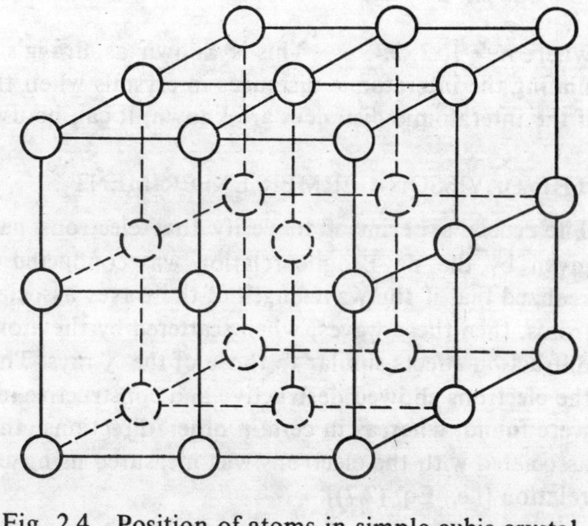

Fig. 2.4 Position of atoms in simple cubic crystal.

When an electromagnetic wave falls on a crystal, each atom of the crystal scatters the wave in all directions. As the atoms in a crystal are periodically arranged, the waves in certain directions interfere constructively, whereas in certain other directions they interfere destructively. Consider a beam of X-rays of wavelength λ falling on a cubic crystal as shown in Fig. 2.5.

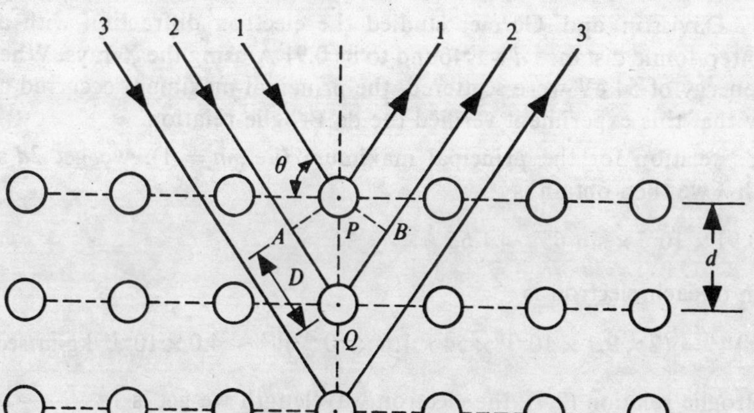

Fig. 2.5 X-ray scattering from crystal.

Rays 1, 2, 3 are scattered by the atoms at the points P, Q, R, respectively, in all possible directions. These scattered rays interfere constructively only when the path difference between any two consecutive rays is an integral multiple of the wavelength and the rays are parallel to each other. From the figure, the path difference between rays 1 and 2 is $AQ + QB = 2PQ \sin \theta = 2d \sin \theta$. Thus, the constructive interference takes place only when

$$2d \sin \theta = n\lambda, \tag{2.9}$$

where $n = 1, 2, 3, \ldots$. This is known as Bragg's law of X-ray diffraction, and is used for finding the interatomic distances in crystals when the wavelength is known. On the other hand, if the interatomic distances are known, it can be used for determining the wavelength.

THE DAVISSON-GERMER EXPERIMENT

The crucial experiment to verify that electrons have wave-like properties with a wavelength given by the de Broglie relation was conducted by C. J. Davisson and L. H. Germer. They realized that if the wavelength of the waves associated with electrons is of the order of angstroms, then these waves, when scattered by the atoms in a crystal, will interfere, and thus show diffraction effects similar to those of the X-rays. This was indeed found to be true. In particular, the electrons showed destructive and constructive interference. In certain directions, no electrons were found, whereas in certain other directions, the number was maximum. The wavelength associated with the electrons was measured using Bragg's law, and this confirmed the de Broglie relation [i.e., Eq. (2.7)].

The wave-like nature of neutrons has also been studied and again it has been found that neutrons obey the de Broglie relation. The wave property has now been established to be a universal attribute of matter.

EXAMPLE 2.6 Davisson and Germer studied the electron diffraction with a nickel crystal for which the interatomic distance d was found to be 0.91 Å using the X-rays. When the electrons with a kinetic energy of 54 eV were scattered, the principal maximum occurred at $\theta = 65°$ (see Fig. 2.5). Show that this experiment verified the de Broglie relation.

Using Bragg's relation for the principal maximum (i.e., $n = 1$), we get $2d \sin \theta = n\lambda = \lambda$. The wavelength λ we then obtain is

$$\lambda = 2 \times 0.91 \times 10^{-7} \times \sin 65° \sim 1.65 \text{ Å}.$$

The momentum of each electron is

$$p = (2mE)^{1/2} = (2 \times 9.1 \times 10^{-31} \times 54 \times 1.6 \times 10^{-19})^{1/2} \sim 4.0 \times 10^{-24} \text{ kg-m/sec}.$$

Using the de Broglie relation (2.7), the electron wavelength we get is

$$\lambda = \frac{6.6 \times 10^{-34}}{4.0 \times 10^{-24}} \sim 1.65 \text{ Å}.$$

We see that this wavelength agrees with that obtained in the diffraction experiment.

EXAMPLE 2.7 Why is the wave-like nature of matter not apparent in our daily experience?

Consider a stone of 0.05 kg being thrown with a velocity of 1 m/sec through a window. The de Broglie wavelength associated with the stone would be

$$h/(mv) = 6.6 \times 10^{-34}/(.05 \times 1) \sim 10^{-32} \text{ m}$$

which is very small as compared to the dimensions of a window. Thus, the diffraction effects would be negligible and the stone will move like a particle. Similarly, a macroscopic phenomenon involves a large momentum (of the order of kg-m/sec), implying that the de Broglie wavelength associated with a macroscopic object is very small. The wave aspect is apparent when we deal with a particle of a very small mass such as an electron and a proton.

QUESTION 2.6 Explain why no wave aspect of a particle (e.g., an electron) can be observed when h (Planck's constant) $\rightarrow 0$.

2.4 WAVE FUNCTION

COMPLEMENTARITY

The concept that light consists of particles (i.e., photons) is necessary to understand the photoelectric effect. On the other hand, the concept that light consists of waves is required to understand the interference and diffraction phenomena. Similarly, the concept that electrons, protons, neutrons, and other types of matter, which are normally considered as particles, have wave-like properties is needed to understand the interference effects they exhibit. These concepts are usually expressed by stating that *all matter has both wave- and particle-like property.*

The particle aspect of matter requires a particle to be 'localized'. A particle is an object occupying a certain definite (small) volume in space at any given time. Its position at this time can also be specified. On the other hand, a wave always extends over a certain region. Thus, the two views basically differ and it is this difference that causes difficulties when we attempt to interpret a given experimental observation in terms of both waves and particles. We use either the particle or wave aspect to understand such an observation in terms of the classical concepts. This was stated as a principle by N. Bohr and is referred to as the principle of complementarity. According to this principle, matter exhibits either the wave or particle aspect. In other words, in a given experiment if matter or radiation exhibits the particle character, then it is impossible to prove the wave character in the experiment, and conversely. The two concepts, i.e., the wave and particle, are complementary.

LOCALIZED WAVES, GROUP VELOCITY

The two seemingly different concepts, namely, of particles and waves, are not necessarily conflicting. We can understand the particle behaviour of matter in terms of 'localized waves' which exist only in a small region of space.

Consider a free particle of momentum p and energy E moving along the positive x-axis. Let us associate with the particle a wave packet, that is, a wave confined to a small region, moving with the same velocity as the particle. A snapshot of the wave packet at a given time is shown in Fig. 2.6. We see that this wave is quite different from a sinusoidal wave which extends over the entire region of space. A wave packet can be constructed by superposing several sinusoidal waves (see Appendix 2). This superposition gives rise to constructive interference of these waves in a small region. The maximum amplitude produced by such a constructive interference moves with a velocity called the group velocity rather than with the phase velocity, i.e., the velocity of an individual sinusoidal wave. The group velocity v_g is given [see Eq. (A2.9) in Appendix 2] by

$$v_g = d\omega/dk, \tag{2.10}$$

and the phase velocity v by $v = \omega/k$, where ω is the angular frequency and $k = 2\pi/\lambda$, λ being the wavelength. We now show that the group velocity of a wave packet is in fact the particle

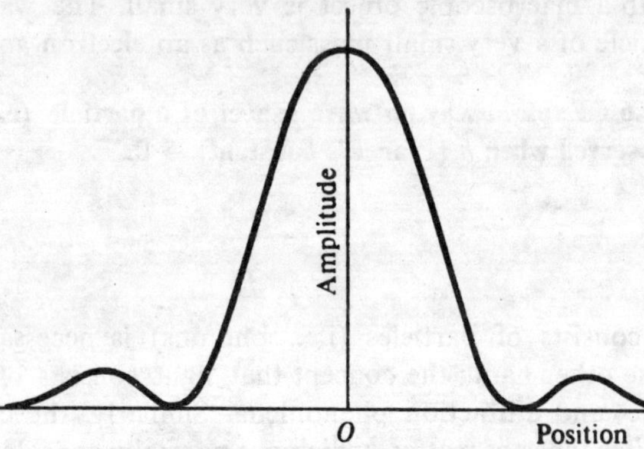

Fig. 2.6 Snapshot of wave packet at given time.

velocity. Using $E = h\nu$ and $p = h/\lambda$, we get

$$v_g = \frac{d(2\pi\nu)}{d(2\pi/\lambda)} = \frac{d(E/h)}{d(p/h)} = \frac{dE}{dp} = \frac{d}{dp}\left(\frac{p^2}{2m}\right) = \frac{p}{m} = v_p,$$

where v_p is the velocity of the particle.

EXAMPLE 2.8 What is the kinetic energy of an electron whose de Broglie wavelength is equal to 5000 Å?

Using the de Broglie relation, we get

$$p = \frac{h}{\lambda} = \frac{6.63 \times 10^{-34}}{5000 \times 10^{-10}} = 1.326 \times 10^{-27} \text{ kg-m/sec.}$$

The kinetic energy then is

$$p^2/(2m) \sim 6 \times 10^{-6} \text{ eV}.$$

It should be noted that to calculate the energy we cannot use here the expression $E = h\nu$ because $\lambda\nu \neq v_p$ (the velocity of an electron).

BORN'S INTERPRETATION

We shall now describe the interference phenomenon using the concept of waves. This would lead to a physical picture of the waves associated with matter. This picture was first proposed by M. Born and forms an important step in the formulation of quantum mechanics.

Let us consider the interference phenomenon observed when light waves pass through two slits. The experimental arrangement is shown in Fig. 2.7. Here, S_1 and S_2 are the two slits receiving the waves from the left-hand side. These slits act as two sources of waves which propagate to the right-hand side and ultimately fall on a screen (S) kept far away. As a result, interference fringes can be seen on the screen, some being completely dark, some bright. Let us recall the essential reasons for the occurrence of such fringes.

The wave amplitude at any point P on the screen is obtained by using the *principle of superposition*. This principle states that the wave amplitude at a given point is the sum of the wave amplitudes due to S_1 and S_2. We can specify the point P by setting up a coordinate system. Let (x_p, y_p, z_p) be the cartesian coordinates of P, and let the wave at P due to the source S_1 at time t be denoted by the amplitude $A_1(x_p, y_p, z_p, t)$. Similarly, $A_2(x_p, y_p, z_p, t)$ denotes the wave at P due to the source S_2 at the time t. The total amplitude at P, according to the principle of superposition, is the sum of the individual amplitudes. Denoting the total amplitude by $A(x_p, y_p, z_p, t)$, we have

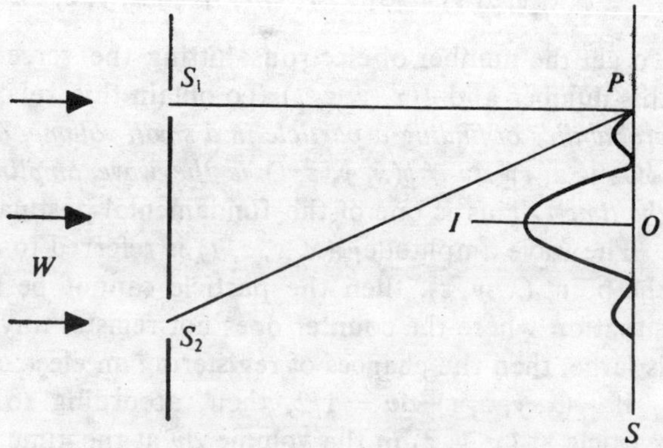

Fig. 2.7 Interference pattern due to two-slit arrangement. S_1, S_2, slits; W, incident wave; I, intensity as function of position on screen S.

$$A(x_p, y_p, z_p, t) = A_1(x_p, y_p, z_p, t) + A_2(x_p, y_p, z_p, t). \tag{2.11}$$

For a light wave, we do not observe the amplitude but the intensity which is the energy supplied per unit area per unit time and is proportional to the $|\text{amplitude}|^2$ [see Eq. (A1.21) in Appendix 1]. Therefore, the intensity I at the point P on the screen for a light wave is given by

$$I \propto |A(x_p, y_p, z_p, t)|^2 \tag{2.12}$$

or

$$I \propto |A_1(x_p, y_p, z_p, t) + A_2(x_p, y_p, z_p, t)|^2. \tag{2.13}$$

If at some point on the screen the two amplitudes A_1 and A_2 are equal in magnitude but opposite in sign, then they cancel each other, and so $I = 0$. This is destructive interference and a dark fringe is observed at this point. On the other hand, if at some other point both A_1 and A_2 have the same sign, then the two amplitudes add to give constructive interference and a bright fringe is seen.

Consider an experiment, similar to the one just described, being performed with matter waves, using, say, electrons. In other words, the light wave incident from the left-hand side (see Fig. 2.7) is now replaced by a beam of electrons with a momentum p. As already noted, the wavelength associated with a matter wave is $\lambda = h/p$. To observe the electrons, electron counters can be kept at various points on the screen S. To observe the interference pattern, we must make several electrons (say, N per second) pass through the slits. The intensity of electrons at any point P on the screen S can then be defined as the number of electrons reaching per second a unit area around P. An interference pattern means that on the screen there are some points where no electrons are received and some other points where many electrons are counted. To explain this pattern using the wave aspect, let $\psi_1(x, y, z, t)$ and $\psi_2(x, y, z, t)$ be the amplitudes of the wave associated with an electron arriving at an arbitrary point (x, y, z) at a time t from the sources S_1 and S_2, respectively. The total amplitude $\psi(x, y, z, t)$, given by the principle of superposition, then is

$$\psi(x, y, z, t) = \psi_1(x, y, z, t) + \psi_2(x, y, z, t). \tag{2.14}$$

To get the number of electrons hitting the screen per unit time, we need a relation between this number and $\psi(x, y, z, t)$. To obtain this relation, Born made an assumption, namely, *the probability of finding a particle in a small volume Δv at a point (x, y, z) at a time t is given by $|\psi(x, y, z, t)|^2 \Delta v$ if $\psi(x, y, z, t)$ is the wave amplitude associated with the particle at (x, y, z) at the time t.* This is one of the fundamental postulates of quantum mechanics.

The wave amplitude $\psi(x, y, z, t)$ is referred to as the *wave function.* If $\psi(x, y, z, t)$ is zero at the point (x, y, z), then the particle cannot be found at this point. This corresponds to the situation where the counter does not register any electron. On the other hand, if $|\psi(x, y, z, t)|^2$ is large, then the chances of registering an electron at this point are high.

If $|\psi(x, y, z, t)|^2 \Delta v = 1/2$, then, according to the postulate, the probability of finding the particle at (x, y, z) in the volume Δv at the time t is $1/2$. This, however, does not mean that the volume has half the particle. Once a given particle, say, an electron, has been detected in a volume, it is there and nowhere else. The physical meaning of the relation $|\psi(x, y, z, t)|^2 \Delta v = 1/2$ is that if the interference experiment is performed a large number of times, say, N, then th number of times a particle will be found in the volume Δv around (x, y, z) at the time t is $N/2$. In the interference experiment using an electron beam we have described, this means that the experiment is conducted by sending several electrons. The wave associated with each such electron passes through both the slits resulting in a wave function $\psi(x, y, z, t)$ given by Eq. (2.14). Thus, each electron has a certain probability $|\psi(x_p, y_p, z_p, t)|^2 \Delta v$ of being found in the volume Δv around the point P on the screen. However, we do not know the point at which a given electron reaches the screen. At a point where $\psi = 0$, the probability of finding an electron is zero, and at a point where $|\psi|^2$ is nonzero, the probability of finding an electron is nonzero. This explains the interference pattern observed for the electrons. If N electrons, each with a momentum p, are sent through the slits, then we have N identical observations. For each such electron, the wave function is $\psi(x_p, y_p, z_p, t)$, and so the number of electrons detected in the volume Δv will be $N|\psi(x_p, y_p, z_p, t)|^2 \Delta v$. It should be noted that the interference pattern does not result from the superposition of the wave functions associated with two different electrons. Instead, it is the wave associated with each electron which, after it passes through the two slits, produces the interference pattern.

EXAMPLE 2.9 The wave function of a particle confined in a box of length L is

$$\psi(x) = (2/L)^{1/2} \sin(\pi x/L)$$

in the region $0 < x < L$ and zero everywhere else. Calculate the probability of finding the particle in the region $0 < x < L/2$. Do the same for a particle with a wave function

$$\psi(x) = [1/(2L)]^{1/2} \exp[-2\pi i(Et - px)/h]$$

in the region $0 < x < 2L$ and zero outside this region.

The problems concern the motion of a particle in one dimension, and hence their answers depend only on x, and require that the volume element Δv in Born's postulate be replaced by the line element Δx.

For the first problem,

$$P(x)\Delta x = \frac{2}{L}\sin^2\left(\frac{\pi x}{L}\right)\Delta x.$$

Hence, the probability of finding the particle in the region $0 < x < L/2$ is

$$P = \int_0^{L/2} P(x)\,dx = \frac{2}{L}\int_0^{L/2} \sin^2\left(\frac{\pi x}{L}\right) dx = 1/2.$$

For the second problem, it should be noted that ψ is a complex number, and therefore we have

$$P(x)\Delta x = |\psi|^2\Delta x = \psi^*\psi\Delta x,$$

where ψ^* is the complex conjugate of ψ. The complex conjugate of $\exp[(-2\pi i/h)(Et - px)]$ is $\exp[(2\pi i/h)(Et - px)]$. Thus, the required probability is

$$P = \int_0^{L/2} \frac{1}{(2L)^{1/2}} \exp\left[\frac{-2\pi i}{h}(Et - px)\right] \frac{1}{(2L)^{1/2}} \exp\left[\frac{2\pi i}{h}(Et - px)\right] dx$$

$$= \frac{1}{2L}\int_0^{L/2} dx = \tfrac{1}{4}.$$

2.5 UNCERTAINTY PRINCIPLE

The fact that matter exhibits both particle-like and wave-like aspects has far-reaching consequences. One such very important result is the uncertainty principle formulated by W. Heisenberg. We shall now discuss the principle.

Both the momentum and position of a particle cannot be measured simultaneously and precisely. Suppose we have a particle whose position and momentum are being measured. If the measurements are carried out repeatedly, it is not necessary that the value of the position or the momentum will be the same. This is because of the probabilistic nature of the wave function associated with the particle. In general, the position or the momentum of a particle will have a spread about a mean value. The order of magnitude of this spread is referred to as the uncertainty in the position or the momentum of the particle. If Δx is the uncertainty in the measurement of the x-component of its position and Δp_x is the uncertainty in the simultaneous measurement of the x-component of its momentum, then, according to the Heisenberg uncertainty principle,

$$\Delta x \Delta p_x \geqslant h/(4\pi). \quad (2.15)$$

We shall not prove this result but discuss the physical reason which leads us to it. Consider the measurement of the x-component of the position of a particle, say, an electron, to an accuracy of Δx by making it pass through a slit of width Δx (see Fig. 2.8). We know that any

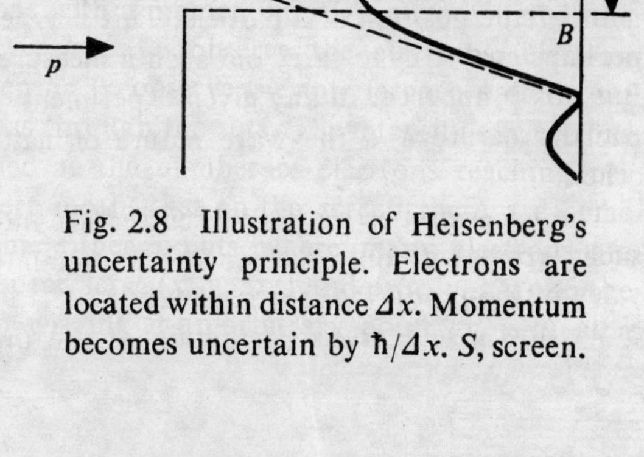

Fig. 2.8 Illustration of Heisenberg's uncertainty principle. Electrons are located within distance Δx. Momentum becomes uncertain by $\hbar/\Delta x$. S, screen.

electron at the time of passing through a slit lies between 0 and Δx. Let us assume that the electron has a momentum only along the y-axis. That is, initially, its x-component of momentum p_x is zero. But does p_x remain zero after it passes through the slit? To answer this question, let us repeatedly carry out the measurement conceptually and see where the electron hits the screen. If p were the initial momentum along the y-axis, then the electrons would have behaved like a wave of wavelength h/p and a diffraction pattern would have been seen on the screen. Also, the electron will not just be confined to the region AB but will spread out. This means that the electron has travelled along the x-axis after leaving the slit and before hitting the screen. In so doing, it has developed the x-components of velocity and momentum. Let us estimate p_x. From the theory of diffraction, we know that the first minimum of the diffraction pattern occurs at an angle $\Delta\theta$ given by $\Delta\theta = \lambda/\Delta x$, the second minimum at $\Delta\theta = 2\lambda/\Delta x$, and so on. On most occasions, the electron is detected within the first peak, and so it is not a bad approximation to assume that the electron is confined within $\Delta\theta = \lambda/\Delta x$. Thus,

$$\Delta p_x = p \sin \Delta\theta \sim p\Delta\theta = p(\lambda/\Delta x).$$

We do not know precisely where a given electron will hit the screen. We only know the probability with which it will be found at the various points on the screen. As such, we do not know the exact values of p_x, and they can lie anywhere between $+p\lambda/\Delta x$ and $-p\lambda/\Delta x$. In other words, the uncertainty Δp_x is $p\lambda/\Delta x$. This uncertainty in p_x is not because of any errors in the measurement but due to the role played by the wave nature of the electron. It is an inherent property of all particles since they all exhibit the wave nature. The magnitude of the uncertainty in p_x is

$$\Delta p_x = p\lambda/\Delta x \qquad \text{or} \qquad \Delta p_x \Delta x \sim p\lambda = h.$$

We see that the product of the uncertainties is not quite the same as in relation (2.15). This is because we have crudely estimated the uncertainties. Our aim was to bring out the physical basis of the uncertainty principle rather than to obtain its correct numerical value. Let us recall the basic point of our argument, namely, in attempting to measure the x-component of the position of the electron, we disturbed its x-component of momentum because of its wave nature. If the uncertainty in position is very large (i.e., large slit width), the uncertainty in momentum will be correspondingly reduced. This is understandable because as the slit width increases, the diffraction effects become less pronounced and the electron essentially falls within the geometrical shadow (AB in Fig. 2.8). On the other hand, if the uncertainty in position is small (i.e., small slit width), the uncertainty in momentum will be large because the diffraction peak becomes broad.

In our foregoing discussion, we have used a particular method for measuring the x-component of the position of a particle. This may lead us to think that a different method would also perhaps enable us to carry out such a measurement without disturbing the particle momentum. But this is not true. In any given experiment, it can be shown that the interplay between the particle nature and the wave nature of matter is such that a measurement of one disturbs the other.

The uncertainties we have discussed are independent of the measurement errors. Therefore, such errors will only add to the uncertainties. Inequality (2.15) indicates that the minimum value of the product $\Delta x \Delta p_x$ is $h/(4\pi)$.

We now give a precise definition of the uncertainty in measurement. Let us consider the measurement of the position of a particle. When a large number, say, N, of measurements is made, we will, in general, get different values of the position, say, $x_1, x_2, \ldots, x_N$. We define the average value $\langle x \rangle$ of the position as

$$\langle x \rangle = \frac{1}{N} \sum_{i=1}^{N} x_i$$

and the uncertainty Δx as

$$\Delta x = [\frac{1}{N} \sum_{i=1}^{N} (x_i - \langle x \rangle)^2]^{1/2}.$$

Similarly, the uncertainty in momentum can be defined as

$$\Delta p = [\frac{1}{N} \sum_{i=1}^{N} (p_i - \langle p \rangle)^2]^{1/2},$$

where $\langle p \rangle$ is the average value of the momentum. With these definitions of Δx and Δp, inequality (2.15) can be proved.

There are uncertainty relations also between the y- and z-component of position and momentum. These are

$$\Delta p_y \Delta y \geqslant h/(4\pi), \qquad \Delta p_z \Delta z \geqslant h/(4\pi),$$

where Δy, Δz are the uncertainties in the y-, z-component of position, and Δp_y, Δp_z are the uncertainties in the y-, z-component of momentum.

There are similar uncertainty relations involving several other pairs of measurable physical quantities. One such relation is the energy-time uncertainty relation. If ΔE is the uncertainty in the measurement of energy and Δt is the time required for measurement, then the product of ΔE and Δt satisfies the relation

$$\Delta E \Delta t \geqslant h/(4\pi). \qquad (2.16)$$

We can verify this relation for a free particle. To do this, consider a particle with momentum p in the x-direction. Since the energy of the particle is $p^2/(2m)$, the uncertainty in its energy is $p\Delta p/m$. Using the position-momentum uncertainty relation (2.15), we get

$$\Delta E = \frac{p}{m} \Delta p \geqslant \frac{p}{m} \frac{h}{4\pi \Delta x} \qquad \text{or} \qquad \Delta E \frac{\Delta x}{p/m} \geqslant \frac{h}{4\pi}.$$

The velocity of the particle is p/m, and $\Delta x/v = \Delta t$. Thus, $\Delta E \Delta t \geqslant h/(4\pi)$. Since we are considering a free particle, we may wonder why its energy is uncertain. This is because the measurement disturbs the particle, leading to an uncertainty in its energy. Though we have verified the uncertainty relation (2.16) in the context of a free particle, it is valid in general. However, a proper explanation of the physics of this relation would be involved, and therefore we shall not attempt it. The correct interpretation of the relation is as follows. If Δt is the time interval characterizing the rate of change of energy of a system, then the uncertainty in the energy of the system is given by

$$\Delta E \geqslant h/(4\pi \Delta t).$$

We shall study the application of this relation in Chapter 4.

EXAMPLE 2.10 Assume that an electron is inside a nucleus of radius 10^{-15} m. From the uncertainty principle, estimate the kinetic energy of the electron.

Since the electron can be anywhere inside the nucleus, the uncertainty $\Delta x \sim 10^{-15}$ m. Using the uncertainty relation $\Delta x \Delta p \sim h/(4\pi)$, we have

$$\Delta p \sim \frac{h}{4\pi \Delta x} = \frac{6.63 \times 10^{-34}}{4\pi \times 10^{-15}} \text{ kg-m/sec.}$$

Since the *magnitude* of the momentum cannot be less than that of the uncertainty, the momentum of the electron will at least be equal to Δp. Thus, the minimum kinetic energy of the electron will be

$$\frac{(\Delta p)^2}{2m} = \frac{(6.63 \times 10^{-19})^2}{2 \times 9.1 \times 10^{-31} \times (4\pi)^2} \text{J} \sim 0.95 \times 10^{10} \text{ eV.}$$

EXAMPLE 2.11 Bullets, each of mass 10^{-3} kg, take 1 sec to reach their target. Considering the bullets as mass points, find the order of magnitude of spread of successive shots at the target under the best possible conditions.

Let the bullets be fired parallel to the x-axis. We will study the spread of the bullets in a direction perpendicular to this, say, the y-axis. To begin with, we do not know either the y-component of their position or the y-component of their momentum. If Δy is zero initially, then, from the uncertainty principle, $\Delta p_y = h/\Delta y = \infty$. In this case, the y-component of momentum of each bullet has an enormous initial uncertainty, and so the bullets will go in all possible directions, that is, the spread in the y-direction, as the bullets hit the target, is extremely large.

If Δp_y is zero initially, then $\Delta y = \infty$, and therefore the initial position of each bullet is completely indeterminate. Thus, the spread in the y-direction will again be very large.

To find the best possible situation (i.e., minimum spread), we assume that at the time each bullet is fired the uncertainty in the y-component of its position is Δy and that in its momentum is $h/\Delta y$. The uncertainty in the y-component of a bullet's position after a time t is

$$\Delta y(t) \sim \Delta y + vt = \Delta y + (p_y/m)t.$$

Since the magnitude of the momentum p_y should at least be equal to the uncertainty, we have

$$\Delta y(t) \sim \Delta y + \Delta p_y t/m = \Delta y + ht/(m\Delta y).$$

At $t = 1$ sec, the spread will be

$$\Delta y(1) = \Delta y + h/(m\Delta y).$$

Minimizing this with respect to Δy, we get

$$1 - \frac{h}{m(\Delta y)^2} = 0, \quad \Delta y = \left(\frac{h}{m}\right)^{1/2}$$

or

$$\Delta y(1) = 2(h/m)^{1/2} = 2\left(\frac{6.6 \times 10^{-34}}{10^{-3}}\right)^{1/2} = 1.6 \times 10^{-15} \text{ m.}$$

The uncertainty principle helps in avoiding any contradiction that may appear to arise because of the presence of both particle-like and wave-like nature of an object. To illustrate this, let us go back to the two-slit interference pattern (Fig. 2.7). The wave associated with the particle goes through both the slits and the superposition of its components causes the interference pattern. If we try to argue the occurrence of the interference pattern in terms of the particle aspect, we run into difficulties as we cannot visualize a particle passing through both the slits. However, we might think that the interference is between two different particles, one coming from S_1 and the other from S_2. But this is not true. The experiment has been performed by using very low intensity beams, ensuring that, at any given time, only one particle on an average was present between the source and the screen. This only shows that the interference pattern is due to a single particle passing through both the slits and giving rise to two components which superpose. In what follows, we shall discover that the uncertainty principle clarifies this situation. We would normally think that the particle will pass through one of two slits. Let us then try to find the slit through which it actually passes. To do this, we need to measure the position of the particle to an accuracy of the order of the width of either slit. But, according to the uncertainty principle, such an attempt would lead to an uncertainty in the momentum, which means that the momentum of the particle would get disturbed. This, in turn, would change the de Broglie wavelength. We shall now show that this change would be so large that it would destroy the interference pattern on the screen.

The experimental condition to observe the interference pattern is that the wavelength associated with the particle be of the order of the interslit separation d. Thus, the initial momentum needed is $p = h/\lambda \sim h/d$. If the particle is located in a slit of dimension a, then the uncertainty in momentum is $|\Delta p| \sim h/a$. Thus,

$$\left|\frac{\Delta p}{p}\right| \sim \frac{d}{a} \gg 1,$$

showing that the uncertainty in the wavelength is much larger than the wavelength originally associated with the particle. As a result, the wave associated with the particle reaching the screen will not have a definite wavelength, and thus no interference pattern will be seen.

We have discussed the apparent difficulty in uniting the particle aspect and the wave aspect in a specific experiment. If a similar analysis is made of all experiments, we would always find that the moment we attempt to locate the particle, the disturbance caused to the particle due to the uncertainty principle would destroy the effect of the wave aspect.

Thus, from our foregoing arguments, we note that the common assumption (based on the experience of the macroscopic world), namely, if a measurement on a system is properly made, the nature of the system remains unchanged, cannot be extended to the microscopic world. In other words, when we 'look at a system', the system gets disturbed and the results we obtain would depend on the way we 'look at the system'. That is, depending on the nature of an experiment, we obtain the wave aspect or the particle aspect.

2.6 SCHROEDINGER'S EQUATION

The probability interpretation of the wave function of a system gives a physical significance to the wave function. However, a complete understanding of the system needs the description of the system as time passes. This means that we must have a method for knowing the wave function of the system as a function of time. We expect the wave function associated with a particle to develop differently under varying circumstances in much the same way as a particle moves differently depending on the nature of the force acting on it. In classical mechanics, the motion of a particle is described by Newton's laws. In quantum mechanics, the development of the wave function with time is given by Schroedinger's equation.

We cannot really derive Schroedinger's equation, just as we cannot derive Newton's equations. Their validity lies in the experimental verification of the results obtained from them. We shall, however, indicate a way to get Schroedinger's equation. It should be noted that the presentation we give represents neither a proof nor a derivation of this equation.

Consider the motion of a free particle (i.e., a particle on which no forces are acting) in one dimension along the x-axis with a momentum p. According to de Broglie's relation, a wave of wavelength $\lambda = h/p$ is associated with the particle. Let us assume this wave to be of the form

$$\psi(x, t) = A \exp[-i(\omega t - kx)] = A[\cos(\omega t - kx) - i \sin(\omega t - kx)], \tag{2.17}$$

where $\omega = 2\pi\nu$ and $k = 2\pi/\lambda$. The wave is travelling along the positive x-axis with a phase velocity $v = \omega/k$ (see Appendix 2). The wave function we have chosen is complex (see Appendix 3). The reason for such a choice rather than just $\cos(\omega t - kx)$ or $\sin(\omega t - kx)$ will become clear in Chapter 3 (see Problem 3.23). Using $E = h\nu$ and $p = h/\lambda$, we get

$$\psi(x, t) = A \exp\left[\frac{-i2\pi}{h}(Et - px)\right] = A \exp\left[\frac{-i}{\hbar}(Et - px)\right], \tag{2.18}$$

where $\hbar$ (pronounced h bar or h cross) $= h/(2\pi)$, a quantity which occurs very often in quantum mechanics. We now have an equation of a wave for a free particle with energy E and momentum p. We have to be careful in the interpretation of this equation. This is because $v = \omega/k = E/p$ is not the velocity of the particle. It represents the phase velocity and not the group velocity. It is the group velocity that has more physical significance. (For a discussion on the concept of group velocity, see Appendix 2.) The group velocity v_g is given by Eq. (2.10), i.e.,

$$v_g = d\omega/dk = p/m, \tag{2.19}$$

which is indeed the velocity of the particle as discussed earlier. We are interested in obtaining a wave equation which will enable us to study the development of the wave function with time. In classical mechanics, Newton's second law when applied to motion in the x-direction gives the differential equation

$$\text{mass} \times \frac{d^2x}{dt^2} = \text{force}. \tag{2.20}$$

This equation enables us to study the change of position of a particle with time. In quantum mechanics too, the wave equation is a differential equation. This differential equation must have as a solution Eq. (2.18), i.e.,

$$\psi(x, t) = A \exp[-(i/\hbar)(Et - px)]$$

for the free particle case. Performing partial differentiation with respect to x on this equation, we get

$$\frac{\partial\psi(x, t)}{\partial x} = A \exp\left[\frac{-i}{\hbar}(Et - px)\right]\left(\frac{ip}{\hbar}\right) = \left(\frac{ip}{\hbar}\right)\psi(x, t)$$

or

$$-i\hbar\frac{\partial\psi(x, t)}{\partial x} = p\psi(x, t). \tag{2.21}$$

Similarly, differentiating Eq. (2.18) with respect to time, we get

$$i\hbar\frac{\partial\psi(x, t)}{\partial t} = E\psi(x, t). \tag{2.22}$$

Equation (2.21) relates $p\psi(x, t)$ with the spatial derivative of ψ, and Eq. (2.22) relates $E\psi(x, t)$ with the time derivative of ψ. To connect these two, we use the relation $E = p^2/(2m)$. Differentiating Eq. (2.21) with respect to x, we have

$$-i\hbar\frac{\partial^2\psi}{\partial x^2} = p\frac{\partial\psi}{\partial x}.$$

Substituting for $\partial\psi/\partial x$ from Eq. (2.21) on the right-hand side of this equation, we obtain

$$-i\hbar\frac{\partial^2\psi}{\partial x^2} = p(\frac{i}{\hbar})p\psi = \frac{i}{\hbar}p^2\psi$$

or

$$(-i\hbar)^2\frac{\partial^2\psi}{\partial x^2} = p^2\psi = 2mE\psi. \tag{2.23}$$

Substituting in Eq. (2.23) for $E\psi$ from Eq. (2.22), we get

$$(-i\hbar)^2\frac{\partial^2\psi}{\partial x^2} = 2mi\hbar\frac{\partial\psi}{\partial t}.$$

Thus, we have

$$i\hbar\frac{\partial\psi}{\partial t} = \frac{-\hbar^2}{2m}\frac{\partial^2\psi(x, t)}{\partial x^2}. \tag{2.24}$$

Since we have obtained Eq. (2.24) by using the explicit form of

$$\psi(x, t) = A \exp [\frac{-i}{\hbar}(Et - px)],$$

the latter clearly satisfies, and hence is a solution of Eq. (2.24). We generalize Eq. (2.24) to a particle acted upon by a force. Then, the energy of the particle will not just be its kinetic energy. If the force is conservative, we can write the total energy as

$$E = p^2/(2m) + V(x), \tag{2.25}$$

where $V(x)$ is the potential energy of the particle.

QUESTION 2.7 For a conservative force, what is the relation between the force and the potential? Why can we not extend the concept of potential to a nonconservative force?

We now make the crucial assumption that Eq. (2.21) and Eq. (2.22) hold for all wave functions and not just for a free particle. This means that

$$E\psi(x, t) = i\hbar \frac{\partial \psi(x, t)}{\partial t} = (\frac{p^2}{2m} + V)\psi(x, t). \tag{2.26}$$

Substituting from Eq. (2.23) for $p^2\psi(x, t)$, we get

$$E\psi(x, t) = i\hbar \frac{\partial \psi(x, t)}{\partial t} = \frac{(-i\hbar)^2}{2m} \frac{\partial^2 \psi(x, t)}{\partial x^2} + V(x)\psi(x, t).$$

This leads to the time-dependent Schroedinger equation in one dimension, namely,

$$\boxed{i\hbar \frac{\partial \psi (x, t)}{\partial t} = \frac{-\hbar^2}{2m} \frac{\partial^2 \psi (x, t)}{\partial x^2} + V(x)\, \psi (x, t).} \tag{2.27}$$

The study of Schroedinger's equation is of utmost importance since it is the basic equation governing the behaviour of matter. The Schroedinger equation along with the Born probability interpretation forms the basis of quantum mechanics.

EXAMPLE 2.12 Consider a particle confined in a box of length L. The potential inside the box is zero. Verify that

$$\psi(x, t) = C_n \exp [-iE_n t/\hbar] \sin (n\pi x/L)$$

is a solution of the Schroedinger equation (2.27), where $E_n = n^2\pi^2\hbar^2/(2mL^2)$.

The Schroedinger equation for a particle inside a box is

$$i\hbar \frac{\partial \psi}{\partial t} = -\frac{\hbar^2}{2m} \frac{\partial^2 \psi}{\partial x^2}$$

for $V(x) = 0$ inside the box. Using the relation

$$\psi(x, t) = C_n \exp (-\frac{iE_n t}{\hbar}) \sin (n\pi x/L),$$

we get

$$i\hbar \frac{\partial \psi}{\partial t} = E_n C_n \exp (-\frac{iE_n t}{\hbar}) \sin (n\pi x/L), \tag{2.28}$$

$$\frac{\partial \psi}{\partial x} = C_n \exp (-\frac{iE_n t}{\hbar}) \cos (n\pi x/L)(n\pi/L).$$

We therefore have

$$\frac{\hbar^2}{2m} \frac{\partial^2 \psi}{\partial x^2} = -\frac{\hbar^2}{2m} C_n \exp (-\frac{iE_n t}{\hbar}) \sin (n\pi x/L)(n\pi/L)^2. \tag{2.29}$$

From Eqs. (2.28) and (2.29), we find that Schroedinger's equation is satisfied if

$$E_n = \frac{\hbar^2}{2m}(\frac{n\pi}{L})^2 = \frac{n^2\pi^2\hbar^2}{2mL^2}.$$

Before we take up the applications of Schroedinger's equation, we shall discuss some important concepts which are essential in dealing with quantum systems.

2.7 OPERATORS, AVERAGE VALUES

OPERATORS

We had stated that $p\psi = -i\hbar(\partial\psi/\partial x)$ is true for a system whose wave function is given by $\psi(x, t)$. We interpret this equation in the following way. An operator p *operates* on ψ and gives a new function $-i\hbar(\partial\psi/\partial x)$. The operation consists of taking the partial derivative with respect to x and multiplying the result so obtained by $-i\hbar$. We write the operator equation as

$$p = -i\hbar(\partial/\partial x). \tag{2.30}$$

The meaning of this equation is that the operation of p on ψ is the same as that of $-i\hbar(\partial/\partial x)$ on ψ, giving $-i\hbar(\partial\psi/\partial x)$. The operator concept is not used for p alone in quantum mechanics. For example, x can also be considered as an operator. When it operates on $\psi(x, t)$, it gives a new function $x\psi(x, t)$. Similarly, E is another operator and its operation on ψ is given by Eq. (2.22). The operator form of E then is

$$E = +i\hbar(\partial/\partial t). \tag{2.31}$$

In fact, in quantum mechanics, we associate an operator with each measurable quantity (see Appendix 4).

EXAMPLE 2.13 The operator p acts on $\psi(x, t) = 3x^2 \exp(-i\omega t)$. Obtain the functions $p\psi$ and $p^2\psi$.

From Eq. (2.30), we have

$$p\psi(x, t) = -i\hbar\frac{\partial}{\partial x}[3x^2 \exp(-i\omega t)] = -i\hbar 6x \exp(-i\omega t).$$

For the operator p^2, we use Eq. (2.30) twice to obtain $p^2\psi$ as

$$p^2\psi(x, t) = p(p\psi) = p[-i\hbar 6x \exp(-i\omega t)] = -i\hbar\frac{\partial}{\partial x}[-i\hbar 6x \exp(-i\omega t)]$$
$$= (-i\hbar)^2 6 \exp(-i\omega t) = -6\hbar^2 \exp(-i\omega t).$$

We specify an operator by specifying its operation on any wave function $\psi(x, t)$. Two operators A and B are the same if the functions $A\psi$ and $B\psi$ are identical for all possible wave functions $\psi(x, t)$.

We can give a meaning to the addition of two operators. For example, $(ax + p)$ is specified by its operation on any function $\psi(x, t)$, i.e.,

$$(ax + p)\psi = ax\psi - i\hbar(\partial\psi/\partial x), \tag{2.32}$$

where a is a constant of appropriate dimensions.

We can also give a meaning to the product of two operators. Taking the two operators x and p, we define the operator xp by the equation

$$xp\psi = x(-i\hbar\frac{\partial\psi}{\partial x}) = -i\hbar x\frac{\partial\psi}{\partial x}. \tag{2.33}$$

Here, first p operates on $\psi(x, t)$ and gives the new function $[-i\hbar(\partial\psi/\partial x)]$ on which x then operates. Had we written px, the order of operations would have then been reversed, i.e.,

$$px\psi = -i\hbar\frac{\partial(x\psi)}{\partial x} = -i\hbar\psi - i\hbar x\frac{\partial\psi}{\partial x}. \tag{2.34}$$

Here, x first operates to give $(x\psi)$ on which p then operates to give $-i\hbar(\partial/\partial x)(x\psi)$. Comparing Eqs. (2.33) and (2.34), we notice that $xp\psi$ is not the same as $px\psi$. From this, we come to a very important conclusion about operator multiplication, namely, *the operator xp is different from the operator px.* The order in which the operators are multiplied is important. In other words, multiplication is *noncommutative.*

From Eqs. (2.33) and (2.34), we have

$$xp\psi - px\psi = (xp - px)\psi = i\hbar\psi.$$

This can be written as an operator equation, namely,

$$xp - px = i\hbar. \tag{2.35}$$

EXAMPLE 2.14 Calculate $(A + B)\psi$, $AB\psi$, and $BA\psi$, where the operators A and B and the wave function ψ are

$$A = 3\frac{d}{dx} + \frac{4}{L}, \qquad B = \frac{x^2}{L^3}, \qquad \psi(x) = \left(\frac{2}{L}\right)^{1/2} \sin\frac{\pi x}{L}.$$

First, consider the operation of A on ψ, i.e.,

$$A\psi = \left(3\frac{d}{dx} + \frac{4}{L}\right)\left(\frac{2}{L}\right)^{1/2} \sin\frac{\pi x}{L} = \left(\frac{2}{L}\right)^{1/2}\left(\frac{3\pi}{L}\cos\frac{\pi x}{L} + \frac{4}{L}\sin\frac{\pi x}{L}\right).$$

Similarly,

$$B\psi = \left(\frac{2}{L}\right)^{1/2}\frac{x^2}{L^3}\sin\frac{\pi x}{L}.$$

Therefore,

$$(A + B)\psi = \left(\frac{2}{L}\right)^{1/2}\left[\frac{3\pi}{L}\cos\frac{\pi x}{L} + \left(\frac{4}{L} + \frac{x^2}{L^3}\right)\sin\frac{\pi x}{L}\right].$$

We now carry out the operation of AB on ψ, i.e.,

$$\begin{aligned} AB\psi &= A(B\psi) = A\left[\left(\frac{2}{L}\right)^{1/2}\frac{x^2}{L^3}\sin\frac{\pi x}{L}\right] \\ &= \left(3\frac{d}{dx} + \frac{4}{L}\right)\left[\left(\frac{2}{L}\right)^{1/2}\frac{x^2}{L^3}\sin\frac{\pi x}{L}\right] \\ &= \left(\frac{2}{L}\right)^{1/2}\left(\frac{6x}{L^3}\sin\frac{\pi x}{L} + \frac{3\pi}{L}\frac{x^2}{L^3}\cos\frac{\pi x}{L} + \frac{4x^2}{L^4}\sin\frac{\pi x}{L}\right). \end{aligned}$$

Similarly,

$$BA\psi = B(A\psi) = B(\frac{2}{L})^{1/2}(\frac{3\pi}{L} \cos \frac{\pi x}{L} + \frac{4}{L} \sin \frac{\pi x}{L})$$

$$= \frac{x^2}{L^3}(\frac{2}{L})^{1/2}(\frac{3\pi}{L} \cos \frac{\pi x}{L} + \frac{4}{L} \sin \frac{\pi x}{L}).$$

It should be noted that $AB\psi \neq BA\psi$. This means that the operators A and B do not commute with each other.

EXAMPLE 2.15 Obtain the operator equation $x^2p - px^2 = 2i\hbar x$.

Consider any function $\psi(x, t)$. We have

$$x^2p\psi = x^2(-i\hbar\frac{\partial\psi}{\partial x}) = -i\hbar x^2\frac{\partial\psi}{\partial x}.$$

On the other hand,

$$px^2\psi = p(x^2\psi) = -i\hbar\frac{\partial}{\partial x}(x^2\psi) = -i\hbar(2x\psi + x^2\frac{\partial\psi}{\partial x}).$$

Thus, $(x^2p - px^2)\psi = 2i\hbar x\psi$. Since we have obtained this equation for an arbitrary wave function ψ, it will be true for all possible $\psi(x, t)$. Hence, we can write the operator equation

$$x^2p - px^2 = 2i\hbar x.$$

AVERAGE VALUE OF POSITION

Let us consider an experiment to measure the position x of a particle with a wave function $\psi(x, t)$. In any given measurement, we will get some specific value of x, say, x_1. On repeating the same measurement by taking care to prepare the same state $\psi(x, t)$, that is, by ensuring that the measurements are carried out under identical conditions, we will, in general, get a different value of position. This is because, given $\psi(x, t)$, we know only the probability of finding the particle in a particular region. In any situation similar to the one we have described, the physical quantity to be measured is the average value of the position. This is done by carrying out the measurements for a sufficiently large number of times, say, N, on identical systems and then taking their average value.

We shall now see how the experimentally obtained average value of x is related to the wave function $\psi(x, t)$. Consider, for example, a wave function of a particle, which is nonvanishing only in two small nonoverlapping regions x_1 to $(x_1 + \Delta x_1)$ and x_2 to $(x_2 + \Delta x_2)$. That is,

$$\begin{aligned}\psi(x, t) &= C_1 \quad \text{if } x_1 < x < (x_1 + \Delta x_1)\\ &= C_2 \quad \text{if } x_2 < x < (x_2 + \Delta x_2),\end{aligned} \tag{2.36}$$

and zero everywhere else. Since the probability of finding the particle in the region x to $(x + \Delta x)$ is proportional to $|\psi(x, t)|^2$, it can be found only in the two regions where $\psi(x, t)$ is not zero. On performing the measurements, the particle will be found either in region 1 $[x_1 < x < (x_1 + \Delta x_1)]$ or in region 2 $[x_2 < x < (x_2 + \Delta x_2)]$.

Let us assume that the particle is found N_1 times in region 1 and N_2 times in region 2. Then,

$$N_1 + N_2 = N, \tag{2.37}$$

and the average position $\langle x \rangle$ is

$$\langle x \rangle = \frac{N_1 x_1 + N_2 x_2}{N}. \tag{2.38}$$

From Born's postulate, the probability of finding the particle in region 1 is $|C_1|^2 \Delta x_1$ and in region 2, it is $|C_2|^2 \Delta x_2$. The volume element Δv used in the postulate has here been replaced by the line element Δx as we are dealing with a one-dimensional situation. Since the particle has been found N_1 times in the interval around x_1 out of N attempts, the probability of it being found in region 1 is N_1/N, so that

$$N_1/N = |C_1|^2 \Delta x_1. \tag{2.39}$$

Similarly,

$$N_2/N = |C_2|^2 \Delta x_2. \tag{2.40}$$

The requirement that the particle can only be in the two regions means that the probability of it being found in region 1 or in region 2 is one. This probability is the sum of the probabilities of it being found in region 1 and region 2. Thus,

$$|C_1|^2 \Delta x_1 + |C_2|^2 \Delta x_2 = 1. \tag{2.41}$$

The expression for the average value, i.e., (2.38), is

$$\langle x \rangle = (N_1/N)x_1 + (N_2/N)x_2 = |C_1|^2 x_1 \Delta x_1 + |C_2|^2 x_2 \Delta x_2. \tag{2.42}$$

QUESTION 2.8 Why is the assumption that regions 1 and 2 are nonoverlapping necessary to obtain Eq. (2.41)?

We can now generalize the foregoing discussion to the situation when $\psi(x, t)$ does not vanish in the nonoverlapping regions $x_1 < x < (x_1 + \Delta x_1), \ldots, x_n < x < (x_n + \Delta x_n)$, where the wave function takes the values $C_1, C_2, \ldots, C_n$. Since the particle is in any one of the n regions, the combined probability of it being found in any of the regions is one, i.e.,

$$|C_1|^2 \Delta x_1 + |C_2|^2 \Delta x_2 + \ldots + |C_n|^2 \Delta x_n = 1$$

or

$$\sum_{i=1}^{n} |C_i|^2 \Delta x_i = 1. \tag{2.43}$$

The average value $\langle x \rangle$ is given by

$$\langle x \rangle = |C_1|^2 \Delta x_1 x_1 + |C_2|^2 \Delta x_2 x_2 + \ldots + |C_n|^2 \Delta x_n x_n = \sum_{i=1}^{n} |C_i|^2 \Delta x_i x_i. \tag{2.44}$$

Finally, if we consider an infinite number of regions by letting $n \to \infty$ and making $\Delta x_i \to 0$, the generalization of Eq. (2.43) is then

$$\int |C|^2 \, dx = \int |\psi|^2 \, dx = 1$$

and that of Eq. (2.44) is

$$\langle x\rangle = \int |C|^2 x\, dx = \int |\psi|^2 x\, dx.$$

To summarize, when $\psi(x, t)$ is nonvanishing and x takes on continuous values from $-\infty$ to $+\infty$, then the sum in Eq. (2.43) is replaced by an integral and we have

$$\int_{-\infty}^{+\infty} |\psi(x, t)|^2\, dx = 1. \tag{2.45}$$

This is referred to as the *normalization condition.* It ensures that the total probability of finding the particle somewhere in the region $-\infty$ to $+\infty$ is one. Similarly, Eq. (2.44) is replaced by

$$\langle x\rangle = \int_{-\infty}^{+\infty} |\psi(x, t)|^2 x\, dx. \tag{2.46}$$

This is usually written, using $|\psi|^2 = \psi^*\psi$, as

$$\langle x\rangle = \int_{-\infty}^{+\infty} \psi^*(x, t) x \psi(x, t)\, dx. \tag{2.47}$$

We can also generalize this to physical quantities other than the position. For example, if we want to find the average value of x^2, then we replace x by x^2 in Eq. (2.47) and get

$$\langle x^2\rangle = \int_{-\infty}^{+\infty} \psi^*(x, t) x^2 \psi(x, t)\, dx. \tag{2.48}$$

QUESTION 2.9 Consider a system consisting of two particles. The wave function of this system is $\psi(x_1, x_2)$, where x_1 and x_2 refer to the positions of particles 1 and 2, respectively. Generalize Born's postulate and the normalization condition for the system.

EXAMPLE 2.16 The wave function of a particle is given by

$$\psi = C \exp(-x^2\alpha^2), \qquad -\infty < x < +\infty,$$

where C and α are constants. Calculate the probability of finding the particle in the region $0 < x < \infty$.

From Born's postulate, the probability of finding the particle in the region $0 < x < \infty$ is

$$P = \int_0^\infty \psi^*\psi\, dx = \int_0^\infty C^* \exp(-\alpha^2x^2) C \exp(-\alpha^2x^2)\, dx$$

$$= |C|^2 \int_0^\infty \exp(-2\alpha^2x^2)\, dx.$$

Using

$$\int_0^\infty \exp(-x^2)\, dx = \pi^{1/2}/2,$$

we get

$$P = \frac{C^2}{2\alpha}\left(\frac{\pi}{2}\right)^{1/2}.$$

We need to impose the normalization condition

$$\int_{-\infty}^{+\infty} |\psi|^2\, dx = \int_{-\infty}^{+\infty} |C|^2 \exp(-2\alpha^2 x^2)\, dx = 1$$

or

$$C^2\left(\frac{\pi}{2}\right)^{1/2}\frac{1}{\alpha} = 1$$

to ensure that there is one particle in the region $-\infty < x < +\infty$. Substituting for C^2 in P, we get $P = \frac{1}{2}$.

EXAMPLE 2.17 A particle is confined in a one-dimensional box of length L. Its wave function is $\psi(x) = C \sin(\pi x/L)$ in the region $0 < x < L$ and $\psi(x) = 0$ outside this region. Calculate the average values of x and x^2.

We first calculate the constant C from the normalization condition

$$\int_{-\infty}^{+\infty} \psi^*\psi\, dx = 1.$$

Substituting in it the wave function, we have

$$\int_0^L |C|^2 \sin^2\left(\frac{\pi x}{L}\right) dx = 1.$$

Evaluating this integral, we get

$$\frac{|C|^2}{2}\left[1 - \frac{\sin(2\pi x/L)}{2\pi/L}\right]_0^L = 1$$

or $|C|^2(L/2) = 1$. Thus, we obtain $|C| = (2/L)^{1/2}$. The average value of x is then

$$\langle x\rangle = \int_{-\infty}^{+\infty} \psi^*(x)x\psi(x)\, dx = |C|^2 \int_0^L \sin\left(\frac{\pi x}{L}\right)x \sin\left(\frac{\pi x}{L}\right) dx$$

$$= |C|^2 \int_0^L \sin^2\left(\frac{\pi x}{L}\right)x\, dx = |C|^2\left(\frac{L^2}{4}\right) = \frac{L}{2}.$$

Similarly, the average value of x^2 is

$$\langle x^2\rangle = |C|^2 \int_0^L \sin^2\left(\frac{\pi x}{L}\right)x^2\, dx = |C|^2\frac{L^3}{2}\left(\frac{1}{3} - \frac{1}{2\pi^2}\right) = L^2\left(\frac{1}{3} - \frac{1}{2\pi^2}\right).$$

It should be noted that $\langle x^2\rangle \neq \langle x\rangle^2$. That is, the average value of the square is not the same as the square of the average value.

We shall now discuss the procedure for obtaining the average values of the measurable physical quantities other than the position, for example, momentum and kinetic energy. In so doing, we shall use the concept, mentioned earlier in this section, of associating an operator with each measurable quantity.

AVERAGE VALUE OF OBSERVABLE

Any physical quantity which can be experimentally measured is referred to as an *observable.* For example, position, momentum, and energy are all observables. In quantum mechanics, we associate an operator with each observable. We list in Table 2.1 some of the physical quantities and the operators associated with them.

TABLE 2.1 Operators associated with some physical quantities

Physical quantity	Operator	Result of operation on ψ
Position	x	$x\psi = x\psi$
Momentum	$-i\hbar\frac{\partial}{\partial x}$	$p\psi = -i\hbar\frac{\partial\psi}{\partial x}$
(Momentum)^2	$(-i\hbar)^2\frac{\partial^2}{\partial x^2}$	$p^2\psi = -\hbar^2\frac{\partial^2\psi}{\partial x^2}$

We generalize Eq. (2.47) by replacing the operator x in it by an arbitrary operator. Let O be this arbitrary operator associated with an observable. Then, the average value of O is

$$\langle O\rangle = \int_{-\infty}^{+\infty} \psi^*(x, t)[O\psi(x, t)]\, dx, \tag{2.49}$$

where $\psi(x, t)$ is a normalized wave function, that is, it satisfies the normalization condition (2.45). When the observable is position, the operator associated with it is x and Eq. (2.49) becomes identical with expression (2.47).

Now, if we want the average value of momentum, we replace in Eq. (2.49) the operator O by the operator associated with momentum and write

$$\langle p\rangle = \int_{-\infty}^{+\infty} \psi^*(x, t)[p\psi(x, t)]\, dx = \int_{-\infty}^{+\infty} \psi^*(x, t)[-i\hbar\frac{\partial}{\partial x}\psi(x, t)]\, dx$$

$$= -i\hbar \int_{-\infty}^{+\infty} \psi^*(x, t)\frac{\partial\psi(x, t)}{\partial x}\, dx. \tag{2.50}$$

It must be noted that we have not derived Eq. (2.49) and Eq. (2.50). These equations are part of the fundamental assumptions in quantum mechanics.

EXAMPLE 2.18 Calculate the average value of p and p^2 for the wave function $\psi(x) = (2/L)^{1/2} \sin(\pi x/L)$ in the region $0 < x < L$ and $\psi(x) = 0$ outside this region.

In any problem such as this one, we should first check whether the given wave function is normalized, that is, if it satisfies Eq. (2.45). In Example 2.17, we noted that the wave function given here is normalized.

The average value of p is then

$$\langle p\rangle = \int_0^L (\frac{2}{L})^{1/2} \sin \frac{\pi x}{L}(-i\hbar\frac{\partial}{\partial x})[(\frac{2}{L})^{1/2} \sin \frac{\pi x}{L}]\, dx$$

$$= \frac{2}{L}(-i\hbar) \int_0^L \sin \frac{\pi x}{L}(\frac{\pi}{L}) \cos \frac{\pi x}{L}\, dx = 0.$$

Further, the average value of p^2 is

$$\langle p^2\rangle = \int_0^L \left(\frac{2}{L}\right)^{1/2} \sin\frac{\pi x}{L}\left(-i\hbar\frac{\partial}{\partial x}\right)\left[-i\hbar\frac{\partial}{\partial x}\left\{\left(\frac{2}{L}\right)^{1/2}\sin\frac{\pi x}{L}\right\}\right] dx$$

$$= (-i\hbar)^2\frac{2}{L}\int_0^L \sin\frac{\pi x}{L}\frac{\partial}{\partial x}\left(\frac{\partial}{\partial x}\sin\frac{\pi x}{L}\right) dx$$

$$= (-i\hbar)^2\frac{2}{L}\int_0^L \sin\frac{\pi x}{L}\frac{\partial^2}{\partial x^2}\sin\frac{\pi x}{L}\, dx$$

$$= \frac{2\hbar^2\pi^2}{L^3}\int_0^L \sin^2\frac{\pi x}{L}\, dx = \frac{\hbar^2\pi^2}{L^2}.$$

We should again like to point out that $\langle p\rangle^2 \neq \langle p^2\rangle$.

2.8 TIME-INDEPENDENT SCHROEDINGER EQUATION

Let us consider the Schroedinger equation (2.27), i.e.,

$$i\hbar\frac{\partial\psi(x, t)}{\partial t} = -\frac{\hbar^2}{2m}\frac{\partial^2\psi(x, t)}{\partial x^2} + V(x)\psi(x, t), \tag{2.51}$$

for a system. We solve this equation by assuming that the wave function separates into a product of two functions, one depending only on x and the other only on t. Thus,

$$\psi(x, t) = X(x)\Phi(t). \tag{2.52}$$

Such a technique of solving a differential equation is widely used and is called the method of separation of variables. Substituting Eq. (2.52) in Eq. (2.51), we get

$$i\hbar X(x)\frac{d\Phi(t)}{dt} = -\frac{\hbar^2}{2m}\Phi(t)\frac{d^2X(x)}{dx^2} + V(x)\Phi(t)X(x). \tag{2.53}$$

Dividing both sides of this equation by $\Phi(t)X(x)$, we have

$$i\hbar\frac{1}{\Phi(t)}\frac{d\Phi(t)}{dt} = -\frac{\hbar^2}{2m}\frac{1}{X(x)}\frac{d^2X(x)}{dx^2} + V(x). \tag{2.54}$$

In this equation, the left-hand side is a function of only time and the right-hand side is a function of only position. This equation would hold for all values of x and t only if both the left- and right-hand side are equal to a constant. To see this, let us vary x. Then the left-hand side remains unchanged as it is a function of only time. Therefore, the right-hand side must also remain unchanged as x changes. In other words, both the sides must be equal to a constant. Thus, we write

$$\frac{i\hbar}{\Phi(t)}\frac{d\Phi(t)}{dt} = A, \tag{2.55}$$

$$-\frac{\hbar^2}{2m}\frac{1}{X(x)}\frac{d^2X(x)}{dx^2} + V(x) = A, \tag{2.56}$$

where A is a constant to be determined. Solving Eq. (2.55), we get

$$\Phi(t) = \Phi(0) \exp\left(\frac{-iAt}{\hbar}\right), \tag{2.57}$$

where $\Phi(0)$ is the value of Φ at the time $t = 0$. We now show that the constant A is nothing but the energy of the system. From Eqs. (2.22) and (2.52), we have

$$i\hbar\frac{\partial\psi(x, t)}{\partial t} = E\psi(x, t) \quad \text{or} \quad i\hbar\frac{\partial[X(x)\Phi(t)]}{\partial t} = EX(x)\Phi(t).$$

Cancelling $X(x)$ on both sides, we get

$$i\hbar\frac{d\Phi(t)}{dt} = E\Phi(t). \tag{2.58}$$

Comparing it with Eq. (2.55), we have $A = E$. Equation (2.57) then becomes

$$\Phi(t) = \Phi(0) \exp(-iEt/\hbar). \tag{2.59}$$

Now that the constant A has been identified with energy E, we can write Eq. (2.56) as

$$\frac{d^2X(x)}{dx^2} + \frac{2m}{\hbar^2}[E - V(x)]X(x) = 0. \tag{2.60}$$

This is called the time-independent Schroedinger equation. It plays a major role in our understanding of atomic, molecular, solid state, and nuclear physics.

The time-independent Schroedinger equation can be obtained only when the potential is independent of time. If the potential is time-dependent, we cannot then separate the wave function as a product of two functions [given by Eq. (2.52)]. Whenever the potential is independent of time, we can obtain the solution of the time part, i.e., $\Phi(t)$, and this solution is always given by Eq. (2.59).

A system described by a wave function of the form

$$\psi(x, t) = \Phi(0) \exp\left(\frac{-iEt}{\hbar}\right)X(x)$$

is in a stationary state. This is because $|\psi|^2$ is independent of time, and so the probability density, that is, the probability per unit length, is independent of time, as we shall now show. From Eqs. (2.52) and (2.59), we have

$$|\psi(x, t)|^2 = \psi^*(x, t)\psi(x, t) = \left[\exp\left(\frac{-iEt}{\hbar}\right)\Phi(0)X(x)\right]^* \exp\left(\frac{-iEt}{\hbar}\right)X(x)\Phi(0)$$

$$= X^*(x) \exp\left(\frac{+iEt}{\hbar}\right) \exp\left(\frac{-iEt}{\hbar}\right)X(x)\Phi^*(0)\Phi(0).$$

We note that the time-dependence cancels and we get

$$|\psi(x, t)|^2 = X^*(x)X(x)\Phi^*(0)\Phi(0).$$

We write this expression as

where $\psi(x) = X(x)\Phi(0)$ satisfies the time-independent Schroedinger equation

$$\frac{d^2\psi(x)}{dx^2} + \frac{2m}{\hbar^2}[E - V(x)]\,\psi(x) = 0. \tag{2.61}$$

We can also express the time-independent Schroedinger equation in an operator form. To do this, we rewrite Eq. (2.61) as

$$\left[\frac{-\hbar^2}{2m}\frac{d^2}{dx^2} + V(x)\right]\psi(x) = E\psi(x).$$

We note that the quantity within the square brackets in the left-hand side is the operator equivalent of $[p^2/(2m) + V(x)]$ which represents energy in the operator form. We represent the energy operator by H to distinguish it from E which denotes the value of energy. Thus, the time-independent Schroedinger equation in the operator form is

$$H\psi = E\psi. \tag{2.62}$$

Here, H is the operator associated with energy and is known as the Hamiltonian. Equation (2.62) is referred to as the energy eigenvalue equation (see Appendix 4).

EXAMPLE 2.19 The wave function of a particle in a stationary state with an energy E_0 at the time $t = 0$ is $\psi(x)$. After how much time will the wave function again be $\psi(x)$?

Since the particle is in a stationary state,

$$\psi(x, t) = \exp\left(\frac{-iE_0t}{\hbar}\right)\psi(x).$$

Suppose the wave function is again $\psi(x)$ at a certain time T. Then,

$$\psi(x, t = T) = \exp(-iE_0T/\hbar)\psi(x) = \psi(x) \quad \text{or} \quad \exp(-iE_0T/\hbar) = 1.$$

The solution of this equation is

$$E_0T/\hbar = 2n\pi \quad \text{or} \quad T = 2n\pi\hbar/E_0,$$

where $n = 0, 1, 2, \ldots$.

SUMMARY

We now summarize our main results:

(1) All matter has both particle and wave aspects.

(2) The physical significance of the wave function $\psi(x, t)$ of a particle is that $|\psi(x, t)|^2\,dx$ is the probability of finding the particle in a small region x to $(x + dx)$ at a time t.

(3) The wave function $\psi(x, t)$ obeys the Schroedinger equation

$$i\hbar\frac{\partial\psi(x, t)}{\partial t} = -\frac{\hbar^2}{2m}\frac{\partial^2\psi(x, t)}{\partial x^2} + V(x)\psi(x, t).$$

(4) Each observable is associated with an operator. For example, the x-component of the momentum operator is $p_x = -i\hbar(\partial/\partial x)$.

(5) If O is an observable (treated as an operator), the average value of O then is

$$\langle O \rangle = \int_{-\infty}^{+\infty} \psi^*(x, t) O \psi(x, t) \, dx.$$

The average value is also known as the expectation value.

(6) For a time-independent potential, $\psi(x, t) = \psi(x) \exp(-iEt/\hbar)$ with $\psi(x)$ satisfying the time-independent Schroedinger equation

$$\frac{d^2\psi(x)}{dx^2} + \frac{2m}{\hbar^2}[E - V(x)]\psi(x) = 0.$$

(7) The normalization condition is

$$\int_{-\infty}^{+\infty} \psi^*(x, t)\psi(x, t) \, dx = \int_{-\infty}^{+\infty} \psi^*(x)\psi(x) \, dx = 1.$$

2.9 SIMPLE ILLUSTRATION—PARTICLE IN BOX

Consider a particle confined in a one-dimensional box of length L (see Fig. 2.9). We shall study the implications of applying the principles of quantum mechanics to such a system. Since the particle is confined in the region $0 < x < L$, the probability of finding it outside this region is zero. This means that $\psi(x, t) = 0$ outside the interval $0 < x < L$.

The particle is free inside the box, that is, no force acts on it. Therefore, $V = 0$ inside the box. The particle is confined inside the box by making $V \to \infty$ at $x = 0$ and $x = L$ (see Fig. 2.9). Since V is independent of time, we use the time-independent Schroedinger equation

$$\frac{d^2\psi(x)}{dx^2} + \frac{2m}{\hbar^2} E\psi(x) = 0, \tag{2.63}$$

where $\psi(x) = \chi(x)\Phi(0)$. The complete wave function is

$$\psi(x, t) = \psi(x) \exp(-iEt/\hbar). \tag{2.64}$$

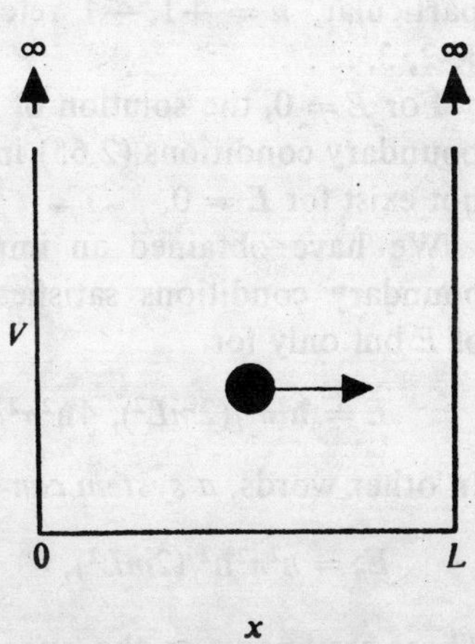

Fig. 2.9 Particle in one-dimensional box.

As $\psi(x, t) = 0$ for $x \leqslant 0$ and $x \geqslant L$, we have on the boundary, that is, at $x = 0$ and $x = L$,

$$\psi(x = 0, t) = \psi(x = L, t) = 0.$$

From Eq. (2.64), we see that

$$\psi(x = 0) = \psi(x = L) = 0. \tag{2.65}$$

We have to solve the Schroedinger equation (2.63) with this boundary condition.

The general solution of Eq. (2.63) for $E \neq 0$ is

$$\psi(x) = A \sin\left[\left(\frac{2mE}{\hbar^2}\right)^{1/2} x\right] + B \cos\left[\left(\frac{2mE}{\hbar^2}\right)^{1/2} x\right]. \tag{2.66}$$

Substituting $x = 0$ in Eq. (2.66) and using Eq. (2.65), we get

$$B = 0. \tag{2.67}$$

Similarly, for $x = L$, we have, using $\psi(L) = 0$,

$$A \sin [(\frac{2mE}{\hbar^2})^{1/2} L] = 0.$$

This implies either $A = 0$ or

$$(\frac{2mE}{\hbar^2})^{1/2} L = n\pi \qquad (n = 0, \pm 1, \pm 2, \ldots). \tag{2.68}$$

If A is zero, then $\psi = 0$, implying there is no particle in the box. Thus, we assume that A is not equal to zero and, from Eq. (2.68), we get

$$E = n^2\pi^2\hbar^2/(2mL^2), \tag{2.6d}$$

where $n = 0, \pm 1, \pm 2, \ldots$. We shall discover later in this section that $n = 0$ does not give an acceptable wave function and $n = -1, -2, \ldots$ give the same wave functions as those obtained with $n = 1, 2, \ldots$, the difference in the two sets of wave functions being only the sign. In particular, $n = +1, -1$ refer to the same physical state. Thus, the allowed values of n are $1, 2, 3, \ldots$.

For $E = 0$, the solution of Eq. (2.63) is $\psi(x) = P + Qx$, where P and Q are constants. The boundary conditions (2.65) imply that both P and Q are zero. Hence, the wave function does not exist for $E = 0$.

We have obtained an important result, namely, the wave function $\psi(x)$ with appropriate boundary conditions satisfies the time-independent Schroedinger equation not for all values of E but only for

$$E = \hbar^2\pi^2/(2mL^2),\ 4\hbar^2\pi^2/(2mL^2), \ldots, n^2\hbar^2\pi^2/(2mL^2).$$

In other words, *a system can take only certain energy values* given by

$$E_n = n^2\pi^2\hbar^2/(2mL^2), \qquad n = 1, 2, \ldots . \tag{2.70}$$

These are known as the energy *eigenvalues*. The lowest energy of the system is not zero but $\hbar^2\pi^2/(2mL^2)$. On the other hand, in classical physics, the lowest energy a particle confined in a box can have is zero.

The integer n in Eq. (2.68) is referred to as the *quantum number*. It specifies the energy and the wave function of the particle confined in the box. The wave function corresponding to the quantum number n, from Eqs. (2.66), (2.67), and (2.68), is

$$\psi_n(x) = A \sin \frac{n\pi x}{L}. \tag{2.71}$$

The constant A is determined from the normalization condition

$$\int_{-\infty}^{+\infty} |\psi_n(x)|^2 \, dx = 1.$$

But $\psi_n(x) = 0$ outside $0 < x < L$. Therefore, this condition becomes

$$\int_0^L |\psi_n(x)|^2 \, dx = 1. \tag{2.72}$$

Substituting Eq. (2.71) in Eq. (2.72), we get

$$A^*A \int_0^L \sin^2 \frac{n\pi x}{L} \, dx = 1 \qquad \text{or} \qquad A^*A = \frac{2}{L}.$$

This is satisfied by $A = (2/L)^{1/2} \exp(i\delta)$, where δ is a phase factor whose value lies between 0 and 2π. Thus, the wave function is

$$\psi_n(x) = (\frac{2}{L})^{1/2} \exp(i\delta) \sin \frac{n\pi x}{L}.$$

Since, in this relation, it is $|\psi_n(x)|^2$ which has a physical significance, the phase factor can be chosen arbitrarily. A value different from the one chosen does not change $\psi^*\psi$ or $\psi^*O\psi$, where O is an operator, and therefore the physical situation. We choose $\delta = 0$ by convention and write

$$\psi_n(x) = (\frac{2}{L})^{1/2} \sin \frac{n\pi x}{L} \tag{2.73}$$

which are the solutions of the time-independent Schroedinger equation. These solutions are referred to as the energy *eigenfunctions*. It should be noted that the wave function corresponding to $n = 0$ is $\psi = 0$. This means that the probability of finding the particle in the box is zero. This is unphysical, and hence is unacceptable. If we choose $n = -m$, then

$$\psi_{-m} = (\frac{2}{L})^{1/2} \sin (\frac{-m\pi x}{L}) = -(\frac{2}{L})^{1/2} \sin (\frac{m\pi x}{L}) = -\psi_m = \exp(i\pi)\psi_m.$$

Thus, we see that choosing $n = -m$ is the same as choosing $n = m$ and $\delta = \pi$. Since any change in δ does not alter the physical situation, $n = \pm m$ correspond to the same state. The complete solution of the time-dependent Schroedinger equation then is

$$\psi_n(x, t) = \psi_n(x) \exp(-iE_n t/\hbar), \tag{2.74}$$

where $n = 1, 2, 3, \ldots$. We can determine the average value of any observable using this wave function.

The most important result of our study in this section is that a system can take only certain values of energy. In other words, the energy of the system is quantized. We shall subsequently see that this is a general property of bound systems and use it to explain several experimental observations in atomic, molecular, and nuclear physics.

QUESTION 2.10 Use the uncertainty principle to show that the lowest energy of a particle confined in a box cannot be zero.

In general, $\psi(x)$ is complex. However, what we measure is the probability density $\psi^*(x)\psi(x) = |\psi|^2$ which is a positive, real quantity. In the illustrative example we have considered, the wave function happens to be real. Figure 2.10 shows a plot of the ground state wave function ψ_1 and the first two excited states ψ_2 and ψ_3.

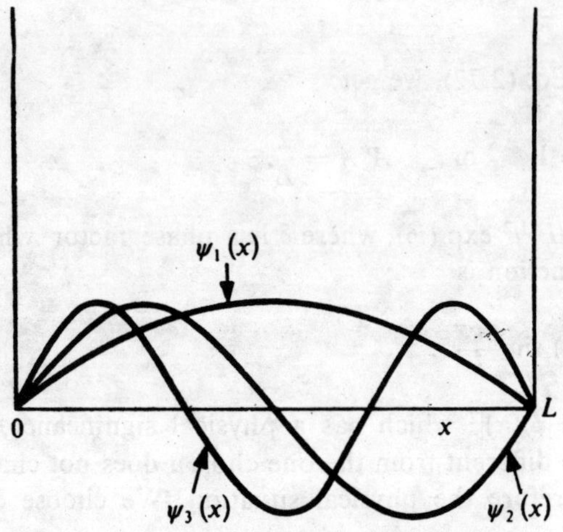

Fig. 2.10 Wave function of particle in box corresponding to ground state ψ_1 and first two excited states ψ_2 and ψ_3.

2.10 GENERALIZATION TO THREE DIMENSIONS

We have so far restricted ourselves to one-dimensional problems. We shall now generalize our discussion to three dimensions so that we can deal with a realistic physical situation. In such a situation, the wave function ψ is a function of x, y, z, and t. The Schroedinger equation (2.27) will then take the form

$$i\hbar\frac{\partial\psi(x, y, z, t)}{\partial t} = -\frac{\hbar^2}{2m}\left[\frac{\partial^2\psi(x, y, z, t)}{\partial x^2} + \frac{\partial^2\psi(x, y, z, t)}{\partial y^2} + \frac{\partial^2\psi(x, y, z, t)}{\partial z^2}\right] + V(x, y, z)\psi(x, y, z, t). \tag{2.75}$$

This is written in a more compact form as

$$-\frac{\hbar^2}{2m}\nabla^2\psi(x, y, z, t) + V(x, y, z)\psi(x, y, z, t) = i\hbar\frac{\partial\psi(x, y, z, t)}{\partial t}, \tag{2.76a}$$

where

$$\nabla^2\psi = \frac{\partial^2\psi(x, y, z, t)}{\partial x^2} + \frac{\partial^2\psi(x, y, z, t)}{\partial y^2} + \frac{\partial^2\psi(x, y, z, t)}{\partial z^2},$$

and the time-independent Schroedinger equation is

$$\nabla^2\psi(x, y, z) + \frac{2m}{\hbar^2}[E - V(x, y, z)]\psi(x, y, z) = 0. \tag{2.76b}$$

The momentum vector in three dimensions will have three components, namely, p_x, p_y, p_z.

Each of these is an operator in quantum mechanics. Generalizing Eq. (2.21), we have

$$p_x\psi(x, y, z, t) = -i\hbar\frac{\partial\psi(x, y, z, t)}{\partial x}, \tag{2.77}$$

$$p_y\psi(x, y, z, t) = -i\hbar\frac{\partial\psi(x, y, z, t)}{\partial y}, \tag{2.78}$$

$$p_z\psi(x, y, z, t) = -i\hbar\frac{\partial\psi(x, y, z, t)}{\partial z}. \tag{2.79}$$

The normalization condition in three dimensions is

$$\int_{-\infty}^{+\infty}\int_{-\infty}^{+\infty}\int_{-\infty}^{+\infty} |\psi(x, y, z, t)|^2\, dx\, dy\, dz = 1. \tag{2.80}$$

Similarly, we can generalize expressions (2.47) and (2.49) for the average values of operators to three dimensions.

EXAMPLE 2.20 Calculate the eigenfunctions and eigenvalues of energy for a particle confined in a two-dimensional square box of length L. Assume that the particle is free in the box [i.e., $V(x, y) = 0$].

As the particle is confined in the box, we have $\psi = 0$ outside the box. The time-independent Schroedinger equation in two dimensions for a particle inside a box is

$$\frac{\partial^2\psi(x, y)}{\partial x^2} + \frac{\partial^2\psi(x, y)}{\partial y^2} + \frac{2m}{\hbar^2}E\psi(x, y) = 0, \tag{2.81}$$

where $V = 0$. Let the box be so placed that the coordinates of its four corners are (0, 0), (0, L), (L, L), and (L, 0). Then, the boundary conditions on the wave functions are

$$\psi(x, y) = 0 \quad \text{if } x \geqslant L \text{ or } y \geqslant L \text{ or } x \leqslant 0 \text{ or } y \leqslant 0. \tag{2.82}$$

The Schroedinger equation (2.81) can be solved by the method of separation of variables, that is, by assuming $\psi(x, y) = R(x)S(y)$, where R is a function of x alone and S is a function of y alone. Then, the Schroedinger equation becomes

$$S(y)\frac{d^2R(x)}{dx^2} + R(x)\frac{d^2S(y)}{dy^2} + \frac{2m}{\hbar^2}ER(x)S(y) = 0. \tag{2.83}$$

Dividing Eq. (2.83) by $R(x)S(y)$, we obtain

$$\frac{1}{R(x)}\frac{d^2R(x)}{dx^2} + \frac{1}{S(y)}\frac{d^2S(y)}{dy^2} + \frac{2m}{\hbar^2}E = 0. \tag{2.84}$$

In this equation, the first term depends only on x and the second term only on y, and the last term is a constant. Such an equation can be satisfied only if each of the terms in it is a constant. Let

$$\frac{1}{S(y)}\frac{d^2S(y)}{dy^2} = C, \tag{2.85}$$

where C is a constant. Substituting this in Eq. (2.84), we get

$$\frac{1}{R(x)}\frac{d^2R(x)}{dx^2} = -\frac{2m}{\hbar^2}E - C = D, \tag{2.86}$$

where D is another constant. Let us first solve Eq. (2.85); it can be rewritten as

$$d^2S/dy^2 - CS = 0. \tag{2.87}$$

If $C > 0$, then $S = A \exp(C^{1/2}y) + B \exp(-C^{1/2}y)$. Using the boundary conditions [see Eq. (2.82)] $\psi(x, y = 0) = \psi(x, y = L) = 0$, we get the relations

$$A + B = 0, \qquad A \exp(C^{1/2}L) + B \exp(-C^{1/2}L) = 0.$$

Combining these two relations, we obtain $A = B = 0$. This implies the wave function is zero everywhere. This in turn means that Eq. (2.85) has no solution if $C > 0$.

If $C = 0$, the solution of Eq. (2.85) is $S = P + Qx$. The boundary conditions (2.82) can be satisfied only when both P and Q are zero. Thus, Eq. (2.85) has again no solution if $C = 0$.

When $C < 0$, we can assume

$$C = -k_y^2, \tag{2.88}$$

where k_y is some real constant. Substituting this in Eq. (2.85), we get

$$d^2S/dy^2 + k_y^2 S = 0. \tag{2.89}$$

The solution of this equation is

$$S(y) = A \sin k_y y + B \cos k_y y. \tag{2.90}$$

Proceeding in a manner similar to that for a particle in a one-dimensional box and using the boundary conditions at $y = 0, L$ on $\psi(x, y)$, we have

$$S(y) = A \sin(\pi n_y y/L), \tag{2.91}$$

where

$$k_y = n_y \pi/L \tag{2.92}$$

with

$$n_y = 1, 2, 3, \ldots . \tag{2.93}$$

Using an argument similar to that for a one-dimensional box, we find that here $n_y = 0$ is unacceptable. Consider now the equation

$$\frac{d^2R(x)}{dx^2} - DR(x) = 0.$$

It is similar to Eq. (2.87), and we find that the boundary conditions (2.82) can be satisfied only if D is negative. Writing

$$D = -k_x^2, \tag{2.94}$$

we get

$$R(x) = A' \sin(\pi n_x x/L), \tag{2.95}$$

where A' is a constant and

$$n_x = 1, 2, 3, \ldots . \tag{2.96}$$

We also have

$$k_x = n_x \pi/L. \tag{2.97}$$

Using Eqs. (2.86), (2.88), and (2.94), we get

$$-2mE/\hbar^2 = C + D = -k_x^2 - k_y^2 \tag{2.98}$$

or

$$E = \frac{\hbar^2}{2m}(k_x^2 + k_y^2) = \frac{\hbar^2\pi^2}{2mL^2}(n_x^2 + n_y^2) \tag{2.99}$$

with $n_x = 1, 2, \ldots$ and $n_y = 1, 2, \ldots$. The allowed energy values are again discrete [see Eq. (2.70)] and not continuous. The wave function corresponding to the energy $\hbar^2\pi^2(n_x^2 + n_y^2)/(2mL^2)$ is given by

$$\psi_{n_x, n_y}(x, y) = \frac{2}{L} \sin \frac{n_x \pi x}{L} \sin \frac{n_y \pi y}{L}. \tag{2.100}$$

The constant $2/L$ can be determined by using the normalization condition

$$\int_0^L \int_0^L \psi^*_{n_x, n_y}(x, y) \psi_{n_x, n_y}(x, y)\, dx\, dy = 1. \tag{2.101}$$

This system has two quantum numbers, namely, n_x and n_y. The ground state (i.e., the lowest energy state) of the system has the energy

$$E = \hbar^2\pi^2/(mL^2) \tag{2.102}$$

which corresponds to $n_x = n_y = 1$. If $n_x = 0$ or $n_y = 0$, this will mean $\psi_{n_x, n_y} = 0$ and would correspond to the situation when there is no particle in the box. For the next state (i.e., first excited state), where $n_x = 2$, $n_y = 1$, the energy is

$$E_1 = \frac{\hbar^2\pi^2}{2mL^2}(2^2 + 1^2) = \frac{5\hbar^2\pi^2}{2mL^2}. \tag{2.103}$$

The wave function corresponding to Eq. (2.103) is $(2/L) \sin(2\pi x/L) \sin(\pi y/L)$. The state with $n_y = 2$, $n_x = 1$ also has the same energy E_1 but the wave function is $(2/L) \sin(\pi x/L) \sin(2\pi y/L)$.

Summarizing, the eigenvalues and the corresponding eigenfunctions of energy for a particle in the two-dimensional box we have are

$$E_{n_x, n_y} = \frac{\hbar^2\pi^2}{2mL^2}(n_x^2 + n_y^2), \qquad \psi_{n_x, n_y} = \frac{2}{L} \sin \frac{n_x \pi x}{L} \sin \frac{n_y \pi y}{L}, \tag{2.104}$$

where the quantum numbers n_x and n_y take only the nonzero positive integral values.

In Example 2.20, we have two different wave functions, namely, $(2/L)\sin(2\pi x/L)\sin(\pi y/L)$ and $(2/L)\sin(\pi x/L)\sin(2\pi y/L)$, corresponding to the same energy. Such a state is referred to as *two-fold degenerate*. In general, if n different wave functions correspond to the same energy, the state is said to be *n-fold degenerate*.

EXAMPLE 2.21 Consider a particle of mass m moving in a one-dimensional potential $V(x)$ and satisfying $V(x) = V(-x)$. If $\psi(x)$ is a stationary state with an energy E, then show that $\psi(-x)$ is also a stationary state with the same energy. Further, if there is no degeneracy associated with the energy E, then show that $\psi(x) = \psi(-x)$ or $\psi(x) = -\psi(-x)$.

The one-dimensional time-independent Schroedinger equation is

$$\frac{d^2\psi(x)}{dx^2} + \frac{2m}{\hbar^2}[E - V(x)]\psi(x) = 0, \tag{2.105}$$

where $\psi(x)$ is the stationary state wave function corresponding to the energy E. Replacing x by $-x$ in Eq. (2.105), we have

$$\frac{d}{d(-x)}\frac{d}{d(-x)}\psi(-x) + \frac{2m}{\hbar^2}[E - V(-x)]\psi(-x) = 0. \tag{2.106}$$

Using the symmetry of the potential $V(x) = V(-x)$, we get

$$\frac{d^2\psi(-x)}{dx^2} + \frac{2m}{\hbar^2}[E - V(x)]\psi(-x) = 0. \tag{2.107}$$

Comparing Eq. (2.107) with Eq. (2.105), we see that $\psi(-x)$ also satisfies the Schroedinger equation (2.105) for the same energy E. If there is no degeneracy associated with this energy, the wave functions $\psi(x)$ and $\psi(-x)$ then cannot be linearly independent. Thus,

$$\psi(x) = C\psi(-x), \tag{2.108}$$

where C is a constant. Substituting $-x$ for x in Eq. (2.108), we get

$$\psi(-x) = C\psi(x). \tag{2.109}$$

Using Eqs. (2.108) and (2.109), we obtain

$$\psi(x) = C\psi(-x) = C^2\psi(x). \tag{2.110}$$

This leads to $C^2 = 1$ or $C = \pm 1$. Thus,

$$\psi(x) = \psi(-x) \tag{2.111}$$

or

$$\psi(x) = -\psi(-x). \tag{2.112}$$

We shall use this result in Example 5.2.

2.11 GENERAL REQUIREMENTS ON WAVE FUNCTION

The solution $\psi(x, y, z)$ of the time-independent Schroedinger equation must satisfy the following conditions:

(1) The wave function must be finite, single-valued, and continuous everywhere.

(2) The three first partial derivatives of the wave function must also be finite, single-valued, and continuous everywhere.

The wave function must be finite and single-valued so that the probability density ($|\psi|^2$) is finite and single-valued. Further, if it is not continuous, then its first derivatives will not be finite. This means that the average value of momentum will not be finite, and therefore will not give a meaningful result. In fact, in order that the average value of momentum is physically reasonable, the first derivatives must be finite and single-valued. Finally, the first derivatives must also be continuous because the second derivatives needed in writing the Schroedinger equation (2.76b) have to be finite. If the partial derivatives are not continuous, then the second derivatives become infinite, and so $(E - V)\psi$ becomes infinite.

Let us consider the solution obtained for a particle in a box in one dimension [see Eq. (2.71)], i.e.,

$$\psi_n(x) = \left(\frac{2}{L}\right)^{1/2} \sin \frac{n\pi x}{L} \qquad (0 < x < L)$$

$$= 0 \qquad \text{(outside).}$$

It should be noticed that the wave function here is finite, single-valued, and continuous everywhere, i.e., $-\infty < x < +\infty$. The first derivative of the wave function is given by

$$\frac{d\psi_n(x)}{dx} = \left(\frac{2}{L}\right)^{1/2} \frac{n\pi}{L} \cos \frac{n\pi x}{L} \qquad (0 < x < L) \tag{2.113}$$

$$= 0 \qquad \text{(outside).}$$

It is finite and single-valued everywhere. However, $d\psi_n(x)/dx$ is not continuous at $x = 0, L$. To see this, consider the point $x = 0$. If we evaluate $d\psi_n/dx$ from $x > 0$, we get, using Eq. (2.113),

$$\lim_{x\to 0^+} \frac{d\psi_n(x)}{dx} = \left(\frac{2}{L}\right)^{1/2} \frac{n\pi}{L}. \tag{2.114}$$

On the other hand, when we approach from $x < 0$, we obtain

$$\lim_{x\to 0^-} \frac{d\psi_n(x)}{dx} = 0. \tag{2.115}$$

This shows that the value of $d\psi_n/dx$ at $x = 0$ depends on the side—left or right—from which we approach $x = 0$. Therefore, $d\psi_n/dx$ is not continuous at $x = 0$. A similar argument can be made for $x = L$. Thus, the second derivative $d^2\psi_n(x)/dx^2$ at $x = 0$ and $x = L$ is infinite which means that, in the Schroedinger equation

$$\frac{d^2\psi_n(x)}{dx^2} + \frac{2m}{\hbar^2}(E_n - V)\psi_n = 0,$$

$(E_n - V)\psi_n$ is infinite. Since ψ_n and E_n are finite, V must be infinite at $x = 0$ and $x = L$. This situation corresponds to a particle confined in a box.

The foregoing analysis shows that $d\psi_n/dx$ is not continuous only because we assumed that $V(x)$ is infinite at $x = 0$ and $x = L$. No physical problem corresponds to such a situation. In other words, for all physical problems, $d\psi_n/dx$ is continuous. The reason we have considered the example of a particle in a box is that the mathematics involved is simple.

2.12 SYSTEM OF TWO INDEPENDENT PARTICLES

We now consider a system of two particles. Such a system would require two coordinates $\boldsymbol{r}_1$ and $\boldsymbol{r}_2$ for the wave function to be specified. We denote this wave function by $\psi(\boldsymbol{r}_1, \boldsymbol{r}_2)$, and shall give its physical interpretation by generalizing Born's analysis of the wave function for a single particle. The probability of finding particle 1 in a volume Δv_1 around $\boldsymbol{r}_1$ and particle 2 in a volume Δv_2 around $\boldsymbol{r}_2$ is given by

$$P(\boldsymbol{r}_1, \boldsymbol{r}_2)\Delta v_1 \Delta v_2 = |\psi(\boldsymbol{r}_1, \boldsymbol{r}_2)|^2 \Delta v_1 \Delta v_2, \tag{2.116}$$

where $P(\boldsymbol{r}_1, \boldsymbol{r}_2)$ is the probability density for the two-particle system. The Schroedinger equation, generalizing Eq. (2.76a), is then

$$-\frac{\hbar^2}{2m_1}\nabla_1^2\psi(\boldsymbol{r}_1, \boldsymbol{r}_2, t) - \frac{\hbar^2}{2m_2}\nabla_2^2\psi(\boldsymbol{r}_1, \boldsymbol{r}_2, t) + V(\boldsymbol{r}_1, \boldsymbol{r}_2, t)\psi(\boldsymbol{r}_1, \boldsymbol{r}_2, t) = i\hbar\frac{\partial\psi(\boldsymbol{r}_1, \boldsymbol{r}_2, t)}{\partial t}, \tag{2.117}$$

where

$$\nabla_1^2\psi(\boldsymbol{r}_1, \boldsymbol{r}_2, t) = \frac{\partial^2\psi(\boldsymbol{r}_1, \boldsymbol{r}_2, t)}{\partial x_1^2} + \frac{\partial^2\psi(\boldsymbol{r}_1, \boldsymbol{r}_2, t)}{\partial y_1^2} + \frac{\partial^2\psi(\boldsymbol{r}_1, \boldsymbol{r}_2, t)}{\partial z_1^2} \tag{2.118}$$

and ∇_2^2 is similar to expression (2.118). If the potential is independent of time, we can write the wave function as

$$\psi(\boldsymbol{r}_1, \boldsymbol{r}_2, t) = \psi(\boldsymbol{r}_1, \boldsymbol{r}_2) \exp(-iEt/\hbar), \tag{2.119}$$

where E is the total energy of the system. The wave function $\psi(\boldsymbol{r}_1, \boldsymbol{r}_2)$ obeys the time-independent Schroedinger equation

$$+\frac{\hbar^2}{2m_1}\nabla_1^2\psi + \frac{\hbar^2}{2m_2}\nabla_2^2\psi + (E - V)\psi = 0. \tag{2.120}$$

EXAMPLE 2.22 Consider a system of two particles of mass m_1 and m_2 confined in a one-dimensional box of length L. The two particles are moving freely so that the energy of the system is the sum of their kinetic energies. What are the allowed energies and the corresponding wave functions for this system?

In one dimension, the time-independent part of the wave function is $\psi(x_1, x_2)$ and it satisfies the Schroedinger equation (with $V = 0$)

$$\frac{\hbar^2}{2m_1}\frac{\partial^2\psi(x_1, x_2)}{\partial x_1^2} + \frac{\hbar^2}{2m_2}\frac{\partial^2\psi(x_1, x_2)}{\partial x_2^2} + E\psi(x_1, x_2) = 0. \tag{2.121}$$

The boundary conditions are

$$\psi(x_1, x_2) = 0 \qquad (x_1 \geqslant L \text{ or } x_1 \leqslant 0;\ x_2 \geqslant L \text{ or } x_2 \leqslant 0). \tag{2.122}$$

Using the method of separation of variables, we write

$$\psi(x_1, x_2) = \phi(x_1)\chi(x_2) \tag{2.123}$$

and separate Eq. (2.121) into the two equations

$$\frac{\hbar^2}{2m_1}\frac{d^2\phi}{dx_1^2} + E_1\phi = 0, \tag{2.124}$$

$$\frac{\hbar^2}{2m_2}\frac{d^2\chi}{dx_2^2} + E_2\chi = 0 \tag{2.125}$$

with

$$E_1 + E_2 = E. \tag{2.126}$$

The wave functions $\phi(x_1)$ and $\chi(x_2)$ satisfy the boundary conditions

$$\phi(x_1) = 0 \qquad (x_1 \geqslant L,\ x_1 \leqslant 0), \tag{2.127}$$

$$\chi(x_2) = 0 \qquad (x_2 \geqslant L,\ x_2 \leqslant 0). \tag{2.128}$$

We see that each of Eqs. (2.124) and (2.125) is identical with the equation for a particle in a box, namely, (2.73), and hence the solutions of the two equations are

$$\phi(x_1) = (\frac{2}{L})^{1/2} \sin \frac{n_1\pi x_1}{L} \qquad (n_1 = 1, 2, \ldots), \tag{2.129}$$

$$\chi(x_2) = (\frac{2}{L})^{1/2} \sin \frac{n_2\pi x_2}{L} \qquad (n_2 = 1, 2, \ldots) \tag{2.130}$$

with

$$E_1 = \frac{\hbar^2\pi^2}{2m_1L^2}n_1^2, \tag{2.131}$$

$$E_2 = \frac{\hbar^2\pi^2}{2m_2L^2}n_2^2. \tag{2.132}$$

Thus, the total allowed energies of the system are

$$E = \frac{\hbar^2\pi^2}{2L^2}(\frac{n_1^2}{m_1} + \frac{n_2^2}{m_2}), \qquad n_1 = 1, 2, 3, \ldots,\ n_2 = 1, 2, 3, \ldots, \tag{2.133}$$

and the wave functions corresponding to it are

$$\psi_{n_1n_2}(x_1, x_2) = \frac{2}{L} \sin \frac{n_1\pi x_1}{L} \sin \frac{n_2\pi x_2}{L}. \tag{2.134}$$

We see that we need two quantum numbers n_1 and n_2 to describe a two-particle system in one dimension.

2.13 SYSTEM OF TWO IDENTICAL PARTICLES

When we have a system of two identical particles, e.g., two electrons and two photons, then the wave function of the system is required to satisfy an important property, namely, if $\psi(\mathbf{r}_1, \mathbf{r}_2, t)$ describes the system at a given time t, then the wave function is either symmetric or antisymmetric under the interchange of particles 1 and 2. That is,

$$\psi(\mathbf{r}_1, \mathbf{r}_2, t) = \psi(\mathbf{r}_2, \mathbf{r}_1, t) \qquad \text{(symmetric)} \tag{2.135}$$

or

$$\psi(\mathbf{r}_1, \mathbf{r}_2, t) = -\psi(\mathbf{r}_2, \mathbf{r}_1, t) \qquad \text{(antisymmetric).} \tag{2.136}$$

If the wave function of the particles is symmetric, then the particles are said to obey the *Bose-Einstein statistics*, and if it is antisymmetric, the particles are said to obey the *Fermi-Dirac statistics*. The former type of particles are known as *bosons* and the latter, *fermions*. Here, we are ignoring spin wave functions which we shall discuss in Chapters 4 and 6.

EXAMPLE 2.23 Consider two identical particles in a one-dimensional box of length L. Assume that the particles are moving independent of each other. For this two-particle system, find the wave functions corresponding to the allowed energies if the particles are (a) bosons and (b) fermions.

The single-particle wave functions for the two particles are

$$\psi_{n_1}(x_1) = \left(\frac{2}{L}\right)^{1/2} \sin\frac{\pi n_1 x_1}{L}, \qquad \psi_{n_2}(x_2) = \left(\frac{2}{L}\right)^{1/2} \sin\frac{\pi n_2 x_2}{L}.$$

From Eq. (2.133) with $m_1 = m_2 = m$, the total energy of the system is

$$E = \frac{\hbar^2\pi^2}{2mL^2}(n_1^2 + n_2^2)$$

and its wave function is

$$\psi_{n_1n_2}(x_1, x_2) = \psi_{n_1}(x_1)\psi_{n_2}(x_2).$$

We note that the wave function

$$\psi_{n_2n_1}(x_1, x_2) = \psi_{n_2}(x_1)\psi_{n_1}(x_2)$$

also describes a state of the system with the same energy, i.e., with

$$E = \frac{\hbar^2\pi^2}{2mL^2}(n_1^2 + n_2^2).$$

We can write the general wave function corresponding to the energy E as a linear combination of the two functions $\psi_{n_1n_2}$ and $\psi_{n_2n_1}$. Thus,

$$\psi(x_1, x_2) = A\psi_{n_1}(x_1)\psi_{n_2}(x_2) + B\psi_{n_1}(x_2)\psi_{n_2}(x_1). \tag{2.137}$$

For bosons, we have

$$\psi(x_1, x_2) = \psi(x_2, x_1) = A\psi_{n_1}(x_2)\psi_{n_2}(x_1) + B\psi_{n_1}(x_1)\psi_{n_2}(x_2). \tag{2.138}$$

Comparing Eqs. (2.137) and (2.138), we find that $A = B$. For fermions, we have

$$\psi(x_1, x_2) = -\psi(x_2, x_1) = -A\psi_{n_1}(x_2)\psi_{n_2}(x_1) - B\psi_{n_1}(x_1)\psi_{n_2}(x_2). \tag{2.139}$$

Comparing Eqs. (2.137) and (2.139), we see that $A = -B$. We determine the constant A using the normalization condition

$$\int_0^L \int_0^L |\psi(x_1, x_2)|^2 \, dx_1 \, dx_2 = 1.$$

Thus, if the particles are bosons and $n_1 \neq n_2$, the wave function is

$$\psi(x_1, x_2) = \frac{1}{\sqrt{2}}[\psi_{n_1}(x_1)\psi_{n_2}(x_2) + \psi_{n_2}(x_1)\psi_{n_1}(x_2)], \tag{2.140}$$

and if the particles are fermions, the wave function is

$$\psi(x_1, x_2) = \frac{1}{\sqrt{2}}[\psi_{n_1}(x_1)\psi_{n_2}(x_2) - \psi_{n_2}(x_1)\psi_{n_1}(x_2)]. \tag{2.141}$$

For fermions, the wave function vanishes when $n_1 = n_2$. This means that the two particles cannot be in the same state. In the case of bosons, the normalized wave function, when $n_1 = n_2$, is

$$\psi(x_1, x_2) = \psi_{n_1}(x_1)\psi_{n_1}(x_2). \tag{2.142}$$

PROBLEMS

2.1 What is the energy of a photon with a wavelength 7000 Å?

2.2 What is the wavelength of a photon with an energy 1 eV?

2.3 Show that the dimensions of Planck's constant are the same as those of angular momentum.

2.4 Estimate the number of photons hitting per second an eye of a person looking at a 100-W bulb from a distance of 10 m (diameter of eyeball = 3 mm).

2.5 Calculate (a) Planck's constant and (b) the work function from the following data obtained from a photoelectric experiment:

Wavelength	*Stopping potential*
2536 Å	1.95 V
3650 Å	0.5 V

2.6 A proton is bound to a heavy nucleus with a binding energy of 1 MeV ($=10^6$ eV). Calculate the wavelength of a photon needed for removing the proton from the nucleus.

2.7 An electron falls from a height of 10 m and passes through a hole of radius 1 cm. To study the motion of the electron after it has passed through the hole, should we apply the wave aspect or the particle aspect?

2.8 Derive the expression for the ratio of the electron wavelength to the photon wavelength for the electrons and photons of equal energy.

2.9 (a) What is the speed of a proton with a de Broglie wavelength 0.1 Å?

(b) What is the electrostatic potential needed to accelerate a proton to the speed obtained in Problem 2.9a?

2.10 Consider two electrons, A and B, initially having the same kinetic energy. The electron A hits a heavy target and emits all its energy in the form of a photon. The electron B, on the other hand, continues to move freely. Which of the two—photon or electron B—will have the larger momentum?

2.11 What is the energy of a photon having the same momentum as that of a 10-MeV proton?

2.12 A single crystal of KCl has a simple cubic structure. Calculate the spacing d between the atoms, given that the molecular weight of KCl is 74.6 and the density of KCl is 1.98 g-cm^{-3}. What would be the angle θ for the first order diffraction if an electron diffraction experiment is performed with this crystal using electrons of energy 50 keV? Compare the angle so obtained with that calculated using photons of energy 50 keV.

2.13 Write an expression for a plane wave travelling along the negative x-axis.

2.14 Write an expression for a plane wave travelling in two dimensions along a direction making an angle of 45° with the positive x-axis.

2.15 The wave function of a particle confined in a cubical box of side L is

$$\psi = \left(\frac{2}{L}\right)^{3/2} \sin \frac{\pi x}{L} \sin \frac{\pi y}{L} \sin \frac{\pi z}{L}.$$

Calculate the probability of finding the particle in the cubical volume $0 \leqslant x \leqslant L/2$, $0 \leqslant y \leqslant L/2$, and $0 \leqslant z \leqslant L/2$.

2.16 Use the uncertainty principle to estimate the energy of a marble of mass 10 g kept in a cubical box of side 10 cm.

2.17 Assume that an atom is confined in a cubical box of side 1 Å. Use the uncertainty principle to estimate its energy.

2.18 Show that the solution of Schroedinger's equation for a particle moving freely (i.e., $V = 0$) in one dimension is

$$\psi = \text{constant} \cdot \exp\left[-\frac{i}{\hbar}(Et - px)\right].$$

2.19 A particle is confined in a one-dimensional box of length L. The box lies in the region $-L/2 \leqslant x \leqslant L/2$. Calculate the energy eigenvalues of the particle and the wave functions corresponding to these eigenvalues. Also, show that these eigenvalues are the same as those obtained in Section 2.9.

2.20 Plot the probability density as a function of x for a particle in a one-dimensional box (see Section 2.9) when the particle is in (a) the ground state, (b) the first excited state, and (c) the second excited state.

2.21 What is the average value of momentum for a system described by

$$\psi(x) = A \exp(ikx) + B \exp(-ikx)$$

in the region $0 \leqslant x \leqslant \pi/k$, and $\psi(x) = 0$ outside this region?

2.22 Consider a particle of mass 10 g and energy $E = 1$ erg confined in a one-dimensional box of length 10 cm. Calculate (a) the approximate quantum number n of this state of the system and (b) the energy required to take the system to its next excited state.

2.23 Consider a particle in a two-dimensional rectangular box of sides L_1 and L_2. Find the allowed energy values of the particle and the wave functions corresponding to these values.

2.24 For a particle in a two-dimensional square box of length L, calculate the expectation values of x^2, $x^2 + y^2$, p_x^2, and $p_x^2 + p_y^2$. Assume that the particle is in the ground state.

2.25 Find the allowed energy values and the wave functions corresponding to these values for a particle of mass m moving in a cuboid of sides a, b, and c.

2.26 Consider a particle confined in a three-dimensional cubical box of side L. Use the method of separation of variables to solve the time-independent Schroedinger equation and obtain, for the particle, (a) the energy eigenvalues, (b) the wave functions corresponding to the energy eigenvalues, and (c) the degeneracy of the first and second excited states.

2.27 Find the allowed energy values and the wave functions corresponding to these values for a system of two particles of masses m_1 and m_2 moving freely in a cuboid of sides a, b, and c $(a > b > c)$.

2.28 In Problem 2.27, assume that the particles are identical (i.e., $m_1 = m_2 = m$). Find the ground state wave function of the system if these particles are (a) bosons, (b) fermions.

3
Some Applications of Quantum Mechanics

3.1 INTRODUCTION

In this chapter, we shall apply the principles of quantum mechanics to some simple systems. In so doing, we shall bring out the general features of quantum physics and use these when dealing with such systems as atoms, molecules, nuclei, and solids.

3.2 BOUND STATE PROBLEMS

Whenever a particle is confined in a finite region of space, we refer to it as a bound particle. Any solid, for instance, is a bound system of many atoms. Also, an atom is a bound system of electrons and the nucleus.

A general feature of a bound particle system is that the wave function of the particle vanishes when we go infinitely far from the region in which it is confined. The reason for this is that the probability of finding it very far from this region is negligibly small. In our study of a particle confined in a box (see Section 2.9), we had $\psi(x) = 0$ outside the box. This is an example of a bound system. In contrast, the wave function of an unbound system or a free particle does not vanish at an infinite distance. For example, a plane wave $\exp[i(kx - \omega t)]$, which represents a free particle, does not vanish as $x \to \pm\infty$.

As we shall soon discover, any bound particle, just as a particle in a box, is allowed to have only certain values of energy. In other words, the energy is quantized, that is, it can take only discrete values. The various allowed values of energy for a system and the corresponding wave functions depend on the nature of the forces binding the system. We shall consider two important examples of bound systems to illustrate these general observations.

3.3 SIMPLE HARMONIC OSCILLATOR (ONE-DIMENSIONAL)

In a simple harmonic oscillator, a particle moves under a force which is proportional to the displacement of the particle from a point (say, the origin) and acts towards the point. If F is the force and x is the displacement, then

$$F = -kx, \tag{3.1}$$

where k is the spring constant. The potential energy of the particle is

$$V = -\int_0^x F\,dx \qquad \text{or} \qquad V(x) = \tfrac{1}{2}kx^2 + \text{constant}.$$

If we take the potential energy V as zero when $x = 0$, then the potential is

$$V(x) = \tfrac{1}{2}kx^2. \tag{3.2}$$

The total energy of the system is then $p^2/(2m) + kx^2/2$, where m is the mass of the particle. As the potential is time-independent, the wave function $\psi(x, t)$ can be written as

$$\psi(x, t) = \exp(-iEt/\hbar)\psi(x) \tag{3.3}$$

with $\psi(x)$ satisfying the Schroedinger equation

$$\frac{d^2\psi}{dx^2} + \frac{2m}{\hbar^2}(E - \tfrac{1}{2}kx^2)\psi = 0, \tag{3.4}$$

where E is the energy of the particle. The boundary condition on the wave function is $\psi(x) = 0$ as $x \to \pm\infty$. The physical reason for this boundary condition, as already stated, is that the particle is confined in a finite region.

The solution of the differential equation (3.4) under the boundary conditions specified is well known. (For the details of solving this equation, see Appendix 13.) Since our aim is to understand the physics underlying the solution of this equation, we shall give here only the solution. Equation (3.4), under the boundary condition

$$\psi(x) \to 0 \quad \text{as} \quad x \to \pm\infty, \tag{3.5}$$

has solutions only for certain values of E. These values of E are given by

$$E_n = (n + \tfrac{1}{2})\hbar\omega, \tag{3.6}$$

where $\omega = (k/m)^{1/2}$ and $n = 0, 1, 2, \ldots$. This means that the allowed energy levels are $\hbar\omega/2$, $3\hbar\omega/2, \ldots$. The wave function $\psi_n(x)$ corresponding to the eigenvalue with energy E_n is a rather complicated expression. Therefore, we give the wave functions for only the ground state and the first excited state:

$$\psi_0(x) = C_0 \exp(-\rho^2/2), \tag{3.7}$$

$$\psi_1(x) = C_1 \exp(-\rho^2/2)\rho, \tag{3.8}$$

where

$$\rho = (km/\hbar^2)^{1/4}x. \tag{3.9}$$

Here, C_0 and C_1 are constants which are evaluated using the normalization condition

$$\int_{-\infty}^{+\infty} |\psi_n|^2\, dx = 1. \tag{3.10}$$

The wave functions $\psi_n(x)$ are expressed in terms of the Hermite polynomials. (For a list of these polynomials, see any standard text on quantum mechanics.) Here we shall not need these wave functions except those given by (3.7) and (3.8).

The allowed energy values of a simple harmonic oscillator along with the simple harmonic oscillator potential are shown in Fig. 3.1a. From this figure, we observe that:

(1) The minimum energy a particle moving in a one-dimensional simple harmonic oscillator potential can have is $\hbar\omega/2$. It cannot be zero because we have applied the boundary condition

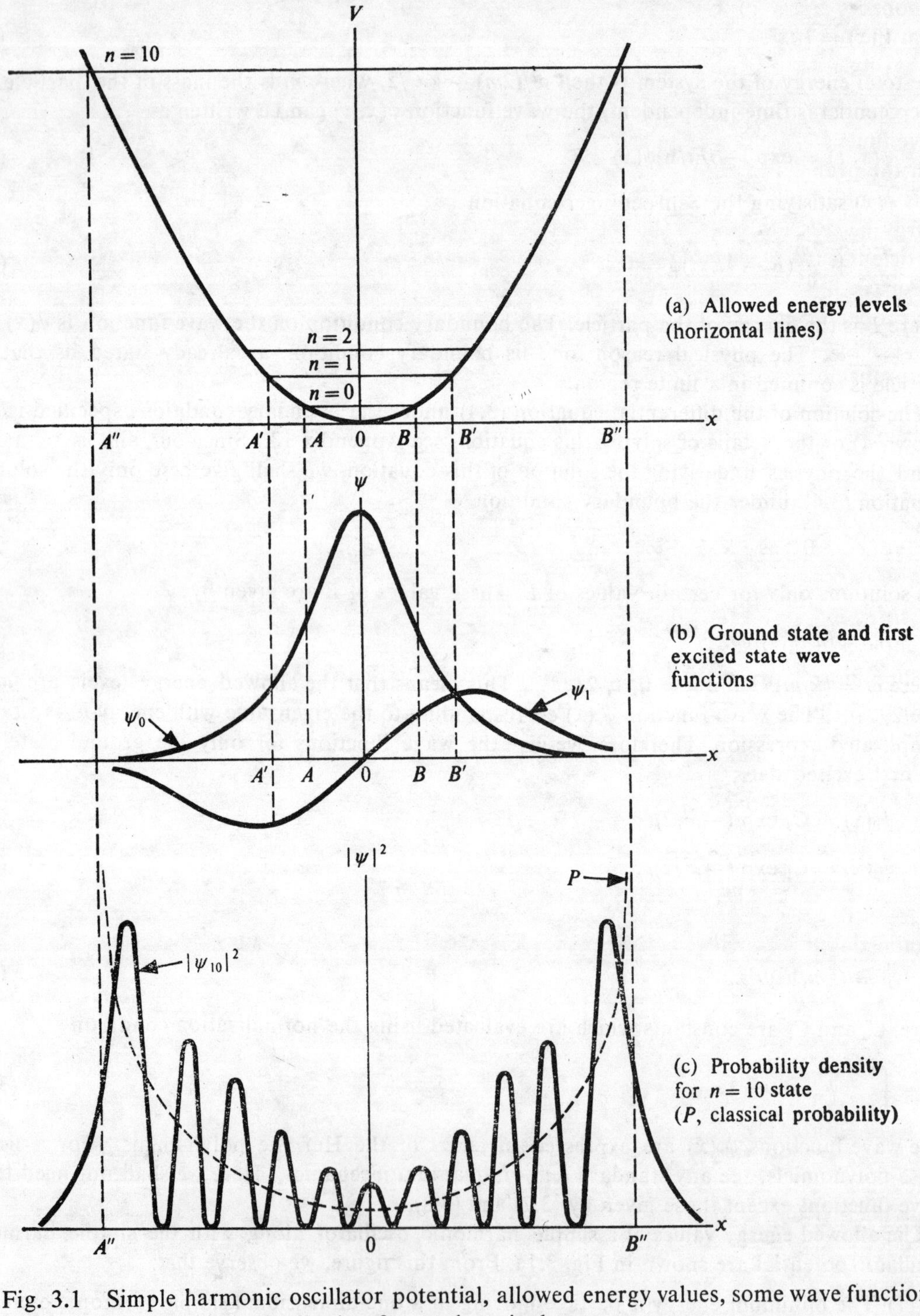

Fig. 3.1 Simple harmonic oscillator potential, allowed energy values, some wave functions, and probability density distribution for tenth excited state.

on the wave function. [The ground state energy is known as the *zero point energy*. The existence of nonzero energy in the minimum energy state (ground state) can be physically understood from the uncertainty principle (see also Example 3.1).]

(2) The allowed energy states are uniformly spaced.

The general shape of the first two eigenfunctions is sketched in Fig. 3.1b. Here, the ground state wave function has a maximum at $x = 0$ and it decreases as $|x|$ increases. It should be noted that the probability of finding the particle for a large value of $|x|$, though small, is not zero. This result is drastically different from that in classical physics where the particle having an energy $\hbar\omega/2$ is confined in the region *AB*. Quantum mechanically, the probability of finding a particle outside the classically allowed region is nonzero.

For the first excited state, the wave function has a node (i.e., it vanishes) at the origin. Thus, the probability of finding the particle at $x = 0$, when it is in the first excited state, is zero. As we go to a higher excited state, the wave function develops more nodes. For a large value of n, the probability density $\psi^*\psi$ exhibits an oscillatory behaviour and the average probability density resembles the one obtained classically (see Problem 3.7). The quantum mechanical probability density for the tenth excited state is shown in Fig. 3.1c.

QUESTION 3.1 Suppose a particle in a simple harmonic potential has an energy $\hbar\omega/2$. What would be the maximum value of x for the particle when the harmonic oscillator is treated classically?

EXAMPLE 3.1 Use the uncertainty principle to estimate the minimum energy of a particle in a simple harmonic potential $kx^2/2$.

Suppose the particle position is fixed precisely at $x = 0$. The application of the uncertainty principle then gives $\Delta x \Delta p \sim \hbar/2$. Since $\Delta x = 0$, Δp is infinite, and so the kinetic energy is infinite. This therefore cannot be the lowest energy state.

To obtain the ground state energy, let us assume that the particle position is uncertain by Δx. Then, from the uncertainty principle, the uncertainty in particle momentum is $\Delta p = \hbar/(2\Delta x)$. Thus, the total energy of the system is of the order of

$$\tfrac{1}{2}k(\Delta x)^2 + \frac{(\Delta p)^2}{2m} = \tfrac{1}{2}k(\Delta x)^2 + \frac{\hbar^2}{8m(\Delta x)^2}.$$

Minimizing this expression with respect to Δx, we get

$$k\Delta x - \frac{\hbar^2}{4m(\Delta x)^3} = 0 \quad \text{or} \quad \Delta x = \left(\frac{\hbar^2}{4mk}\right)^{1/4}.$$

Substituting the value of Δx in the expression for total energy, we get

$$E = \tfrac{1}{2}k\left(\frac{\hbar^2}{4mk}\right)^{1/2} + \frac{\hbar^2}{8m}\left(\frac{4mk}{\hbar^2}\right)^{1/2} = \frac{\hbar}{2}\left(\frac{k}{m}\right)^{1/2} = \frac{\hbar\omega}{2}.$$

The uncertainty principle can give us only an estimate and not the precise value of the ground state energy. In this example, we have accidentally obtained the correct numerical value.

EXAMPLE 3.2 A simple pendulum of mass 100 g and length 1 m has an amplitude of oscillation of 1 cm. Evaluate (a) the energy of this system, treating the system classically, (b) the approximate value of the quantum number n so that it leads to the same energy value as that obtained in (a), and (c) the energy separation between the adjacent allowed states.

(a) The energy of the pendulum at its extreme point of oscillation is entirely its potential energy since its kinetic energy is zero. So,

$$E = mgl(1 - \cos\theta_0) \sim \frac{mgl}{2}\theta_0^2,$$

where m and l are the mass and the length of the pendulum and θ_0 is the angle made by the pendulum with the vertical when it is at its extreme position. In our example, $\theta_0 = 1/100 = 10^{-2}$ rad, $m = 0.1$ kg, and $l = 1$ m. Thus, the energy of the system is

$$E = \frac{0.1 \times 9.8 \times 1 \times (10^{-2})^2}{2} = 4.9 \times 10^{-5} \text{ J}.$$

(b) Since the pendulum is a simple harmonic oscillator, its allowed energy values in quantum mechanics are

$$E = (n + \tfrac{1}{2})\hbar\omega,$$

where

$$\omega = \frac{2\pi}{T} = \left(\frac{g}{l}\right)^{1/2} = \left(\frac{9.8}{1}\right)^{1/2} \sim 3.1 \text{ rad/sec}.$$

The approximate value of the quantum number can be obtained by equating this expression to the energy obtained classically, i.e.,

$$E = (n + \tfrac{1}{2})\hbar \times 3.1 = 4.9 \times 10^{-5},$$

$$n + \tfrac{1}{2} = \frac{4.9 \times 10^{-5}}{1.05 \times 10^{-34} \times 3.1}, \qquad n \sim 1.5 \times 10^{29}.$$

(c) The separation between the adjacent energy levels is

$$E_{n+1} - E_n = [(n + 1) + \tfrac{1}{2}]\hbar\omega - (n + \tfrac{1}{2})\hbar\omega = \hbar\omega \sim 1.05 \times 10^{-34} \times 3.1 \sim 3.3 \times 10^{-34} \text{ J}.$$

It should be noted that the energy separation is extremely small as compared to the energy of the pendulum. In fact, it is impossible to observe such a small energy change and, for all practical purposes, the energy of the pendulum can be treated as a continuous variable. Therefore, the classical concepts are well suited for understanding the mechanics of a pendulum.

We have observed that the quantum number of the pendulum is very large. This is a particular case of a general result that the classical systems are the limits of quantum systems when the quantum numbers become very large.

EXAMPLE 3.3 Consider a particle of mass m in the two-dimensional potential $V = \frac{1}{2}k(x^2 + y^2)$. Find the allowed energy levels (eigenvalues) of this system.

The Schroedinger equation here is

$$i\hbar\frac{\partial\psi(x, y, t)}{\partial t} = -\frac{\hbar^2}{2m}\left[\frac{\partial^2\psi(x, y, t)}{\partial x^2} + \frac{\partial^2\psi(x, y, t)}{\partial y^2}\right] + \tfrac{1}{2}k(x^2 + y^2)\psi(x, y, t). \qquad (3.11)$$

Since the given potential is time-independent, we separate the wave function as

$$\psi(x, y, t) = \exp(-iEt/\hbar)\psi(x, y). \tag{3.12}$$

The wave function $\psi(x, y)$ obeys the time-independent Schroedinger equation

$$\frac{\partial^2\psi(x, y)}{\partial x^2} + \frac{\partial^2\psi(x, y)}{\partial y^2} + \frac{2m}{\hbar^2}[E - \tfrac{1}{2}k(x^2 + y^2)]\psi(x, y) = 0. \tag{3.13}$$

The boundary condition on the wave function is

$$\psi(x, y) \to 0 \quad \text{as} \quad x \to \pm\infty \quad \text{or} \quad y \to \pm\infty. \tag{3.14}$$

If the potential can be written as the sum $V(x) + V(y)$, the method of separation of variables is applicable (see Problem 3.10). So, Eq. (3.13) can be solved by assuming

$$\psi(x, y) = \Phi(x)X(y). \tag{3.15}$$

Substituting this in Eq. (3.13) and dividing the result so obtained by $\Phi(x)X(y)$, we get

$$\frac{1}{\Phi(x)}\frac{d^2\Phi(x)}{dx^2} + \frac{1}{X(y)}\frac{d^2X(y)}{dy^2} + \frac{2m}{\hbar^2}[E - \tfrac{1}{2}k(x^2 + y^2)] = 0 \tag{3.16}$$

or

$$\frac{1}{\Phi(x)}\frac{d^2\Phi(x)}{dx^2} + \frac{2m}{\hbar^2}(E - \tfrac{1}{2}kx^2) = -\frac{1}{X(y)}\frac{d^2X(y)}{dy^2} + \frac{2m}{\hbar^2}\frac{k}{2}y^2. \tag{3.17}$$

It should be noted that in Eq. (3.17) the left-hand side depends only on x, and the right-hand side only on y. Since this is valid for all values of x and y, we have

$$\frac{1}{\Phi(x)}\frac{d^2\Phi(x)}{dx^2} + \frac{2m}{\hbar^2}(E - \tfrac{1}{2}kx^2) = -\frac{1}{X(y)}\frac{d^2X(y)}{dy^2} + \frac{2m}{\hbar^2}\frac{k}{2}y^2 = A, \tag{3.18}$$

where A is a constant. Equation (3.18) gives two equations, namely,

$$\frac{d^2\Phi}{dx^2} + \frac{2m}{\hbar^2}(E - \tfrac{1}{2}kx^2)\Phi = A\Phi, \tag{3.19}$$

$$\frac{d^2X}{dy^2} - \frac{2m}{\hbar^2}\tfrac{1}{2}ky^2X = -AX. \tag{3.20}$$

Equations (3.19) and (3.20) can be rewritten as

$$\frac{d^2\Phi}{dx^2} + \frac{2m}{\hbar^2}(E_1 - \tfrac{1}{2}kx^2)\Phi = 0, \tag{3.21}$$

$$\frac{d^2X}{dy^2} + \frac{2m}{\hbar^2}(E_2 - \tfrac{1}{2}ky^2)X = 0, \tag{3.22}$$

where

$$E_1 = E - \hbar^2A/(2m), \tag{3.23}$$

$$E_2 = \hbar^2A/(2m). \tag{3.24}$$

The boundary conditions on $\Phi(x)$ and $X(y)$ follow from the boundary condition (3.14) and Eq. (3.15), and are given by

$$\Phi(x) \to 0 \quad \text{as} \quad x \to \pm\infty, \tag{3.25}$$

$$X(y) \to 0 \quad \text{as} \quad y \to \pm\infty. \tag{3.26}$$

We note that Eq. (3.21) is similar to Eq. (3.4) and the boundary condition (3.25) is similar to the boundary condition (3.5). Thus, the allowed energies are

$$E_1 = (n_1 + \tfrac{1}{2})\hbar\omega, \tag{3.27}$$

where $n_1 = 0, 1, 2, \ldots$ and $\omega = (k/m)^{1/2}$. Similarly, Eq. (3.22) and the boundary condition (3.26) give $E_2 = (n_2 + \frac{1}{2})\hbar\omega$, where $n_2 = 0, 1, 2, \ldots$. The energy E of the particle, from Eqs. (3.23) and (3.24), is

$$E = E_1 + E_2 = (n_1 + n_2 + 1)\hbar\omega. \tag{3.28}$$

This energy is allowed only discrete values, and now depends on two quantum numbers n_1 and n_2.

The energies of the ground state and the first two excited states of the particle are

$$E_0 = \hbar\omega \qquad (n_1 = n_2 = 0),$$

$$E_1 = 2\hbar\omega \qquad (n_1 = 1, n_2 = 0 \text{ or } n_1 = 0, n_2 = 1),$$

$$E_2 = 3\hbar\omega \qquad (n_1 = 2, n_2 = 0 \text{ or } n_1 = n_2 = 1 \text{ or } n_1 = 0, n_2 = 2).$$

As is clear, the ground state is nondegenerate, whereas the first excited state is two-fold degenerate and the second excited state three-fold degenerate.

QUESTION 3.2 How many quantum numbers are needed to describe a particle moving in a three-dimensional harmonic oscillator potential?

3.4 PARTICLE IN SPHERICALLY SYMMETRIC POTENTIAL

Consider a particle moving in three dimensions under the influence of a force exerted by another particle located at the origin. Let us assume that the force can be represented by a potential $V(\mathbf{r})$, where $\mathbf{r}$ is the separation between the two particles. A potential is said to be spherically symmetric if

$$V(\mathbf{r}) = V(|\mathbf{r}|) = V(r). \tag{3.29}$$

In other words, the potential depends on the distance between the two particles alone and not on the direction of r. In our analysis, we shall use the polar coordinates r, θ, ϕ (see Appendix 5) for solving the Schroedinger equation, the reason being that the potential depends only on r and not on θ and ϕ. This, as we shall see, simplifies the Schroedinger equation. Were we to use the cartesian coordinates, V then would be a function of all the three coordinates x, y, and z. A force giving rise to a spherically symmetric potential is quite common and is known as the *central force*. An example of such a force is the force between two particles, one with a positive charge and the other carrying a negative charge, where the potential is given by

$$V(r) = q_1 q_2/(4\pi\epsilon_0 r), \tag{3.30}$$

r being the distance between the two particles, and q_1 and q_2 the respective charges. Clearly, this potential here does not depend on the angles θ and ϕ.

We shall first study the general case of motion of a particle in a spherically symmetric potential of which the Coulomb potential is a special case, and then discuss the quantum mechanics of a particle moving in a Coulomb potential.

QUESTION 3.3 Is the potential between two magnetic dipoles central?

QUESTION 3.4 A particle is moving in a potential $V = (k_1x^2 + k_2y^2 + k_3z^2)/2$. Is the potential spherically symmetric?

The time-independent Schroedinger equation for a particle in a spherically symmetric potential $V(r)$ is given by

$$\nabla^2\psi(r, \theta, \phi) + \frac{2m}{\hbar^2}[E - V(r)]\psi(r, \theta, \phi) = 0. \tag{3.31}$$

Substituting for the operator ∇^2 in spherical polar coordinates (see Appendix 5), we get

$$\frac{1}{r^2}\frac{\partial}{\partial r}\left(r^2\frac{\partial\psi}{\partial r}\right) + \frac{1}{r^2 \sin\theta}\frac{\partial}{\partial\theta}\left(\sin\theta\frac{\partial\psi}{\partial\theta}\right) + \frac{1}{r^2\sin^2\theta}\frac{\partial^2\psi}{\partial\phi^2} + \frac{2m}{\hbar^2}[E - V(r)]\psi = 0. \tag{3.32}$$

Since V is independent of θ and ϕ and depends only on r, we use the method of separation of variables. In Eq. (3.32), separating from ψ the r-part and the (θ, ϕ)-part, we obtain

$$\psi(r, \theta, \phi) = R(r)\Omega(\theta, \phi). \tag{3.33}$$

Substituting this in Eq. (3.32) and dividing the result obtained by $R(r)\Omega(\theta, \phi)/r^2$, we have

$$\frac{1}{R(r)}\frac{d}{dr}\left(r^2\frac{dR}{dr}\right) + \frac{1}{\Omega(\theta, \phi)}\frac{1}{\sin\theta}\frac{\partial}{\partial\theta}\left[\sin\theta\frac{\partial\Omega(\theta, \phi)}{\partial\theta}\right]$$
$$+\frac{1}{\Omega(\theta, \phi)}\frac{1}{\sin^2\theta}\frac{\partial^2\Omega(\theta, \phi)}{\partial\phi^2} + \frac{2mr^2}{\hbar^2}[E - V(r)] = 0. \tag{3.34}$$

Here, the second and third terms are independent of r, whereas the first and last terms are independent of θ and ϕ. Using the now familiar technique of separation of variables, we get the two equations

$$\frac{1}{\Omega(\theta, \phi)}\left[\frac{1}{\sin\theta}\frac{\partial}{\partial\theta}\left(\sin\theta\frac{\partial\Omega(\theta, \phi)}{\partial\theta}\right) + \frac{1}{\sin^2\theta}\frac{\partial^2\Omega(\theta, \phi)}{\partial\phi^2}\right] = C, \tag{3.35}$$

$$-\frac{1}{R(r)}\frac{d}{dr}\left(r^2\frac{dR}{dr}\right) - \frac{2mr^2}{\hbar^2}[E - V(r)] = C, \tag{3.36}$$

where C is a constant. We note that Eq. (3.35) does not contain any term dependent on the potential energy. This is always true of all spherically symmetric potentials. Thus, the solution of Eq. (3.35), which gives the angular part of the wave function, is the same for all such potentials. As we shall see, this equation is related to the angular momentum. Before we can do this, let us discuss the angular momentum quantum mechanically.

ANGULAR MOMENTUM

In classical mechanics, the angular momentum $\boldsymbol{L}$ of a particle about the origin is defined as

$$L = r \times p, \tag{3.37}$$

where r and p are the position and momentum vectors, respectively.

QUESTION 3.5 Does the angular momentum vector depend on the choice of the origin of the coordinate axes?

The concept of angular momentum is useful in a situation where the angular momentum is conserved. A particle moving in a spherically symmetric potential is one such situation. To see this, let us differentiate Eq. (3.37) with respect to time. Then,

$$\frac{dL}{dt} = \frac{dr}{dt} \times p + r \times \frac{dp}{dt}.$$

Here, the first term on the right-hand side is $v \times p = v \times mv = 0$. Thus,

$$\frac{dL}{dt} = r \times \frac{dp}{dt} = r \times F. \tag{3.38}$$

For a central potential, since V is a function of only the distance from the origin, the equipotentials are spheres with the origin as the centre. Since the direction of the force is normal to the equipotential surface, it is radial for a central potential. In other words, the force is either towards or away from the centre. This means

$$F \propto r, \qquad dL/dt = r \times F \propto r \times r = 0$$

or L is a constant vector.

The angular momentum is an observable and, in quantum mechanics, we associate an operator with it. This is done by using the momentum operator

$$p = p_x\hat{i} + p_y\hat{j} + p_z\hat{k} = -i\hbar\left(\hat{i}\frac{\partial}{\partial x} + \hat{j}\frac{\partial}{\partial y} + \hat{k}\frac{\partial}{\partial z}\right) = -i\hbar\nabla, \tag{3.39}$$

where

$$\nabla = \hat{i}\frac{\partial}{\partial x} + \hat{j}\frac{\partial}{\partial y} + \hat{k}\frac{\partial}{\partial z}, \tag{3.40}$$

and is referred to as the gradient operator. Substituting Eq. (3.39) in Eq. (3.37), we get

$$L = -i\hbar r \times \nabla. \tag{3.41}$$

Equation (3.41) is a vector equation whose x-, y-, and z-component can be written as

$$L_x = -i\hbar\left(y\frac{\partial}{\partial z} - z\frac{\partial}{\partial y}\right), \qquad L_y = -i\hbar\left(z\frac{\partial}{\partial x} - x\frac{\partial}{\partial z}\right), \qquad L_z = -i\hbar\left(x\frac{\partial}{\partial y} - y\frac{\partial}{\partial x}\right). \tag{3.42}$$

The square of the angular momentum operator is defined as

$$L^2 = L_x^2 + L_y^2 + L_z^2. \tag{3.43}$$

Operators (3.42) and (3.43) can be written in terms of the spherical polar coordinates r, θ, ϕ (see Appendix 5) as

$$L_x = i\hbar(\sin\phi\frac{\partial}{\partial\theta} + \cot\theta\cos\phi\frac{\partial}{\partial\phi}),$$

$$L_y = -i\hbar(\cos\phi\frac{\partial}{\partial\theta} - \cot\theta\sin\phi\frac{\partial}{\partial\phi}), \tag{3.44}$$

$$L_z = -i\hbar\frac{\partial}{\partial\phi},$$

$$L^2 = -\hbar^2[\frac{1}{\sin\theta}\frac{\partial}{\partial\theta}(\sin\theta\frac{\partial}{\partial\theta}) + \frac{1}{\sin^2\theta}\frac{\partial^2}{\partial\phi^2}]. \tag{3.45}$$

We observe that Eq. (3.35) can be rewritten, using Eq. (3.45), as

$$L^2\Omega = -\hbar^2[\frac{1}{\sin\theta}\frac{\partial}{\partial\theta}(\sin\theta\frac{\partial\Omega}{\partial\theta}) + \frac{1}{\sin^2\theta}\frac{\partial^2\Omega}{\partial\phi^2}] = -\hbar^2 C\Omega(\theta, \phi),$$

$$L^2\Omega(\theta, \phi) = \lambda\Omega(\theta, \phi), \tag{3.46}$$

where $\lambda = -\hbar^2 C$. This differential equation is the eigenvalue equation for the operator L^2. Here the eigenfunctions are $\Omega(\theta, \phi)$ and the eigenvalues are λ (see Appendix 14). Equation (3.46) can be solved by the method of separation of variables by substituting in it

$$\Omega(\theta, \phi) = \Theta(\theta)\Phi(\phi). \tag{3.47}$$

This gives rise to two differential equations, one for $\Theta(\theta)$ and another for $\Phi(\phi)$; these are

$$\frac{1}{\Phi(\phi)}\frac{d^2\Phi(\phi)}{d\phi^2} = -[\frac{1}{\Theta(\theta)}\sin\theta\frac{d}{d\theta}(\sin\theta\frac{d\Theta}{d\theta}) + \frac{\lambda}{\hbar^2}\sin^2\theta] = \alpha, \tag{3.48}$$

where α is a constant. Equation (3.46) often occurs in many physical problems and its solutions are expressed in terms of spherical harmonics. Spherical harmonics are standard functions, similar to the trigonometric functions, and can be found in any good handbook. We shall first describe and then use these spherical harmonics.

Spherical Harmonics

The eigenvalue equation (3.46) for the operator L^2 has solutions for only certain values of λ. In other words, the constant λ in the equation

$$L^2\Omega(\theta, \phi) = \lambda\Omega(\theta, \phi) \tag{3.49}$$

can take only discrete values. If λ is not equal to one of these discrete values, then there exists no single-valued continuous function $\Omega(\theta, \phi)$ which satisfies Eq. (3.46). These discrete values are known as the *eigenvalues of the operator* L^2 and are given by the equation

$$\lambda = \hbar^2 l(l+1), \tag{3.50}$$

where $l = 0, 1, 2, 3, \ldots$ is referred to as the *orbital quantum number*. Thus,

$$\begin{aligned} \lambda &= 0 \qquad (l = 0) \\ &= 2\hbar^2 \qquad (l = 1) \\ &= 6\hbar^2 \qquad (l = 2). \end{aligned} \tag{3.51}$$

The solutions (i.e., spherical harmonics) of Eq. (3.46) corresponding to these eigenvalues (see

Appendix 14) are

$$\Omega(\theta, \phi) = Y_0^0(\theta, \phi) = \left(\frac{1}{4\pi}\right)^{1/2} \qquad (l = 0), \tag{3.52}$$

$$\Omega(\theta, \phi) = Y_1^1(\theta, \phi) = -\left(\frac{3}{8\pi}\right)^{1/2} \sin\theta \exp(i\phi)$$

$$\Omega(\theta, \phi) = Y_1^0(\theta, \phi) = \left(\frac{3}{4\pi}\right)^{1/2} \cos\theta \qquad (l = 1), \tag{3.53}$$

$$\Omega(\theta, \phi) = Y_1^{-1}(\theta, \phi) = \left(\frac{3}{8\pi}\right)^{1/2} \sin\theta \exp(-i\phi)$$

$$\Omega(\theta, \phi) = Y_2^2(\theta, \phi) = \tfrac{1}{4}\left(\frac{15}{2\pi}\right)^{1/2} \sin^2\theta \exp(2i\phi)$$

$$\Omega(\theta, \phi) = Y_2^1(\theta, \phi) = -\left(\frac{15}{8\pi}\right)^{1/2} \sin\theta \cos\theta \exp(i\phi)$$

$$\Omega(\theta, \phi) = Y_2^0(\theta, \phi) = \left(\frac{5}{4\pi}\right)^{1/2} \frac{3\cos^2\theta - 1}{2} \qquad (l = 2). \tag{3.54}$$

$$\Omega(\theta, \phi) = Y_2^{-1}(\theta, \phi) = \left(\frac{15}{8\pi}\right)^{1/2} \sin\theta \cos\theta \exp(-i\phi)$$

$$\Omega(\theta, \phi) = Y_2^{-2}(\theta, \phi) = \tfrac{1}{4}\left(\frac{15}{2\pi}\right)^{1/2} \sin^2\theta \exp(-2i\phi)$$

In general, for a given l, there are $(2l + 1)$ solutions denoted by

$$Y_l^l(\theta, \phi),\ Y_l^{l-1}(\theta, \phi),\ \ldots,\ Y_l^{-l}(\theta, \phi) \tag{3.55}$$

or

$$\Omega(\theta, \phi) = Y_l^{m_l}(\theta, \phi), \tag{3.56}$$

where, for a fixed l, m_l takes the $(2l + 1)$ integer values

$$m_l = l, l - 1, l - 2, \ldots, -l.$$

We note that the two numbers l and m_l are needed to specify a given solution.

The spherical harmonics $Y_l^{m_l}$ are normalized functions, that is, they satisfy the condition

$$\int_{\theta=0}^{\pi} \int_{\phi=0}^{2\pi} Y_l^{m_l^*}(\theta, \phi) Y_l^{m_l}(\theta, \phi) \sin\theta \, d\theta \, d\phi = 1. \tag{3.57}$$

EXAMPLE 3.4 Obtain the solution for $\Phi(\phi)$ by solving the differential equation (3.48).

From Eq. (3.48), the equation for $\Phi(\phi)$ is

$$\frac{1}{\Phi(\phi)} \frac{d^2\Phi(\phi)}{d\phi^2} = \alpha$$

or

$$\frac{d^2\Phi(\phi)}{d\phi^2} - \alpha\Phi(\phi) = 0. \tag{3.58}$$

The general solution of this equation is

$$\Phi(\phi) = A \exp(\alpha^{1/2}\phi) + B \exp(-\alpha^{1/2}\phi), \tag{3.59}$$

where A and B are constants and $(\alpha)^{1/2}$ is chosen to be real. We now apply the physical condition that the wave function is single-valued everywhere, i.e.,

$$\Phi(\phi) = \Phi(\phi + 2\pi), \tag{3.60}$$

because (r, θ, ϕ) and $(r, \theta, \phi + 2\pi)$ correspond to the same point in space. This condition leads to

$$A \exp[\alpha^{1/2}(\phi + 2\pi)] + B \exp[-\alpha^{1/2}(\phi + 2\pi)] = A \exp(\alpha^{1/2}\phi) + B \exp(-\alpha^{1/2}\phi).$$

Differentiating this equation with respect to ϕ, we get

$$\alpha^{1/2}\{A \exp[\alpha^{1/2}(\phi + 2\pi)] - B \exp[-\alpha^{1/2}(\phi + 2\pi)]\} = \alpha^{1/2}[A \exp(\alpha^{1/2}\phi) - B \exp(-\alpha^{1/2}\phi)]$$

or

$$A \exp[\alpha^{1/2}(\phi + 2\pi)] - B \exp[-\alpha^{1/2}(\phi + 2\pi)] = A \exp(\alpha^{1/2}\phi) - B \exp(-\alpha^{1/2}\phi).$$

Adding and subtracting the foregoing two equations, we obtain

$$A \exp[\alpha^{1/2}(\phi + 2\pi)] = A \exp(\alpha^{1/2}\phi),$$

$$B \exp[-\alpha^{1/2}(\phi + 2\pi)] = B \exp(-\alpha^{1/2}\phi).$$

These can be satisfied only if $\exp(\pm 2\pi\alpha^{1/2}) = 1$ or $2\pi(\alpha^{1/2}) = 2\pi i m$, where m takes the values $0, \pm 1, \pm 2, \ldots$. Substituting for $\alpha^{1/2}$ in Eq. (3.59), we get

$$\Phi(\phi) = A \exp(im\phi), \tag{3.61}$$

where $m = \pm 1, \pm 2, \ldots$. This is the ϕ-dependent part of the spherical harmonics if we identify $m = m_l$. We can write the spherical harmonics as

$$Y_l^{m_l}(\theta, \phi) = \Theta(\theta) \exp(im_l\phi). \tag{3.62}$$

Quantization of Angular Momentum

We have seen that the operator L^2 associated with the angular momentum takes only discrete values [given by (3.49) and (3.50)]. In a similar manner, energy takes only discrete values. For example, when

$$\Omega(\theta, \phi) = Y_1^{m_l}(\theta, \phi), \tag{3.63}$$

where $m_l = 1, 0, -1$, L^2 takes the value $l(l + 1)\hbar^2 = 2\hbar^2$ for $l = 1$. This means that the magnitude of the angular momentum vector given by $L = (L^2)^{1/2}$ takes the value $2^{1/2}\hbar$. In other words, the length of the angular momentum vector is $\hbar[l(l + 1)]^{1/2}$. On the other hand, as we know, in classical mechanics, the length is not quantized and can take arbitrary values. The length is specified by the *orbital quantum number l.*

QUESTION 3.6 What will be the magnitude of the angular momentum vector for $\Omega(\theta, \phi) = Y_2^{m_2}(\theta, \phi)$, $m_2 = 2, 1, 0, -1, -2$?

Space Quantization

There are $(2l + 1)$ wave functions $Y_l^{m_l}(\theta, \phi)$, $m_l = l, l-1, l-2, \ldots, -l$, all of which correspond to the same magnitude of angular momentum. The number m_l distinguishes the various wave functions and is known as the *azimuthal quantum number.* It determines the z-component of the angular momentum vector in much the same way as the orbital quantum number determines the length of the angular momentum vector. In order to see this, let us evaluate $L_z\Omega(\theta, \phi)$, where $\Omega(\theta, \phi) = Y_l^{m_l}(\theta, \phi)$, using the operator associated with L_z [see Eqs. (3.44)]. Then,

$$L_z\Omega(\theta, \phi) = -i\hbar\frac{\partial\Omega(\theta, \phi)}{\partial\phi}. \tag{3.64}$$

Substituting for $\Omega(\theta, \phi)$ from Eq. (3.47), we have

$$L_z\Omega(\theta, \phi) = L_zY_l^{m_l}(\theta, \phi) = -i\hbar\frac{\partial}{\partial\phi}Y_l^{m_l}(\theta, \phi). \tag{3.65}$$

In order to evaluate the right-hand side of Eq. (3.65), we need only the ϕ-dependent part of $Y_l^{m_l}(\theta, \phi)$. Thus, Eq. (3.65) becomes, using Eq. (3.62),

$$L_zY_l^{m_l}(\theta, \phi) = -i\hbar\Theta(\theta)\frac{d}{d\phi}\exp(im_l\phi) = m_l\hbar\Theta(\theta)\exp(im_l\phi) = m_l\hbar Y_l^{m_l}(\theta, \phi). \tag{3.66}$$

The eigenvalues of the operator associated with L_z are $m_l\hbar$. This means that the z-component of the angular momentum is also quantized and takes on integral values in units of $\hbar$. The maximum value m_l can have is l. This is consistent with the fact that the magnitude of a component of a vector is always smaller than the length of the vector.

The foregoing lengthy mathematical arguments can be summarized in terms of a simple physical picture. We shall describe this picture, using an explicit case, namely, $l = 2$. The total angular momentum vector has a length $6^{1/2}\hbar$. The z-component of the angular momentum can take the values $2\hbar$, $\hbar$, 0, $-\hbar$, and $-2\hbar$. This means that the angle between the z-axis and the angular momentum vector can take only certain specific values (see Fig. 3.2). Till now, we have considered only the z-component of the angular momentum vector. To complete the description of the picture, we need to also consider the other two components of the angular momentum vector, i.e., L_x and L_y. The average value of these components for a particle described by the wave function $Y_l^{m_l}(\theta, \phi)$, it so turns out, is zero. That is,

$$\langle L_x\rangle = \int Y_l^{m_l*}(\theta, \phi)L_xY_l^{m_l}(\theta, \phi)\sin\theta\, d\theta\, d\phi = 0,$$

$$\langle L_y\rangle = \int Y_l^{m_l*}(\theta, \phi)L_yY_l^{m_l}(\theta, \phi)\sin\theta\, d\theta\, d\phi = 0.$$

In other words, if L_x and L_y are measured a large number of times, their mean value is found to be zero. This result can also be understood in terms of the uncertainty relations existing between the components L_x, L_y, and L_z of angular momentum. These relations are that no two components of angular momentum can be simultaneously measured with infinite precision. Since in our foregoing analysis the component L_z of the particle with a wave function $Y_l^{m_l}(\theta, \phi)$ is precisely known to be $m_l\hbar$, the x- and y-component cannot be fixed. In particular, $Y_l^{m_l}(\theta, \phi)$ are not the eigenfunctions of the operators L_x and L_y. The values of L_x and L_y keep on fluctuating around the average value zero. The only exception to this is the trivial case where $l = 0$, $m_l = 0$. In this situation, $L_x Y_0^0 = L_y Y_0^0 = L_z Y_0^0 = 0$, and hence the wave function Y_0^0 is a simultaneous eigenfunction of all the three components of angular momentum.

The quantization of angular momentum can also be visualized in terms of a vector diagram. In such a diagram, the angular momentum vector traces a cone about the z-axis. This is because the angle between the z-axis and the vector is fixed. For different values of m_l, the angle of the cone will vary as shown in Fig. 3.2. The result that the angular momentum vector can take only certain fixed directions in space with respect to the z-axis is referred to as *space quantization.*

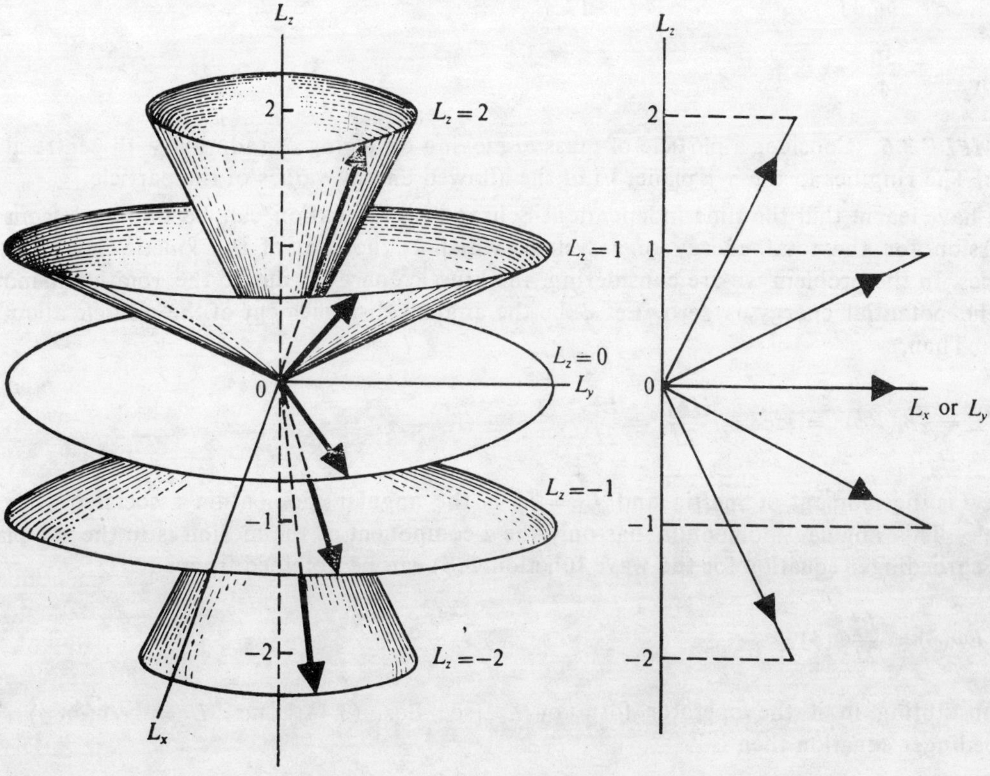

Fig. 3.2 Precession of angular momentum vector with $l = 2$ and its possible orientations with respect to z-axis.

EXAMPLE 3.5 Find the average value of the z-component of angular momentum, L_z, when $\Omega(\theta, \phi) = Y_1^1(\theta, \phi)$.

Since the wave function is $Y_1^1(\theta, \phi)$, we have $l = 1$ and $m_l = 1$. Substituting these quantum numbers in Eq. (3.65), we get

$$L_z\Omega(\theta, \phi) = -i\hbar\frac{\partial}{\partial\phi}Y_1^1(\theta, \phi) = -i\hbar\frac{\partial}{\partial\phi}[-(\frac{3}{8\pi})^{1/2} \sin\theta \exp(i\phi)] = -\hbar(\frac{3}{8\pi})^{1/2} \sin\theta \exp(i\phi).$$

The average value of L_z then is [see Eq. (A5.31) in Appendix 5]

$$\langle L_z\rangle = \iint \Omega^*(\theta, \phi)[L_z\Omega(\theta, \phi)] \sin\theta \, d\theta \, d\phi$$

$$= \iint [-(\frac{3}{8\pi})^{1/2} \sin\theta \exp(i\phi)]^*[-\hbar(\frac{3}{8\pi})^{1/2} \sin\theta \exp(i\phi)] \sin\theta \, d\theta \, d\phi$$

$$= \hbar\frac{3}{8\pi}\int_{\theta=0}^{\pi}\int_{\phi=0}^{2\pi} \sin\theta \exp(-i\phi) \sin\theta \exp(i\phi) \sin\theta \, d\theta \, d\phi$$

$$= \frac{3\hbar}{8\pi}\iint \sin^3\theta \, d\theta \, d\phi = \frac{3\hbar}{8\pi}2\pi\int_{\theta=0}^{\pi} \sin^3\theta \, d\theta$$

$$= \frac{3\hbar}{4} \times \tfrac{4}{3} = \hbar.$$

EXAMPLE 3.6 Consider a particle of mass m moving on a ring of radius R with centre at the origin. The ring lies in the x-y plane. Find the allowed energy values of the particle.

We have learnt that the time-independent Schroedinger equation can be obtained from the expression for the total energy of a system, which is the sum of the kinetic and potential energies. In the problem we are considering, the kinetic energy is due to the rotational motion and the potential energy is zero. Let ϕ be the angular displacement of the particle about the x-axis. Then,

$$E = \tfrac{1}{2}m(R\dot{\phi})^2 = \tfrac{1}{2}I\dot{\phi}^2 = \frac{(I\dot{\phi})^2}{2I} = \frac{L_z^2}{2I},$$

where I is the moment of inertia and $L_z = I\dot{\phi}$ is the angular momentum associated with the particle. This angular momentum has only the z-component as the motion is in the x-y plane. The Schroedinger equation for the wave function $\Phi(\phi)$ can be obtained from

$$E\Phi(\phi) = \frac{L_z^2}{2I}\Phi(\phi)$$

by substituting in it the operator form of L_z [see Eqs. (3.44)], i.e., $L_z = -i\hbar(\partial/\partial\phi)$. The Schroedinger equation then is

$$\frac{1}{2I}(-i\hbar\frac{\partial}{\partial\phi})(\frac{-i\hbar\partial}{\partial\phi})\Phi(\phi) - E\Phi(\phi) = 0$$

or

$$\frac{-i\hbar}{2I}(\frac{-i\hbar\partial}{\partial\phi})\frac{\partial}{\partial\phi}\Phi(\phi) - E\Phi(\phi) = 0$$

or

$$\frac{\hbar^2}{2I}\frac{d^2\Phi(\phi)}{d\phi^2} + E\Phi(\phi) = 0$$

or

$$\frac{d^2\Phi(\phi)}{d\phi^2} + (\frac{2I}{\hbar^2}E)\Phi(\phi) = 0.$$

The general solution of this equation is

$$\Phi(\phi) = A \cos (\frac{2IE}{\hbar^2})^{1/2}\phi + B \sin (\frac{2IE}{\hbar^2})^{1/2}\phi.$$

We now apply the physical requirement (i.e., boundary condition) that the wave function is single-valued, that is, $\Phi(\phi + 2\pi) = \Phi(\phi)$. We can show, as in Example 3.4, that this condition leads to

$$\sin [(\frac{2IE}{\hbar^2})^{1/2}(\phi + 2\pi)] = \sin (\frac{2IE}{\hbar^2})^{1/2}\phi,$$

$$\cos [(\frac{2IE}{\hbar^2})^{1/2}(\phi + 2\pi)] = \cos (\frac{2IE}{\hbar^2})^{1/2}\phi.$$

These equations imply

$$(2IE/\hbar^2)^{1/2}2\pi = 2m\pi,$$

where $m = 0, \pm1, \pm2, \ldots$. Thus, $E = \hbar^2 m^2/(2I)$. The allowed energy values are then $0, \hbar^2/(2I), 4\hbar^2/(2I), \ldots$.

QUESTION 3.7 The energy $E = 0$ is not allowed for a particle in a box. Why can we not apply the same argument in Example 3.6 to eliminate the solution $E = 0$?

Physical Significance of Eigenfunctions

We have seen that the spherical harmonics are the eigenfunctions of the operators L^2 and L_z, i.e.,

$$L^2 Y_l^{m_l}(\theta, \phi) = l(l + 1)\hbar^2 Y_l^{m_l}(\theta, \phi), \tag{3.67}$$

$$L_z Y_l^{m_l}(\theta, \phi) = m_l\hbar Y_l^{m_l}(\theta, \phi). \tag{3.68}$$

However, the spherical harmonics are not the eigenfunctions of the operators L_x and L_y. For example,

$$L_x Y_1^0(\theta, \phi) = i\hbar(\sin \phi\frac{\partial}{\partial\theta} + \cot \theta \cos \phi\frac{\partial}{\partial\phi})(\frac{3}{4\pi})^{1/2} \cos \theta$$

$$= i\hbar \sin \phi\frac{\partial}{\partial\theta}[(\frac{3}{4\pi})^{1/2} \cos \theta] = -(\frac{3}{4\pi})^{1/2}i\hbar \sin \theta \sin \phi. \tag{3.69}$$

We see that

$$L_x Y_1^0(\theta, \phi) \neq \text{constant} \times Y_1^0(\theta, \phi). \tag{3.70}$$

This means that if the angular part of the wave function of a system is given by $Y_l^{m_l}(\theta, \phi)$, then the system has a definite z-component of the angular momentum, namely, $m_l\hbar$. In particular, if the system is described by the wave function $Y_1^1(\theta, \phi)$, then the value of the z-component of the angular momentum when measured is $\hbar$. However, this is not true of the other two components of angular momentum, i.e., L_x and L_y. The average value of these components is zero. It should be noted that a given measurement of L_x or L_y may not be zero, but it is the average of a large number of measurements which is zero. In general, the repeated measurement of an observable gives the same value only if the wave function describing the system is an eigenfunction of the operator associated with the observable. If the wave function is not an eigenfunction of the operator associated with the observable, then the repeated measurement of the observable will not lead to the same value. As we know, it is only the average value of such measurements which has a physical significance. If the wave function is an eigenfunction of an operator, say, O, and has an eige nvalue λ, then the average value of the observable is λ, because all the measurements of the observable described by the operator O will yield the same value λ. We can express this mathematically as follows. If

$$O\psi = \lambda\psi, \tag{3.71}$$

then

$$\langle O \rangle = \int \psi^* O\psi \, dv = \int \psi^* \lambda\psi \, dv = \lambda \int \psi^*\psi \, dv = \lambda. \tag{3.72}$$

We shall illustrate the foregoing ideas by considering a system described by the wave function $Y_1^0(\theta, \phi)$. Here, the value of the z-component of angular momentum, whenever it is measured, will be zero. On the other hand, only the average value of the x-component of angular momentum will be zero. However, a given measurement of the x-component, in general, may not be zero.

We might wonder what is so special about the direction (i.e., z-axis) that makes the component L_z along this direction take only the integral values. The fact is that there is no preferred axis. For example, if the component of angular momentum of a system is measured in any direction, say, $\hat{n}$, then the values it can take will only be the integral multiples of $\hbar$. Let us assume that the value of such a measurement is $\hbar$. The system has now been prepared with the component of angular momentum in the direction $\hat{n}$ with a value $\hbar$. Effectively, this means that all subsequent measurements of the component of angular momentum for this system along $\hat{n}$ will give us the value $\hbar$. If now we measure the component of angular momentum along a direction perpendicular to $\hat{n}$, then the values of angular momentum we obtain would also be the integral multiples of $\hbar$. However, if repeated measurements are carried out on a system prepared with $\hbar$ as the component of angular momentum along the $\hat{n}$-direction, the average value of the component of angular momentum in the perpendicular direction will then be zero.

RADIAL PART OF SCHRÖEDINGER'S EQUATION

We now go back to Eq. (3.36) with the knowledge that the constant $C = -l(l+1)$. Thus,

$$-\frac{1}{R(r)}\frac{d}{dr}\left(r^2\frac{dR}{dr}\right) - \frac{2mr^2}{\hbar^2}[E - V(r)] = -l(l+1)$$

or

$$\frac{1}{r^2}\frac{d}{dr}\left(r^2\frac{dR}{dr}\right)+\left\{\frac{2m}{\hbar^2}[E-V(r)]-\frac{l(l+1)}{r^2}\right\}R=0. \tag{3.73}$$

As we know, the complete three-dimensional wave function, replacing $\Omega(\theta, \phi)$ in Eq. (3.33) by the spherical harmonics, is

$$\psi(r, \theta, \phi)=R(r)Y_l^{m_l}(\theta, \phi). \tag{3.74}$$

As already stated, the boundary condition the wave function of a bound state system has to satisfy is that the wave function must vanish when $r \rightarrow \infty$. This boundary condition, from Eq. (3.74), is

$$R(r) \rightarrow 0 \quad \text{as} \quad r \rightarrow \infty. \tag{3.75}$$

We have to solve Eq. (3.73) under condition (3.75). An explicit solution of this equation is possible only when $V(r)$ is known. However, in what follows, we shall make a few remarks on the nature of its solutions.

The solutions of Eq. (3.73) exist only for certain allowed values of E in much the same way as the solutions of the Schroedinger equation for a particle in a one-dimensional box exist only for certain allowed values of energy. These energy values depend, in general, on l since Eq. (3.73) depends on l. For each l, we may have several allowed energy levels, say, $E_1^l, E_2^l, \ldots, E_n^l$. Labelling the radial solutions corresponding to these levels as $R_1^l, R_2^l, \ldots, R_n^l$, we can write the eigenfunctions corresponding to E_n as

$$\psi_{n,\,l,\,m_l}=R_n^l(r)Y_l^{m_l}(\theta, \phi). \tag{3.76}$$

Since m_l is allowed to take the values from $-l$ to $+l$ in integral steps, we have $(2l+1)$ wave functions, each differing in the value of m_l, corresponding to the same energy. Thus, the states are $(2l+1)$-fold degenerate. In order to specify these states, we need three quantum numbers n, l, and m_l. This is because we are dealing with a three-dimensional system having three degrees of freedom.

Let us now consider a particle moving in a Coulomb potential given by $V(r)=-e^2/(4\pi\epsilon_0 r)$. Equation (3.73) will then be

$$\frac{1}{r^2}\frac{d}{dr}\left(r^2\frac{dR}{dr}\right)+\left[\frac{2m}{\hbar^2}\left(E+\frac{e^2}{4\pi\epsilon_0 r}\right)-\frac{l(l+1)}{r^2}\right]R=0. \tag{3.77}$$

This equation can be solved by any of the known techniques (see Appendix 15). We shall state only the final result which, it so turns out, is that the allowed values of E, for $E<0$, are

$$E_n=-\frac{me^4}{2\hbar^2n^2(4\pi\epsilon_0)^2}$$

with n taking positive integral values excluding zero. Further, l is restricted to be less than or equal to $n-1$. In other words, there are n different values of l (i.e., $0, 1, \ldots, n-1$), each having the same energy E. It should be noted that the energy E_n is negative, indicating that the particle is bound. Since the energy of a particle at an infinite distance from the centre of the force is necessarily positive (i.e., the potential energy is zero), the particle is confined in a finite region. We note that the energy of a particle in a Coulomb potential does not depend on l.

The three quantum numbers, n, l, and m_l required to specify the state of the particle are called the *principal*, *orbital*, and *azimuthal quantum numbers*, respectively.

3.5 SCATTERING PROBLEM IN ONE DIMENSION

In our study of the bound state problems so far, we have applied the boundary condition that the wave function vanishes as $x \to \pm\infty$. This boundary condition gives us the general result that a particle has only discrete values of energy. However, there are physical situations which do not correspond to this boundary condition.

As we shall learn in subsequent chapters, the phenomenon of scattering of particles by matter gives us a great deal of information about the atomic and nuclear structure. In an experiment to observe this phenomenon, a beam of particles impinges on a target and is scattered by the atoms and the nuclei of the target. To understand the scattering of microscopic particles such as electrons and protons, it is necessary to study the phenomenon quantum mechanically. In what follows, we shall do this in one dimension and, in the process, highlight the main results. These results, as we shall realize, are drastically different from those expected from classical physics.

The incident beam consists of free particles, each having a definite momentum and energy. After scattering, each particle will have a different momentum and energy. Let a particle in the beam be described by a wave function

$$\psi(x, t) = A \exp\left[\frac{i}{\hbar}(px - Et)\right].$$

There is a mathematical difficulty in normalizing this wave function. Consider the normalization condition (2.45), i.e.,

$$\int_{-\infty}^{+\infty} \psi^*\psi \, dx = 1.$$

This leads to

$$|A|^2 \int_{-\infty}^{+\infty} dx = 1$$

which can be satisfied only if $A = 0$. This, of course, is reasonable, for, if a particle is spread over the entire space ($-\infty$ to $+\infty$), the probability of finding it in any finite region is zero. To ensure that the wave function is not zero, we assume the particle to be contained in a large box of length L. We then have

$$|A|^2 \int_0^L dx = 1 \quad \text{or} \quad |A|^2 = \frac{1}{L}.$$

All the physical results for an unbound system are obtained by letting $L \to \infty$. The wave function

$$\psi(x, t) = \frac{1}{(L)^{1/2}} \exp\left[\frac{i}{\hbar}(px - Et)\right]$$

can be associated with a beam of particles separated by a distance L. To see this, we note that the probability of finding a particle in the region $(0, L)$, $(L, 2L)$, ... is one. If L is very large, we can assume the particles to be independent of one another (i.e., free). Thus, the constant A is related to the flux of the incident beam of particles.

We shall learn later in this section that the physically meaningful results are independent of the constant A. This is because, for any quantity of physical interest, we can express the result as a ratio where the constant A, which depends on L, cancels.

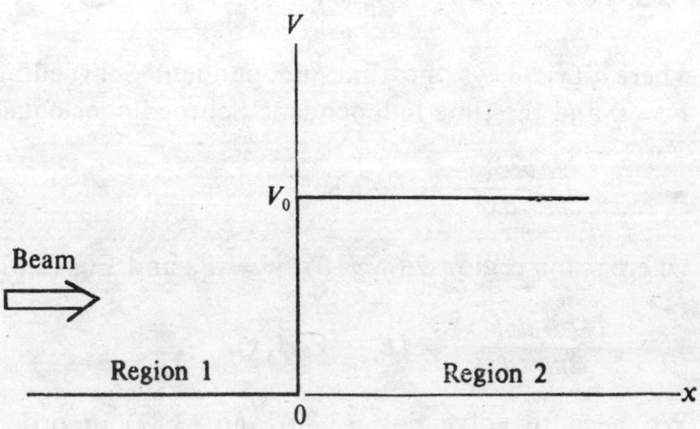

Fig. 3.3 Potential step.

BARRIER PENETRATION

Let us see how a beam of free particles incident from the left (i.e., $x < 0$) behaves when it enters a region ($x > 0$) where the potential V_0 is constant. This is shown in Fig. 3.3.

QUESTION 3.8 Describe an experimental set-up to make an electron encounter the potential shown in Fig. 3.3.

The particles of the beam have an initial kinetic energy E. We shall consider two situations, namely, $E < V_0$ and $E > V_0$, separately as these lead to different physical situations. Our discussion is confined to one dimension to keep the mathematics simple.

(1) $E < V_0$ Let us first use classical physics to analyze this situation. On the left (i.e., region 1 in Fig. 3.3), a particle of the beam is free and its total energy is E. The particle does not have sufficient energy to cross the potential barrier at $x = 0$, and so it bounces back. Mathematically, the energy of the particle, if it enters the region $x > 0$, is $p_1^2/(2m) + V_0$, where p_1 is the momentum of the particle in the region $x > 0$. Since the energy is conserved,

$$p_1^2/(2m) + V_0 = E. \tag{3.78}$$

If this equation has to be valid, then $p_1^2/(2m)$ must be negative. This is clearly unphysical since the kinetic energy cannot be negative. Therefore, the particle cannot enter the region $x > 0$. As the particle has energy E, its momentum is $\pm(2mE)^{1/2}$. Since the particle cannot be in region 2 and it cannot stop, it must turn back at $x = 0$. Now that the direction of motion of the particle has changed, its momentum is $-(2mE)^{1/2}$. In other words, the particle bounces back.

QUESTION 3.9 At $x = 0$, why cannot the particle come to rest instead of bouncing back?

We now analyze the situation quantum mechanically. Let us associate a wave function $\psi(x, t)$ with the particle. This wave function has to be obtained by solving the Schroedinger equation (2.27). Once $\psi(x, t)$ is known, we can obtain the physical properties of a system by using the

Born interpretation. As we know, the potential does not depend on time, and so we can write

$$\psi(x, t) = \exp\left(-\frac{iEt}{\hbar}\right)\psi(x), \tag{3.79}$$

where $\psi(x)$ obeys the time-independent Schroedinger equation (2.61). For region 1 ($x < 0$), $V = 0$ and the time-independent Schroedinger equation becomes

$$-\frac{\hbar^2}{2m}\frac{d^2\psi(x)}{dx^2} = E\psi(x), \tag{3.80}$$

whereas for region 2 ($x > 0$), $V = V_0$ and Eq. (2.61) becomes

$$-\frac{\hbar^2}{2m}\frac{d^2\psi(x)}{dx^2} = (E - V_0)\psi(x). \tag{3.81}$$

We need to solve Eqs. (3.80) and (3.81) in order to discuss the quantum behaviour of the particle.

From Eq. (3.80), we have

$$\frac{d^2\psi}{dx^2} + k^2\psi = 0 \tag{3.82}$$

with

$$k^2 = 2mE/\hbar^2. \tag{3.83}$$

The solutions of these equations are well known, and we can write their most general solution as

$$\psi = A \exp(ikx) + B \exp(-ikx) \quad \text{for } x < 0, \tag{3.84}$$

where A and B are constants to be determined. We rewrite Eq. (3.81) as

$$d^2\psi/dx^2 - \alpha^2\psi = 0, \tag{3.85}$$

where

$$\alpha^2 = 2m(V_0 - E)/\hbar^2. \tag{3.86}$$

The solution of Eq. (3.85) for $V_0 > E$ is

$$\psi(x) = C \exp(-\alpha x) + D \exp(\alpha x) \quad \text{for } x > 0, \tag{3.87}$$

where C and D are again constants to be determined.

Equations (3.84) and (3.87) are the solutions of the Schroedinger equation (2.61) in the two regions $x < 0$ and $x > 0$. The constants A, B, C, and D occurring in these solutions have yet to be determined. We shall evaluate these by imposing some physical requirements on the wave functions [Eqs. (3.84) and (3.87)].

A wave function must be finite everywhere, otherwise the probability of finding the particle becomes infinite which is unphysical. The wave function for $x < 0$ [i.e., Eq. (3.84)] is finite everywhere. This is because both $\exp(ikx)$ and $\exp(-ikx)$ are finite for all x. Thus, the finiteness condition has no restriction on A or B. On the other hand, the wave function for $x > 0$ [i.e., Eq. (3.87)] has a term $\exp(-\alpha x)$ and a term $\exp(\alpha x)$. The former term is finite for all $x > 0$, whereas $\exp(\alpha x)$ becomes infinite as $x \rightarrow +\infty$. It is the term $\exp(\alpha x)$ that makes the wave function infinite as $x \rightarrow \infty$, and this is unacceptable. Thus, the coefficient of this term must be zero. This leads to the condition $D = 0$.

The wave function given by Eq. (3.84) is valid in the region $x < 0$, whereas that given by Eq. (3.87) is valid in the region $x > 0$. $x = 0$ is the boundary between these two regions, and can be reached from either of the regions. We impose the restriction that $\psi(x = 0)$ obtained from region 1 is identical with $\psi(x = 0)$ obtained from region 2. This ensures that the wave function at $x = 0$ is unique. If this were not so, the probability density $|\psi(x = 0)|^2$ at $x = 0$ would have taken two different values, one when $x = 0$ is approached from region 1 and the other when $x = 0$ is approached from region 2. Mathematically, this physical condition means that the wave function is single-valued and continuous everywhere. The value of $\psi(x = 0)$, when evaluated from Eq. (3.84), is

$$\psi(x = 0) = [A \exp(ikx) + B \exp(-ikx)]_{x=0} = A + B, \tag{3.88}$$

and when evaluated from Eq. (3.87), using $D = 0$, it is

$$\psi(x = 0) = [C \exp(-\alpha x)]_{x=0} = C. \tag{3.89}$$

Thus, the single-valuedness of ψ implies

$$A + B = C. \tag{3.90}$$

There is one more condition the wave function must satisfy, namely, $d\psi/dx$ must be continuous and single-valued. If this were not so, $d^2\psi/dx^2$ would have been infinite, implying $2m(E - V)\psi$ is infinite. But both ψ and $(2m/\hbar^2)(E - V)$ are finite, and hence $d^2\psi/dx^2$ is finite. This explains why $d\psi/dx$ must be continuous. For region 1, $d\psi/dx$, from Eq. (3.84), is

$$d\psi/dx = Aik \exp(ikx) - Bik \exp(-ikx), \qquad x < 0, \tag{3.91}$$

and for region 2, from Eq. (3.87), it is

$$d\psi/dx = -C\alpha \exp(-\alpha x), \qquad x > 0. \tag{3.92}$$

The continuity of $d\psi/dx$ demands that Eqs. (3.91) and (3.92) give the same value for the boundary point $x = 0$. Thus,

$$Aik - Bik = -C\alpha. \tag{3.93}$$

From Eqs. (3.90) and (3.93), we solve for B and C in terms of A. We thus get

$$B = \frac{k - i\alpha}{k + i\alpha} A, \tag{3.94}$$

$$C = \frac{2k}{k + i\alpha} A. \tag{3.95}$$

The most surprising result is that $C \neq 0$. This means that the probability of finding the particle in the region $x > 0$ is not zero. This contradicts the result we obtained classically, namely, the particle cannot enter region 2, that is, the probability of finding the particle in region 2 is zero. As we shall see in our subsequent study, the experimental result agrees with that arrived at quantum mechanically.

The wave function in the region $x > 0$ is of the form $\exp(-\alpha x)$ which decreases rapidly as x increases. Therefore, the probability of finding the particle in the region $x > 1/\alpha$ is quite small.

EXAMPLE 3.7 Use the time-dependent wave function to show that all the particles that are incident from the left ($x < 0$)—see Fig. 3.3—are reflected back when $E < V_0$.

The potential here is time-independent. Therefore, we can write, as in Example 2.18, the time-dependent wave function as

$$\psi(x, t) = \exp\left(-\frac{iEt}{\hbar}\right)\psi(x).$$

For $x < 0$,

$$\psi(x, t) = \exp\left(-\frac{iEt}{\hbar}\right)[A \exp(ikx) + B \exp(-ikx)]$$

$$= A \exp\left[\frac{i}{\hbar}(kx - Et)\right] + B \exp\left[-\frac{i}{\hbar}(kx + Et)\right].$$

On the right-hand side of this equation, the first term, as we already know, represents a wave travelling to the right with an amplitude A and the second term a wave travelling to the left (see Problem 2.13) with an amplitude B. We can therefore write

$$\psi(x, t) = \psi_{\text{incident}} + \psi_{\text{reflected}},$$

where

$$\psi_{\text{incident}} = A \exp\left[\frac{i}{\hbar}(kx - Et)\right], \qquad \psi_{\text{reflected}} = B \exp\left[-\frac{i}{\hbar}(kx + Et)\right].$$

The ratio R of the probability densities of the reflected and the incident waves, using

$$\left|\exp\left\{\pm\frac{i}{\hbar}(kx + Et)\right\}\right|^2 = 1,$$

is

$$R = \frac{|\psi_{\text{reflected}}|^2}{|\psi_{\text{incident}}|^2} = \frac{|B|^2}{|A|^2}.$$

The ratio R is known as the reflection coefficient. Using Eq. (3.94), we have

$$R = \frac{|B|^2}{|A|^2} = \frac{|k - i\alpha|^2}{|k + i\alpha|^2} = \frac{(k - i\alpha)(k + i\alpha)}{(k + i\alpha)(k - i\alpha)} = 1.$$

This shows that all the particles that are incident from the left are reflected back.

It should be noted that the correct way of defining the reflection coefficient is to take the ratio of the number of particles reflected per unit time and the number of particles incident per unit time. In so doing, the result obtained, as we shall note in our discussion following Example 3.8, is the same as that in the problem we have just considered. Since all the particles are turning back, we might wonder how we can find any particle at all in the region $x > 0$.

The answer to this is that there exists a finite probability for the particles to first enter the region $x > 0$ and then turn back. The demonstration of this result using the principles of quantum mechanics is beyond the scope of this text.

(2) $E > V_0$ Classically, we expect the particle to travel into the region $x > 0$ but with a reduced momentum. From Eq. (3.78), the momentum of the particle is $p_1 = [2m(E - V_0)]^{1/2}$.

Quantum mechanically, we have in the region $x < 0$ the same solution [i.e., (3.84)] as when $E < V_0$, namely, $\psi = A \exp(ikx) + B \exp(-ikx)$. In the region $x > 0$, we have

$$d^2\psi/dx^2 + k_1^2\psi = 0 \tag{3.96}$$

with

$$(2m/\hbar^2)(E - V_0) = k_1^2. \tag{3.97}$$

The general solution of Eq. (3.96) is

$$\psi = C \exp(ik_1x) + D \exp(-ik_1x). \tag{3.98}$$

In each of Eqs. (3.84) and (3.98), the first term represents a wave travelling from the left and the second term, a wave travelling from the right (see Example 3.7). In region 1, the first term denotes the incoming beam of particles and the second term, the reflected beam. The first term in region 2 represents the part of the incident beam transmitted to region 2. In this region, there is no beam incident from the right. Such a beam could have existed only if some particles had been sent to the left from $x = +\infty$. Since we are sending the particles only from the left to the right (see Fig. 3.3), the coefficient of the wave incident from the right in region 2 must be zero. In other words,

$$D = 0. \tag{3.99}$$

Matching the boundary conditions for ψ and $d\psi/dx$, we obtain

$$C = \frac{2k}{k + k_1}A, \qquad B = \frac{k - k_1}{k + k_1}A. \tag{3.100}$$

We can calculate the reflection coefficient as in Example 3.7. We thus get

$$R = \left|\frac{k - k_1}{k + k_1}\right|^2. \tag{3.101}$$

This (quantum mechanical) result, once again, contradicts the one obtained classically, namely, the particles are not reflected back when $E > V_0$. The experimental observations support the quantum mechanical result.

QUESTION 3.10 Is the de Broglie wavelength of the electron in region 2 different from that in region 1?

EXAMPLE 3.8 A beam with a flux of N particles per second is incident from the left with an energy $E > V_0$. Calculate the number of particles transmitted per unit time.

The reflection coefficient R is given by Eq. (3.101), i.e.,

$$R = \left|\frac{B}{A}\right|^2 = \left|\frac{k - k_1}{k + k_1}\right|^2.$$

Therefore, the number of particles reflected per unit time is

$$NR = N\frac{(k - k_1)^2}{(k + k_1)^2}.$$

The particles which are not reflected are transmitted. Hence, the number of transmitted particles is

$$N - N \times \frac{(k - k_1)^2}{(k + k_1)^2} = N[1 - \frac{(k - k_1)^2}{(k + k_1)^2}] = \frac{4kk_1N}{(k + k_1)^2}.$$

From Example 3.8, the transmission coefficient T can be calculated and is $T = 4kk_1/(k + k_1)^2$. The transmission coefficient can also be calculated directly from its definition which is

$$T = \frac{\text{number of particles transmitted per unit time}}{\text{number of particles incident per unit time}} = \frac{N_t}{N_i}.$$

If v_i is the velocity of the incident particles, then

$$N_i = v_i \times (\text{number of incident particles in unit length}).$$

The number of incident particles in unit length is

$$\int_0^1 |A \exp [\frac{i}{\hbar}(kx - Et)]|^2 \, dx = |A|^2.$$

Thus, $N_i = v_i|A|^2$. Similarly, $N_t = v_t|C|^2$, where v_t is the velocity of the transmitted particles. Substituting N_i and N_t in the expression for T, we get

$$T = \frac{v_t|C|^2}{v_i|A|^2} = \frac{v_t}{v_i}\frac{4k^2}{(k + k_1)^2}.$$

Using

$$\frac{v_t}{v_i} = (\frac{\hbar k_1}{m})/(\frac{\hbar k}{m}) = \frac{k_1}{k},$$

we obtain $T = 4kk_1/(k + k_1)^2$.

The reflection coefficient should, as noted in our discussion following Example 3.7, be calculated using the fluxes of the reflected and incident beams. However, since the velocities v_r and v_i of the reflected and incident particles are the same, the reflection coefficient is given by

$$R = \frac{v_r}{v_i}\frac{|B|^2}{|A|^2} = \frac{|B|^2}{|A|^2}.$$

This justifies our earlier calculation of the reflection coefficient [see Eq. (3.101)].

3.6 TUNNELING

The fact that a particle can cross a potential barrier even when its energy is less than the barrier height has very interesting consequences, one of which is the tunneling of a particle through a potential barrier such as the one shown in Fig. 3.4. Assume that particles of energy E are incident from the left ($x < 0$) and that $E < V_0$. Classically, such particles will be reflected back and can never appear in the region $x > a$. However, this is not true quantum mechanically. We shall now show this by solving the Schroedinger equation.

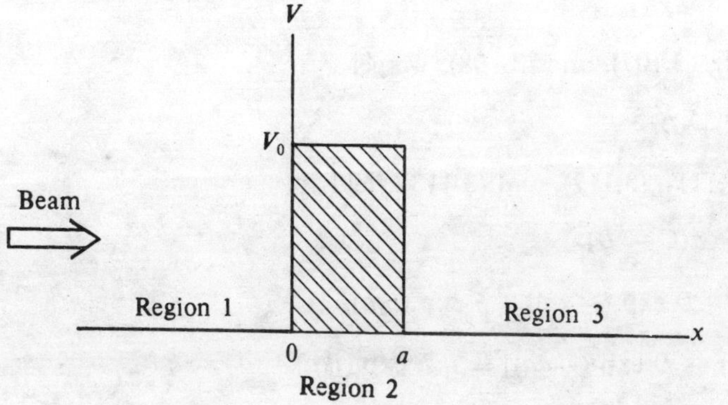

Fig. 3.4 Potential barrier.

The potential $V(x)$ is given by

$$\begin{aligned} V(x) &= 0 \quad (x < 0) \quad &\text{(region 1)} \\ &= V_0 \quad (0 < x < a) \quad &\text{(region 2)} \\ &= 0 \quad (a < x) \quad &\text{(region 3).} \end{aligned} \tag{3.102}$$

The time-independent Schroedinger equations for these regions are

$$d^2\psi/dx^2 + k^2\psi = 0 \qquad \text{(region 1),} \tag{3.103}$$

$$d^2\psi/dx^2 - \alpha^2\psi = 0 \qquad \text{(region 2),} \tag{3.104}$$

$$d^2\psi/dx^2 + k^2\psi = 0 \qquad \text{(region 3),} \tag{3.105}$$

where

$$k^2 = 2mE/\hbar^2, \qquad \alpha^2 = (2m/\hbar^2)(V_0 - E). \tag{3.106}$$

The solutions of these equations are

$$\psi_1(x) = A \exp(ikx) + B \exp(-ikx) \qquad \text{(region 1),} \tag{3.107}$$

$$\psi_2(x) = C \exp(\alpha x) + D \exp(-\alpha x) \qquad \text{(region 2),} \tag{3.108}$$

$$\psi_3(x) = F \exp(ikx) + G \exp(-ikx) \qquad \text{(region 3).} \tag{3.109}$$

Since no particles are incident from the right, there can be no wave of the form $\exp(-ikx)$ in region 3. This means that the coefficient G in Eq. (3.109) must be zero. We have to ensure that each wave function and its derivative are continuous at $x = 0$ and $x = a$. That is,

$$\psi_1(x)|_{x=0} = \psi_2(x)|_{x=0}, \tag{3.110}$$

$$\left.\frac{d\psi_1(x)}{dx}\right|_{x=0} = \left.\frac{d\psi_2(x)}{dx}\right|_{x=0}, \tag{3.111}$$

$$\psi_2(x)|_{x=a} = \psi_3(x)|_{x=a}, \tag{3.112}$$

$$\frac{d\psi_2(x)}{dx}\bigg|_{x=a} = \frac{d\psi_3(x)}{dx}\bigg|_{x=a}. \tag{3.113}$$

From Eqs. (3.110), (3.107), and (3.108), we get

$$A + B = C + D. \tag{3.114}$$

Similarly, Eqs. (3.111), (3.112), and (3.113) yield

$$ik(A - B) = \alpha(C - D), \tag{3.115}$$

$$C \exp(\alpha a) + D \exp(-\alpha a) = F \exp(ika), \tag{3.116}$$

$$\alpha[C \exp(\alpha a) - D \exp(-\alpha a)] = ikF \exp(ika). \tag{3.117}$$

QUESTION 3.11 The coefficient of exp (αx) in region 2 is not zero in Example 3.8, whereas in Section 3.5 we argued that it is zero for the potential step (Fig. 3.3). Why?

The transmitted wave in region 3 is given by $F \exp(ikx)$, and $|F \exp(ikx)|^2$ is proportional to the probability of finding the particle in this region. It is straightforward but lengthy to calculate F in terms of A. Therefore, we give only the final formula, namely,

$$F = \frac{A \exp(-ik_1 a)}{\cosh \alpha a - \frac{i}{2}(\frac{\alpha}{k} - \frac{k}{\alpha}) \sinh \alpha a}. \tag{3.118}$$

We see that $F \neq 0$ and has a finite value. The transmission coefficient is given by

$$T = |\frac{F}{A}|^2 = \frac{1}{\cosh^2 \alpha a + \frac{1}{4}(\alpha/k - k/\alpha)^2 \sinh^2 \alpha a}. \tag{3.119}$$

This phenomenon of a particle being able to cross a barrier even when its energy is less than the barrier height is known as quantum mechanical tunneling. The particle tunnels through the barrier to the other side. The tunnel diodes used in electronic circuits are based on this phenomenon. The tunneling effect forms the basis for understanding the emission of alpha-particles (i.e., nucleus of the He atom) from heavy nuclei. We shall study more about this phenomenon in Chapter 9.

EXAMPLE 3.9 Consider an electron impinging on a rectangular potential barrier of height $V_0 = 5$ eV and thickness $a = 10^{-10}$ m. If the kinetic energy of the electron is 2.5 eV, calculate the transmission coefficient.

We need to evaluate αa to calculate the transmission coefficient. Using Eq. (3.119), we get

$$\alpha a = [\frac{2m}{\hbar^2}(V_0 - E)]^{1/2} a \tag{3.120}$$

$$= (\frac{2 \times 9.1 \times 10^{-31} \times 2.5 \times 1.6 \times 10^{-19}}{(1.05)^2 \times 10^{-68}})^{1/2} \times 10^{-10} \sim 0.8,$$

$$\frac{\alpha}{k} = \left(\frac{V_0 - E}{E}\right)^{1/2} = 1, \qquad \frac{k}{\alpha} = \left(\frac{E}{V_0 - E}\right)^{1/2} = 1.$$

Thus,

$$T = \frac{1}{(\cosh 0.8)^2} = \frac{1}{(1.34)^2} = 0.6.$$

We see that the transmission probability is quite large.

When $E > V_0$, the method of obtaining the wave function is similar to that when $E < V_0$. The solutions in the three regions are

$$\begin{aligned} \psi_1 &= A \exp(ikx) + B \exp(-ikx), \\ \psi_2 &= C \exp(ik_1x) + D \exp(-ik_1x), \\ \psi_3 &= E \exp(ikx) + F \exp(-ikx), \end{aligned} \tag{3.121}$$

where

$$k^2 = 2mE/\hbar^2, \qquad k_1^2 = 2m(E - V_0)/\hbar^2. \tag{3.122}$$

We can get the transmission coefficient directly by replacing in Eq. (3.119) α by ik_1 because $k_1^2 = -\alpha^2$. Thus, using the relations

$$\cosh \alpha a = \cosh ik_1a = \cos ka, \qquad \sinh \alpha a = \sinh ik_1a = i \sin k_1a,$$

we get

$$T = \frac{1}{\cos^2 k_1a + \frac{1}{4}(k_1/k - k/k_1)^2 \sin^2 k_1a}.$$

We note that if the energy of the incident particle is such that $k_1a = n\pi$, then $T = 1$, showing that the incident wave is transmitted, as if the barrier were not present. Further, for this value of energy, there is no reflected wave in region 1. Such effects are actually observed in the scattering of electrons in an atom, and in the scattering of protons and neutrons by a nucleus.

3.7 PROBABILITY CURRENT*

ONE-DIMENSIONAL

When discussing the scattering phenomenon, we used particle flux (defined as the number of particles crossing a point per unit time) to calculate the transmission and reflection coefficients. We define the transmission coefficient T as

$$T = \frac{\text{transmitted flux}}{\text{incident flux}},$$

and the reflection coefficient R as

$$R = \frac{\text{reflected flux}}{\text{incident flux}}.$$

*A student taking an introductory course may skip this discussion.

The incident, transmitted, and reflected particle fluxes can be calculated from the wave function, using the concept of probability current which we shall now introduce.

Consider a beam of particles moving towards the right. For a single particle, the wave function associated with it moves to the right with the same velocity as that of the particle. As such, the probability of finding the particle in a given region changes with time. We can also express this by saying that a current, i.e., probability current, is travelling to the right. At a time t, let the probability current at x be $S(x, t)$ which denotes the 'amount of probability' crossing the point x per unit time from the left to the right. Let us take a region between x_1 and x_2 ($x_2 > x_1$) and study the change in the probability of finding the particle in this region with time. Now, the probability current $S(x_1, t)$ at x_1 represents the 'amount of probability' entering per unit time the region and $S(x_2, t)$ at x_2 the 'amount of probability' leaving per unit time the region. The rate of increase in probability in the region is

$$dP(t)/dt = S(x_1, t) - S(x_2, t), \tag{3.123}$$

where $P(t)$ is the probability of finding the particle in the region at the time t. If $\psi(x, t)$ is the wave function of the particle, then we know that the probability of finding the particle between x_1 and x_2 is

$$P(t) = \int_{x_1}^{x_2} \psi^*(x, t)\psi(x, t)\, dx. \tag{3.124}$$

The rate of change in probability $dP(t)/dt$ can be calculated from Eq. (3.124) as

$$\frac{dP(t)}{dt} = \frac{d}{dt}\left[\int_{x_1}^{x_2} \psi^*(x, t)\psi(x, t)\, dx\right] = \int_{x_1}^{x_2} \left[\frac{\partial\psi^*(x, t)}{\partial t}\psi(x, t) + \psi^*(x, t)\frac{\partial\psi(x, t)}{\partial t}\right] dx. \tag{3.125}$$

The time derivatives of ψ and ψ^* can be obtained from the Schroedinger equation

$$\frac{\partial\psi(x, t)}{\partial t} = \frac{1}{i\hbar}\left[-\frac{\hbar^2}{2m}\frac{\partial^2\psi(x, t)}{\partial x^2} + V\psi(x, t)\right]$$

and its complex conjugate

$$\frac{\partial\psi^*(x, t)}{\partial t} = \frac{1}{-i\hbar}\left[-\frac{\hbar^2}{2m}\frac{\partial^2\psi^*(x, t)}{\partial x^2} + V\psi^*(x, t)\right].$$

It should be noted that the potentials are assumed to be real. Inserting the foregoing two equations in Eq. (3.125) and simplifying, we get

$$\frac{dP(t)}{dt} = \frac{\hbar}{2mi}\int_{x_1}^{x_2} \left[\frac{\partial^2\psi^*(x, t)}{\partial x^2}\psi(x, t) - \psi^*(x, t)\frac{\partial^2\psi(x, t)}{\partial x^2}\right] dx. \tag{3.126}$$

We can integrate this equation by using the identity

$$\frac{\partial^2\psi^*(x, t)}{\partial x^2}\psi(x, t) - \psi^*(x, t)\frac{\partial^2\psi(x, t)}{\partial x^2} \equiv \frac{\partial}{\partial x}\left[\frac{\partial\psi^*(x, t)}{\partial x}\psi(x, t) - \psi^*(x, t)\frac{\partial\psi(x, t)}{\partial x}\right].$$

This identity can be derived by observing that the terms $(\partial\psi^*/\partial x)(\partial\psi/\partial x)$ cancel out on its right-hand side. Substituting the identity in Eq. (3.126), we get

$$\frac{dP(t)}{dt} = \frac{\hbar}{2mi}\int_{x_1}^{x_2} \frac{\partial}{\partial x}\left[\frac{\partial \psi^*(x, t)}{\partial x}\psi(x, t) - \psi^*(x, t)\frac{\partial \psi(x, t)}{\partial x}\right] dx$$

$$= \frac{\hbar}{2mi}\left[\frac{\partial \psi^*(x, t)}{\partial x}\psi(x, t) - \psi^*(x, t)\frac{\partial \psi(x, t)}{\partial x}\right]_{x_1}^{x_2}$$

$$= -\frac{\hbar}{2mi}\left[\left\{\frac{\partial \psi^*(x, t)}{\partial x}\psi(x, t) - \psi^*(x, t)\frac{\partial \psi(x, t)}{\partial x}\right\}_{x=x_1} - \left\{\frac{\partial \psi^*(x, t)}{\partial x}\psi(x, t) - \psi^*(x, t)\frac{\partial \psi(x, t)}{\partial x}\right\}_{x=x_2}\right]. \quad (3.127)$$

Comparing Eqs. (3.123) and (3.127), we see that

$$S(x, t) = -\frac{\hbar}{2mi}\left[\frac{\partial \psi^*(x, t)}{\partial x}\psi(x, t) - \psi^*(x, t)\frac{\partial \psi(x, t)}{\partial x}\right]$$

or

$$S(x, t) = \frac{\hbar}{m}\ \mathrm{Im}\left[\psi^*\frac{\partial \psi(x, t)}{\partial x}\right]. \quad (3.128)$$

This expression provides a way to calculate the probability current from the wave function.

EXAMPLE 3.10 Calculate the probability current for a plane wave.

The wave function for a plane wave is

$$\psi(x, t) = A \exp\left[\frac{i}{\hbar}(px - Et)\right].$$

To calculate the probability current, we shall use Eq. (3.128). Substituting for ψ^*, the complex conjugate of $\psi(x, t)$, i.e.,

$$\psi^*(x, t) = A^* \exp\left[-\frac{i}{\hbar}(px - Et)\right],$$

in Eq. (3.128), we get

$$S(x, t) = -\frac{\hbar}{2mi}\left[|A|^2\left(\frac{-ip}{\hbar}\right) - |A|^2\frac{(ip)}{\hbar}\right] = |A|^2\frac{p}{m} = |A|^2 v,$$

where v is the velocity of the particle associated with the wave. We have used this expression for calculating the particle flux (see the discussion following Example 3.8). We note that this probability current expression is analogous to particle current density nv, where n is the density of the particles and v is their velocity.

THREE-DIMENSIONAL

We can generalize the concept of probability current to three dimensions. Here, the vector probability current density $\boldsymbol{S}(\boldsymbol{r}, t)$ is defined as the 'probability flow' per unit area per unit time, where the normal to the area is along S. The x-component of the vector is

$$S_x(\mathbf{r}, t) = -\frac{\hbar}{2mi}\left[\frac{\partial\psi^*(\mathbf{r}, t)}{\partial t}\psi(\mathbf{r}, t) - \psi^*(\mathbf{r}, t)\frac{\partial\psi(\mathbf{r}, t)}{\partial t}\right]. \tag{3.129}$$

The expressions for the other components of the vector are similar.

3.8 QUALITATIVE FEATURES OF BOUND STATE WAVE FUNCTIONS*

We shall now discuss the qualitative features of the wave functions associated with the stationary states of a particle moving in an arbitrary one-dimensional potential well, say, a potential of the shape shown in Fig. 3.5. The potential tends to zero as $x \to \pm\infty$ and is attractive

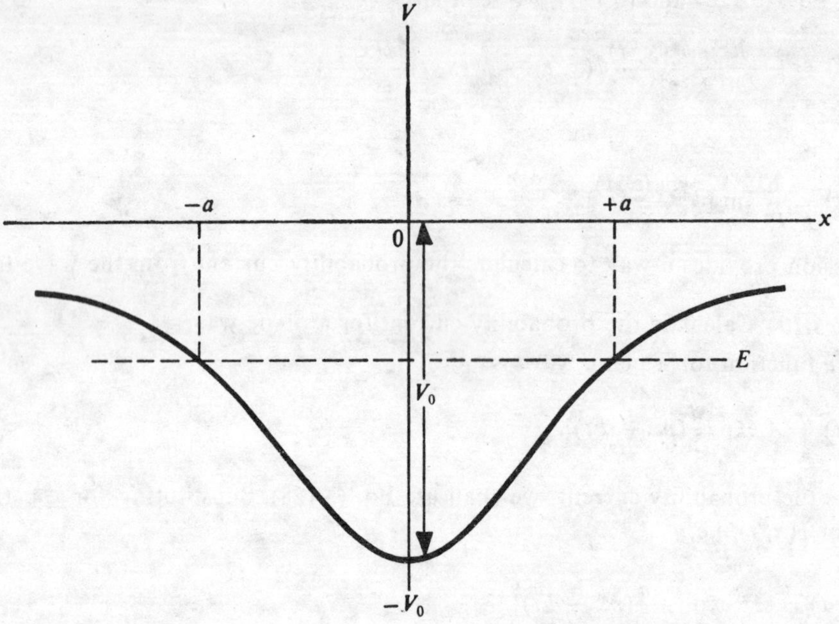

Fig. 3.5 One-dimensional well. The potential is attractive and vanishes as $x \to \pm\infty$. The classically allowed region for particle with energy E is $-a \leqslant x \leqslant +a$.

everywhere ($V < 0$). The wave functions corresponding to a definite energy E (i.e., stationary states) can be broadly classified as bound states ($E < 0$) and scattering (unbound) states ($E > 0$). In the former, the particle is bound and we use the boundary condition that $\psi(x) \to 0$ as $x \to \pm\infty$. On the other hand, the scattering states do not have any such restriction.

Let us first focus our attention on the bound states. The energy E of the particle is between $-V_0$ and 0, as shown in Fig. 3.5. In our discussion on the solutions of the Schroedinger equation for such potentials as one-dimensional box and harmonic oscillator, we had noted that the allowed energy values are quantized if the particle is confined in a region. This result is general

*A student taking an introductory course may skip this discussion.

and is valid for any arbitrary potential. We shall explain it using the qualitative features of the wave functions.

Let us write the Schroedinger equation in the form

$$\frac{d^2\psi/dx^2}{\psi} = \frac{2m}{\hbar^2}[V(x) - E]. \tag{3.130}$$

Also, let us assume that the wave function ψ is real. This, as shown in Appendix 16, is always possible. We consider the following two regions:

(1) The classically allowed region, i.e., $-a < x < a$. In this region, $E > V$ and, in the classical description, the motion of the particle is restricted within it.

(2) The classically forbidden region, i.e., $x > a$, $x < -a$. In this region, $E < V$, and thus this region corresponds to the particle having a negative kinetic energy.

We note that ψ'' (i.e., the second derivative of ψ), given by Eq. (3.130), has the same sign as ψ in the classically forbidden region, whereas it has the sign opposite to that of ψ in the classically allowed region. As we know, the second derivative of a function determines the curvature of the function. If ψ'' is positive, the curve will be concave upwards. This is easy to see because the slope (ψ') of such a curve first becomes less negative as x increases, then goes to zero, and finally increases, and therefore ψ'', which is the rate of change of ψ', is always positive (see curve 2 in Fig. 3.6a and curve 1 in Fig. 3.6b). Similarly, for a curve which is concave down-

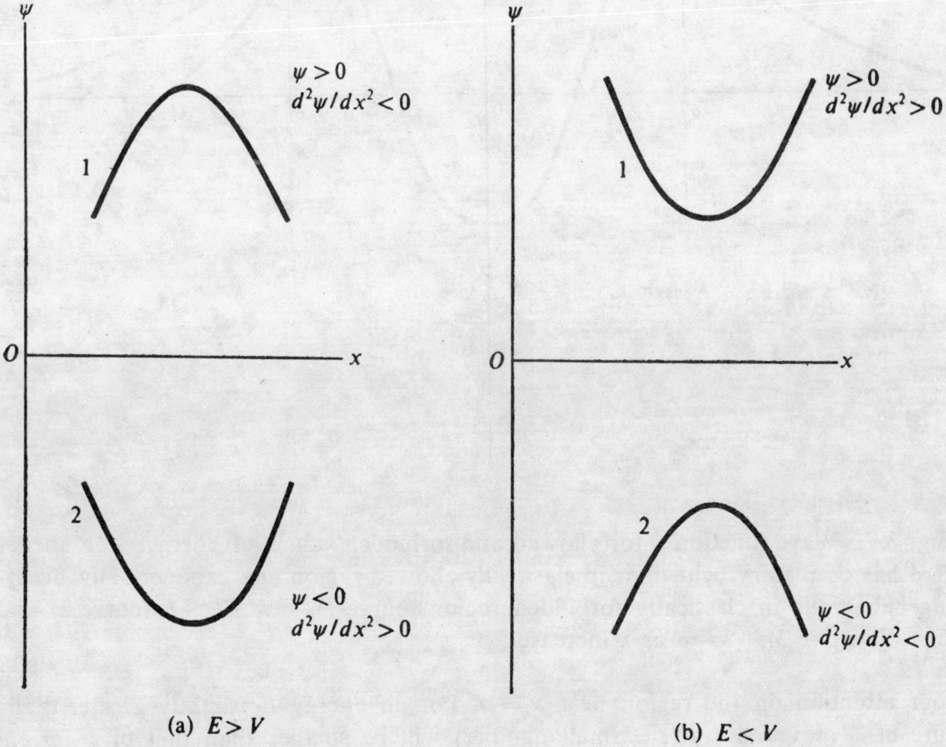

Fig. 3.6 Behaviour of wave function when $E > V$ and $E < V$.

wards, the value of the slope always decreases as x increases, and thus ψ'' is always negative (see curve 1 in Fig. 3.6a and curve 2 in Fig. 3.6b).

The Schroedinger equation which relates ψ'' to ψ determines the general behaviour of ψ. In the classically allowed region, $\psi'' \propto -\psi$, i.e., the wave function is concave downwards for $+|\psi|$ and concave upwards for $-|\psi|$. It is for this reason that the behaviour of the wave function is oscillatory in the region $-a < x < a$ (see Fig. 3.7). On the other hand, in the classically forbidden region, $\psi'' \propto \psi$, i.e., if, to start with, ψ is positive, the function will be concave upwards.

We can now understand why we do not get a physically acceptable solution for any energy value, indicating that the energy is quantized. Let us consider a wave function which is physically acceptable, that is, it satisfies the boundary conditions that $\psi \to 0$ as $x \to \pm\infty$ as shown by curve 1 in Fig. 3.7. This wave function corresponds to an allowed energy, say, E. We

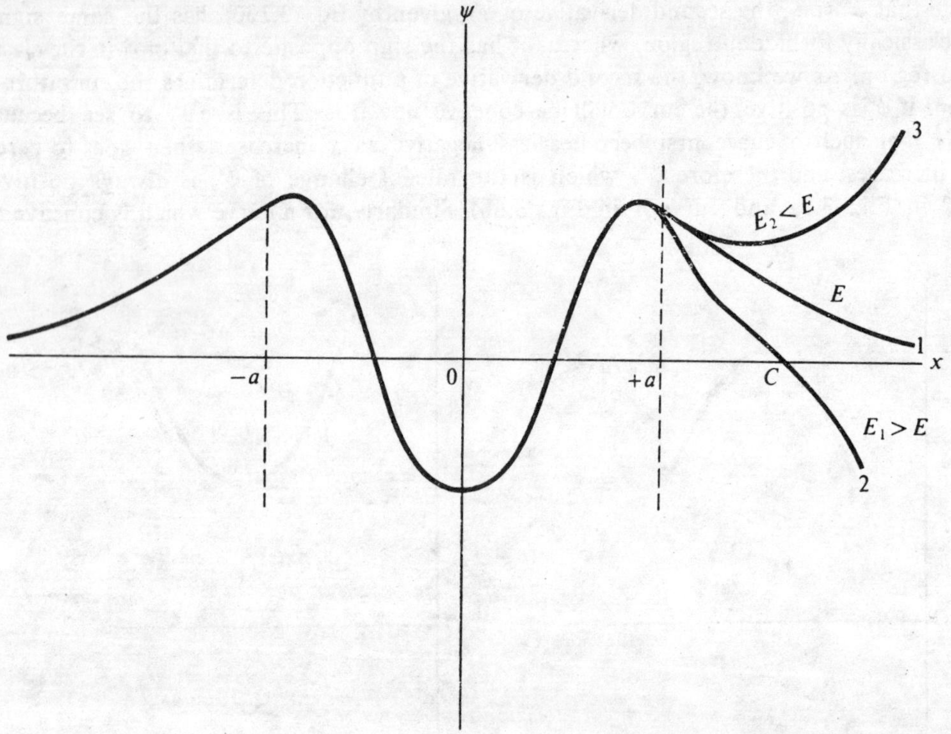

Fig. 3.7 Wave function ψ for allowed and forbidden values of energy. For energy E, ψ has oscillatory behaviour in classically allowed region and exponentially decaying behaviour in classically forbidden region (curve 1). For $E_1 > E$ (curve 2) and $E_2 < E$ (curve 3), $\psi \to \infty$ as x increases.

focus our attention on the region near $x = a$. For an energy E_1 slightly greater than E, the curvature of ψ at $x = a + \epsilon$ (ϵ = small number) will be smaller than that of ψ for energy E because $V - E > V - E_1$. Hence, the wave function which would have otherwise approached the x-axis asymptotically will now cross the x-axis at some point, say, point C of curve 2 in

Fig. 3.7. As soon as ψ crosses the axis (i.e., ψ becomes negative), the curvature of ψ changes its sign and ψ moves away from the x-axis as already discussed. Similarly, for an energy E_2 slightly smaller than E, the curvature of ψ at $x = a + \epsilon$ will be larger than that of ψ for energy E, and hence the curve will move away from the x-axis and approach infinity (see curve 3 in Fig. 3.7).

We have discussed the matching of the wave function for $x \to +\infty$. A similar argument holds for $x \to -\infty$. A physically acceptable bound state wave function must satisfy the boundary conditions both at $x = +\infty$ and $x = -\infty$.

A scattering state (where $E > 0$) does not have the restriction that the wave function must vanish at $x = \pm\infty$. Thus, E can take continuous values for such a state.

PROBLEMS

3.1 Use the uncertainty principle to estimate the ground state energy of a particle in a potential $V = \lambda x^4$.

3.2 Find the energy levels (i.e., eigenvalues) and the first two wave functions for a particle in a potential $V = \frac{1}{2}kx^2 + C$, where C is a constant. Use Eqs. (3.7), (3.8), and (3.9) for finding the wave functions.

3.3 Find the energy eigenvalues for a particle in a potential $V = \frac{1}{2}kx^2 + ax$, where a and k are constants.

3.4 Sketch as a function of x the probability density for the ground state and the first excited state of a particle in a one-dimensional simple harmonic oscillator potential.

3.5 Find the energy levels of a particle of mass m moving in a potential

$$V = \tfrac{1}{2}kx^2 \qquad (x > 0)$$
$$= \infty \qquad (x < 0).$$

3.6 Show that Eqs. (3.7) and (3.8) are the solutions of the Schroedinger equation (3.4).

3.7 A particle with energy $E = (21/2)\hbar\omega$ is moving in a one-dimensional simple harmonic potential. Calculate classically the probability density $P(x)$ between the two turning points, and compare it with the quantum mechanical result shown in Fig. 3.1c.
[*Hint*: Classically, the probability of finding the particle in a region dx is directly proportional to the time the particle spends in the region.]

3.8 Find the energy eigenvalues of a particle moving in a potential $V = k(x^2 + y^2 + z^2)/2$. What are the degeneracies of the first and second excited states of the particle?

3.9 Find the energy eigenvalues of a particle moving in a potential $V = \frac{1}{2}k_1x^2 + \frac{1}{2}k_2y^2$.

3.10 Consider a particle in a two-dimensional space under the action of a potential $V(x, y) = V_1(x) + V_2(y)$. Assuming that the allowed energy values of the particle are

(a) $E_0, E_1, E_2, \ldots$ if it moves in one dimension under the action of $V_1(x)$ alone, and

(b) $F_0, F_1, F_2, \ldots$ if it moves in one dimension under the action of $V_2(y)$ alone,

show that its allowed energy values are

$$(E_0 + F_0), (E_1 + F_0), (E_2 + F_0), \ldots, (E_0 + F_1), (E_1 + F_1), \ldots .$$

3.11 Show that the operator relation $L_xL_y - L_yL_x = i\hbar L_z$ holds.

3.12 Calculate the average value of L_z when the wave function $\Omega(\theta, \phi)$ is (a) Y_2^2, (b) $(Y_1^1 + Y_1^0)/(2)^{1/2}$.

3.13 Two particles, each of mass m, are rigidly connected by a light rod of length l (dumbbell). The dumbbell is rotating about its centre of mass and the motion is confined to a plane. Find the allowed energy values of this system.

3.14 Find the energy levels of a particle of mass m moving freely on the curved surface of a cylinder of radius R and length L. Also find the wave functions corresponding to these energy levels.

3.15 The operator associated with L^2 is given by Eq. (3.45). Show that $L^2Y_0^0 = 0$ and $L^2Y_2^1 = 6\hbar^2Y_2^1$.

3.16 Draw a diagram showing the space quantization for $l = 3$.

3.17 Show that the average value of each of L_x and L_y is zero when the angular part of the wave function is given by

$$Y_1^1(\theta, \phi) = -(\frac{3}{8\pi})^{1/2} \sin\theta \exp(i\phi).$$

3.18 Derive Eqs. (3.118) and (3.119).

3.19 A particle of mass 1 g travelling with a velocity of 1 cm-sec^{-1} encounters a potential barrier $V_0 = 10^{-7}$ J. Calculate the penetration depth $1/\alpha$ [see Eq. (3.86)] of this particle.

3.20 Consider a beam of particles of energy $E > V_0$ (see Fig. 3.3) incident from the right. Show that under this potential (see figure) the beam is reflected and transmitted. Calculate the reflection coefficient.

3.21 Show that the transmission coefficient [given by Eq. (3.119)] can be reduced to

$$T \sim 16\frac{E}{V_0}(1 - \frac{E}{V_0}) \exp(-2\alpha a)$$

when $\alpha a \gg 1$.

3.22 Consider the scattering of electrons (of mass m_e) by a potential well V, where

$$V = 0 \qquad (x < 0,\ x > a),$$

$$V = -V_0 \qquad (0 < x < a),$$

V_0 being positive. Calculate the reflection and transmission coefficients. At what energy of the electrons will there be no reflection?

3.23 Consider a particle whose wave function is real and is such that $\psi(x) \to 0$ as $x \to \pm\infty$. Show that the average value of momentum of the particle is zero. (When the momentum of a particle is not zero, we cannot choose a real wave function.)

4

Atomic Physics

4.1 INTRODUCTION

The concept that matter consists of discrete particles (atoms) is very old. The concrete evidence supporting it emerged during the last century from the investigations of chemical and electrical properties of substances. The study of chemistry led to the concepts of atomic and molecular weight and Avogadro number. The Avogadro number gives an approximate idea of the dimensions of an atom, namely, the volume occupied by an atom or a molecule is about 10^{-29} m^3. The kinetic theory of gases supported the atomistic view of matter and was successful in explaining many physical properties of gases. This theory also led to the conclusion that the volume of an atom or a molecule is approximately 10^{-29} m^3. In other words, the size (radius) of an atom is approximately 10^{-10} m.

> **QUESTION 4.1 Estimate roughly the volume occupied by an aluminium atom. (The atomic weight of aluminium = 27 and the density of aluminium = 2.7×10^3 kg/m^3.)**

The theoretical ideas about the shape of the atom were developed during the beginning of this century. The study of electrical discharge through tubes indicated that electrons are common to an atom and are the fundamental building blocks of matter. As electrons are negatively charged particles and the atom as a whole is neutral, it is clear that some parts of the atom must be positively charged. These parts must therefore contain most of the atomic mass since the total mass of the electrons in an atom is very small as compared to that of the atom.

4.2 THE RUTHERFORD MODEL

The modern concept of the atom is based on the model E. Rutherford proposed in 1911. According to this model, an atom consists mainly of a nucleus occupying a very small region of space and containing all the positive charge. The nucleus is surrounded by negatively charged electrons. The major mass of the atom is contributed by the nucleus. The space between the nucleus and the electrons is empty. He arrived at this model from his study of the scattering of charged particles. In such a study, a beam of particles is sent through a thin sheet of matter (target) and the scattering of the particles is observed. The scattered particles help us interpret the structure of the atoms in the target. To understand how this is achieved, let us use a simple analogy. Consider a box containing some unknown material. One way to get some idea about the nature of the material is to fire bullets through the box and observe their path on the other side of the box. If the bullets get deviated by a very small angle (or do not deviate at all), then it means that the box contains some soft material because the bullets have easily passed through it. On the other hand, if some of the bullets get deflected by a large angle, it then indicates that the box has a hard object which has not let these bullets pass through it. The number of such deflected bullets gives us an estimate of the density and size of the object.

In his experiment, Rutherford used the alpha-particles in place of bullets. The alpha-particles are the nuclei of the helium atoms and are emitted by a radioactive source. The target he used was a very thin (10^{-5} m thick) sheet of gold. He observed the scattered α-particles on a fluorescent screen. A schematic diagram of the experimental set-up is given in Fig. 4.1. His observa-

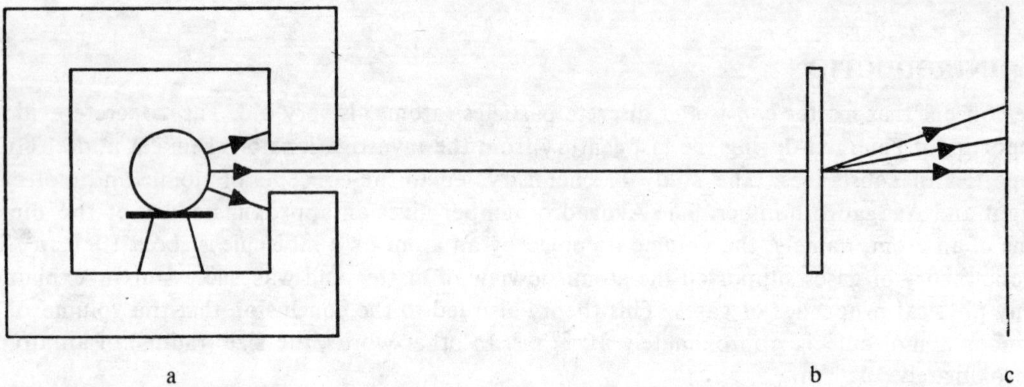

Fig. 4.1 Schematic diagram for scattering. a, α-particle radioactive source in lead chamber; b, gold foil; c, fluorescent screen.

tions showed that most of the α-particles went straight through the gold sheet but some were deflected by large angles. In fact, a few α-particles were completely deflected backwards.

The α-particles, being very much heavier than electrons, could not have been deflected by the electrons. This made Rutherford conclude that the positively charged heavy nuclei must have made the α-particles deflect. Since most α-particles had not been deflected, he further concluded that the positively charged nucleus occupies a very small part of the atom and that the electrons surround the nucleus.

4.3 SPECTRA OF ATOMS

The light emitted by the filament in an electric bulb appears white. When this emitted light is analyzed, its wavelength is found to vary continuously. On the other hand, when an electric discharge is passed through a gas (e.g., sodium vapour), the wavelengths of the light obtained are well defined. Such characteristic wavelengths vary from one gas to another. In fact, helium was discovered to be a constituent of the sun by its characteristic emissions even before it was found on the earth.

The characteristic spectrum of the light emitted by an atom is studied by means of a spectrometer. This instrument consists of a prism or a diffraction grating. When the light beam from a gas is passed through a slit in the spectrometer and made to fall on the prism or the grating, it splits into its wavelength spectrum. This spectrum can be recorded on a photographic plate. It shows a discrete set of lines (each line being the image of the slit of the spectrometer), called the *spectral lines*. The study of the spectra of elements has been of great value in understanding their structure and properties.

Of all the atoms, the hydrogen atom has the simplest spectrum. It consists of two parts, namely, a continuous part and a discrete part (see Fig. 4.2). The presence of this discrete spectrum is

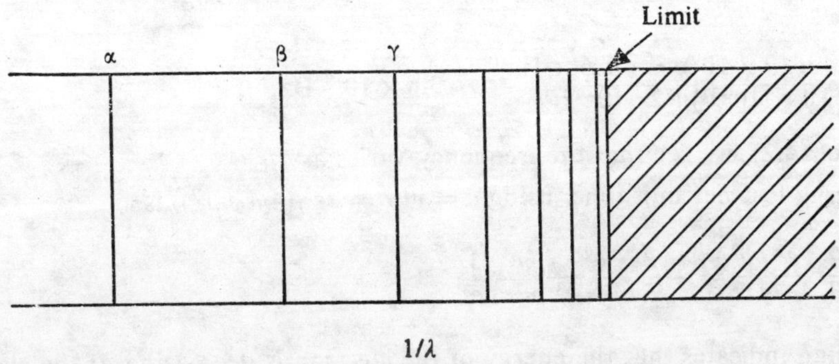

Fig. 4.2 Spectrum of hydrogen atom. Shaded region represents continuous part; lines α, β, γ are discrete wavelengths.

very difficult to understand in terms of classical physics. The reason is as follows. The hydrogen atom has an electron which is attracted by the positively charged nucleus. Let us assume that the electron moves in a circular orbit around the nucleus. According to the electromagnetic theory, if a charged particle moves with a frequency ν in a circular orbit, the light it would emit (electromagnetic radiation) would also have the same frequency ν. In other words, classical physics predicts the emission of light by the particle. However, the electron while emitting light will lose energy, and so will move closer to the nucleus, thus changing the orbit radius. As a result, the frequency of circular motion, and therefore the frequency of radiation, will change. This, according to the classical picture, means that we can expect only a continuous spectrum.

QUESTION 4.2 The electrons in an atom cannot be stationary. Why?

EXAMPLE 4.1 Consider an electron moving around a nucleus of charge $2e$ in a circular orbit of radius 10^{-10} m.

(a) What would be the initial frequency of the light emitted by the electron?

(b) How does the frequency of the emitted light vary with the orbit radius?

(c) How long will the electron take to reach the nucleus (radius of the nucleus = 10^{-14} m)?

Assume that an electron undergoing an acceleration a radiates an energy $(2/3)[1/(4\pi\epsilon_0)][2e^2a^2/(3c^3)]$ J/sec, where c is the velocity of light. This is a result from the classical electromagnetic theory.

The centripetal force for the circular motion of the electron is provided by the Coulomb attraction of the nucleus. If v is the speed of the electron and r the radius of the circular orbit, then

$$\frac{mv^2}{r} = \frac{2e^2}{4\pi\epsilon_0 r^2} \tag{4.1}$$

or $v = [2e^2/(4\pi\epsilon_0 mr)]^{1/2}$. The frequency of the electron moving around the nucleus is

$$f = \frac{v}{2\pi r} = \left(\frac{2e^2}{4\pi\epsilon_0 mr}\right)^{1/2}\frac{1}{2\pi r}. \tag{4.2}$$

(a) The initial frequency ν of the light emitted is

$$\nu = \left(\frac{2}{4\pi\epsilon_0 m}\right)^{1/2} \frac{e}{2\pi r^{3/2}} \tag{4.3}$$

$$= \frac{1.414 \times (9 \times 10^9)^{1/2} \times 1.6 \times 10^{-19}}{(9.1 \times 10^{-31})^{1/2} 2\pi (10^{-10})^{3/2}} \sim 3.6 \times 10^{15} \text{ Hz}.$$

(b) From Eq. (4.2), we see that the frequency varies as $1/r^{3/2}$.

(c) The energy lost per unit time, using $a = v^2/r = 2e^2/(4\pi\epsilon_0 m r^2)$, is

$$\frac{dE}{dt} = -\frac{2}{3}\frac{e^2}{4\pi\epsilon_0}\frac{a^2}{c^3} = -\frac{2}{3}\frac{e^2}{(4\pi\epsilon_0)^3}\frac{4e^4}{m^2 r^4}\frac{1}{c^3}. \tag{4.4}$$

The negative sign indicates that the energy of the electron is decreasing. If the electron is in a circular orbit of radius r, the total energy is

$$E = \text{kinetic energy} + \text{potential energy} = \tfrac{1}{2}mv^2 - 2e^2/(4\pi\epsilon_0 r). \tag{4.5}$$

Using Eq. (4.1), we get

$$\tfrac{1}{2}mv^2 = e^2/(4\pi\epsilon_0 r). \tag{4.6}$$

Substituting this equation in Eq. (4.5), we have

$$E = -e^2/(4\pi\epsilon_0 r). \tag{4.7}$$

Differentiating Eq. (4.7), the rate of change of energy we obtain is

$$\frac{dE}{dt} = \frac{e^2}{4\pi\epsilon_0 r^2}\frac{dr}{dt}. \tag{4.8}$$

Equating this with Eq. (4.4), we get

$$\frac{e^2}{4\pi\epsilon_0}\frac{1}{r^2}\frac{dr}{dt} = -\frac{8}{3}\frac{e^6}{(4\pi\epsilon_0)^3}\frac{1}{m^2c^3}\frac{1}{r^4},$$

$$\frac{dr}{dt} = -\frac{8}{3}\left(\frac{1}{4\pi\epsilon_0}\right)^2\frac{e^4}{m^2c^3}\frac{1}{r^2} \tag{4.9}$$

or

$$r^2\,dr = -\frac{8}{3}\left(\frac{1}{4\pi\epsilon_0}\right)^2\frac{e^4}{m^2c^3}\,dt.$$

Integrating the foregoing equation from $r = 10^{-10}$ m at $t = 0$ to $r = 10^{-14}$ m at $t = T$, we get

$$\int_{10^{-10}}^{10^{-14}} r^2\,dr = -\frac{8}{3}\left(\frac{1}{4\pi\epsilon_0}\right)^2\frac{e^4}{m^2c^3}\int_0^T dt, \tag{4.10}$$

$$\left.\frac{r^3}{3}\right|_{10^{-10}}^{10^{-14}} = \frac{10^{-42} - 10^{-30}}{3} = -\tfrac{8}{3}\left(\frac{1}{4\pi\epsilon_0}\right)^2\frac{e^4}{m^2c^3}T. \tag{4.11}$$

The approximate time taken by the electron to reach the nucleus is then

$$T \sim \tfrac{3}{8}(4\pi\epsilon_0)^2\left(\frac{m^2c^3}{e^4}\right)10^{-30} \qquad (\text{since } 10^{-30} \gg 10^{-42})$$

$$\sim \tfrac{3}{8}\left(\frac{4\pi\epsilon_0\hbar c}{e^2}\right)^2\left(\frac{mc}{\hbar}\right)^2\frac{10^{-30}}{c} \sim \tfrac{3}{8}\,\frac{1}{\alpha^2}\,\frac{1}{\lambda\!\!\!^{-}_c{}^2}\,\frac{10^{-30}}{c},$$

where $\alpha = e^2/(4\pi\epsilon_0\hbar c) = 137$ is the fine structure constant, $\lambda_c[=h/(mc)]$ is the Compton wavelength of the electron, and $\lambda\!\!\!^{-}_c = \hbar/(mc)$. Thus,

$$T \sim \tfrac{3}{8} \times 137 \times 137 \times \frac{1}{(2.4 \times 10^{-12})^2} \times \frac{1 \times 10^{-30}}{3 \times 10^8} \sim 4 \times 10^{-12} \text{ sec.}$$

So we see that the electron reaches the nucleus in a very short time. In other words, the atom collapses to the size of the nucleus. This contradicts the experimental result that the size of an atom is 10^{-10} m.

J. J. Balmer and several other workers empirically found that the discrete part of the spectrum of the hydrogen atom could be written in terms of wave number $(1/\lambda)$ in a simple form as

$$\frac{1}{\lambda} = R\left(\frac{1}{m^2} - \frac{1}{n^2}\right), \tag{4.12}$$

where λ is the wavelength of the light emitted, R is Rydberg's constant $(=1.096776 \times 10^7 \text{ m}^{-1})$, and m, n take integral values (with $m < n$) such as

$$\begin{aligned} m &= 1, \quad n = 2, 3, 4, \ldots, \\ m &= 2, \quad n = 3, 4, 5, \ldots, \\ m &= 3, \quad n = 4, 5, 6, \ldots . \end{aligned} \tag{4.13}$$

EXAMPLE 4.2 From Eq. (4.12), calculate the wavelength of the radiation emitted if $m = 2$ and $n = 3$.

The wavelength is given by

$$\frac{1}{\lambda} = R(\tfrac{1}{4} - \tfrac{1}{9}) \sim 1.097 \times 10^7 \times \tfrac{5}{36} \text{ m}^{-1}.$$

Thus,

$$\lambda = \frac{36}{5 \times 1.097} \times 10^{-7} \sim 6.563 \times 10^{-7} \text{ m} \sim 6563 \text{ Å}.$$

The spectrum of the hydrogen atom was explained by Bohr, using a simple model. We shall first describe his model since it (1) has certain physical features which are easy to understand and (2) satisfactorily explains the atomic structure of hydrogen. Thereafter, we shall treat the hydrogen atom quantum mechanically to explain most of the results obtained from Bohr's model. In so doing, we shall bring out some important differences between the two models, and thus see that the model formulated quantum mechanically is more acceptable.

4.4 BOHR'S MODEL OF HYDROGEN ATOM

The hydrogen atom consists of an electron and a nucleus with a positive charge (equal and opposite to that of the electron). The nucleus is very heavy (about 1800 times the electron). Bohr assumed that the electron goes around the nucleus in a circular orbit in the same manner as the planets move around the sun. The nucleus, being heavy, is thought to be at rest. Such a model does not explain the stability of the atom since, as noted in Example 4.1, the electron loses energy while emitting electromagnetic radiation and falls into the nucleus. Bohr circumvented this problem by assuming that the electron does not radiate only when it moves in certain circular orbits and hence forms stable orbits, i.e., it can only be in these stable orbits. In other words, the orbits are quantized. He found the radii of these stable orbits by assuming that *the angular momentum of the electron is an integral multiple of* $\hbar$. The radiation is emitted only when the electron makes a transition from one stable orbit to another. The energy carried by this radiation is the difference in the energy of the electron in the two orbits. Bohr also assumed Einstein's relation, namely, Eq. (2.4). We now summarize his assumptions, now known as Bohr's postulates:

(1) The electron in a hydrogen atom moves in a circular orbit around the nucleus, the centripetal force being provided by the Coulomb attraction between the electron and the nucleus.

(2) An allowed orbit of the electron is one in which the orbital angular momentum L satisfies the relation

$$L = mvr = nh/(2\pi) = n\hbar, \tag{4.14}$$

where v is the speed of the electron and r is the orbit radius.

(3) The electron does not radiate as long as it is in an orbit satisfying relation (4.14). Further, it emits radiation only when it jumps from an allowed orbit with an energy E_i to another allowed orbit with an energy E_f. The frequency ν of the emitted radiation is given by

$$h\nu = E_i - E_f. \tag{4.15}$$

Bohr proposed his model in 1913, long before the principles of quantum mechanics had been formulated. As we shall subsequently discover, this model correctly explains many properties of the hydrogen atom. There are, however, some properties that cannot be understood in terms of this simple model. We shall discuss these properties in Section 4.8, where we shall also study the hydrogen atom using the Schroedinger equation. Since this model correctly explains several properties and provides a simple picture of the hydrogen atom, it is advisable that we understand Bohr's postulates in terms of the principles of quantum mechanics.

Consider an electron moving in a circular orbit of radius r with a constant speed v. The wavelength of the de Broglie wave associated with the electron is

$$\lambda = h/(mv). \tag{4.16}$$

Combining this with Bohr's second postulate [i.e., Eq. (4.14)], we get

$$2\pi r = nh/(mv) = n\lambda, \tag{4.17}$$

where n is a positive integer. According to Eq. (4.17), the circumference of the orbit is an integral multiple of the wavelength associated with the electron. This means that the wave at any point will be in phase with the wave that would result after the electron completes one revolution. Such waves would interfere constructively, leading to a standing wave. In other words, a stable orbit would result (see Fig. 4.3). On the other hand, if the orbit circumference

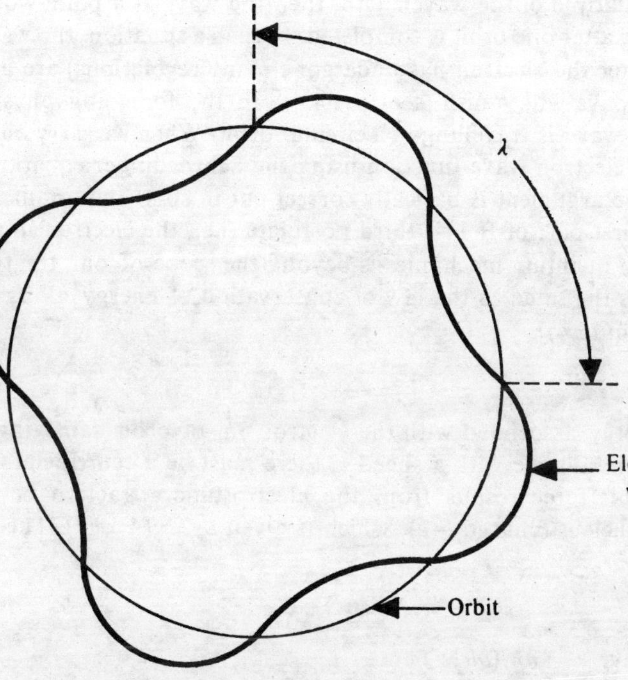

Fig. 4.3 Constructive interference of electron wave.

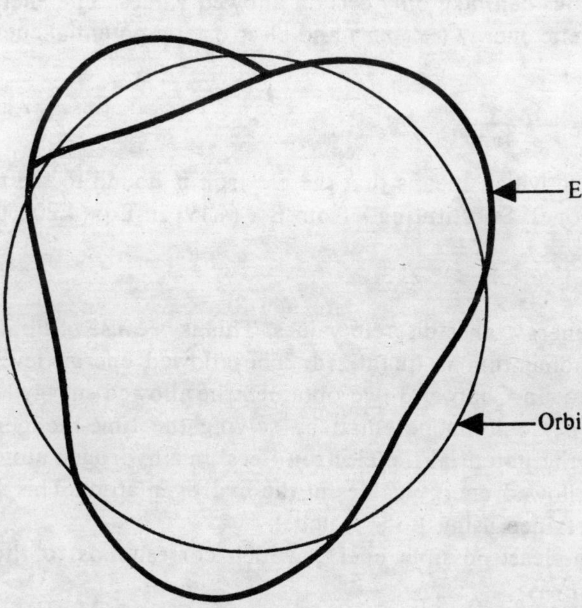

Fig. 4.4 Example of wave not satisfying Bohr condition $mvr = n\hbar$.

were not an integral multiple of the wavelength, then the wave at a point would not be in phase with the wave resulting after one orbit is completed. In such a situation, the wave train will, when the waves (resulting after the electron has undergone many revolutions) are added to it, cancel itself and the electron wave will vanish (see Fig. 4.4). In the foregoing physical argument, we have assumed that the wave is travelling in a circular orbit. When we carry out the mathematical analysis of finding the electron wave function using the Schroedinger equation (see Section 4.8), we shall realize that the argument is basically correct but in some important aspects it is wrong.

The analysis of the first part of Bohr's third postulate (i.e., the electron in a stable orbit does not radiate) in terms of quantum mechanics is beyond the scope of our treatment. The second part of this postulate is the same as the law of conservation of energy along with the Einstein relation $E = h\nu$ [i.e., Eq. (2.4)].

ENERGY LEVELS

We now derive the energy associated with the electron in an orbit satisfying Eq. (4.14). Since the electron is moving in a circle with a speed v, there must be a centripetal force of magnitude mv^2/r acting on it. This force results from the electrostatic attraction between the electron (charge $-e$) and the nucleus (charge $+e$), which is given by $e^2/(4\pi\epsilon_0 r^2)$. Therefore,

$$\frac{mv^2}{r} = \frac{e^2}{4\pi\epsilon_0 r^2}. \tag{4.18}$$

From Eq. (4.14), we have $v = n\hbar/(mr)$. Thus,

$$r = \frac{4\pi\epsilon_0 n^2\hbar^2}{me^2} \qquad (n = 1, 2, 3, \ldots). \tag{4.19}$$

Clearly then, the orbit radius can take only certain allowed values. The energy of the hydrogen atom is in the form of kinetic energy ($=\frac{1}{2}mv^2$) and electrostatic potential energy [$=-e^2/(4\pi\epsilon_0 r)$], and so its total energy is

$$E = \tfrac{1}{2}mv^2 - \frac{e^2}{4\pi\epsilon_0 r} = -\frac{1}{2}\frac{e^2}{4\pi\epsilon_0 r}. \tag{4.20}$$

The energy here is negative which means that the electron is bound to the nucleus (to separate the two, work has to be done). Substituting r from Eq. (4.19) in Eq. (4.20), the energy we get is

$$E_n = -\frac{me^4}{(4\pi\epsilon_0)^2 2\hbar^2}\left(\frac{1}{n^2}\right), \qquad n = 1, 2, 3, \ldots. \tag{4.21}$$

We see that the allowed energy takes discrete values. This is because of Bohr's second postulate, namely, orbital angular momentum is quantized. The allowed energy levels of the hydrogen atom are shown in Fig. 4.5. In Chapter 3, we obtained the allowed energy levels [see Eq. (3.72)] for a particle moving in a Coulomb potential by solving the time-independent Schroedinger equation. This, in fact, is the potential the electron 'sees' in a hydrogen atom, and so the energy levels (3.72) refer to the allowed energy values of the hydrogen atom. This is in perfect agreement with result (4.21) obtained using Bohr's model.

The ground state has the least possible energy which corresponds to the state with $n = 1$. The ground state energy E_1 is

$$E_1 = -\frac{me^4}{(4\pi\epsilon_0)^2 2\hbar^2} \sim -13.6 \text{ eV}.$$

The radius of the Bohr orbit (a_0) for the ground state, from Eq. (4.19), is

$$a_0 = (4\pi\epsilon_0)\frac{\hbar^2}{me^2} \sim 9.53 \times 10^{-10} \text{ m}.$$

It is convenient to write these expressions for energy and radius in terms of the fine structure

Fig. 4.5 Allowed energy levels (horizontal lines) of hydrogen atom. Corresponding energy values (far right) and transitions (arrows) giving rise to spectrum of hydrogen atom are also shown.

constant (dimensionless)

$$\alpha = \frac{e^2}{4\pi\epsilon_0\hbar c} \sim \frac{1}{137} \tag{4.22}$$

and $mc^2 \sim 511$ keV. (The constant α appears frequently in atomic physics.) Thus,

$$E_1 = -\frac{mc^2}{2}\frac{e^4}{(4\pi\epsilon_0)^2\hbar^2c^2} = -\frac{mc^2}{2}\alpha^2 \sim -13.6 \text{ eV}, \tag{4.23a}$$

$$a_0 = \frac{\hbar^2 4\pi\epsilon_0}{me^2} = (\frac{\hbar}{mc})\frac{4\pi\epsilon_0\hbar c}{e^2} = (\frac{\hbar}{mc})\frac{1}{\alpha}. \tag{4.23b}$$

For an orbit, we have

$$r_n = a_0 n^2 = \frac{\hbar}{mc}\frac{n^2}{\alpha}, \tag{4.23c}$$

$$E_n = -\frac{mc^2\alpha^2}{2n^2} = \frac{E_1}{n^2}. \tag{4.23d}$$

We note from equations (4.23c) and (4.23d) that the magnitude of energy decreases as $1/n^2$ and the orbit radius increases as n^2. This result is understandable physically because an electron which is far away from the nucleus is less bound to the nucleus than the one nearer the nucleus.

The discrete spectrum of the hydrogen atom observed experimentally follows from Bohr's third postulate and the energy of the allowed orbits [given by Eq. (4.21)]. If an electron makes a transition from an orbit of energy E_m to an orbit of energy E_n, the frequency of the radiation emitted is

$$\nu = \frac{E_m - E_n}{h} = \frac{mc^2\alpha^2}{2h}\left(\frac{1}{n^2} - \frac{1}{m^2}\right), \qquad n < m,$$

or

$$\frac{1}{\lambda} = R\left(\frac{1}{n^2} - \frac{1}{m^2}\right), \tag{4.24}$$

where R is the Rydberg constant and is given by

$$R = mc\alpha^2/(4\pi\hbar). \tag{4.25}$$

It should be noted that Eq. (4.24) is the same as formula (4.12) obtained experimentally by Balmer and other workers. The value of the Rydberg constant given by Eq. (4.25) is equal to 1.09737×10^7 m^{-1} which agrees well with that determined experimentally [see Eq. (4.12)]. Thus, Bohr's model satisfactorily explains the spectra of the hydrogen atom.

EXAMPLE 4.3 What is the energy, momentum, and wavelength of the photon emitted by a hydrogen atom when the atom makes a transition from the ($n = 2$)- to the ($n = 1$)-state? Assume that the atom is initially at rest.

The energy carried by the photon can be obtained by calculating the difference $E_1 - E_2$ from Eq. (4.23d). Thus, $E = 10.2$ eV. The momentum of the photon is

$$p = E/c = 5.4 \times 10^{-27} \text{ kg-m/sec.}$$

The wavelength of the photon is $\lambda = h/p \sim 1210$ Å.

In Bohr's model, the nucleus is assumed to be infinitely heavy, and thus considered to be at rest. However, as we know, when a photon is emitted, the atom recoils because it has a finite mass. The effects of this recoil can be easily incorporated into the Bohr model (see Example 4.4).

QUESTION 4.3 What is wrong in assuming that the nucleus does not recoil?

EXAMPLE 4.4 How would the results of Example 4.3 be modified when we take into account the recoil of the nucleus in a hydrogen atom?

The initial momentum of a hydrogen atom is zero since the atom is at rest. The law of conservation of momentum demands that the total final momentum of the atom be also zero. Thus, if the photon carries a momentum p_1, the atom has to carry a momentum $-p_1$. The energy of the final system (photon and recoil atom) is the sum of the photon energy (cp_1) and the kinetic energy of the hydrogen atom, i.e.,

$$E_{\text{final}} = cp_1 + \frac{(-p_1)^2}{2M},$$

where M is the mass of the hydrogen atom. This energy must be equal to the energy given out by the hvdrogen atom when it makes the transition from the ($n = 2$)- to the ($n = 1$)-state. This energy as calculated in Example 4.3 is 10.2 eV. Thus,

$$cp_1 + p_1^2/(2M) = 10.2 \text{ eV}.$$

Solving for p_1, we get

$$p_1 = M[-c \pm (c^2 + \frac{20.4 \times 1.6 \times 10^{-19}}{M})^{1/2}] = Mc[-1 \pm 1(1 + \frac{20.4 \times 1.6 \times 10^{-19}}{Mc^2})^{1/2}].$$

Here, Mc^2 is the rest energy of the hydrogen atom and is approximately 940 MeV (see Chapter 8). Therefore,

$$\frac{20.4}{Mc^2} \sim \frac{20.4 \text{ eV}}{940 \times 10^6 \text{ eV}} \sim 2.2 \times 10^{-8}$$

which is very small as compared to unity. In the expression for p_1, the square root can be expanded and, keeping only the first three terms in this expression, we can write

$$p_1 = Mc[-1 \pm \{1 + \tfrac{1}{2}\frac{20.4}{Mc^2} - \tfrac{1}{8}(\frac{10.2}{Mc^2})^2\}].$$

The two solutions of this equation are

$$p_1 \sim -2Mc$$

$$\sim \frac{10.2}{c}[1 - \tfrac{1}{4}(\frac{20.4}{Mc^2})].$$

The magnitude of the momentum cannot be negative. Thus, the first solution is unphysical and we get, from the second solution,

$$p_1 = \frac{10.2}{c}(1 - \tfrac{1}{4} \times 2.2 \times 10^{-8}).$$

Comparing this with the result, i.e., $p_1 = 10.2/c$, we obtained in Example 4.3, we see that the momentum value here is less by 5×10^{-7}%. This is also true for the energy of the photon. The reason the energy and momentum values are less here is that the atom, when it recoils, carries away with it some of the energy and momentum. It should be observed that as $M \to \infty$, we recover the results obtained in Example 4.3.

QUESTION 4.4 In a hydrogen atom, what is the angular momentum of the electron in the ground state according to Bohr's model?

Ionization Energy

As already noted, the radius of the electron orbit in a hydrogen atom can take only discrete values, giving rise to different energy states of the atom. Also, the lowest state of the hydrogen atom corresponds to $n = 1$, its energy being

$$-\frac{me^4}{(4\pi\epsilon_0)^2}\frac{1}{2\hbar^2} = -\frac{mc^2}{2}\alpha^2 \sim -13.6 \text{ eV}. \tag{4.26}$$

The fact that the energy is negative means that the electron is bound to the nucleus. To separate the two, work has to be done. The total energy needed to do this so that the electron and the nucleus are infinitely far apart and also at rest is 13.6 eV. This energy is known as the ionization energy of the hydrogen atom in the ground state and is also the binding energy of the hydrogen atom.

> **QUESTION 4.5 What will be the ionization energy of a hydrogen atom if the atom is in its first excited state?**

Continuous Spectrum

As noted in Section 3.8, if an electron is unbound, its energy can take continuous values. When a free electron having a kinetic energy E jumps into an orbit of a hydrogen atom with an energy E_n, the energy released is $E - E_n$. This energy is released by the emission of a photon of frequency given by

$$\nu = \frac{E - E_n}{h} = \frac{1}{h}[E - (-\frac{mc^2}{2} \times \frac{\alpha^2}{n^2})] = \frac{1}{h}(E + \frac{mc^2\alpha^2}{2n^2}). \tag{4.27}$$

Since E can take continuous values, the frequency spectrum of the emitted photon is also continuous. This explains the continuous part of the spectrum.

4.5 THE FRANCK-HERTZ EXPERIMENT

The energy levels of an atom can be experimentally determined by measuring the energy required to excite the atom from its ground state to higher states, that is, the energy needed to put the electron into higher orbits. This energy can be supplied by bombarding the atom with electrons or photons. Using electrons as the bombarding particles, J. Franck and G. Hertz performed a simple experiment, and provided one of the earliest confirmations of the quantization of the energy levels of atoms. Figure 4.6 shows a schematic diagram of the apparatus they used. The vapour of the atoms to be bombarded is filled in a tube containing a filament F, grid G, and plate P. The filament emits electrons which can be accelerated by varying the potential V and applying it between the filament and the grid. The grid is a wire mesh and allows the electrons to pass through. The plate is maintained at a potential slightly negative than that at the grid. In this experiment, the current I in the ammeter is measured as a function of the voltage V. A typical curve showing the variation of current with voltage is also depicted in Fig. 4.6.

As the voltage V increases, the electron energy goes up, and so the electrons can overcome the retarding potential E to reach the plate P. This gives rise to a current in the ammeter, which initially increases as V increases. At some value of $V (= V_1)$, the electrons have sufficient energy to excite the atom in its ground state to its next allowed state. In so doing, the electrons lose energy, and therefore the number of electrons reaching the plate decreases. This gives rise to a peak in the *I-V* curve. When V is further increased, the electrons gain enough energy to

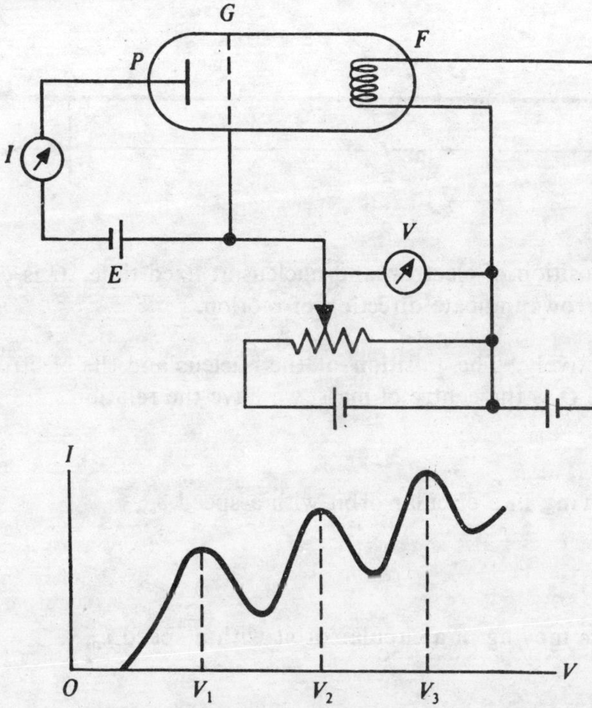

Fig. 4.6 Franck-Hertz experiment. Curve shows variation of plate current I with grid voltage V.

excite the atom to its next higher states. In Fig. 4.6, each peak corresponds to an excited state of the atom. The positions of the peaks give the allowed energy states of the atom.

> **QUESTION 4.6 What will be the energy difference (in electron volts) between the ground state and the first excited state of an atom if the first peak in the *I-V* curve is observed at 4.9 V?**

4.6 EFFECT OF FINITE MASS OF NUCLEUS

We have so far assumed that the nucleus remains fixed in space while the electron is moving in a circular orbit around it. Strictly speaking, the nucleus also moves as it experiences the electrostatic force exerted on it by the electron. The movement of the nucleus is much smaller because the nucleus is much heavier than the electron. The effect of this slight motion of the nucleus can be observed if the experiment we have described is accurately carried out. If the effect of the finite mass of the nucleus has also to be taken into account, it is then necessary that we modify the results we earlier obtained using Bohr's model.

Consider a hydrogen atom at rest. Let the mass of the electron and the nucleus be m and M, respectively. Since there are no external forces acting on the system, the centre of mass of the hydrogen atom will remain at rest. Both the nucleus and the electron will move in circles about their centre of mass (see Fig. 4.7). Let r_e and r_n be the radii of the circular orbits of the electron

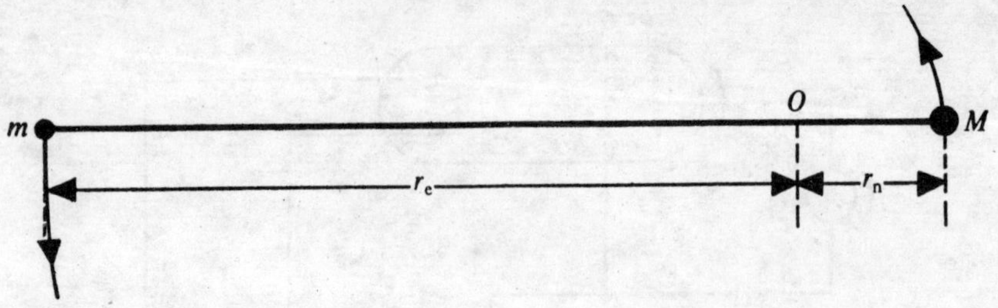

Fig. 4.7 Position of electron and nucleus at fixed time. O is centre of mass and arrows indicate direction of motion.

and the nucleus, respectively. The position of the nucleus and the electron at some instant is shown in Fig. 4.7. Since O is the centre of mass, we have the relation

$$mr_e = Mr_n. \tag{4.28}$$

Since the electron is moving in a circular orbit with a speed v_e,

$$\frac{mv_e^2}{r_e} = \frac{e^2}{4\pi\epsilon_0(r_e + r_n)^2}. \tag{4.29}$$

Similarly, for the nucleus moving in a circular orbit with a speed v_n,

$$\frac{Mv_n^2}{r_n} = \frac{e^2}{4\pi\epsilon_0(r_e + r_n)^2}. \tag{4.30}$$

As the centre of mass does not move, the angular velocities of the electron and the nucleus are the same. Thus,

$$\frac{v_e}{r_e} = \frac{v_n}{r_n}. \tag{4.31}$$

According to Bohr's second postulate, the orbital angular momentum of an electron is an integral multiple of $\hbar$. Since the nucleus is also moving in a circular orbit, we modify the postulate and state that the total angular momentum of the system (including the orbital angular momentum of the nucleus) is an integral multiple of $\hbar$. Thus,

$$L_{\text{total}} = mv_e r_e + Mv_n r_n = n\hbar \qquad (n = 1, 2, 3, \ldots). \tag{4.32}$$

If ω is the angular velocity of the electron, Eq. (4.29) then becomes

$$mr_e\omega^2 = \frac{e^2}{4\pi\epsilon_0(r_e + r_n)^2} = \frac{e^2}{4\pi\epsilon_0 r^2}, \tag{4.33}$$

where r is the distance between the electron and the nucleus, i.e.,

$$r = r_e + r_n. \tag{4.34}$$

Equations (4.28) and (4.34) give

$$r_e = \frac{M}{m + M}r. \tag{4.35}$$

Substituting this in Eq. (4.33), we get

$$mr_e\omega^2 = \frac{mM}{m+M}r\omega^2 = \frac{e^2}{4\pi\epsilon_0 r^2}. \tag{4.36}$$

QUESTION 4.7 Prove Eq. (4.31) and use it to show that Eq. (4.36) can be obtained from Eq. (4.30).

The total angular momentum of the hydrogen atom then is

$$L_{\text{total}} = (mr_e^2 + Mr_n^2)\omega = [\frac{mM^2 + Mm^2}{(M+m)^2}]r^2\omega = \frac{mM}{m+M}r^2\omega. \tag{4.37}$$

Thus, the quantization condition becomes

$$\frac{mM}{m+M}\omega r^2 = n\hbar. \tag{4.38}$$

The total energy of the system is

$$E = \tfrac{1}{2}mv_e^2 + \tfrac{1}{2}Mv_n^2 - \frac{e^2}{4\pi\epsilon_0 r} \tag{4.39}$$

$$= \tfrac{1}{2}(mr_e^2 + Mr_n^2)\omega^2 - \frac{e^2}{4\pi\epsilon_0 r} = \frac{1}{2}\frac{mM}{m+M}r^2\omega^2 - \frac{e^2}{4\pi\epsilon_0 r}. \tag{4.40}$$

We rewrite Eqs. (4.36), (4.38), and (4.40), using $\mu = mM/(m+M)$, as

$$\mu\omega^2 r = e^2/(4\pi\epsilon_0 r^2), \tag{4.41}$$

$$\mu\omega r^2 = n\hbar, \tag{4.42}$$

$$E = \tfrac{1}{2}\mu r^2\omega^2 - e^2/(4\pi\epsilon_0 r). \tag{4.43}$$

Equations (4.41), (4.42), (4.43) become identical with Eqs. (4.18), (4.14), (4.20) (with $\omega = v/r$) if in the latter set of equations we replace the mass of the electron m by μ. Thus, we see that the effect of the finite mass of the nucleus is equivalent to replacing in the former set of equations the mass of the electron m by a slightly reduced value, namely, $m[M/(m+M)]$, known as the reduced mass. The reduced mass μ is less than m by a factor $1/(1 + m/M)$. To calculate the allowed orbital radii and energy levels, we can proceed in a manner similar to that followed in Section 4.4, except that now m is replaced by μ. We thus get

$$r_n = \frac{4\pi\epsilon_0\hbar^2}{\mu e^2}n^2 = \frac{\hbar}{\mu c}\frac{n^2}{\alpha}, \tag{4.44}$$

$$E_n = -\frac{\mu e^4}{(4\pi\epsilon_0)^2}\frac{1}{2\hbar^2}(\frac{1}{n^2}) = -\frac{\mu c^2\alpha^2}{2n^2}, \tag{4.45}$$

$$\frac{1}{\lambda} = R_\mu(\frac{1}{n^2} - \frac{1}{m^2}), \tag{4.46}$$

where

$$R_\mu = \frac{\mu c\alpha^2}{4\pi\hbar} = \frac{\mu R}{m}. \tag{4.47}$$

Comparing the set of Eqs. (4.44)–(4.46) with Eqs. (4.23c), (4.23d), (4.24), we find that μ has replaced m in the former set. Substituting the numerical values in Eq. (4.47), we obtain

$$R_\mu = 1.096810 \times 10^7 \text{ m}^{-1}, \tag{4.48}$$

which is in much better agreement with the experimental value (i.e., 1.096776×10^7 m^{-1}).

QUESTION 4.8 If the reduced mass is used in calculating the energy levels, is it necessary to take into account the recoil effect of the atom as done in Example 4.4 to obtain the exact energy of the photon?

EXAMPLE 4.5 A positron is a particle with a mass identical with that of an electron but with a charge equal in magnitude and opposite in sign to that of the electron. In a positronium atom, the electron is bound to the positron. In other words, the nucleus is replaced by the positron. Calculate the allowed energy values and the ionization energy of the positron, using Bohr's model.

Since, as we have noted, the mass of the nucleus (positron) is equal to that of the electron, we need to use the reduced mass of positronium for evaluating the energy levels [i.e., Eq. (4.45)]

$$E_n = -\mu c^2\alpha^2/(2n^2)$$

with $\mu = mM/(m + M) = m/2$ since $M = m$. Thus,

$$E_n = -\frac{6.8}{n^2} \text{ eV}.$$

It should be noted that the energy of each of these allowed states is half the energy of the corresponding state of a hydrogen atom with an infinitely heavy nucleus. Therefore, the ionization energy is 6.8 eV.

4.7 HYDROGEN-LIKE ATOM

An atom of atomic number Z contains Z electrons moving around the nucleus with a charge Ze. If $Z - 1$ electrons are removed, then, as in a hydrogen atom, there will only be a single electron moving around the nucleus in the atom except that the nuclear charge would be Ze instead of e. It is straightforward to calculate the allowed energy values of such an atom. The energy values $E_{Z,n}$ are given by the formula

$$E_{Z,n} = -Z^2E_n, \tag{4.49}$$

where E_n are the allowed energy values of the hydrogen atom [see Eq. (4.45)]. The wavelength of the spectrum of an ionized atom is

$$\frac{1}{\lambda} = RZ^2\left(\frac{1}{m^2} - \frac{1}{n^2}\right), \tag{4.50}$$

where

$$m = 1, \quad n = 2, 3, 4, \ldots; \qquad m = 2, \quad n = 3, 4, 5, \ldots; \qquad m = 3, \quad n = 4, 5, 6, \ldots;$$

and so on.

EXAMPLE 4.6 **Calculate the frequency of the electron in the *n*-th orbit of a hydrogen-like atom with a nuclear charge *Ze*.**

From the expressions

$$\frac{mv^2}{r} = \frac{Ze^2}{4\pi\epsilon_0 r^2}, \qquad mvr = n\hbar,$$

we get

$$r_n = \frac{4\pi\epsilon_0\hbar^2}{me^2}\frac{n^2}{Z} = \frac{a_0 n^2}{Z} \qquad \text{(where } a_0 \text{ is the Bohr radius),}$$

$$v_n = \frac{Ze^2}{4\pi\epsilon_0\hbar}\left(\frac{1}{n}\right) = \frac{Z\alpha c}{n}.$$

The frequency of the electron in the *n*-th orbit then is

$$\nu_n = \frac{v_n}{2\pi r_n} = \frac{Z^2\alpha c}{2\pi a_0}\left(\frac{1}{n^3}\right).$$

Using

$$\frac{\alpha}{a_0} = \left(\frac{e^2}{4\pi\epsilon_0 c\hbar}\right)\frac{me^2}{4\pi\epsilon_0\hbar^2} = \frac{mc\alpha^2}{\hbar},$$

we get

$$\nu_n = \frac{Z^2 mc^2\alpha^2}{2\pi\hbar}\left(\frac{1}{n^3}\right). \tag{4.51}$$

We should like to again emphasize that Bohr's picture of the hydrogen atom is only a model and not a theory. It is a reasonably good model, giving a simple account of the atom. The development of quantum mechanics has however given us a theory, using which we can understand the properties of atoms. As already noted, the main assumption [i.e., Eq. (4.14)] in Bohr's model incorporates in it a quantum concept, namely, quantization of angular momentum. We shall now treat the hydrogen atom in terms of quantum mechanics. In other words, we shall use the Schroedinger equation to study the hydrogen atom.

4.8 QUANTUM MECHANICAL TREATMENT OF HYDROGEN ATOM

In a hydrogen atom, the potential energy of the electron in the presence of the nucleus is $-e^2/(4\pi\epsilon_0 r)$. Thus, the Schroedinger equation (3.31) for the motion of the electron is

$$\nabla^2\psi + \frac{2m}{\hbar^2}\left(E + \frac{e^2}{4\pi\epsilon_0}\frac{1}{r}\right)\psi = 0. \tag{4.52}$$

As we can see, the potential is spherically symmetric. Thus, as indicated in Section 3.3, it is convenient to use the spherical polar coordinates. Writing $\nabla^2\psi$ in the spherical polar coordinates, we get

$$\frac{1}{r^2}\frac{\partial}{\partial r}\left(r^2\frac{\partial\psi}{\partial r}\right) + \frac{1}{r^2\sin\theta}\frac{\partial}{\partial\theta}\left(\sin\theta\frac{\partial\psi}{\partial\theta}\right) + \frac{1}{r^2\sin^2\theta}\frac{\partial^2\psi}{\partial\phi^2} + \frac{2m}{\hbar^2}\left(E + \frac{e^2}{4\pi\epsilon_0 r}\right)\psi = 0. \tag{4.53}$$

As we have already learnt, we can separate Eq. (4.53) into a radial part and an angular part. For a spherically symmetric (central) potential, the solution of the angular part, as we know, can be given in terms of the spherical harmonics $Y_l^m(\theta, \phi)$. The solution of Eq. (4.53) is then of the form

$$\psi(r, \theta, \phi) = R_n^l(r) Y_l^{m_l}(\theta, \phi), \qquad m_l = -l \text{ to } +l;\ l = 0, 1, 2, \ldots . \tag{4.54}$$

The radial part $R_n^l(r)$ of the wave function here satisfies the equation [see Eq. (3.71)]

$$\frac{1}{r^2}\frac{d}{dr}\left(r^2 \frac{dR_n^l}{dr}\right) + \frac{2m}{\hbar^2}\left[E + \frac{e^2}{4\pi\epsilon_0 r} - \frac{\hbar^2}{2m}\frac{l(l+1)}{r^2}\right] R_n^l = 0. \tag{4.55}$$

This differential equation has the physical boundary condition (3.69), i.e.,

$$R(r) \to 0 \quad \text{as} \quad r \to \infty.$$

Since our intention here is to understand the physics of the solution of this differential equation, we make no effort to solve it. (For the algebraic details, see Appendix 15.) As discussed in Chapter 3, the boundary condition can be satisfied only when E takes the values

$$E = E_n = -\frac{me^4}{2\hbar^2(4\pi\epsilon_0)^2}\frac{1}{n^2} = -\frac{mc^2\alpha^2}{2}\left(\frac{1}{n^2}\right), \qquad n = 1, 2, \ldots . \tag{4.56}$$

As can be noted, these values are the same as those obtained in Bohr's model (4.23d). The quantum mechanical treatment also gives the wave function $R_n^l(r) Y_l^{m_l}(\theta, \phi)$ associated with the energy level E_n. The wave function helps us calculate the probability of finding the electron in a given volume. It also enables us to compute the average value of various physically measurable quantities such as radius, angular momentum, and magnetic moment. Therefore, the knowledge of the wave function is necessary to get a detailed picture of the hydrogen atom. Before we discuss the wave functions of a hydrogen atom, we state their following general properties:

(1) For a fixed value of n, there are several allowed values of l. In particular, l can take the values $0, 1, 2, \ldots, (n-1)$. For example, when $n = 3$, $l = 0, 1, 2$.

(2) For a fixed value of l, m_l takes on $(2l+1)$ values, i.e., $-l, -(l-1), \ldots, 0, \ldots, +(l-1), l$.

The three numbers n, l, m_l completely specify the wave function of the electron in a hydrogen atom. These are known as the principal, orbital (or azimuthal), and magnetic quantum numbers, respectively. The reason the quantum numbers are three is that we need three coordinates to specify the electron in the atom.

QUESTION 4.9 How many quantum numbers are required to describe a particle moving on a spherical surface of a fixed radius?

Degeneracy

The principal quantum number n of the ground state is 1. Since the orbital quantum number l can have only the values $0, 1, \ldots, (n-1)$, in the ground state $l = 0$. Also, the magnetic quantum number m_l in the ground state has only one allowed value, namely, $m_l = 0$. Thus, the ground state is specified by

$$n = 1, \quad l = 0, \quad m_l = 0. \tag{4.57}$$

This state is nondegenerate.

For the first excited state, $n = 2$, and therefore the allowed values of l are 0 and 1. For $l = 0$, $m = 0$, and for $l = 1$, m_l can have three possible values, namely, 0 and ± 1. Thus, the possible first excited states can be specified by the following quantum numbers:

$$\begin{aligned} n &= 2, \quad l = 0, \quad m_l = 0, \\ n &= 2, \quad l = 1, \quad m_l = 1, \\ n &= 2, \quad l = 1, \quad m_l = 0, \\ n &= 2, \quad l = 1, \quad m_l = -1. \end{aligned} \tag{4.58}$$

QUESTION 4.10 For a hydrogen atom in the ground state, what is the value of the (a) square of angular momentum? (b) *z*-component of angular momentum?

Since the energy depends only on the principal quantum number n [see Eq. (4.56)], each of the foregoing four states corresponds to the same energy. Hence, the first excited state is four-fold degenerate.

The wave function corresponding to a state is denoted by ψ_{n, l, m_l}. Thus, the ground state wave function is $\psi_{1,0,0}$, and the four wave functions associated with the first excited state are $\psi_{2,0,0}$, $\psi_{2,1,1}$, $\psi_{2,1,0}$, and $\psi_{2,1,-1}$. Although each of these four wave functions corresponds to the same energy, they describe the different states of the hydrogen atom. For example, for the z-component of angular momentum, $\psi_{2,1,-1}$ has a value $-\hbar$ and $\psi_{2,1,0}$ has the value zero.

EXAMPLE 4.7 Show that the degeneracy of a state with the principal quantum number n is n^2, ignoring the spin quantum number of the electron (see Section 4.12 for a discussion on electron spin).

When the principal quantum number is n, the allowed values of l are $0, 1, 2, \ldots, n-1$ and, for each such value, there are $(2l+1)$ allowed values, namely, $-l, -l+1, \ldots, l-1, l$, of m_l. So, the number of allowed states for a given value of n is

$$\begin{aligned} 0 + 3 + \ldots + [2(n-1) + 1] &= \sum_{l=0}^{n-1} (2l + 1) = 2 \sum_{l=0}^{n-1} l + \sum_{l=0}^{n-1} 1 \\ &= \frac{2(n-1)(n)}{2} + n = n^2. \end{aligned}$$

As we shall subsequently discover (see Section 4.12), when the spin is taken into account, the degeneracy of a state with the principal quantum number n is $2n^2$.

WAVE FUNCTIONS

The ground state wave function is given by

$$\psi_{1,0,0}(r, \theta, \phi) = R_1^0(r) Y_0^0(\theta, \phi) = \left(\frac{1}{\pi a_0^3}\right)^{1/2} \exp\left(-\frac{r}{a_0}\right), \tag{4.59}$$

where a_0 is the Bohr radius. This wave function depends only on r and not on θ and ϕ.

The four wave functions corresponding to the first excited state, as specified by the n, l, m_l values, are

$$\psi_{2,0,0}(r, \theta, \phi) = \frac{1}{4(2\pi a\ )^{1/2}}\left(2 - \frac{r}{a_0}\right) \exp\left(-\frac{r}{2a_0}\right), \tag{4.60}$$

$$\psi_{2,1,1}(r, \theta, \phi) = \frac{1}{8(\pi a_0^3)^{1/2}} \frac{r}{a_0} \exp\left(-\frac{r}{2a_0}\right) \sin\theta \exp(i\phi), \tag{4.61}$$

$$\psi_{2,1,0}(r, \theta, \phi) = \frac{1}{4(2\pi a_0^3)^{1/2}} \frac{r}{a_0} \exp\left(-\frac{r}{2a_0}\right) \cos\theta, \tag{4.62}$$

$$\psi_{2,1,-1}(r, \theta, \phi) = \frac{1}{8(\pi a_0^3)^{1/2}} \frac{r}{a_0} \exp\left(-\frac{r}{2a_0}\right) \sin\theta \exp(-i\phi). \tag{4.63}$$

We note that $\psi_{2,0,0}(r, \theta, \phi)$ is spherically symmetric, that is, it depends only on r. These wave functions are normalized. In other words, they satisfy the condition

$$\int |\psi(r, \theta, \phi)|^2\, dv = 1, \tag{4.64}$$

where the integration is carried out over the entire space. As noted in Chapter 2, this physical condition ensures that the total probability of finding the electron in the entire space is unity.

In terms of the principles of quantum mechanics, the wave function of a system completely describes the state of the system. In other words, we can obtain all the physical properties of the system once we know the wave function.

4.9 QUANTUM MECHANICAL PICTURE OF HYDROGEN ATOM

We can get a picture of the hydrogen atom by calculating the probability of finding the electron in the space around the nucleus.

PROBABILITY DENSITY

The probability of finding the electron in a small volume Δv is

$$P(r, \theta, \phi)\Delta v = |\psi(r, \theta, \phi)|^2 \Delta v, \tag{4.65}$$

where $P(r, \theta, \phi)$ is the probability per unit volume or the probability density. The volume element Δv in the spherical polar coordinates is expressed (see Appendix 5) as

$$\Delta v = r^2 \sin\theta\, dr\, d\theta\, d\phi. \tag{4.66}$$

Thus, we have

$$P(r, \theta, \phi)\Delta v = |\psi(r, \theta, \phi)|^2 r^2 \sin\theta\, dr\, d\theta\, d\phi. \tag{4.67}$$

The probability $P(r)\, dr$ of finding the electron between r and $(r + dr)$, irrespective of the coordinates θ, ϕ, is obtained by integrating Eq. (4.67) over the coordinates θ, ϕ. Thus,

$$P(r)\, dr = \int_0^{2\pi} d\phi \int_0^{\pi} d\theta \sin\theta\, |\psi(r, \theta, \phi)|^2 r^2\, dr. \tag{4.68}$$

For the ground state of the hydrogen atom, we have

$$|\psi(r, \theta, \phi)|^2 = \frac{1}{\pi a_0^3} \exp\left(-\frac{2r}{a_0}\right). \tag{4.69}$$

Substituting this in Eq. (4.68), we get the probability $P(r)\,dr$ as

$$P(r)\,dr = \int_0^{2\pi} d\phi \int_0^{\pi} d\theta \sin\theta \frac{1}{\pi a_0^3} \exp\left(-\frac{2r}{a_0}\right) r^2\,dr \tag{4.70}$$

$$= 2\pi \int_0^{\pi} d\theta \sin\theta \frac{1}{\pi a_0^3} \exp\left(-\frac{2r}{a_0}\right) r^2\,dr = [2\pi(-\cos\theta)]\Big|_0^{\pi} \frac{1}{\pi a_0^3} \exp\left(-\frac{2r}{a_0}\right) r^2\,dr$$

$$= \left[\frac{1}{\pi a_0^3} \exp\left(-\frac{2r}{a_0}\right)\right] 4\pi r^2\,dr. \tag{4.71}$$

It should be noted that here $4\pi r^2\,dr$ is the volume of a spherical shell of radius r and thickness dr. Hence, Eq. (4.71) is the probability density times the volume element, and can be written as

$$P(r)\,dr = |\psi_{1,0,0}|^2 4\pi r^2\,dr. \tag{4.72}$$

This simple interpretation in terms of ψ is possible only when ψ is independent of θ and ϕ. In general, we have to use Eq. (4.68) to evaluate $P(r)\,dr$. The radial probability density $P(r)$, from Eq. (4.71), is

$$P(r) = \frac{4}{a_0^3} \exp\left(-\frac{2r}{a_0}\right) r^2. \tag{4.73}$$

It should be noted that the total probability of finding the electron between $r = 0$ and $r = \infty$ is

$$\int_0^{\infty} P(r)\,dr$$

$$= \frac{4}{a_0^3} \int_0^{\infty} \exp\left(-\frac{2r}{a_0}\right) r^2\,dr = 1 \tag{4.74}$$

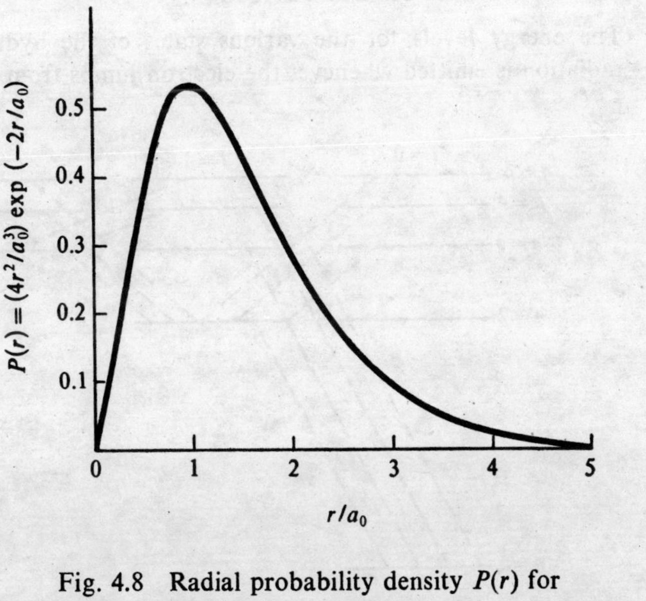

Fig. 4.8 Radial probability density $P(r)$ for state $n = 1$ of hydrogen atom.

as it should be. A graph between $P(r)$ and r is shown in Fig. 4.8. As can be seen, $P(r)$ is large in the region around $r = a_0$ and is maximum at $r = a_0$.

EXAMPLE 4.8 Calculate the probability of finding the electron in the region $a_0/2 < r < 2a_0$ in a hydrogen atom in the ground state.

The probability is given by

$$\int_{a_0/2}^{2a_0} P(r)\,dr = \int_{a_0/2}^{2a_0} \frac{4}{a_0^3} \exp\left(-\frac{2r}{a_0}\right) r^2\,dr = 0.68.$$

Average Radius

From Fig. 4.8, we see that the electron does not have a definite orbit around the nucleus, but has a probability distribution around $r = a_0$. We define the average radius $\langle r \rangle$ of the hydrogen atom as the expectation value of r in the state $\psi_{1,0,0}$, i.e., as

$$\langle r \rangle = \iiint |\psi_{1,0,0}|^2 r r^2\, dr \sin\theta\, d\theta\, d\phi \tag{4.75}$$

$$= \int_0^\infty P(r) r\, dr. \tag{4.76}$$

Substituting for $P(r)$ from Eq. (4.73), we get

$$\langle r \rangle = \frac{4}{a_0^3} \int_0^\infty r^3 \exp\left(-\frac{2r}{a_0}\right) dr = \frac{3a_0}{2}. \tag{4.77}$$

The maximum value of $P(r)$ and the average radius $\langle r \rangle$ indicate that the probability of finding the electron at a distance of $\sim a_0$ $(0.5 \times 10^{-10}$ m) from the nucleus is large.

HYDROGEN ATOM SPECTRUM

The energy levels for the various states of the hydrogen atom are shown in Fig. 4.9. The radiation is emitted whenever the electron jumps from one state to another. Since a discussion

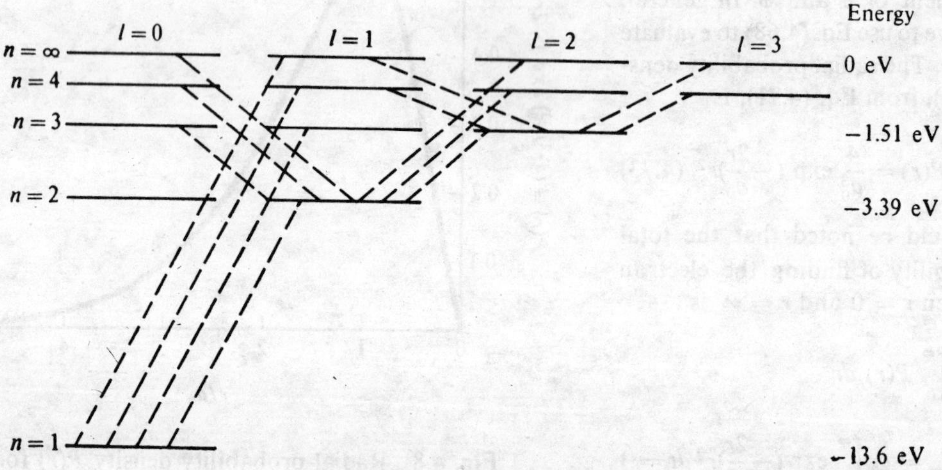

Fig. 4.9 Energy levels of hydrogen atom and corresponding n, l values. Allowed transitions obey $l_i - l_f = \Delta l = \pm 1$.

on the dynamics of the emission process is beyond the scope of this text, we state only the result of the quantum mechanical treatment of this dynamics, namely, the transition of the electron from one state to another by emission of a photon takes place only when the two states differ in their value l by 1. Thus, if l_i and l_f are the initial and final orbital angular momenta of the initial and final states, then

$$l_i - l_f = +1 \tag{4.78}$$

$$= -1. \tag{4.79}$$

Equations (4.78), (4.79) constitute a selection rule for a transition. The frequency of the photon emitted is obtained by using the usual formula

$$\nu = \frac{E_i - E_f}{h}, \tag{4.80}$$

where E_i and E_f are the initial and final energies of the electron. The selection rule gives the allowed transitions, and thus the spectrum of the hydrogen atom (see Fig. 4.9).

4.10 COMPARISON WITH BOHR'S MODEL

Now that we have the quantum mechanical picture of the hydrogen atom, it will be interesting to compare it with the results obtained from Bohr's semiclassical model.

The energy levels of the hydrogen atom calculated using quantum mechanics are the same as those obtained using Bohr's model. In the quantum mechanical picture, the electron does not have a definite radius. However, the most probable distance between the electron and the nucleus for the ground state is exactly a_0, the Bohr radius.

The angular momentum of the ground state, according to Bohr's model, can be obtained by putting $n = 1$ in Eq. (4.14) and its value is $\hbar$. In quantum mechanics, the ground state wave function is $R_1^0 Y_0^0(\theta, \phi)$. If the angular part of the wave function is $Y_l^{m_l}(\theta, \phi)$, then, as we learnt in Chapter 3 [see Eq. (3.50)], the magnitude of the angular momentum is $[l(l+1)]^{1/2}\hbar$. Thus, the ground state, where $l = 0$, has no angular momentum at all. This result contradicts the one given by Bohr's model. However, the quantum mechanical picture, it has been shown, is correct.

Further, Bohr's model cannot predict the degeneracy of states, whereas in the quantum mechanical treatment, this is possible. As we shall learn in Section 4.13, the degeneracy concept plays an important role in making us understand the periodic table of elements.

Let us now compare the orbit radius r given by the Bohr model with the average value of r obtained using the wave function ψ_{n,l,m_l}. The average value of r is given by

$$\langle r_{n,l,m_l} \rangle = n^2 a_0 [1 + \tfrac{1}{2}\{1 - \frac{l(l+1)}{n^2}\}]. \tag{4.81}$$

The average radius depends only on n and l but not on m_l. The radius obtained from the Bohr model (4.19) is

$$r_n = n^2 a_0. \tag{4.82}$$

The difference then between $\langle r_{n,l,m_l} \rangle$ and r_n is

$$\langle r_{n,l,m_l} \rangle - r_n = \frac{n^2 a_0}{2}[1 - \frac{l(l+1)}{n^2}]. \tag{4.83}$$

This difference is minimum when $l = n - 1$. This means that the picture of the hydrogen atom we obtain from $\psi_{n,n-1,m_l}$ is closest to that we get from Bohr's model. We illustrate this by studying the probability distribution of the electron shown in Fig. 4.10. The probability distribution in the state $n = 1$, $l = 0$ is spherically symmetric. For the electron in the state $n = 3$, $l = 2$, $m_l = \pm 2$, the probability of finding the electron is maximum in the x-y plane, i.e., the electron is most likely to be found in a ring lying in this plane. This picture of the electron

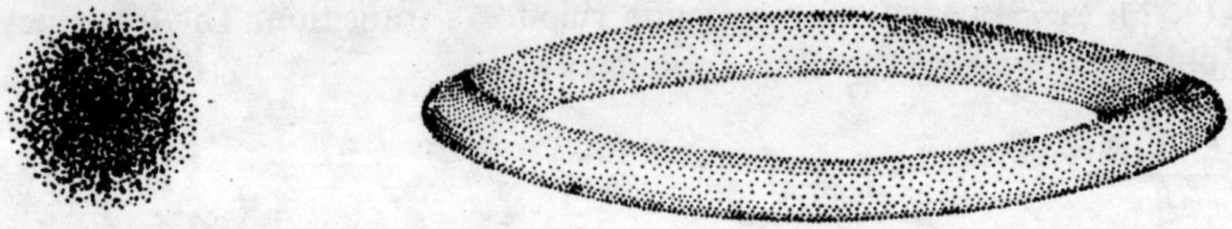

$n=1,\ l=0,\ m_l=0$ $n=3,\ l=2,\ m_l=\pm 2$

Fig. 4.10 Probability distribution of electron in three dimensions for two states.

closely resembles that we have from the Bohr model. For the electron in other states, e.g., $n = 3$, $l = 1$, $m_l = \pm 1$, the probability distribution is not confined in a plane.

QUESTION 4.11 In which quantum mechanical state is the difference $\langle r_{n,l,m_l} \rangle - r_n$ a maximum?

4.11 BOHR'S CORRESPONDENCE PRINCIPLE

The Bohr correspondence principle states that in the limit of large quantum numbers, the results obtained from quantum mechanics tend to become the same as those arrived at from classical physics. As we have already seen, a classical system, e.g., a simple pendulum (see Example 3.2), in fact corresponds to a quantum system with a very large quantum number. This is one example of the correspondence principle. Let us now find out how the correspondence principle is applied to the hydrogen atom. In order to keep the arguments simple, we shall use Bohr's model instead of the quantum theory. When the electron makes a transition, it emits a photon having a frequency given by Eq. (4.15). The frequency of the photon is distinct from the frequency of the electron in its orbit. But classically, the two frequencies are expected to be equal. In Bohr's model, the number of revolutions per second the electron makes (see Example 4.6) is

$$\nu_n = \frac{Z^2mc^2\alpha^2}{2\pi\hbar}\frac{1}{n^3}. \tag{4.84}$$

Let us consider the frequency of the emitted radiation when the transition takes place from a state of quantum number n_1 to a state of quantum number n_2. Then, from Eq. (4.23),

$$\nu_{\text{transition}} = \frac{Z^2mc^2\alpha^2}{4\pi\hbar}\left(\frac{1}{n_1^2} - \frac{1}{n_2^2}\right) \tag{4.85}$$

$$= \frac{Z^2mc^2\alpha^2}{2\pi\hbar}\,\frac{(n_2 - n_1)(n_2 + n_1)}{2n_1^2n_2^2}. \tag{4.86}$$

As $n_2 > n_1$,

$$\tfrac{1}{2}\left(\frac{n_1 + n_2}{n_1^2n_2^2}\right) = \tfrac{1}{2}\left(\frac{1}{n_1n_2^2} + \frac{1}{n_2n_1^2}\right) > \tfrac{1}{2}\left(\frac{1}{n_1^3} + \frac{1}{n_2^3}\right) > \frac{1}{n_2^3}. \tag{4.87}$$

Similarly, we have

$$\tfrac{1}{2}\left(\frac{n_1 + n_2}{n_1^2n_2^2}\right) < \frac{1}{n_1^3}. \tag{4.88}$$

Equations (4.85) and (4.88), for a transition $n_1 - n_2 = 1$, lead to

$$\frac{Z^2mc^2\alpha^2}{2\pi\hbar}\left(\frac{1}{n_2^3}\right) < \frac{Z^2mc^2\alpha^2}{2\pi\hbar}\frac{(n_2+n_1)}{2n_1^2n_2^2} < \frac{Z^2mc^2\alpha^2}{2\pi\hbar}\left(\frac{1}{n_1^3}\right). \tag{4.89}$$

From Eqs. (4.84) and (4.86), we get

$$\nu_{n_2} < \nu_{\text{transition}} < \nu_{n_1 = n_2 - 1}. \tag{4.90}$$

In other words, the frequency of the emitted photon is between the frequency of the electron in the initial and final orbits. Let us assume that the initial orbit has $n_2 \gg 1$. Then,

$$\nu_{n_2} \sim \nu_{n_1} \sim \nu_{\text{transition}}. \tag{4.91}$$

According to Eq. (4.91), for a large quantum number, the frequency of the emitted photon is equal to the frequency of the electron in its orbit. Thus, as we see, in the limit of large quantum numbers, the classical and quantum physics give the same result.

4.12 SPIN OF ELECTRON

We shall now consider the experimental evidence for a very important physical property of the electron. An electron possesses a certain intrinsic angular momentum, the value of which is $\hbar/2$. This angular momentum is as much a property of an electron as are its mass and charge. The intrinsic angular momentum of an electron is referred to as its spin. The evidence for the spin of the electron was first found while studying the spectra of many-electron atoms. We shall not discuss the historic development of this concept. Instead, we shall present the salient features of the spin quantum number and describe an experiment to demonstrate that an electron has an intrinsic angular momentum (spin) given by $\hbar/2$.

The electron spin, like the orbital angular momentum, is a vector quantity. Just as L_x, L_y, and L_z are the three components of angular momentum, so too we have three components S_x, S_y, and S_z of spin angular momentum $\boldsymbol{S}$. Thus, we can write

$$\boldsymbol{S} = S_x\hat{i} + S_y\hat{j} + S_z\hat{k}, \tag{4.92}$$

where $\hat{i}$, $\hat{j}$, and $\hat{k}$ are the unit vectors along the x-, y-, and z-axis.

The orbital angular momentum of a particle is due to its motion and is given by $\boldsymbol{r} \times \boldsymbol{p}$. In quantum mechanics, the operators L_x, L_y, L_z, and L^2, given by Eqs. (3.42) and (3.43), are associated with it. On the other hand, the spin angular momentum cannot be written as $\boldsymbol{r} \times \boldsymbol{p}$. Instead, it can be expressed only in terms of relativistic quantum mechanics. In what follows, we shall state only the salient points necessary to understand the basic differences between the orbital angular and spin angular momenta. The spin angular momentum does not have any differential operators, equivalent to those given by Eqs. (3.42) and (3.43), associated with it. However, it has three abstract operators S_x, S_y, and S_z associated with it. These operators satisfy the following so-called commutation relations:

$$S_xS_y - S_yS_x = i\hbar S_z, \quad S_yS_z - S_zS_y = i\hbar S_x, \quad S_zS_x - S_xS_z = i\hbar S_y. \tag{4.93}$$

The orbital angular momentum operators also satisfy relations (4.93)—see Problem 3.11. Using Eqs. (4.93), we can show that the operators S^2 and S_z satisfy

$$S^2\chi_s^{m_s} = \hbar^2 s(s+1)\chi_s^{m_s}, \quad S_z\chi_s^{m_s} = \hbar m_s\chi_s^{m_s}, \tag{4.94}$$

where $s = 0, \frac{1}{2}, 1, \ldots$, and $m_s = -s, -s+1, \ldots, +s$. Also, $\chi_s^{m_s}$ are the spin eigenfunctions. We shall not go into the details of their explicit forms.

For an electron, the value of $s = 1/2$ which means that

$$S^2\chi_{1/2}^{m_s} = \tfrac{3}{4}\hbar^2\chi_{1/2}^{m_s}, \tag{4.95}$$

where $m_s = 1/2$ or $-1/2$. Further,

$$S_z\chi_{1/2}^{1/2} = \frac{\hbar}{2}\chi_{1/2}^{1/2}, \quad S_z\chi_{1/2}^{-1/2} = -\frac{\hbar}{2}\chi_{1/2}^{-1/2}. \tag{4.96}$$

The electron spin has two orientations, of which one has the z-component $\hbar/2$ and the other, $-\hbar/2$ (see Fig. 4.11).

The other important difference is that the orbital quantum number l takes only integral values, whereas the spin quantum number can also take half-integral values. The reason for this difference is that the boundary condition $\psi(r, \theta, \phi) = \psi(r, \theta, \phi + 2\pi)$, as already noted, is applicable to the orbital motion. There is no such boundary condition on the wave function associated with the spin angular momentum.

In a hydrogen atom, when the electron is in the ground state, the energy of the electron is almost independent of the direction of the z-component of spin. This means that the ground state really consists of two states, one corresponding to $m_s = \frac{1}{2}$ and the other to $m_s = -\frac{1}{2}$.

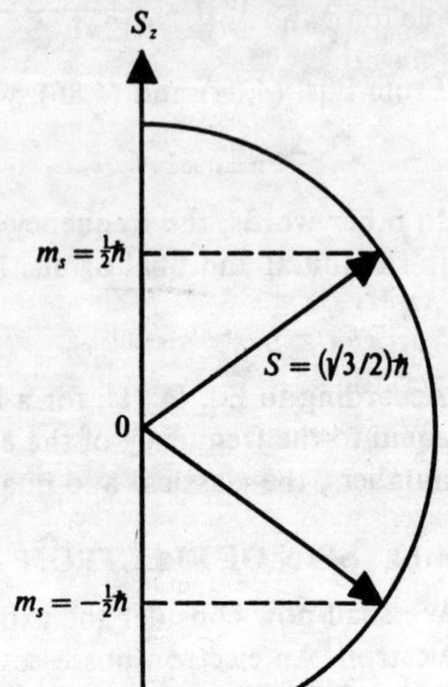

Fig. 4.11 Two states of angular momentum vector ($m_s = \hbar/2$, $m_s = -\hbar/2$).

SPIN AND MAGNETIC MOMENT

If an electron has a spin angular momentum, it is expected to also have an intrinsic magnetic moment. This can be understood as follows. Assume that the intrinsic angular momentum is due to the rotation of the electron about an axis similar to the rotation of the earth about its axis. Since the electron is charged, this rotation gives rise to currents which produce a magnetic field. From the electromagnetic theory, we know that the magnetic field due to a current loop at a large distance from the loop is equivalent to the magnetic field produced by a small magnet kept at the centre of the loop. The magnetic field produced by the current loop can therefore be considered to arise from

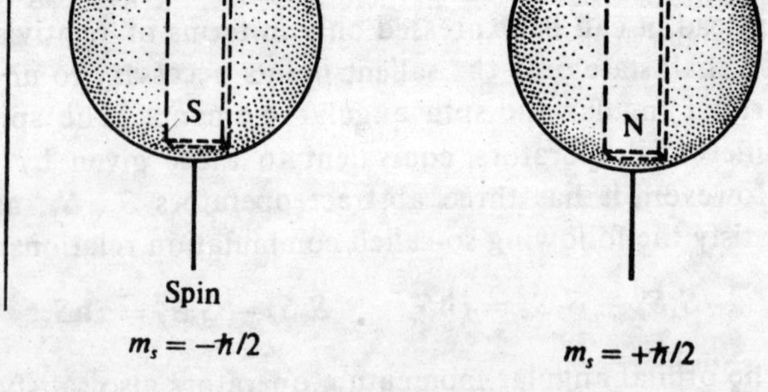

Fig. 4.12 Spin and magnetic moment of electron.

a small magnetic dipole. Thus, an electron has an intrinsic magnetic moment associated with it (see Fig. 4.12). The direction of the magnetic moment vector, that is, the direction from the south pole to the north pole, is opposite to that of the intrinsic angular momentum because the electron is negatively charged.

QUESTION 4.12 If the spin quantum number of an electron is $m_s = -\frac{1}{2}$, in which direction will the magnetic moment point, towards the positive or the negative *z*-axis?

EXAMPLE 4.9 Calculate the orbital magnetic moment of an electron in the first orbit of a hydrogen atom using Bohr's model. Compare it with the average value of the magnetic moment operator in the ground state of the hydrogen atom.

When a charged particle moves, a magnetic field is generated around it. Let us consider a charge e moving in a circle of radius r with a constant speed v. The magnetic moment μ associated with such a motion is $\mu = neA$, where n is the number of revolutions per second and A is the area of the circle. The direction of the magnetic moment is given by the right-hand rule and is perpendicular to the plane of the circle. Thus,

$$\mu = \frac{v}{2\pi r} e\pi r^2 = \frac{evr}{2} = \frac{e}{2m} mvr = \frac{e}{2m} L,$$

where $L = mvr$ is the magnitude of the angular momentum of the particle whose mass is m. Since the direction of angular momentum is the same as that of magnetic moment, we can write the vector equation

$$\boldsymbol{\mu}_L = \frac{e}{2m} \boldsymbol{L}.$$

This relation is true, in general, although we have derived it for a simple case.

In Bohr's model, we have, for the hydrogen atom in the ground state,

$$\mu = \mu_B = \frac{e}{2m} mvr = \frac{e\hbar}{2m} = 9.27 \times 10^{-24} \text{ J/weber-m}^{-2}.$$

In atomic physics, μ_B is used as the unit of magnetic moment. It is referred to as one *Bohr magneton.*

In order to calculate the average value of the magnetic moment in the ground state of the hydrogen atom, we need the operator associated with the magnetic moment. This operator is

$$\boldsymbol{\mu}_L = \frac{e}{2m} \boldsymbol{L},$$

where L is also to be treated as an operator. The average value of μ_L is

$$\langle \boldsymbol{\mu}_L \rangle = \frac{e}{2m} \int \psi^*_{1,0,0}(r, \theta, \phi) \boldsymbol{L} \psi_{1,0,0}(r, \theta, \phi)\, dv.$$

It should be noted that the ground state wave function $\psi_{1,0,0}$ does not depend on θ and ϕ. On the other hand, L_x, L_y, L_z are the differential operators and depend only on θ, ϕ [see Eq. (3.44)]. Thus, $\boldsymbol{L}\psi_{1,0,0} = 0$ which leads to $\langle \mu_L \rangle = 0$. This quantum mechanical result has also been experimentally confirmed.

The foregoing example illustrates that a charged particle moving in a circular orbit possesses a magnetic moment. This is one instance of a general result from electrodynamics, namely, if a charged body has an angular momentum $\boldsymbol{L}$, then it is associated with a magnetic moment $\boldsymbol{\mu}$ given by

$$\boldsymbol{\mu}_L = \frac{q}{2m}\boldsymbol{L}, \tag{4.97}$$

where q and m are the charge and the mass of the body, respectively. In Example 4.9, the angular momentum of the electron in the first orbit, according to Bohr's model, is $\hbar$, and thus the magnetic moment is $e\hbar/(2m)$. The direction of $\boldsymbol{\mu}_L$ is opposite to that of $\boldsymbol{L}$ because the charge of the electron is negative.

In quantum mechanics, the magnetic moment, since it is an observable, has to be associated with an operator. This operator is given by Eq. (4.97). As we know, only one component of L can be fixed, and usually it is the z-component that is chosen. Thus, only the z-component of magnetic moment can be fixed, and so $(\mu_L)_z$ can have the values

$$\frac{q}{2m}m_l\hbar \qquad (m_l = 0, \pm 1, \pm 2, \ldots, \pm l). \tag{4.98}$$

In order to calculate the intrinsic magnetic moment of an electron due to the current loop caused by the rotation of the electron, we need relativistic quantum mechanics which, as already stated, is beyond the scope of this text. The magnetic moment μ_S of an electron is given by

$$\boldsymbol{\mu}_S = -g_s\left(\frac{e}{2m}\right)\boldsymbol{S}, \tag{4.99}$$

where e, as already stated, is the magnitude of the electric charge. It should be noted that this expression is similar to Eq. (4.97), except for the factor g_s. (g_s is called the gyromagnetic ratio for spin or simply the spin factor g. Experimentally, its value has been found to be very close to 2.) Since, again, we can fix only one component of the magnetic moment vector, we choose its z-component. Using Eqs. (4.94) and (4.99), we find that the z-component of $\boldsymbol{\mu}_S$ can take only the values

$$g_s\frac{e}{2m}\hbar m_s \qquad (m_s = +\tfrac{1}{2} \text{ or } -\tfrac{1}{2}). \tag{4.100}$$

Thus, the z-component of the spin magnetic moment of the electron can take only two values.

THE STERN-GERLACH EXPERIMENT

As we have already seen, a magnetic moment is associated with the spin of an electron. This can also be demonstrated by the Stern-Gerlach experiment. The presence of the magnetic moment can be detected by placing an electron in a magnetic field. If the field B is uniform, then, as we know, on the magnetic moment there is no net force but only a torque $\boldsymbol{\tau} = \boldsymbol{\mu} \times \boldsymbol{B}$. However, if the magnetic field is nonuniform, then there is a force acting on the magnetic moment. Thus, if the electron placed in a nonuniform magnetic field experiences a force, it means the electron has a magnetic moment.

The force acting on the magnetic moment can be calculated as follows. Assume that the electron is a small magnet, its north and south poles being separated by a distance d (see Fig. 4.13).

The magnetic field vector, being nonuniform, changes with position, and so we write it as

$$B(x, y, z) = B_x(x, y, z)\hat{i} + B_y(x, y, z)\hat{j} + B_z(x, y, z)\hat{k}.$$

Let us consider the situation where the magnet is lying along the z-axis, with its south pole at the origin. The magnetic moment is md, where m is the magnetic charge, and it points from S to N. In Fig. 4.13, we have indicated the z-component of the forces acting on the magnet. The force on the south pole is along the negative z-axis and its magnitude is $mB_z(0, 0, 0)$. The force on the north pole is along the positive z-axis and has a magnitude $mB_z(0, 0, d)$. Since d is small, we can expand $B_z(0, 0, d)$, using the Taylor series, as

$$B_z(0, 0, d) = B_z(0, 0, 0) + d\frac{\partial B_z}{\partial z}\bigg|_{x=y=z=0} + \ldots \quad (4.101)$$

The net force on the magnet along the z-axis is

$$F = md\frac{\partial B_z}{\partial z}\bigg|_{x=y=z=0} = \mu\frac{\partial B_z}{\partial z}\bigg|_{x=y=z=0}, \quad (4.102)$$

where μ is the z-component of the magnetic moment. Since the z-component of electron spin

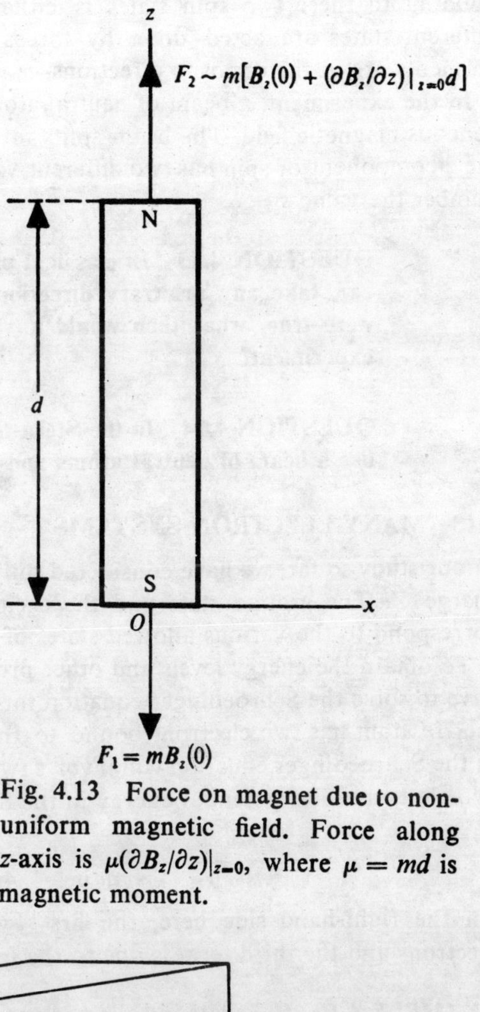

Fig. 4.13 Force on magnet due to nonuniform magnetic field. Force along z-axis is $\mu(\partial B_z/\partial z)|_{z=0}$, where $\mu = md$ is magnetic moment.

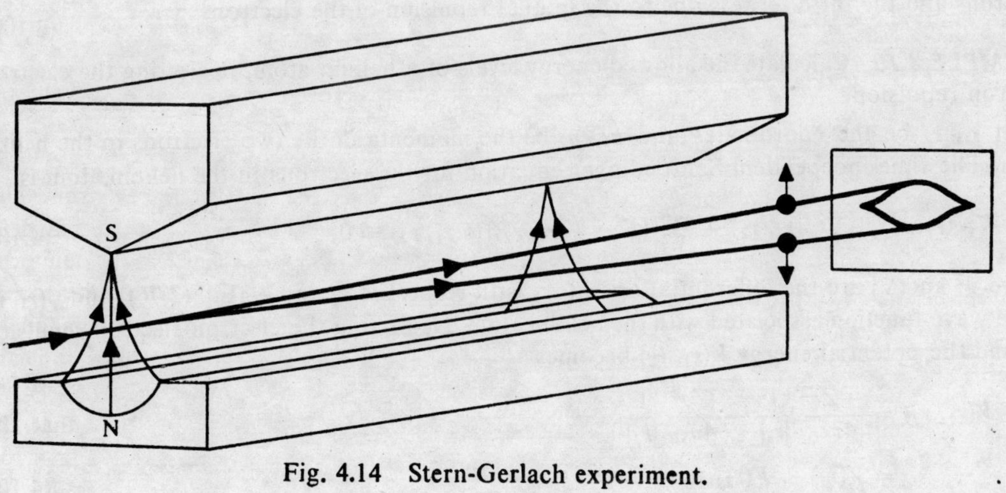

Fig. 4.14 Stern-Gerlach experiment.

can take only one of two values $+\hbar/2$ and $-\hbar/2$, the magnetic moment of the electron corresponding to these two spin states is either towards the positive or negative z-axis. The two different states are acted upon by forces in the opposite directions because the magnetic moments are in the opposite directions.

In the experiment, a beam of neutral atoms, say, hydrogen, is sent through a highly inhomogeneous magnetic field. The beam splits into two parts (see Fig. 4.14), thus demonstrating that the z-component of spin has two different values. It is this that makes us assign the spin quantum number the value $s = \frac{1}{2}$.

QUESTION 4.13 In classical physics, the spin and the magnetic moment can take any arbitrary direction with respect to the magnetic field. If this were true, what then would have been the outcome of the Stern-Gerlach experiment?

QUESTION 4.14 In the Stern-Gerlach experiment, why is it necessary to use a beam of neutral atoms and not of ions?

4.13 MANY-ELECTRON SYSTEMS

In our study so far, we have considered only the system of one electron bound to a nucleus of charge Z. The ground state and the excited states of such an atom (single-electron system) correspond to the various allowed states of the electron.

To obtain the energy levels and other properties of an atom containing many electrons, we have to solve the Schroedinger equation for many electrons bound to the nucleus. For example, the He atom has two electrons bound to the nucleus. Therefore, the wave function occurring in the Schroedinger equation will involve two coordinates $\mathbf{r}_1$, $\mathbf{r}_2$ representing the positions of the two electrons. The potential energy in the Schroedinger equation is the sum of three terms, i.e.,

$$V(\mathbf{r}_1, \mathbf{r}_2) = -\frac{Ze^2}{4\pi\epsilon_0}\frac{1}{|\mathbf{r}_1|} - \frac{Ze^2}{4\pi\epsilon_0}\frac{1}{|\mathbf{r}_2|} + \frac{e^2}{4\pi\epsilon_0}\frac{1}{|\mathbf{r}_1 - \mathbf{r}_2|}. \tag{4.103}$$

On the right-hand side here, the first two terms correspond to the attraction between the electrons and the third term is due to the mutual repulsion of the electrons.

EXAMPLE 4.10 Calculate the allowed energy levels of a helium atom, neglecting the electron-electron repulsion.

Let $\mathbf{r}_1$, $\mathbf{r}_2$ be the coordinates and $\mathbf{p}_1$, $\mathbf{p}_2$ be the momenta of the two electrons in the helium atom. The time-independent Schroedinger equation for the electrons in the helium atom is

$$\nabla_1^2\psi(\mathbf{r}_1, \mathbf{r}_2) + \nabla_2^2\psi(\mathbf{r}_1, \mathbf{r}_2) + \frac{2m}{\hbar^2}[\{E - V(\mathbf{r}_1, \mathbf{r}_2)\}]\psi(\mathbf{r}_1, \mathbf{r}_2) = 0, \tag{4.104}$$

where ∇_1^2 and ∇_2^2 are the differential operators with respect to $\mathbf{r}_1$ and $\mathbf{r}_2$ [Eq. (2.76)] and $\psi(\mathbf{r}_1, \mathbf{r}_2)$ is the wave function associated with the helium atom. Neglecting the electron-electron repulsion, we find the potential energy $V(\mathbf{r}_1, \mathbf{r}_2)$ becomes

$$V(\mathbf{r}_1, \mathbf{r}_2) = \frac{-Ze^2}{4\pi\epsilon_0}\frac{1}{|\mathbf{r}_1|} - \frac{Ze^2}{4\pi\epsilon_0}\frac{1}{|\mathbf{r}_2|} \tag{4.105}$$

$$= V(\mathbf{r}_1) + V(\mathbf{r}_2). \tag{4.106}$$

Substituting this in Eq. (4.104), we see that we can use the method of separation of variables to solve the resulting Schroedinger equation because we have separated the potential energy into two terms, one depending only on $\mathbf{r}_1$ and the other only on $\mathbf{r}_2$. At this stage, we should like to point out that if we had not neglected the electron-electron repulsion, we would have not been able to use the method of separation of variables. Writing

$$\psi(\mathbf{r}_1, \mathbf{r}_2) = \phi(\mathbf{r}_1)\chi(\mathbf{r}_2), \tag{4.107}$$

we get two equations, one depending only on $\mathbf{r}_1$ and the other only on $\mathbf{r}_2$. These are

$$\nabla_1^2\phi(\mathbf{r}_1) + \frac{2m}{\hbar^2}[E_1 - V(\mathbf{r}_1)]\phi(\mathbf{r}_1) = 0, \tag{4.108}$$

$$\nabla_2^2\chi(\mathbf{r}_2) + \frac{2m}{\hbar^2}[E_2 - V(\mathbf{r}_2)]\chi(\mathbf{r}_2) = 0. \tag{4.109}$$

The total energy of the He atom is

$$E = E_1 + E_2. \tag{4.110}$$

Equations (4.108) and (4.109) are just the Schroedinger equations for a Coulomb potential, and, as we have already learnt, the energies E_1 and E_2 are given by

$$E_1 = -\frac{Z^2mc^2\alpha^2}{2}\left(\frac{1}{n_1^2}\right), \tag{4.111}$$

$$E_2 = -\frac{Z^2mc^2\alpha^2}{2}\left(\frac{1}{n_2^2}\right), \tag{4.112}$$

where $n_1 = 1, 2, 3, \ldots$ and $n_2 = 1, 2, 3, \ldots$. Thus, the energy of the helium atom is

$$E = -\frac{Z^2mc^2\alpha^2}{2}\left[\frac{1}{n_1^2} + \frac{1}{n_2^2}\right]. \tag{4.113}$$

This result is physically understandable because the motion of each electron in the helium atom is independent of that of the other since we have neglected the electron-electron repulsion. Therefore, the energy of the helium atom is the sum of the energies of the two electrons that move independently under the Coulomb attraction of the nucleus. The quantum numbers n_1 and n_2 are the principal quantum numbers of the two electrons. The ground state of the helium atom is given by $n_1 = 1$, $n_2 = 1$, and its energy, from Eq. (4.113) with $Z = 2$, is

$$E = -4mc^2\alpha^2 \sim -109 \text{ eV}.$$

Similarly, we can obtain the other allowed energy values of the helium atom.

The ground state energy of the helium atom determined experimentally is ~ -80 eV. Comparing this with the value obtained in the foregoing example, we see that if the electron-electron repulsion is neglected, we get a rough estimate of the energy levels of the helium atom.

QUESTION 4.15 What will be the energy required to remove an electron from a helium atom if the electron-electron repulsion is neglected?

We can also solve the Schroedinger equation for a three- or more electron system in the same way as done for the two-electron system if we ignore the interaction between the electrons. The allowed energy of an atom will then be given by the sum of the energies of the individual electrons. The energy of an individual electron is $-Z^2mc^2\alpha^2/(2n^2)$, where n is the principal quantum number of the electron. However, this approximation gives a completely wrong result for a many-electron system. For example, the ground state energy of a lithium atom, assuming that each of the three electrons in it is in the state $n = 1$, is

$$-9\frac{mc^2\alpha^2}{2}\left(\frac{1}{1^2}+\frac{1}{1^2}+\frac{1}{1^2}\right) = -9\times 13.6\times 3 = -367 \text{ eV}.$$

From this, the energy required to remove one electron from the lithium atom works to be about 122 eV (see Problem 4.22). But the experimental value of this ionization energy is about 5 eV. The reason for this discrepancy is that we have in our calculations ignored an important physical effect. This effect is due to the Pauli exclusion principle.

PAULI'S EXCLUSION PRINCIPLE

The Pauli exclusion principle states that *no two electrons can be in the same quantum state.* In other words, if one electron is in a specific quantum state, the other electron must then be in a different quantum state. The Pauli exclusion principle naturally results from the fact that the electrons obey the Fermi-Dirac statistics. This is discussed in Chapter 6.

Let us now consider the quantum states of the electron in a hydrogen atom. A quantum state is specified by the

(1) principal quantum number $n\,[=1, 2, 3, \ldots]$,

(2) azimuthal quantum number $l\,[=0, 1, \ldots, (n-1)]$,

(3) magnetic quantum number $m_l\,[=-l, (-l+1), \ldots, +l]$, and

(4) spin quantum number $m_s\,[=-\frac{1}{2}$ or $\frac{1}{2}]$.

In our solution of the Schroedinger equation for the hydrogen atom, we discussed in detail the quantum numbers n, l, m_l. These three quantum numbers were required because we described the electron using the three coordinates r, θ, ϕ. The fourth quantum number, namely, m_s, represents the intrinsic angular momentum an electron possesses. Therefore, to completely describe the state of an electron, this additional quantum number is also needed. The quantum number m_s helps label the two possible states of the electron spin.

EXAMPLE 4.11 Specify all the quantum numbers of a hydrogen atom in its ground state and first excited state.

The ground state is two-fold degenerate and the quantum numbers of these two states are

$$n = 1,\quad l = 0,\quad m_l = 0,\quad m_s = -\tfrac{1}{2},$$

$$n = 1,\quad l = 0,\quad m_l = 0,\quad m_s = +\tfrac{1}{2}.$$

The first excited state is eight-fold degenerate and the quantum numbers of these eight states are

$$n = 2,\quad l = 0,\quad m_l = 0,\quad m_s = -\tfrac{1}{2},$$

$$n = 2,\quad l = 0,\quad m_l = 0,\quad m_s = +\tfrac{1}{2},$$

$$n = 2,\quad l = 1,\quad m_l = -1,\quad m_s = -\tfrac{1}{2},$$

$$n = 2, \quad l = 1, \quad m_l = -1, \quad m_s = +\tfrac{1}{2},$$

$$n = 2, \quad l = 1, \quad m_l = 0, \quad m_s = -\tfrac{1}{2},$$

$$n = 2, \quad l = 1, \quad m_l = 0, \quad m_s = +\tfrac{1}{2},$$

$$n = 2, \quad l = 1, \quad m_l = +1, \quad m_s = -\tfrac{1}{2},$$

$$n = 2, \quad l = 1, \quad m_l = +1, \quad m_s = +\tfrac{1}{2}.$$

In view of the results we have obtained and the Pauli exclusion principle, we see that only two electrons can occupy the ground state and the first excited state can accommodate a maximum of eight electrons.

The degeneracy of each state of a hydrogen atom increases by 2 when the spin of the electron is also taken into account. In general, the degeneracy associated with a state with the principal quantum number n becomes $2n^2$.

EXAMPLE 4.12 Consider a one-dimensional box of length L. What is the lowest energy of a system consisting of (a) two electrons in the box? (b) three electrons in the box? Ignore the Coulomb interaction between the electrons.

The energy levels for a particle in a box are given by Eq. (2.70), i.e.,

$$E = \frac{\hbar^2 \pi^2}{2mL^2} n^2.$$

(a) When there are two electrons in the box, the energy of the system is the sum of the energies of the two individual electrons. Thus,

$$E = \frac{\hbar^2 \pi^2}{2mL^2}(n_1^2 + n_2^2), \qquad n_1 = 1, 2, 3, \ldots, \; n_2 = 1, 2, 3, \ldots,$$

where n_1, n_2 are the quantum numbers of the two electrons. The minimum energy is obtained by assigning the lowest possible values to n_1 and n_2, that is, $n_1 = n_2 = 1$. Then,

$$E = \frac{\hbar^2 \pi^2}{2mL^2} \times 2 = \frac{\hbar^2 \pi^2}{mL^2}.$$

The quantum numbers of the two electrons have to be different to satisfy the Pauli exclusion principle. If one of the electrons has a spin quantum number $m_s = +\frac{1}{2}$, then the other has to have $m_s = -\frac{1}{2}$.

(b) When there are three electrons in the box, we have to also take into account the energy of the third electron to get the total energy of the system. Since we are interested in obtaining the lowest energy of the system, we would want the third electron to occupy the lowest allowed energy state, i.e., $n = 1$. But this is not possible since the Pauli exclusion principle must also be satisfied. This is because if the third electron also occupies the $(n = 1)$-state, then its quantum numbers would be either $n = 1$, $m_s = -\frac{1}{2}$ or $n = 1$, $m_s = +\frac{1}{2}$. But the states corresponding to these two quantum numbers have already been occupied by the other two electrons. Therefore, the third electron has to occupy the next lowest allowed energy state which corresponds to $n = 2$. Thus, the total energy of the system (when $n_1 = 1$, $n_2 = 1$, $n_3 = 2$) is

$$E = \frac{\hbar^2\pi^2}{2mL^2}(n_1^2 + n_2^2 + n_3^2) = \frac{3\hbar^2\pi^2}{mL^2}.$$

QUESTION 4.16 List the possible quantum numbers for the third electron in Example 4.12.

We can qualitatively understand a many-electron atom if we take into account the exclusion principle and ignore the mutual Coulomb interaction between the electrons. When the Coulomb interaction is neglected, we can treat each electron in the atom as an independent particle moving around the nucleus (see Example 4.10). The energy of each such electron is given by its motion around the nucleus, that is, it is the same as that of the electron in a hydrogen-like atom. The total energy of the many-electron atom in this approximation is then the sum of the energies of the individual electrons.

The general properties of all elements are systematized by the periodic table. We shall now use the quantum mechanical concepts of allowed energy levels and the Pauli exclusion principle to understand the periodic table.

4.14 PERIODIC TABLE

When we arrange the elements in the order of increasing atomic number, we find the elements with similar physical and chemical properties reoccur at some regular intervals. In other words, the properties of elements are periodic functions of the atomic number Z. The complete periodic table is shown in Fig. 4.15. The rows are called *periods* and the eighteen columns are called *groups*. The elements in a group have similar properties. We note that, in each period, the first element is an alkali atom, the second an alkaline earth, and the last a rare gas. A series of transition elements in each period (after the third period) appear between groups IIA and IIIA. The transition elements in a period considerably resemble one another.

We now build up the periodic table by starting with the hydrogen atom and adding electrons to appropriate energy levels keeping in mind the Pauli exclusion principle.

The ground state of the hydrogen atom is the one in which the electron is in the state $n = 1$, $l = 0$. Suppose we wish to add one more electron to the same state (i.e., $n = 1$, $l = 0$). According to Pauli's principle, this can be done only if at least one of the quantum numbers of the electron to be added is different from the quantum numbers of the first electron. This is possible since the spin of an electron has two quantum numbers $m_s = \pm\frac{1}{2}$. Therefore, if the first electron has $m_s = +\frac{1}{2}$, the second one can take $m_s = -\frac{1}{2}$. Thus, we can have two electrons in the state $n = 1$, $l = 0$. To make the system neutral, we increase the charge of the nucleus to 2. This then constitutes the He atom. The electronic structure of the He atom is schematically shown in Fig. 4.16.

Let us now consider the lithium atom which contains three electrons. Since, according to the Pauli exclusion principle, we cannot put more than two electrons in the state $n = 1$, the ground state of this three-electron system will have the following configuration (see Fig. 4.16): two electrons in the state $n = 1$ and one electron in the state $n = 2$.

If we proceed in a similar manner, we can then accommodate at the most 8 electrons in the state $n = 2$ (see Example 4.10). This means that the eleventh electron must go to the state $n = 3$.

1s	1 H																		2 He
2s	3 Li	4 Be											2p	5 B	6 C	7 N	8 O	9 F	10 Ne
3s	11 Na	12 Mg											3p	13 Al	14 Si	15 P	16 S	17 Cl	18 Ar
4s	19 K	20 Ca	21 Sc	22 Ti	23 V	24 Cr	25 Mn	26 Fe	27 Co	28 Ni	29 Cu	30 Zn	4p	31 Ga	32 Ge	33 As	34 Se	35 Br	36 Kr
5s	37 Rb	38 Sr	39 Y	40 Zr	41 Nb	42 Mo	43 Tc	44 Ru	45 Rh	46 Pd	47 Ag	48 Cd	5p	49 In	50 Sn	51 Sb	52 Te	53 I	54 Xe
6s	55 Cs	56 Ba	57 La	72 Hf	73 Ta	74 W	75 Re	76 Os	77 Ir	78 Pt	79 Au	80 Hg	6p	81 Tl	82 Pb	83 Bi	84 Po	85 At	86 Rn
7s	87 Fr	88 Ra	89 Ac										7p						
	s^1	s^2	d^1	d^2	d^3	d^4	d^5	d^6	d^7	d^8	d^9	d^{10}		p^1	p^2	p^3	p^4	p^5	p^6

4f Lanthanides	58 Ce	59 Pr	60 Nd	61 Pm	62 Sm	63 Eu	64 Gd	65 Tb	66 Dy	67 Ho	68 Er	69 Tm	70 Yb	71 Lu
5f Actinides	90 Th	91 Pa	92 U	93 Np	94 Pu	95 Am	96 Cm	97 Bk	98 Cf	99 Es	100 Fm	101 Md	102 No	103 Lw
	f^1	f^2	f^3	f^4	f^5	f^6	f^7	f^8	f^9	f^{10}	f^{11}	f^{12}	f^{13}	f^{14}

Fig. 4.15 Periodic table of elements.

The configuration of the Na atom which has 11 electrons is also shown in Fig. 4.16.

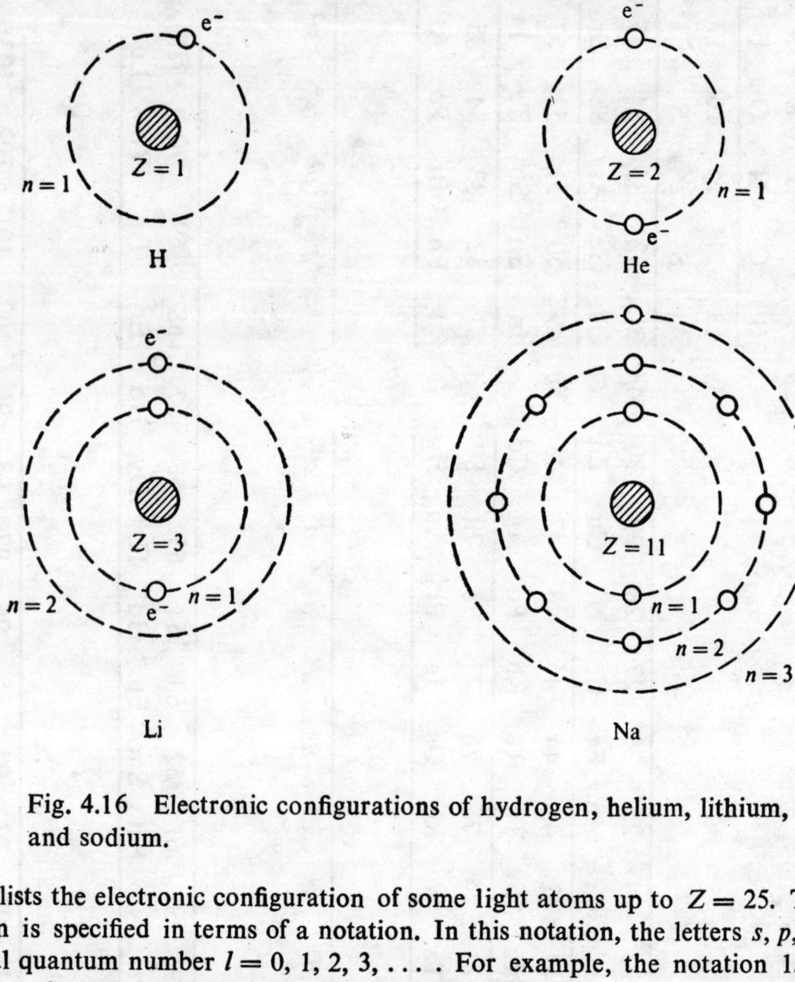

Fig. 4.16 Electronic configurations of hydrogen, helium, lithium, and sodium.

Table 4.1 lists the electronic configuration of some light atoms up to $Z = 25$. The electronic configuration is specified in terms of a notation. In this notation, the letters $s, p, d, f, \ldots$ refer to the orbital quantum number $l = 0, 1, 2, 3, \ldots$. For example, the notation $1s^2$ means that the state $n = 1$, $l = 0$ has two electrons. Further, $1s^2 2s^2 2p^6$ means that the $1s$-state ($n = 1, l = 0$) has 2 electrons, the $2s$-state ($n = 2, l = 0$) has 2 electrons, and the $2p$-state ($n = 2, l = 1$) has 6 electrons.

Thus, we note that the electronic configuration of an atom is obtained by systematically filling its allowed states with electrons, simultaneously taking into account the Pauli principle. The ordering of the allowed levels is

$$1s, 2s, 2p, 3s, 3p, 4s, 3d, 4p, 5s, 4d, 5p, 6s, 5d, 4f, \ldots.$$

In the light of our discussion in Section 4.13, the ordering of the allowed levels should be

$$1s, 2s, 2p, 3s, 3p, 3d, \ldots,$$

that is, the levels should depend only on the principal quantum number n and therefore be arranged in the increasing order of n. This is because, according to the method we have applied in calculating the energy of an atom, the energy depends only on n (see Fig. 4.9). But in reality the energy of a many-electron atom does also depend slightly on the quantum number l, and this is the reason in the ordering of levels $4s$ occurs before $3d$, $5s$ before $4d$, and so on.

TABLE 4.1 Electronic configurations of some elements

Z	Element	Configuration
1	Hydrogen	$1s^1$
2	Helium	$1s^2$
3	Lithium	$1s^2 2s^1$
4	Beryllium	$1s^2 2s^2$
5	Boron	$1s^2 2s^2 2p^1$
6	Carbon	$1s^2 2s^2 2p^2$
10	Neon	$1s^2 2s^2 2p^6$
11	Sodium	$1s^2 2s^2 2p^6 3s^1$
12	Magnesium	$1s^2 2s^2 2p^6 3s^2$
13	Aluminium	$1s^2 2s^2 2p^6 3s^2 3p^1$
17	Chlorine	$1s^2 2s^2 2p^6 3s^2 3p^5$
18	Argon	$1s^2 2s^2 2p^6 3s^2 3p^6$
19	Potassium	$1s^2 2s^2 2p^6 3s^2 3p^6 4s^1$
20	Calcium	$1s^2 2s^2 2p^6 3s^2 3p^6 4s^2$
21	Scandium	$1s^2 2s^2 2p^6 3s^2 3p^6 4s^2 3d^1$
25	Manganese	$1s^2 2s^2 2p^6 3s^2 3p^6 4s^2 3d^5$

The electronic configuration of the elements is also indicated in the periodic table (Fig. 4.15). For example, in period III, ($3s$, $3p$) shows that the configuration of the eight elements, from Na to Ar, is obtained by filling the eight states starting from $3s^1$ to $3s^2 3p^6$ (see Table 4.1). We note that when the ten $3d$-states are filled, we get the transition elements of group IV, and when the ten $4d$-states are filled, we obtain the transition elements of group V. Similarly, by filling the $4f$-states, we obtain the fourteen elements of lanthanide series in period VI, and by filling the $5f$-states, we get the fourteen elements of the actinide series in period VII.

Ionization Energy

A plot of ionization energy as a function of the atomic number Z is shown in Fig. 4.17. As can be noticed, the ionization energy of the alkali atoms (Li, Na, K, . . .) is small, whereas that of the halogens (F, Cl, Br, . . .) and the rare gases (He, Ne, Ar, . . .) is large. In fact, the ionization energy increases as we move from the elements of each period in group I to group VIII. This can be physically explained as follows.

Consider the Na atom which has one valence electron in the $3s$-orbit. This electron is the farthest from the nucleus. Of all the electrons in the atom, it therefore is the least bound. The inner electrons ($1s^2$, $2s^2$, $2p^6$), on the other hand, are very close to the nucleus. Hence, the net positive charge the valence electron 'sees' is considerably less than the charge of the nucleus. If the shielding of the nucleus were complete, the valence electron would see only a positive charge of one unit. However, it is not so because, as we know, the electron orbits (i.e., probability

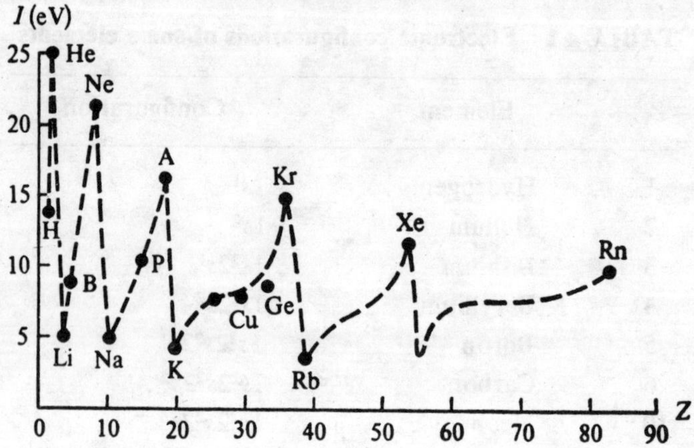

Fig. 4.17 Ionization energy I as function of atomic number Z. Dashed line indicates periodicity.

density distribution of electrons in various allowed states) are complicated. The detailed calculation of the shielding effect shows that the electron in the ($n = 3$)-state of the Na atom sees an effective positive charge (Z_{eff}) $\sim$ 1.9. Therefore, the ionization energy of the valence electron, using Eq. (4.26), is

$$Z_{eff}^2 \frac{mc^2\alpha^2}{2n^2} \sim Z_{eff}^2 \frac{13.6 \text{ eV}}{n^2} \sim 5.4 \text{ eV}.$$

In the elements of a higher group, e.g., Cl and Ar, the number of valence electrons in the $3s$- and $3p$-orbit is more than that in Na. In Cl, the number of valence electrons is seven. For each such electron, $Z_{eff} \sim 3$. Therefore, the ionization energy, since it is proportional to Z_{eff}^2, is large for the Cl atom. A similar argument holds for the argon atom.

Stability of Rare Gases

As we know, the rare gases He, Ne, Ar, Kr, Xe, Rn are chemically inert. We shall explain this by means of an example. Consider the three consecutive elements in the periodic table, namely, Cl, Ar, and K. Their electronic configurations are given in Table 4.1. The ionization energies of these elements are 13 eV, 15.7 eV, and 4.3 eV, respectively. Therefore, it is more difficult to remove a valence electron from the Cl or Ar atom than from the K atom. A chemical reaction involves the transfer of an electron from one atom to another. Thus, an atom is chemically active if it can easily donate or accept an electron. The K atom is chemically active because it releases the valence electron and requires a little energy for doing so. On the other hand, the Cl and Ar atoms cannot donate an electron since their valence electrons are tightly bound. However, the Cl atom can easily accept an electron. This is because its outer shell ($3p$) can accommodate one additional electron. Thus, chlorine is chemically active because it can participate in a chemical reaction as an electron acceptor. Argon, on the other hand, can neither accept nor donate an electron easily. It is thus chemically inert.

4.15 X-RAYS

An inner electron of a many-electron system is rather well bound to the nucleus. To remove an electron from the ($n = 1$)-state of an atom, an energy of the order of a few thousand electron volts (keV) is needed.

EXAMPLE 4.13 Calculate approximately the energy needed to remove an electron from the ($n = 1$)-state of a sodium atom.

The electron in the ($n = 1$)-state of a sodium atom, being the innermost electron, sees the entire nuclear charge of 11 units. From Eq. (4.49), we have

$$E = -Z^2\frac{mc^2\alpha^2}{2} = -121 \times 13.6 \text{ eV} \sim -1.6 \text{ keV}.$$

This is the energy required to remove an electron from the ($n = 1$)-state of the sodium atom.

The removal of an electron from the state $n = 1$ leaves a hole in the state. (The orbit $n = 1$ is also referred to as the K-shell.) Therefore, an electron from the orbit $n = 2$ can now undergo a transition to the K-shell. During such a transition in sodium, the photon emitted has an energy 1.2 keV and a wavelength ~10Å. This photon is in the X-ray region, indicating that the atom emits X-rays. An electron from states higher than the ($n = 2$)-state can also undergo a transition to the K-shell, resulting in X-rays of varying energy. These discrete X-ray energies are called the K-lines as the transition is to the K-shell. If the transition is to the ($n = 2$)-shell (after an electron has been removed from this shell), we get the so-called L-lines. In X-rays, the orbits $n = 1, 2, 3, \ldots$ are referred to as the shells $K, L, M, \ldots$. The K-lines consist of the lines K_α, K_β, ... The K_α-line corresponds to the transition from the state $n = 2$ to the state $n = 1$, the K_β-line to the transition from the state $n = 3$ to the state $n = 1$, and so on. These transitions are shown in Fig. 4.18.

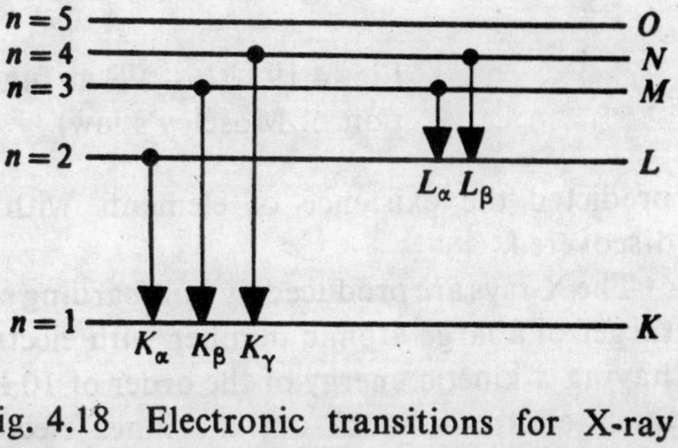

Fig. 4.18 Electronic transitions for X-ray lines.

We can calculate the frequency of the K_α-line if we assume that an electron in the L-shell sees a charge $(Z - 1)e$ on the nucleus. This assumption is valid because the charge Ze of the nucleus is screened by the single electron left in the K-shell (i.e., state $n = 1$). The frequency of the K_α-line, ν_{K_α}, from Eq. (4.50), then is

$$\nu_{K_\alpha} = \frac{c}{\lambda} = cR(Z-1)^2\left(\frac{1}{1^2} - \frac{1}{2^2}\right) = \frac{3}{4}cR(Z-1)^2. \tag{4.114}$$

Therefore, the plot between $\nu_{K_\alpha}^{1/2}$ and Z should be a straight line, as shown in Fig. 4.19. The verification that the frequency of the K_α-line depends on the atomic number of an element was furnished by H. G. J. Moseley and is known as Moseley's law. His investigations in 1913 successfully explained the X-ray line spectra. By using his law, he established a relation between the nuclear charge of an atom and the place of the atom in the periodic table. In particular, he showed that it is the nuclear charge (atomic number) and not the atomic weight of an element that determines its position in the periodic table. He also found gaps in the periodic table and

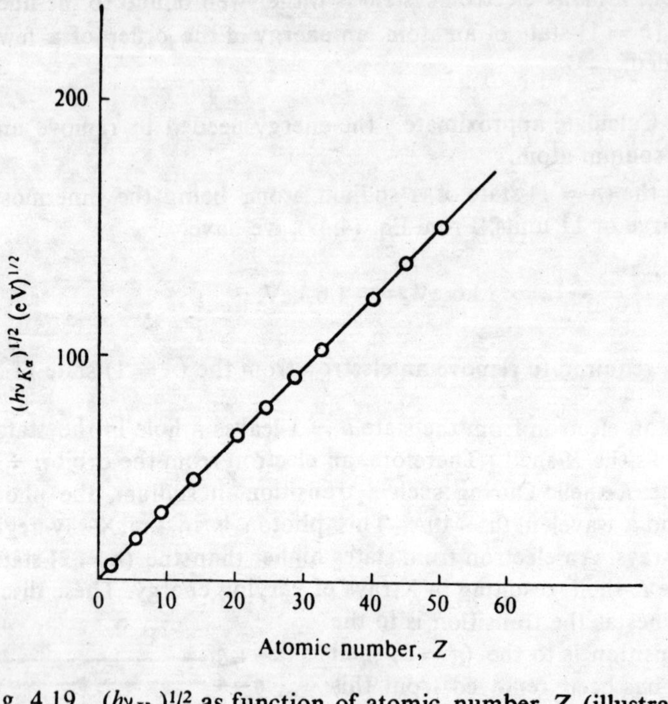

Fig. 4.19 $(h\nu_{K_\alpha})^{1/2}$ as function of atomic number Z (illustration of Moseley's law).

predicted the existence of elements with $Z = 43$, 61, 72, and 75, which were subsequently discovered.

The X-rays are produced by bombarding some target of a large atomic number with electrons having a kinetic energy of the order of 10 keV. Such electrons knock out the inner electrons of the target atom, giving rise to photons, as already explained. These high energy electrons, after they hit the target, get decelerated. A decelerated charge, as we know from classical electrodynamics, emits radiation having a continuous spectrum. A typical X-ray spectrum from a tungsten target is shown in Fig. 4.20. As can be seen, the line spectra here are superposed on a continuous spectrum.

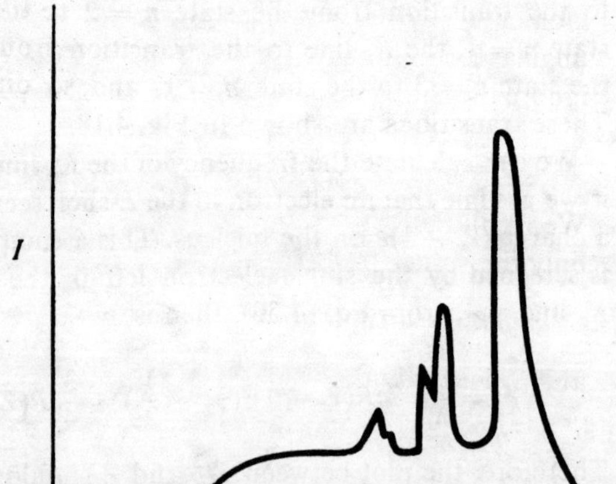

Fig. 4.20 Intensity I of X-ray spectrum from tungsten target as function of wavelength.

EXAMPLE 4.14 Calculate the energies required to create a K-vacancy and an L-vacancy in a copper atom and obtain the wavelength of the K_α-line.

The energy needed to create a vacancy in the K-shell [i.e., ($n = 1$)-orbit] is

$$Z^2 \times 13.6 \text{ eV} = 29^2 \times 13.6 \text{ eV} \sim 11.6 \text{ keV}.$$

The energy required to create a vacancy in the L-shell [i.e., ($n = 2$)-orbit], assuming that $Z_{\text{eff}} = 27$, is

$$(13.6/4) \times 27^2 \sim 2.5 \text{ keV}.$$

We might have in the normal course expected the energy of K_α-radiation to be the difference of the two energies we have calculated. But this is not true because when an electron is making a transition from the L- to the K-shell, there is already a vacancy in the K-shell. The correct method of calculating the wavelength of K_α-line is to use Eq. (4.114). In this way,

$$\lambda = \frac{4}{3} \frac{1}{R(Z-1)^2} \sim 1.5 \text{ Å}.$$

Applications of X-rays

The wavelength of X-ray photons is of the same order as the spacing of atoms in a crystalline solid. Thus, the diffraction of X-rays from a crystal is used for studying the structure of the crystal. In Chapter 2, we discussed Bragg's law which describes the diffraction of X-rays by a crystal. The structure of a complex organic molecule can also be analyzed using X-rays because such a molecule has a periodic structure like a crystal. X-ray analysis plays an important role in the study of the structure of proteins, DNA, RNA, and such other macromolecules. Also, as we know, X-rays are extremely useful in medical diagnosis and treatment.

4.16 TOTAL ANGULAR MOMENTUM OF HYDROGEN ATOM

We know that, for an isolated system, the angular momentum is conserved. Consider, for example, a hydrogen atom in which the electron is in an allowed energy state. The angular momentum of the hydrogen atom is the sum of the orbital angular momentum ($\boldsymbol{L}$) and the spin angular momentum ($\boldsymbol{S}$) associated with the electron. To keep our discussion simple, we shall ignore the spin associated with the nucleus. We denote the total angular momentum by $\boldsymbol{J}$. Then,

$$\boldsymbol{J} = \boldsymbol{L} + \boldsymbol{S}. \tag{4.115}$$

We recall that the vectors $\boldsymbol{L}$ and $\boldsymbol{S}$ are space-quantized (see Sections 3.3 and 4.1), L_z can have only the values [see Eq. (3.66)]

$$m_l\hbar = -l\hbar, -(l-1)\hbar, \ldots, (l-1)\hbar, l\hbar,$$

and S_z has only the values [see Eqs. (4.94)]

$$m_s\hbar = -s\hbar, -(s-1)\hbar, \ldots, (s-1)\hbar, s\hbar.$$

The lengths of $\boldsymbol{L}$ and $\boldsymbol{S}$ are $[l(l+1)]^{1/2}\hbar$ and $[s(s+1)]^{1/2}\hbar$, respectively.

It can be shown that the total angular momentum $\boldsymbol{J}$ has the following properties:

(1) J_z can have only the values

$$m_j\hbar = -j\hbar, -(j-1)\hbar, \ldots, (j-1)\hbar, j\hbar. \tag{4.116}$$

(2) The magnitude of the vector $\boldsymbol{J}$ is given by

$$|\boldsymbol{J}| = [j(j+1)]^{1/2}\hbar, \tag{4.117}$$

where j is the quantum number which determines the allowed values of $|\boldsymbol{J}|$ and J_z.

The quantum number j is related to the quantum numbers l and s. The possible values of j, for given values of l and s, are

$$l+s,\ l+s-1, \ldots, |l-s+1|,\ |l-s|. \tag{4.118}$$

EXAMPLE 4.15 What are the possible values of the quantum numbers j and m_j for a hydrogen atom in the ground state (i.e., $n = 1$, $l = 0$) and the $2p$-state (i.e., $n = 2$, $l = 1$)?

We know that the spin quantum number s for an electron is $\frac{1}{2}$. For the ground state, the allowed values of j, from Eq. (4.118), are $j = |0 + \frac{1}{2}|$ to $j = |0 - \frac{1}{2}|$ or $j = \frac{1}{2}$. Therefore, there is only one allowed value of j for the hydrogen atom in the ground state. From Eq. (4.116), the allowed values of m_j are $m_j = -\frac{1}{2}$, $m_j = +\frac{1}{2}$. Thus, the total angular momentum of the hydrogen atom in the ground state is given by

$$j = \tfrac{1}{2},\ m_j = -\tfrac{1}{2} \quad \text{or} \quad j = \tfrac{1}{2},\ m_j = +\tfrac{1}{2}.$$

In other words, the magnitude of the angular momentum vector $\boldsymbol{J}$ is $(3/2)^{1/2}\hbar$ and its z-component can be either $-\hbar/2$ or $+\hbar/2$.

When the hydrogen atom is in the $2p$-state, the value of $l = 1$ and $s = \frac{1}{2}$. Thus, from Eq. (4.117), we have

$$j = |1 + \tfrac{1}{2}| \text{ to } j = |1 - \tfrac{1}{2}|$$

or

$$j = \tfrac{3}{2} \quad \text{or} \quad j = \tfrac{1}{2}.$$

Using Eq. (4.116), we find that, for $j = 3/2$,

$$m_j = -\tfrac{3}{2}, -\tfrac{1}{2}, +\tfrac{1}{2}, +\tfrac{3}{2}$$

and, for $j = 1/2$,

$$m_j = -\tfrac{1}{2}, \tfrac{1}{2}.$$

Thus, the various allowed quantum numbers (j, m_j) are

$$(\tfrac{3}{2}, -\tfrac{3}{2}), (\tfrac{3}{2}, -\tfrac{1}{2}), (\tfrac{3}{2}, \tfrac{1}{2}), (\tfrac{3}{2}, \tfrac{3}{2}), (\tfrac{1}{2}, -\tfrac{1}{2}), (\tfrac{1}{2}, +\tfrac{1}{2}).$$

The total angular momentum of an atom consisting of several electrons can be obtained by generalizing the foregoing ideas.

SPECTROSCOPIC NOTATION

We now describe the notation used in atomic spectroscopy. In this notation, the energy levels are labelled as

$$n^{(2s+1)}l_j, \tag{4.119}$$

where n is the principal quantum number, s is the total spin angular momentum quantum number, j is the total angular momentum quantum number, and l is the orbital angular momentum quantum number written as $s, p, d, f, \ldots$ to denote $0, 1, 2, 3, \ldots$. Thus, $1^2s_{1/2}$ is the ground state and $2^2s_{1/2}$, $2^2p_{1/2}$, $2^2p_{3/2}$ are the first three excited states of the hydrogen atom. Further, $(2s + 1)$ is known as the multiplicity factor and, for a one-electron system, it is 2. The degeneracy of the states is given by $(2j + 1)$. Thus, the ground state is two-fold degenerate (i.e., $j = \frac{1}{2}$). The first excited state has the principal quantum number $n = 2$. The states $2^2s_{1/2}$ and $2^2p_{1/2}$ are two-fold degenerate, whereas the state $2^2p_{3/2}$ is four-fold degenerate. In the simplest description of the hydrogen atom, the energy depends only on the principal quantum number and not on l or j. Thus, each of the eight states $2^2s_{1/2}$, $2^2p_{1/2}$, and $2^2p_{3/2}$ corresponds to the same energy.

The addition of spin and orbital angular momentum to get the total angular momentum $\boldsymbol{J}$ can be visualized in terms of a physical picture (see Fig. 4.21). The total angular momentum $\boldsymbol{J}$ is conserved for an isolated system. In quantum mechanics, this implies that J^2 and one of its components, say, J_z, each has a definite value which remains constant with time as long as no torque acts on the system. Once the system is in a definite state having fixed J^2 and J_z, the average value of each of J_x and J_y is zero. We can observe this in a manner similar to that discussed (see Section 3.3) for orbital angular momentum (see Fig. 3.2). Here, $\boldsymbol{J}$, instead of $\boldsymbol{L}$, traces a cone about the z-axis. At a given time, $\boldsymbol{L}$ and $\boldsymbol{S}$ add up to give $\boldsymbol{J}$. The component of each of $\boldsymbol{L}$ and $\boldsymbol{S}$ along $\boldsymbol{J}$ remains fixed in order that the magnitude of $\boldsymbol{J}$ remains fixed. Thus, $\boldsymbol{L}$ and $\boldsymbol{S}$ precess around $\boldsymbol{J}$, and $\boldsymbol{J}$ precesses around the z-axis. The picture we have drawn is useful in understanding the qualitative features of the behaviour of angular momentum vector.

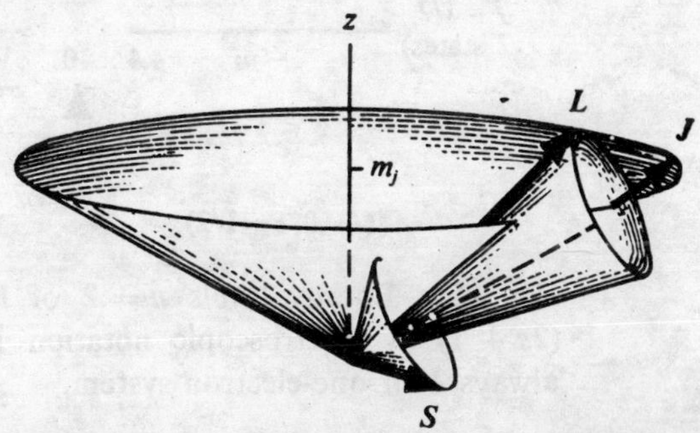

Fig. 4.21 Addition of angular momenta $\boldsymbol{L}$ and $\boldsymbol{S}$ for obtaining total angular momentum $\boldsymbol{J}$. $\boldsymbol{L}$ and $\boldsymbol{S}$ precess around $\boldsymbol{J}$, and $\boldsymbol{J}$ precesses around z-axis.

4.17 FINE STRUCTURE

It has been observed that in an atom the different l-states with the same principal quantum number n have slightly varying energies. Consider, as an example, the eight states of the hydrogen atom corresponding to $n = 2$. The energies of these states are shown in Fig. 4.22. The two states $2p_{3/2}$ and $2p_{1/2}$ are referred to as a doublet. The separation between the states $2s_{1/2}$ and $2p_{1/2}$ is known as the Lamb shift and is about one-tenth the doublet separation. In here, we shall discuss only the energy difference between the states $2p_{1/2}$ and $2p_{3/2}$.

SPIN-ORBIT COUPLING

The energy separation in the doublet $2p_{1/2}$ and $2p_{3/2}$ is due to the magnetic forces present in the hydrogen atom. An electron, as we have already learnt, carries a magnetic moment. As it moves in an electric field, it sees a magnetic field and its magnetic moment interacts with the magnetic field. The strength of the interaction depends on the direction of the magnetic moment. Since the direction of the magnetic moment depends on the direction of the spin of the electron, the interaction is spin-dependent. This causes the energy difference between the levels $j = 3/2$ and $j = 1/2$.

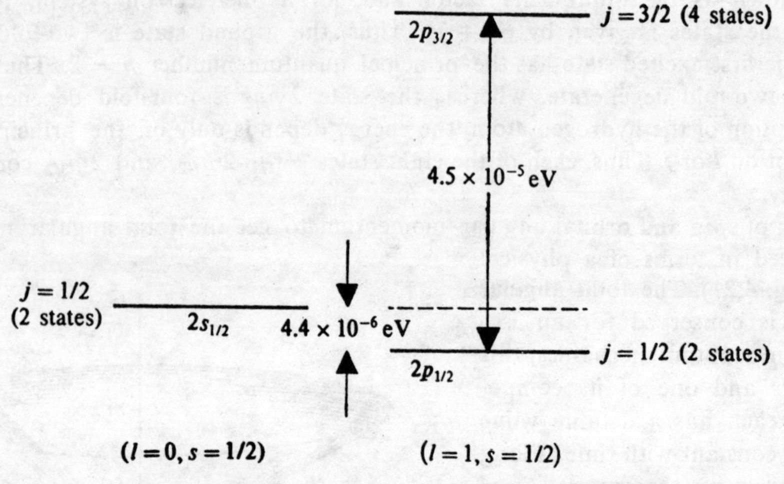

Fig. 4.22 Energy levels $n = 2$ of hydrogen atom. Multiplicity factor $(2s + 1)$ in spectroscopic notation is not shown because its value is always 2 for one-electron system.

The mathematical calculation to exactly determine the separation energy is outside the scope of our treatment. Therefore, we shall follow a simplified approach to compute it. If the magnetic moment of the electron is $\boldsymbol{\mu}$, then the interaction energy E_M due to the magnetic field B is

$$E_M = -\boldsymbol{\mu} \cdot \boldsymbol{B}. \tag{4.120}$$

Assume that the electron is moving in a circular orbit around a proton with a speed v_e. In the rest frame of the electron, we see a proton moving in a circular orbit of radius r in the direction opposite to that of the electron with a speed $v_p (= v_e)$. This produces a current of value $e[v_p/(2\pi r)]$ in the circular loop of radius r. The magnetic field due to such a current is

$$B = \frac{\mu_0 I}{2r} = \frac{\mu_0 e v_p}{4\pi r^2} = \mu_0 \epsilon_0 E v_e, \tag{4.121}$$

where E is the electric field given by $E = e/(4\pi\epsilon_0 r^2)$. The direction of the magnetic field points out of the paper as shown in Fig. 4.23. We can write $\boldsymbol{B}$ in terms of the angular momentum vector $\boldsymbol{L}$ as

$$\boldsymbol{B} = \mu_0 \epsilon_0 E \frac{\boldsymbol{r} \times \boldsymbol{v}_e}{r} = \frac{\mu_0 \epsilon_0 E \boldsymbol{L}}{mr}. \tag{4.122}$$

Using $E = -\partial V/\partial r$, where V is the electrostatic potential, and $\mu_0 \epsilon_0 = 1/c^2$, we get

$$\boldsymbol{B} = -\frac{\mu_0 \epsilon_0}{mr} \frac{\partial V}{\partial r} \boldsymbol{L} = -\frac{1}{mc^2 r} \frac{\partial V}{\partial r} \boldsymbol{L}. \tag{4.123}$$

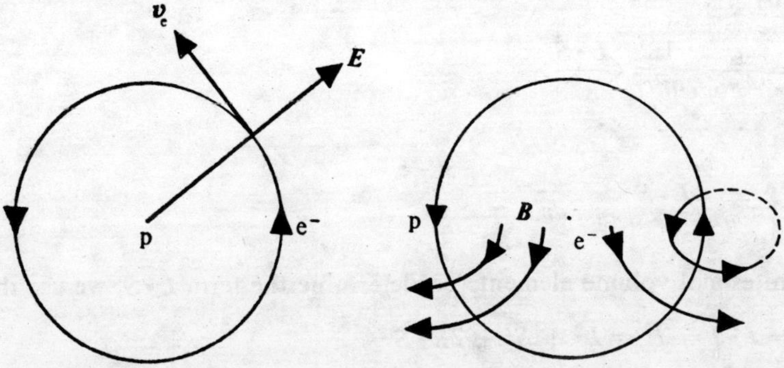

Fig. 4.23 Motion of electron around proton as 'seen' by proton, and motion of proton as 'seen' by electron. Magnetic field lines are also shown.

Substituting this in Eq. (4.120), we obtain

$$E_M = \frac{1}{mc^2 r}\frac{\partial V}{\partial r}\boldsymbol{\mu} \cdot \boldsymbol{L}.$$

Using Eq. (4.99) for the magnetic moment vector, we have

$$E_M = -\frac{ge}{2m^2c^2}\frac{1}{r}\frac{\partial V}{\partial r}\boldsymbol{L} \cdot \boldsymbol{S}. \tag{4.124}$$

A very important but a subtle effect, known as Thomas precession, due to the special theory of relativity changes this result drastically. This changed result is

$$E_M = -\frac{e}{2m^2c^2 r}(g-1)\frac{\partial V}{\partial r}\boldsymbol{L} \cdot \boldsymbol{S}. \tag{4.125}$$

The value of g is approximately equal to 2 (the almost exact value is $g = 2.0023$) and we get, to a good approximation,

$$E_M = -\frac{e}{2m^2c^2 r}\frac{\partial V}{\partial r}\boldsymbol{L} \cdot \boldsymbol{S}. \tag{4.126}$$

Using

$$V = \frac{1}{4\pi\epsilon_0}\frac{e}{r},$$

the value of E_M we obtain is

$$E_M = \frac{e^2}{4\pi\epsilon_0 2m^2c^2r^3}\boldsymbol{L} \cdot \boldsymbol{S}. \tag{4.127}$$

The total energy of the hydrogen atom is the sum of the energy given by Eqs. (4.56) and (4.127). If the electron is in an orbit described by the wave function $\psi_{n,l,j}$, then the magnetic energy is given by

$$(E_M)_{n,l,j} = \frac{e^2}{4\pi\epsilon_0}\frac{1}{2m^2c^2}\langle\frac{\boldsymbol{L}\cdot\boldsymbol{S}}{r^3}\rangle, \tag{4.128}$$

where

$$\langle\frac{\boldsymbol{L}\cdot\boldsymbol{S}}{r^3}\rangle = \int \psi^*_{n,l,j}\frac{\boldsymbol{L}\cdot\boldsymbol{S}}{r^3}\psi_{n,l,j}\,dv, \tag{4.129}$$

dv being an infinitesimal volume element. To determine the term $\boldsymbol{L}\cdot\boldsymbol{S}$, we use the identity

$$(\boldsymbol{L}+\boldsymbol{S})^2 = \boldsymbol{J}\cdot\boldsymbol{J} = J^2 = L^2 + S^2 + 2\boldsymbol{L}\cdot\boldsymbol{S}$$

or

$$\boldsymbol{L}\cdot\boldsymbol{S} = \frac{J^2 - L^2 - S^2}{2}. \tag{4.130}$$

For the state $\psi_{n,l,j}$, we know the values of J^2, L^2, and S^2, and therefore the value of $\boldsymbol{L}\cdot\boldsymbol{S}$ in this state is

$$(\boldsymbol{L}\cdot\boldsymbol{S})_{n,l,j} = \hbar^2\left[\frac{j(j+1) - l(l+1) - s(s+1)}{2}\right]. \tag{4.131}$$

Thus, Eq. (4.128) becomes

$$(E_M)_{n,l,j} = \frac{e^2\hbar^2}{4\pi\epsilon_0}\frac{1}{2m^2c^2}\frac{j(j+1) - l(l+1) - s(s+1)}{2}\int \psi^*_{n,l,j}\frac{1}{r^3}\psi_{n,l,j}\,dv. \tag{4.132}$$

The evaluation of the integral here is tedious, and therefore we give only the final result which is

$$(E_M)_{n,l,j} = \frac{e^2\hbar^2}{4\pi\epsilon_0}\frac{1}{2m^2c^2}\frac{1}{n^3}\frac{1}{(l+1)(l+\frac{1}{2})a_0^3}\left[\frac{j(j+1) - l(l+1) - s(s+1)}{2}\right] \tag{4.133}$$

for $l \neq 0$. For $l = 0$, this result is zero.

QUESTION 4.17 Give a physical argument to prove that $(E_M)_{n,l=0,j=1/2}$ is zero.

The energy difference in the doublet $j = l + \frac{1}{2}$ and $j = l - \frac{1}{2}$ is the energy due to spin-orbit coupling (known as L-S coupling) and is given by

$$E_{L\text{-}S} = (E_M)_{n,l,j=l+1/2} - (E_M)_{n,l,j=l-1/2} = \frac{1}{8\pi\epsilon_0 n^3 l(l+1)}\frac{e^2\hbar^2}{m^2c^2a_0^3} = \frac{\alpha^4 mc^2}{2n^3 l(l+1)}. \tag{4.134}$$

It is easy to verify that, for $n = 2$, $l = 1$, this splitting is 4.53×10^{-5} eV.

4.18 LINE WIDTH

In an atom, when an electron makes a transition from an excited state to its ground state, it emits a photon. The energy of the photon is the difference in the energy of the two states.

However, this energy difference is uncertain by an amount governed by the energy-time uncertainty principle (see Section 2.5). Therefore, the energy of the emitted photon is not fixed, but has a spread, known as the natural width. In the example that follows, we shall estimate this width.

EXAMPLE 4.16 The typical lifetime of an excited state of an atom is $\sim 10^{-8}$ sec. If the energy difference between the excited and ground states of the atom is 1 eV, calculate $\Delta\nu/\nu$, where ν is the frequency and $\Delta\nu$ the uncertainty in the frequency of the emitted photon.

We shall use the uncertainty principle $\Delta E \Delta t \geqslant h/(4\pi)$ in our calculations. Taking Δt to be the lifetime of the excited state, we see that

$$\Delta E \geqslant \frac{h}{4\pi \times 10^{-8}} \sim \frac{6.6 \times 10^{-34}}{4 \times 3.14 \times 10^{-8}} \sim 5.3 \times 10^{-27}.$$

Thus,

$$\frac{\Delta\nu}{\nu} = \frac{\Delta E}{E} \sim \frac{5.3 \times 10^{-27}}{1.6 \times 10^{-19}} \sim 3.3 \times 10^{-8}.$$

The uncertainty in the frequency broadens the lines in the frequency spectrum. In addition, effects such as Doppler broadening increase the width of such lines.

4.19 ATOMS IN MAGNETIC FIELD*

When an atom is placed in a magnetic field, the number of spectral lines observed increases. This is because, in general, each spectral line splits into many lines in the presence of the magnetic field. This effect was first studied by P. Zeeman at the end of the last century and is called the Zeeman effect. The effect can be fully explained in terms of quantum mechanics, and provides another proof that an electron has a spin and a magnetic moment.

The Zeeman effect is due to the interaction of the magnetic moment of the atom with the magnetic field. Let us first consider the hydrogen atom, ignoring the spin of the electron. Then, the magnetic moment of the atom is due only to the orbital angular momentum of the electron [see Eq. (4.97)] and is given by

$$\boldsymbol{\mu}_L = -\frac{e}{2m}\boldsymbol{L}. \tag{4.135}$$

When the atom is in a magnetic field $\boldsymbol{B}$, it experiences a torque ($\boldsymbol{\tau} = \boldsymbol{\mu}_L \times \boldsymbol{B}$) which causes a change in the angular momentum vector $\boldsymbol{L}$. The rate of change of angular momentum is given by

$$\frac{d\boldsymbol{L}}{dt} = \boldsymbol{\tau} = -\frac{e}{2m}\boldsymbol{L} \times \boldsymbol{B} = -\frac{\mu_B}{\hbar}\boldsymbol{L} \times \boldsymbol{B}, \tag{4.136}$$

where μ_B is the Bohr magneton (see Example 4.9). We shall presently see that the angular momentum vector $\boldsymbol{L}$ precesses around $\boldsymbol{B}$ with an angular frequency $\mu_B B/\hbar$. First, we take the scalar product of Eq. (4.136) with $\boldsymbol{B}$. This gives

$$\frac{\boldsymbol{B} \cdot d\boldsymbol{L}}{dt} = -\frac{\mu_B}{\hbar}(\boldsymbol{L} \times \boldsymbol{B}) \cdot \boldsymbol{B} = 0.$$

*A student taking an introductory course may skip this discussion.

Since $\boldsymbol{B}$ is a constant vector,

$$\boldsymbol{B}\cdot\frac{d\boldsymbol{L}}{dt}=\frac{d(\boldsymbol{B}\cdot\boldsymbol{L})}{dt}=0.$$

Integrating, we get

$$\boldsymbol{B}\cdot\boldsymbol{L}=\text{constant}. \tag{4.137}$$

Similarly, we note that $\boldsymbol{L}\cdot(d\boldsymbol{L}/dt)=0$ which gives

$$L^2=\text{constant}. \tag{4.138}$$

Equations (4.137) and (4.138) show that the component of $\boldsymbol{L}$ along $\boldsymbol{B}$ and L^2 remain constant. If we choose $\boldsymbol{B}$ along the z-axis, then L_z and $L_x^2+L_y^2(=L^2-L_z^2)$ are constants. Further, from Eq. (4.136), assuming $\boldsymbol{B}=B\hat{k}$,

$$\frac{dL_x}{dt}=-\frac{\mu_B}{\hbar}BL_y, \tag{4.139}$$

$$\frac{dL_y}{dt}=+\frac{\mu_B}{\hbar}BL_x. \tag{4.140}$$

Differentiating Eq. (4.139), we get

$$\frac{d^2L_x}{dt^2}=-\frac{\mu_B B}{\hbar}\frac{dL_y}{dt}=-\left(\frac{\mu_B B}{\hbar}\right)^2L_x$$

or

$$d^2L_x/dt^2+\omega_B^2L_x=0, \tag{4.141}$$

where $\omega_B=\mu_B B/\hbar$ is called the *Larmor frequency*, and its value is $1.41\times10^{10}B\ \text{sec}^{-1}$, where B is in weber/m^2. Similarly, we get an equation for L_y, namely,

$$d^2L_y/dt^2+\omega_B^2L_y=0. \tag{4.142}$$

Equations (4.141) and (4.142) are the familiar equations of simple harmonic motion and we know that their solutions are

$$L_x=A_x\sin(\omega_Bt+\phi_1),\qquad L_y=A_y\sin(\omega_Bt+\phi_2). \tag{4.143}$$

Substituting these in Eq. (4.139), we get

$$\omega_BA_x\cos(\omega_Bt+\phi_1)=-\omega_BA_y\sin(\omega_Bt+\phi_2). \tag{4.144}$$

This equation implies

$$A_x=A_y=A,\qquad \phi_1-\phi_2=\pi/2.$$

The final solutions, with $\phi_2=0$, are

$$L_x=A\cos\omega_Bt,\qquad L_y=A\sin\omega_Bt,\qquad L_z=\text{constant}. \tag{4.145}$$

Thus, we see that $\boldsymbol{L}$ precesses around the magnetic field (z-axis).

The additional energy of the atom in an external magnetic field is

$$E_M = -\boldsymbol{\mu}_L \cdot \boldsymbol{B} = \frac{e}{2m}\boldsymbol{L} \cdot \boldsymbol{B} = \frac{eB}{2m}L_z \tag{4.146}$$

since we have chosen $\boldsymbol{B}$ to be along the z-axis. The allowed values of L_z are $\hbar m_l$ ($|m_l| < l$). Thus, the total energy of the hydrogen atom in a magnetic field, ignoring the electron spin, is

$$E = -\tfrac{1}{2}\frac{mc^2\alpha^2}{n^2} + \frac{eB}{2m}m_l\hbar. \tag{4.147}$$

We note that the levels having the same l but different m_l are not degenerate. As a result, there is a corresponding change in the energy spectrum of the hydrogen atom.

We shall now consider the effect of the spin of the electron on the energy of the hydrogen atom. The magnetic moment of the atom arises because of the orbital magnetic moment and intrinsic spin of the electron. From Eqs. (4.97) and (4.99), we know that, for a one-electron system,

$$\boldsymbol{\mu}_L = -\frac{e}{2m}g_l\boldsymbol{L}, \qquad \boldsymbol{\mu}_S = -\frac{e}{2m}g_s\boldsymbol{S},$$

where $g_l = 1$ and $g_s = 2$. The total magnetic moment of the atom then is

$$\boldsymbol{\mu} = \boldsymbol{\mu}_L + \boldsymbol{\mu}_S = -\frac{e}{2m}(g_l\boldsymbol{L} + g_s\boldsymbol{S}). \tag{4.148}$$

The magnetic moment vectors $\boldsymbol{\mu}_L$, $\boldsymbol{\mu}_S$, and $\boldsymbol{\mu}$ are shown in Fig. 4.24. The orbital and spin angular momenta precess around $\boldsymbol{J}$. Thus, $\boldsymbol{\mu}_L$, $\boldsymbol{\mu}_S$, and $\boldsymbol{\mu}$ also precess around $\boldsymbol{J}$. We shall assume that the total angular momentum $\boldsymbol{J}$ precesses around $\boldsymbol{B}$ in the same manner as $\boldsymbol{L}$ precesses around $\boldsymbol{B}$. We shall also assume that the precession frequency of $\boldsymbol{J}$ around $\boldsymbol{B}$ is much smaller than the precession frequency of $\boldsymbol{L}$ and $\boldsymbol{S}$ around $\boldsymbol{J}$. Later in this section, we shall state and discuss the condition to justify this assumption. In view of these assumptions, we need to consider the component of magnetic moment only along the direction $\boldsymbol{J}$ to compute the energy of the atom. The other components of the magnetic moment average out to zero as $\boldsymbol{\mu}$ precesses very fast around $\boldsymbol{J}$. Therefore, we define a vector $\boldsymbol{\mu}_J$ with its direction along $\boldsymbol{J}$ and magnitude given by the projection of $\boldsymbol{\mu}$ along $\boldsymbol{J}$. Thus, we can write

$$\boldsymbol{\mu}_J = |\boldsymbol{\mu}_J|\boldsymbol{J}, \qquad |\boldsymbol{\mu}_J| = \frac{\boldsymbol{\mu} \cdot \boldsymbol{J}}{|\boldsymbol{J}|}. \tag{4.149}$$

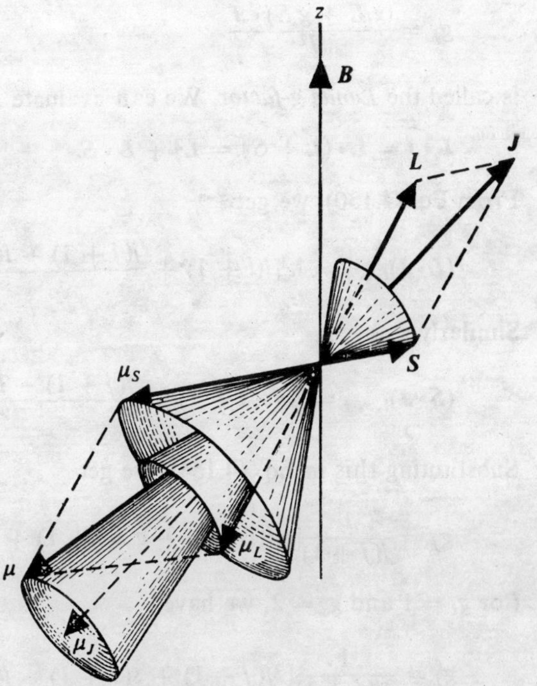

Fig. 4.24 Addition of $\boldsymbol{\mu}_L$ and $\boldsymbol{\mu}_S$ for obtaining $\boldsymbol{\mu}$. As $\boldsymbol{L}$ and $\boldsymbol{S}$ precess around $\boldsymbol{J}$, vectors $\boldsymbol{\mu}_L$, $\boldsymbol{\mu}_S$, and $\boldsymbol{\mu}$ precess around $\boldsymbol{J}$.

Combining these two equations, we get

$$\boldsymbol{\mu}_J = \frac{(\boldsymbol{\mu} \cdot \boldsymbol{J})\boldsymbol{J}}{J^2}. \tag{4.150}$$

Using Eq. (4.148), we obtain

$$\boldsymbol{\mu} \cdot \boldsymbol{J} = -\frac{e}{2m}(g_l \boldsymbol{L} + g_s \boldsymbol{S}) \cdot \boldsymbol{J}. \tag{4.151}$$

Thus,

$$\boldsymbol{\mu}_J = -\frac{e}{2m}\frac{(g_l \boldsymbol{L} + g_s \boldsymbol{S}) \cdot \boldsymbol{J}}{J^2}\boldsymbol{J} \tag{4.152}$$

$$= -\frac{e}{2m} g_j \boldsymbol{J}, \tag{4.153}$$

where

$$g_j = \frac{(g_l \boldsymbol{L} + g_s \boldsymbol{S}) \cdot \boldsymbol{J}}{J^2} \tag{4.154}$$

is called the *Lande g-factor*. We can evaluate $\boldsymbol{L} \cdot \boldsymbol{J}$ as follows:

$$\boldsymbol{L} \cdot \boldsymbol{J} = \boldsymbol{L} \cdot (\boldsymbol{L} + \boldsymbol{S}) = L^2 + \boldsymbol{L} \cdot \boldsymbol{S}. \tag{4.155}$$

From Eq. (4.130), we get

$$(\boldsymbol{L} \cdot \boldsymbol{J})_{l,s,j} = \hbar^2\left[l(l+1) + \frac{j(j+1) - l(l+1) - s(s+1)}{2}\right]. \tag{4.156}$$

Similarly,

$$(\boldsymbol{S} \cdot \boldsymbol{J})_{l,s,j} = \hbar^2\left[s(s+1) + \frac{j(j+1) - l(l+1) - s(s+1)}{2}\right]. \tag{4.157}$$

Substituting this in Eq. (4.154), we get

$$g_j = \frac{1}{j(j+1)}\left[g_l l(l+1) + g_s s(s+1) + (g_l + g_s)\frac{j(j+1) - l(l+1) - s(s+1)}{2}\right]. \tag{4.158}$$

For $g_l = 1$ and $g_s = 2$, we have

$$g_j = \frac{1}{2j(j+1)}[3j(j+1) + s(s+1) - l(l+1)]. \tag{4.159}$$

The magnetic energy of the hydrogen atom is then given by

$$E_M = -\boldsymbol{\mu}_j \cdot \boldsymbol{B} = \frac{e}{2m} g_j \boldsymbol{J} \cdot \boldsymbol{B} = \frac{eB}{2m} g_j J_z \tag{4.160}$$

because we have taken B along the z-axis. Thus,

$$E_M = \frac{e\hbar}{2m} g_j m_j B = \mu_B g_j m_j B \qquad (-j \leqslant m_j \leqslant j), \tag{4.161}$$

μ_B being the Bohr magneton. The total energy of the hydrogen atom in a magnetic field B along the z-direction (ignoring the fine structure) is then

$$E_{n,l,j} = -\frac{1}{2}\frac{mc^2\alpha^2}{n^2} + g_j\mu_B B m_j. \tag{4.162}$$

Thus, we see that the presence of a magnetic field destroys the degeneracy of states, leading to additional transitions and increase in the number of spectral lines. We shall illustrate this through the example that follows.

EXAMPLE 4.17 Find the energies of the spectral lines emitted in the transitions $2p_{1/2} \rightarrow 1s_{1/2}$ when a hydrogen atom is placed in a magnetic field.

The quantum numbers of the $(1s_{1/2})$-state are $l = 0$, $s = \frac{1}{2}$, and $j = \frac{1}{2}$, and $g_j = 2$ [see Eq. (4.159)]. Thus, this energy level splits into two levels. The energy of each such level changes due to the presence of the magnetic field. This change is given by

$$\Delta_s = E_M = +\frac{e\hbar}{2m}2m_jB.$$

Thus, the energy changes in the two levels are

$$\Delta_s = \pm\frac{e\hbar}{2m}B \qquad (+ \text{ for } m_j = \tfrac{1}{2}, - \text{ for } m_j = -\tfrac{1}{2})$$

$$= \pm\mu_B B.$$

For the $(2p_{1/2})$-level, we have $l = 1$, $s = \frac{1}{2}$, and $j = \frac{1}{2}$, and

$$g_j = \frac{3\cdot\frac{3}{4} + \frac{3}{4} - 2}{\frac{3}{2}} = \frac{2}{3}.$$

The energy shift is given by

$$\Delta_p = E_M = \frac{e\hbar}{2m}\frac{m_j}{2}B.$$

Thus, the energy shifts in the two levels (see Fig. 4.25) are

$$\Delta_p = \pm\frac{e\hbar}{2m}\frac{B}{2} \qquad (+ \text{ for } m_j = \tfrac{1}{2}, - \text{ for } m_j = -\tfrac{1}{2})$$

$$= \pm\frac{\mu_B B}{2}.$$

If the transition energy in the absence of the magnetic field is E_0 (= 10.2 eV), then the transition energies in the presence of the magnetic field are $E_0 - 3\Delta/2$, $E_0 - \Delta/2$, $E_0 + \Delta/2$, $E_0 + 3\Delta/2$, where $\Delta = \mu_B B$. As can be noticed, a single spectral line splits into four lines.

In our discussion following Eq. (4.148), we had assumed that the precession frequency ω_j of $\boldsymbol{J}$ around $\boldsymbol{B}$ is much smaller than the precession frequency of $\boldsymbol{L}$ and $\boldsymbol{S}$ around $\boldsymbol{J}$. We now give the condition to justify this assumption. The precession frequency of $\boldsymbol{J}$ around $\boldsymbol{B}$ is directly proportional to the magnitude of $\boldsymbol{B}$ since the torque on the atom is proportional to $\boldsymbol{B}$. The spin-orbit splitting (discussed in Section 4.17) can be interpreted in terms of the precession of $\boldsymbol{L}$ and

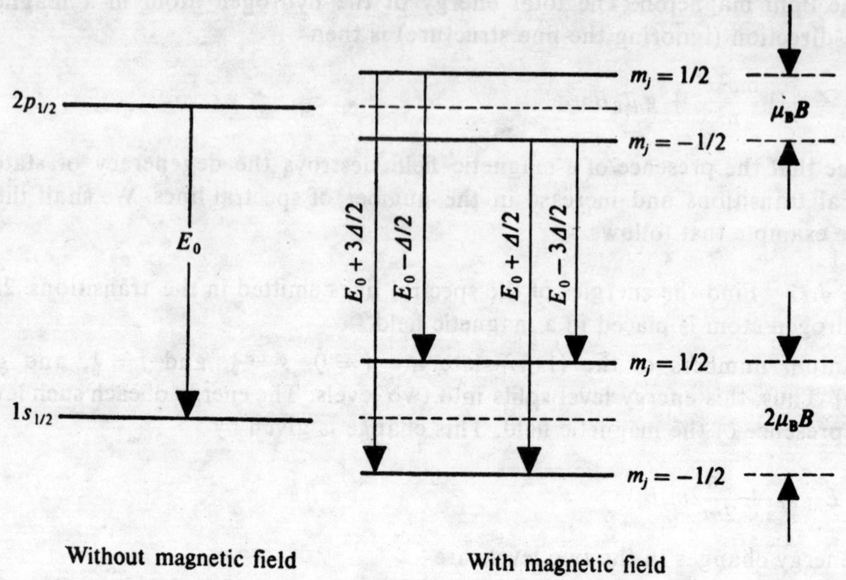

Fig. 4.25 Splitting of energy levels $2p_{1/2}$ and $1s_{1/2}$ in presence of magnetic field. Transitions are also shown.

S around ***J***, as we shall now show. The orbital motion of the electron produces an internal magnetic field and the electron experiences a torque because of its magnetic moment due to spin. As a result, the vector ***S*** precesses around the total angular momentum vector ***J*** which is a constant of motion. The vector ***L*** also precesses around ***J*** because of the condition that the sum ***L*** + ***S*** must be equal to ***J***. The energy of the spin-orbit interaction is the energy associated with the precession of ***L*** and ***S*** around ***J***. The precession frequency $\omega_{L\text{-}S}$ for the doublet, from Eq. (4.134), is given by

$$\omega_{L\text{-}S} = \frac{E_{L\text{-}S}}{\hbar} = \frac{\alpha^4 mc^2}{2n^2 l(l+1)\hbar}. \tag{4.163}$$

Thus, the condition for our assumption to be valid is

$$\omega_j \ll \omega_{L\text{-}S}. \tag{4.164}$$

The separation of the levels $2p_{1/2}$ and $2p_{3/2}$ in the hydrogen atom is 4.5×10^{-5} eV (see Fig. 4.21). Thus, the precession frequency $\omega_{L\text{-}S}$ is

$$\omega_{L\text{-}S} = \frac{4.5 \times 10^{-5}}{\hbar} \text{ eV} \sim 7 \times 10^{10} \text{ sec}^{-1}.$$

Since the precession frequency ω_j is expected to be of the same order of magnitude as the Larmor frequency $\omega_B (= 1.41 \times 10^{10} B \text{ sec}^{-1})$, our assumption is valid for a magnetic field much less than 1 weber/m^2.

The Zeeman effect we have discussed is valid for a weak magnetic field. For a strong magnetic field, the effect is referred to as the Paschen-Back effect.

PROBLEMS

4.1 According to J. J. Thomson, a sphere with the positive charge uniformly spread over it and the electron moving inside it describes the model of an atom.

(a) Show that the force acting on the electron at a point r from the centre of the sphere is $F = -\rho e r/(3\pi\epsilon_0)$, and the motion of the electron is simple harmonic (ρ is the charge density in the sphere).

(b) Apply the classical theory to calculate the frequency of light emitted. Assume the radius of the sphere to be 10^{-10} m.

4.2 Calculate the shortest and longest wavelengths of (a) the Lyman series (i.e., the transitions in which the final state is $n = 1$) and (b) the Balmer series (i.e., the transitions in which the final state is $n = 2$).

4.3 In a hydrogen atom, how much energy must be supplied (a) to put the electron in the ($n = 6$)-state from its ground state? (b) to ionize the atom from the ($n = 3$)-state?

4.4 Apply Bohr's quantization condition to a proton-electron system to calculate the energy of the system by taking into account only the gravitational attraction. What will be the percentage error in the energy of the hydrogen atom obtained by neglecting the gravitational interaction?

4.5 For a μ-mesic hydrogen atom, calculate (a) the energy levels, taking into account the finite mass of the nucleus, (b) the percentage error in the energy, treating the nucleus as fixed, and (c) the radius. A μ-mesic atom is one in which a μ-meson goes around the nucleus. A μ-meson is a particle with a charge $-e$ and mass 207 times that of an electron.

4.6 A deuterium atom consists of an electron going around the nucleus whose charge is $+e$ and mass is twice that of a proton. Calculate the difference between the wavelengths of a deuterium atom and a hydrogen atom for the transition from the ($n = 2$)- to the ($n = 1$)-state.

4.7 A hydrogen atom is in its ground state and is at rest. What should be the energy of the photon needed to take it to the first excited state (a) when the recoil of the atom is neglected? (b) when the recoil of the atom is taken into account?

4.8 Use the Bohr model to calculate the kinetic energy, the potential energy, and the magnetic moment of an electron in the n-th orbit of a hydrogen atom.

4.9 A hydrogen atom, initially in the ($n = 4$)-state, finally attains the ground state. In this process, what will be the possible wavelengths of the electromagnetic waves emitted?

4.10 The hydrogen gas is bombarded with electrons of energy 12.10 eV. What will be the frequencies of the spectral lines emitted?

4.11 Consider the singly-ionized systems $({}^3_2\text{He})^+$ and $({}^4_2\text{He})^+$. Which of these two atoms would require lesser energy to be excited to the ($n = 2$)-state? Calculate the difference in the excitation energy of the two atoms in electron volts. The mass of ${}^3_2$He is (approximately) three times, whereas the mass of ${}^4_2$He is (approximately) four times that of a hydrogen atom.

4.12 Verify by direct substitution in the Schroedinger equation (4.52) for the hydrogen atom that $\psi_{1,0,0}$ and $\psi_{2,0,0}$ [given by Eqs. (4.59) and (4.60)] are its solutions.

4.13 Verify the normalization condition (4.64) for $\psi_{2,0,0}$ [given by Eq. (4.60)].

4.14 Calculate the average value of r^2 for a hydrogen atom in the ground state.

4.15 Calculate the average value of the potential energy of a hydrogen atom in the ground state. Use the result so obtained to find the average value of the kinetic energy of the hydrogen atom in the ground state.

4.16 Consider a one-dimensional box of length L with a particle of mass m having an energy $E_n = \hbar^2\pi^2n^2/(2mL^2)$. Calculate the average value of x and x^2 (a) quantum mechanically, (b) classically. When $n \gg 1$, show that the values of x and x^2 obtained quantum mechanically become identical with those worked out classically. (This is an example of Bohr's correspondence principle.)

4.17 The Stern-Gerlach experiment is performed by using an atomic hydrogen beam. Assume that the length of the magnet is 50 cm and the field gradient $\Delta B/\Delta Z = 0.8$ weber-m^{-3}. Also, let the velocity of the incident atoms be 500 m-sec^{-1} and let the width of the splitting beam just at the end of the magnet be 10^{-3} m. Calculate the magnetic moment of a hydrogen atom.

4.18 In the Stern-Gerlach experiment, a beam of neutral nickel atoms splits into nine components. What is the angular momentum of a nickel atom in its ground state?

4.19 Consider a three-dimensional cubical box of length L. Ten electrons are required to be put into it. What is the lowest energy state available for the system (neglect the Coulomb repulsion effects)?

4.20 If an electron in the state $n_x = n_y = n_z = 1$ is removed from the system in Problem 4.19, what will be the possible frequencies of the electromagnetic waves that will be emitted when the system attains the ground state?

4.21 Write down the electronic configuration, i.e., the quantum numbers of the electrons, of (a) aluminium and (b) zinc.

4.22 Estimate the ionization energy of a lithium atom, assuming that the electron-electron repulsion is negligible and ignoring the Pauli exclusion principle.

4.23 Calculate the wavelength of the K_α-line for a uranium atom.

4.24 Calculate the minimum energy of the electrons needed to produce a K_α-line from a molybdenum target.

5

Molecular Physics

5.1 INTRODUCTION

In Chapters 1–4, we saw that the electrons in an atom are bound to the nucleus by the electrostatic attractive force. Also, we noted that the Schroedinger equation, together with the Pauli exclusion principle, gives us a good description of an atom. We now focus our attention to this question of how the atoms, which are electrically neutral, combine to form molecules. The formation of molecules and the reactions amongst the molecules form the subject matter of physical chemistry. We shall see in this chapter that the basic principles of valence and bonding can be explained in terms of quantum mechanics. The discipline of quantum mechanics plays an important role in unifying chemistry and physics. In fact, it has been possible to predict the structure of complex molecules and their reactions, using the concepts of quantum mechanics.

5.2 MOLECULAR BONDING

We know from the study of physical and chemical properties of molecules that the atoms combine either by *ionic* bond or by *covalent* bond to form molecules. For example, in the formation of a NaCl crystal, ionic bonding is dominant, whereas in an H_2 molecule, covalent bonding is prevalent.

IONIC BOND

In an ionic bond between two atoms, one or more electrons are transferred from one atom to the other, making the two atoms oppositely charged. Thus, in the formation of a NaCl molecule, one electron from the sodium atom is transferred to the chlorine atom, making sodium a positively charged ion (Na^+) and chlorine a negatively charged ion (Cl^-). The electrostatic attraction between the two ions binds them to form the NaCl molecule. The description of the ionic bond we have given raises a number of questions, for example: Why is only one electron transferred from the Na atom to the Cl atom and not two or more? Why don't the ions Na^+ and Cl^- fall into each other because of the electrostatic attraction?

In Section 4.14, we learnt that the ionization energy varies with atomic number (see Fig. 4.17). We recall that a sodium atom has a low ionization potential because the valence electron in it effectively 'sees' only a charge $+e$ and not $+11e$. The inner electrons in the atom screen the nuclear charge, and hence the binding energy of the valence electron in the 3*s*-state is small. On the other hand, the valence electrons in an element having the 3*s*-shell and 3*p*-shell almost filled (e.g., chlorine) are strongly bound. The number of valence electrons in a chlorine atom is seven, and so each of these electrons sees a charge of about $+7e$. Since these electrons are strongly bound, the ionization energy of chlorine is large. In fact, a chlorine atom can accommodate an additional electron to form a Cl^- ion. This is because the attraction of this additional electron to the nucleus is more than the electron-electron repulsion force. We now consider the binding of a NaCl molecule.

The ionization energy of a Na atom is 5.1 eV. In other words, the valence electron becomes free from the Na atom, i.e.,

$$\text{Na} + 5.1 \text{ eV} \rightarrow \text{Na}^+ + e^-, \tag{5.1}$$

when an energy of 5.1 eV is supplied to it. On the other hand, we know that a free electron gets absorbed by a chlorine atom to form a Cl^- ion and an energy of 3.8 eV is released. The energy so released is referred to as *electron affinity*. Thus, the electron affinity of Cl is 3.8 eV and can be expressed as

$$e^- + \text{Cl} \rightarrow \text{Cl}^- + 3.8 \text{ eV}. \tag{5.2}$$

Consider a system consisting of the Na and Cl atoms separated by a distance R. If we ignore the electrostatic potential energy between the ions (i.e., Na^+ and Cl^-), then the energy required to transfer an electron from the Na atom to the Cl atom is the difference between the ionization energy of the Na atom and the electron affinity of the Cl atom. The potential energy V as a function of R is shown in Fig. 5.1. We shall discuss the essential features of this curve. The potential energy $V(R)$ between the Na^+ and Cl^- ions for a large internuclear distance behaves,

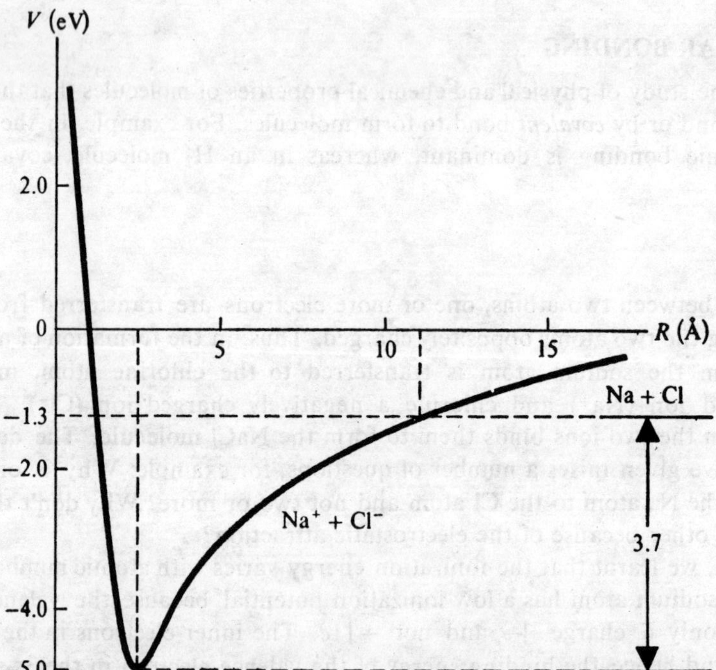

Fig. 5.1 Potential energy of Na^+Cl^- as function of internuclear separation R.

when R is very much greater than the atomic radii, as

$$V(R) = -\frac{e^2}{4\pi\epsilon_0}\frac{1}{R} \tag{5.3}$$

because the two ions can be considered as point charges when they are far apart. As R becomes very small, much less than the internuclear distance, the internuclear repulsion due to the positive charge of the Na and Cl nuclei increases and so does their energy. Thus, the potential energy is large and positive for small R. The potential energy $V(R)$ is minimum, i.e., -5 eV, when $R \sim 2.4$ Å. We can express this quantitatively as

$$\text{Na} + \text{Cl} \rightarrow \underset{\substack{\leftarrow \infty \rightarrow \\ \text{infinite} \\ \text{distance}}}{\text{Na}^+ + \text{Cl}^-} - 1.3 \text{ eV} \tag{5.4}$$

$$\rightarrow \underset{\leftarrow R \rightarrow}{\text{Na}^+ + \text{Cl}^-} - 1.3 \text{ eV} - V(R). \tag{5.5}$$

Thus, Na + Cl is bound in the form of $Na^+ + Cl^-$, with a binding energy of $5 - 1.3 = 3.7$ eV. The potential energy of $Na^+ + Cl^-$ at a large distance is taken to be zero [see Eq. (5.3)]. Therefore, the energy of Na + Cl is -1.3 eV [see reaction (5.4)].

A sodium atom cannot transfer two electrons unless a large amount of energy is supplied to the atom. This is because the second electron of the atom, being in the ($n = 2$)-orbit, is rather well bound. Further, a chlorine atom cannot accept two electrons. This is because the second electron cannot enter the $3p$-shell since the electrons in this shell are not well bound. Thus, it is not energetically favourable to have a system such as $Na^{++}Cl^{--}$.

EXAMPLE 5.1 In Fig. 5.1, evaluate R_0, the internuclear distance beyond which the system $Na^+ + Cl^-$ is unstable.

Let us assume that R_0 is far greater than the interatomic distance ($\sim$1 Å). As we will discover, the result we obtain is consistent with this assumption.

For R very much greater than the interatomic distance,

$$V(R) = -e^2/(4\pi\epsilon_0 R).$$

If $V(R_0) = -1.3$ eV, then both the systems $Na^+ + Cl^-$ (with the ions at a distance R_0) and Na + Cl have the same energy. Further, when $R > R_0$, then $V(R_0) > -1.3$ eV, and $Na^+ + Cl^-$ becomes unstable. Thus,

$$V(R_0) = -e^2/(4\pi\epsilon_0 R_0) = -1.3 \text{ eV} \quad \text{or} \quad R_0 \sim 11 \text{ Å}.$$

COVALENT BOND

When two like atoms join together, the bond cannot be ionic. The two atoms being identical can either receive or donate electrons to form a molecule. In such a molecule, the two atoms share the electrons, i.e., are held together by a covalent bond.

An atom can join with only a limited number of other atoms to form a molecule. For example, one H atom can bind with only one other H atom, whereas nitrogen can combine with three H atoms. In chemistry, this can be explained in terms of the concept of valence.

We shall first consider the simplest molecular system, i.e., hydrogen molecular ion (H_2^+). It consists of two protons and an electron. We assume that in a stable hydrogen molecular ion, the protons, being much heavier, do not move. The potential energy of the electron under the electrostatic attraction of the two protons (see Fig. 5.2) is

$$V(r) = -\frac{e^2}{4\pi\epsilon_0|r - r_1|} - \frac{e^2}{4\pi\epsilon_0|r - r_2|}. \tag{5.6}$$

If E is the energy of the electron, then the Schroedinger equation for the electron is

$$\nabla^2\psi(r) + \frac{2m}{\hbar^2}[(E + \frac{e^2}{4\pi\epsilon_0})(\frac{1}{|r - r_1|} + \frac{1}{|r - r_2|})]\psi(r) = 0. \tag{5.7}$$

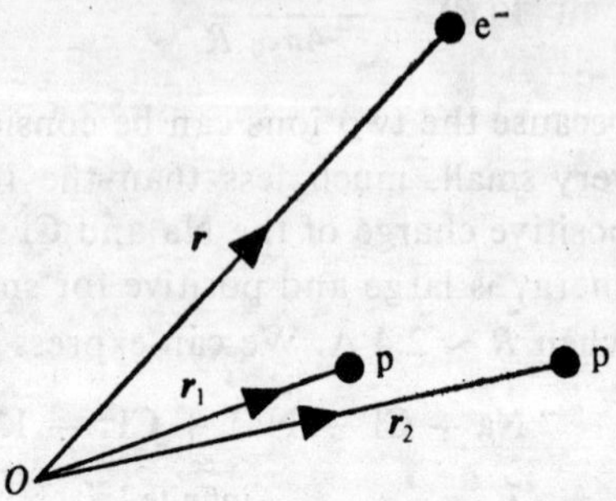

Fig. 5.2 Coordinates of electron and two protons in H_2^+ ion.

The energy E is not the total energy of the molecule since we have not included in Eq. (5.7) the potential energy between the two protons. If E_t is the total energy of H_2^+, then

$$E_t = E + \frac{e^2}{4\pi\epsilon_0|r_1 - r_2|}. \tag{5.8}$$

When we solve for E and ψ, we see, from Eq. (5.7), that they depend on $\boldsymbol{r}_1$ and $\boldsymbol{r}_2$, and the total energy depends on the interproton distance $|\boldsymbol{r}_1 - \boldsymbol{r}_2|$.

The interproton distance for which the energy is minimum gives a stable configuration. If there is no minimum for the energy, a stable configuration cannot exist.

The exact solution of Eq. (5.7) is complicated and we shall therefore not discuss it. Instead, we shall confine ourselves to physical arguments. The solution of the Schroedinger equation (5.7) obtained numerically has justified these physical arguments.

EXAMPLE 5.2 Consider the motion of a particle of mass m confined in a region $a < |x| < (a + b)$ and in a potential

$$V(x) = V_0 \qquad [|x| < a]$$
$$= 0 \qquad [a < |x| < (a + b)]. \tag{5.9}$$

Find the energy of the particle when it is in the ground state and the first excited state.

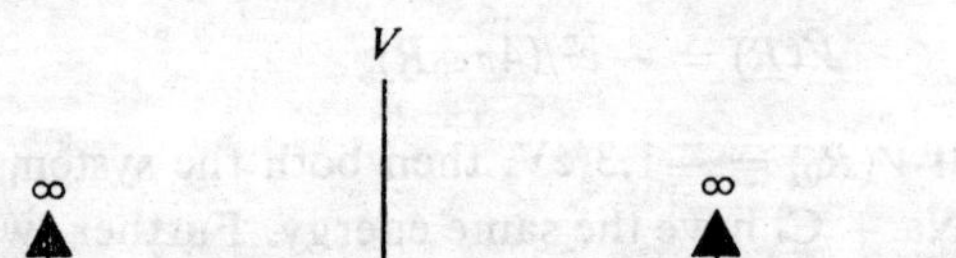

Fig. 5.3 Potential $V(x)$.

We note that the given potential (see Fig. 5.3) is an even function, that is,

$$V(x) = V(-x). \tag{5.10}$$

Thus, we know (see Example 2.21) that the wave function must be either even (symmetric), i.e.,

$$\psi(x) = \psi(-x), \tag{5.11}$$

or odd (antisymmetric), i.e., $\psi(x) = -\psi(-x)$.

From the experience we have so far gathered in our study of the various systems (for example, particle in a box and simple harmonic oscillator), we can safely assume the wave function of the ground state to be even and that of the first excited state to be odd. This assumption simplifies the application of the boundary conditions in that these conditions need be applied only at one of two points, namely, $x = a$ and $x = -a$. The application of the boundary conditions at the other point then follows.

The Schroedinger equation for the particle is

$$\frac{d^2\psi}{dx^2} + \frac{2m}{\hbar^2}E\psi = 0 \qquad [(-a-b) < x < -a],$$

$$\frac{d^2\psi}{dx^2} + \frac{2m}{\hbar^2}(E - V_0)\psi = 0 \qquad [-a < x < +a], \tag{5.12}$$

$$\frac{d^2\psi}{dx^2} + \frac{2m}{\hbar^2}E\psi = 0 \qquad [a < x < (a+b)].$$

The solutions corresponding to Eqs. (5.12)—see Problem 5.4—are

$$\begin{aligned} \psi(x) &= A \sin k(x + a + b) && [(-a-b) < x < -a] \\ &= B \exp(\alpha x) + C \exp(-\alpha x) && [-a < x < +a] \\ &= D \sin k(x - a - b) && [a < x < (a+b)], \end{aligned} \tag{5.13}$$

where $k^2 = 2mE/\hbar^2$ and $\alpha^2 = 2m(V_0 - E)/\hbar^2$. In solutions (5.13), we have imposed the condition $\psi(a+b) = \psi(-a-b) = 0$. The energy levels of the particle and the corresponding wave functions can now be obtained by applying the boundary conditions at $x = \pm a$.

To obtain the energy of the particle in its ground state, we impose the boundary condition $\psi(x) = \psi(-x)$ on the wave function. We thus get

$$B = C, \quad A = -D. \tag{5.14}$$

Applying the boundary condition that ψ and $d\psi/dx$ are continuous at $x = a$, we obtain

$$\begin{aligned} B[\exp(\alpha a) + \exp(-\alpha a)] &= -D \sin kb, \\ B\alpha[\exp(\alpha a) - \exp(-\alpha a)] &= Dk \cos kb \end{aligned} \tag{5.15}$$

or

$$\alpha \tanh \alpha a = -k \cot kb. \tag{5.16}$$

The solution of Eq. (5.16) gives the allowed energy values of the particle. Of these, the lowest value yields the energy of the ground state. This energy, calculated numerically, is plotted in Fig. 5.4 as a function of a We notice that it increases as a increases from 0 to ∞. If a is large, the problem reduces to that of a particle in the two regions $(-a-b) < x < -a$ and $(a+b) > x > a$, the regions being connected by a potential barrier. Classically, if the particle were in the region $(-a-b) < x < -a$ with an energy less than V_0, then it would have stayed in this region. However, quantum mechanically, the particle should have a certain probability of tunneling through the potential to the region $a < x < (a+b)$. This tunneling probability is less if V_0 or a is large. In such a situation, the particle can be imagined to be almost confined to one region. As a decreases, the probability of the particle tunneling through to the other region increases. In other words, the wave function spreads. From the uncertainty principle, we know that if a particle occupies a large region, its energy tends to be low. Thus, when $a \to 0$, the particle is in a region of width $2b$ and its ground state energy is $\hbar^2\pi^2/[2m(2b)^2]$. This explains qualitatively the behaviour of energy as a function of a (see Fig. 5.4).

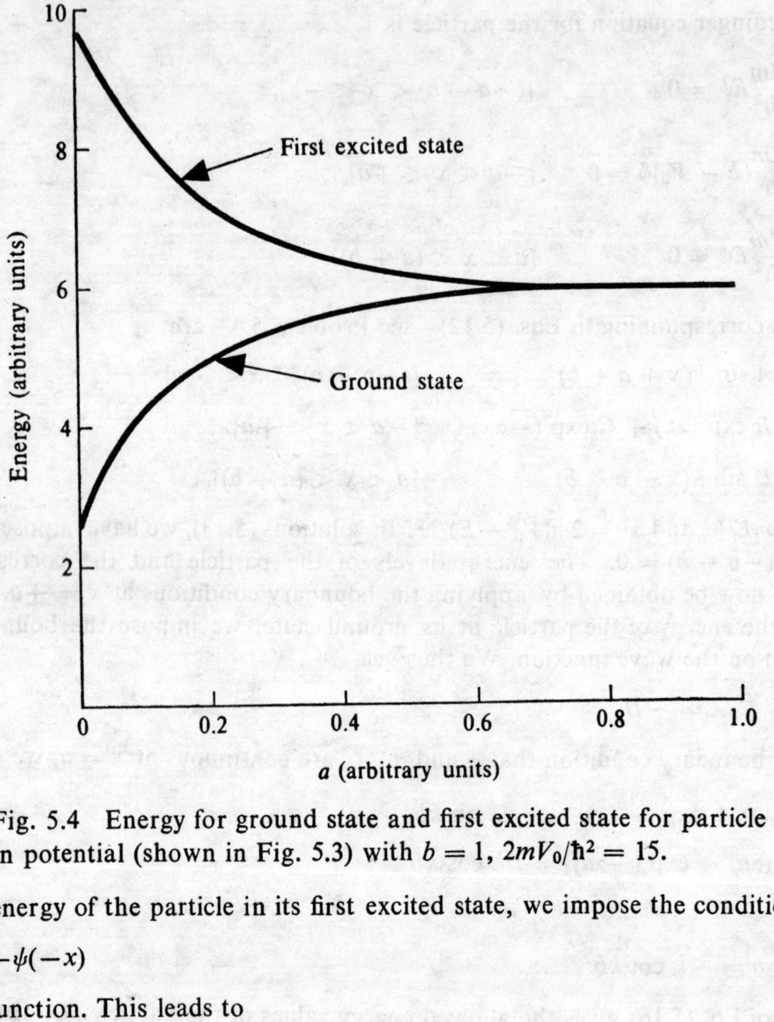

Fig. 5.4 Energy for ground state and first excited state for particle in potential (shown in Fig. 5.3) with $b = 1$, $2mV_0/\hbar^2 = 15$.

To get the energy of the particle in its first excited state, we impose the condition

$$\psi(x) = -\psi(-x) \tag{5.17}$$

on the wave function. This leads to

$$B = -C, \qquad A = D. \tag{5.18}$$

Matching the wave function and its derivative at $x = a$, we get

$$\begin{aligned} B[\exp(\alpha a) - \exp(-\alpha a)] &= -D \sin kb, \\ B\alpha[\exp(\alpha a) + \exp(-\alpha a)] &= Dk \cos kb \end{aligned} \tag{5.19}$$

or

$$\alpha \coth \alpha a = -k \cot kb. \tag{5.20}$$

The variation of the energy of the first excited state as a function of a is also shown in Fig. 5.4.

The wave functions associated with the ground state and the first excited state are depicted in Fig. 5.5.

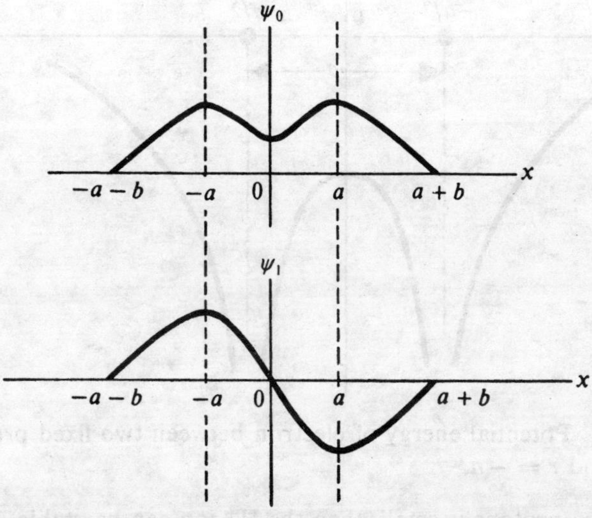

Fig. 5.5 Wave functions corresponding to ground state and first excited state.

QUESTION 5.1 Use the uncertainty principle to explain why the energy of a particle tends to increase when the volume of confinement of the particle decreases.

We shall now consider the motion of the electron in the electrostatic potential due to the two fixed protons (see Fig. 5.6). From this figure, it is clear that the potential is symmetric about the origin, and therefore the ground state wave function is even (symmetric), i.e.,

$$\psi(\boldsymbol{r}) = \psi(-\boldsymbol{r}). \tag{5.21}$$

The electron can be thought of as being attached to one atom for some time and then tunneling to the other atom through the potential barrier. The tunneling time depends on a. If a is very large, then the tunneling time is also very large and we can assume that at any given time the electron is attached to one of the protons. Such a system then corresponds to a system consisting of a hydrogen atom and a proton, the distance between the two being a. The energy of the electron in this system is the same as that of the electron in the hydrogen atom. On the other hand, if a is small, the electron is shared by both the protons, and hence the electron wave function spreads. In such a situation, the uncertainty in the position of the electron is large, i.e., the electron has a low kinetic energy. Thus, the total energy of the electron in H_2^+ is less than that of the electron in the system $H + H^+$. The total energy of H_2^+ is the sum of the energy of the electron and the Coulomb energy between the two protons. If the Coulomb

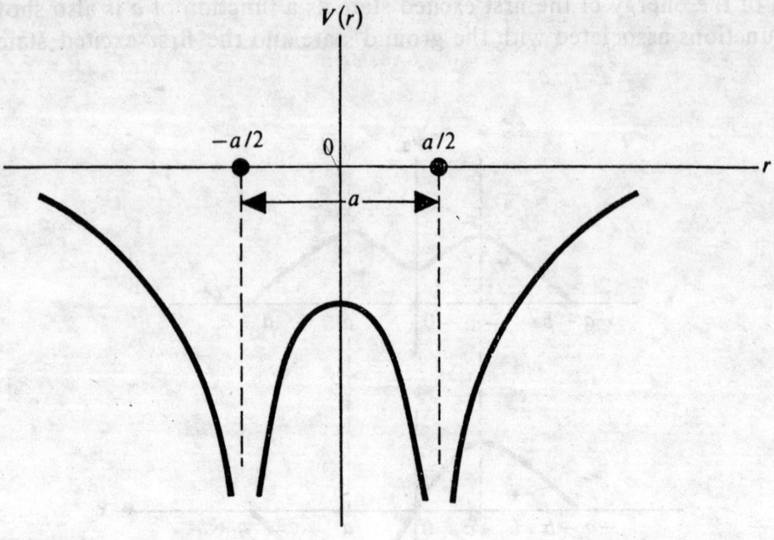

Fig. 5.6 Potential energy of electron between two fixed protons at $r = a$ and $r = -a$.

energy between the two protons is small, then the H_2^+ ion can be stable. Figure 5.7 shows the total energy as a function of the internuclear distance R, obtained by a detailed calculation for

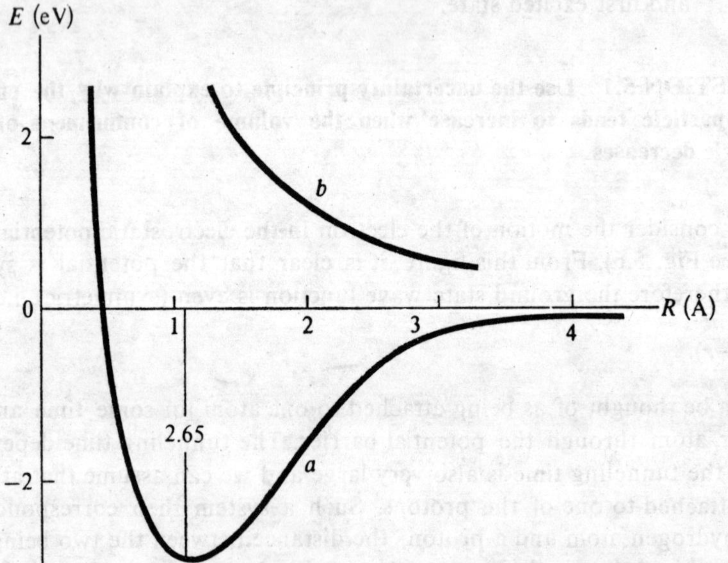

Fig. 5.7 Total energy for H_2^+ ion as function of internuclear distance for symmetric state (curve a) and antisymmetric state (curve b).

the symmetric (even) and antisymmetric (odd) states. We find that the minimum for energy occurs only for the symmetric state which corresponds to the ground state. The total energy associated with the antisymmetric state does not have a minimum, and hence only the ground state of H_2^+ is stable.

In Fig. 5.7, we can also see that the potential energy of the electron in the symmetric state is lower than that in the antisymmetric state. The antisymmetric wave function vanishes midway between the two protons, whereas the symmetric wave function does not. Thus, the electron density midway between the protons is larger for the symmetric state. The electron midway between the two protons has a lower potential energy because it experiences the effect of both the protons.

THE HYDROGEN MOLECULE

The ground state of the H_2^+ system can accommodate an additional electron, with the spins of the two electrons being antiparallel in accordance with the Pauli exclusion principle. This is the configuration of the hydrogen molecule H_2. In chemistry, the one-electron wave function $\psi_{n,l,m}$ is referred to as an orbital. Both the electrons are in the same orbital but their spin quantum numbers m_s are different. With two electrons, the binding energy of H_2 is twice that of the H_2^+ ion, that is, 5.3 eV as compared to 2.65 eV for H_2^+. However, the electron-electron repulsion reduces the strength of the bond between the two hydrogen atoms and the actual binding energy is 4.72 eV. The separation between the two protons, called the bond length, is 0.74 Å. Again, the antisymmetric state does not form a bound state.

The sharing of the valence electrons establishes a bond between the atoms and leads to the formation of a molecule. The number of electrons that can be shared depends on the allowed energy levels of the participating atoms. Thus, the valence concept used in chemistry finds a natural explanation in quantum mechanics.

5.3 MOLECULAR SPECTRA

The transition between the various allowed energy states of a molecule with the emission or absorption of a photon gives rise to a molecular spectrum. The principle underlying this phenomenon is the same as that in atomic spectra. If E_i and E_f are the energies of the initial and final states involved in the transition, then the frequency of the photon is given by

$$\nu = \frac{E_f - E_i}{h} \quad \text{if } E_f > E_i \quad \text{(absorption)},$$
$$\nu = \frac{E_i - E_f}{h} \quad \text{if } E_i > E_f \quad \text{(emission)}. \tag{5.22}$$

The spectrum of a molecule is more complicated than that of an atom because a molecule has many more energy levels. This is mainly due to the fact that a molecule has many more degrees of freedom. Also, in a molecule, the electrons are no longer in a central potential, and therefore the method of separating the angular motion from the radial motion is not applicable. This means that the orbital angular momentum is not conserved, making it impossible to label the energy states using the orbital angular momentum quantum number. Further, the nuclear motion in a molecule cannot be neglected as in an atom. In a molecule, the nuclei can rotate about its centre of mass, or vibrate about their equilibrium position, or execute a complex motion which could be both vibrational and rotational. Thus, it is very difficult to solve the

Schroedinger equation for such a complicated system. However, it turns out that the simple models of molecules that have been proposed explain their spectra quite satisfactorily. We shall therefore discuss only these models.

The excited energy levels for a molecule of a fixed electronic configuration arise from

(1) the rotation of the molecule as a whole and

(2) the vibration of the atoms in the molecule about their mean position.

As we shall presently see, the energy levels arising from (1) are separated by about 0.001 eV, whereas those arising from (2) are separated by 0.1 eV. The separation in the electronic excited states in a molecule resulting from the excited states of atoms is of the order of a few electron volts or more.

The energy of a molecule results from a superposition of all the three types of levels, namely, electronic, vibrational, and rotational. In particular, if the electrons in the atoms remain in the same quantum state, then those atoms that vibrate about their mean position give the vibrational energy levels. Each such vibrational level has several energy levels arising from the various rotational states available to the molecule. Some of the energy levels of a typical molecule are shown in Fig. 5.8.

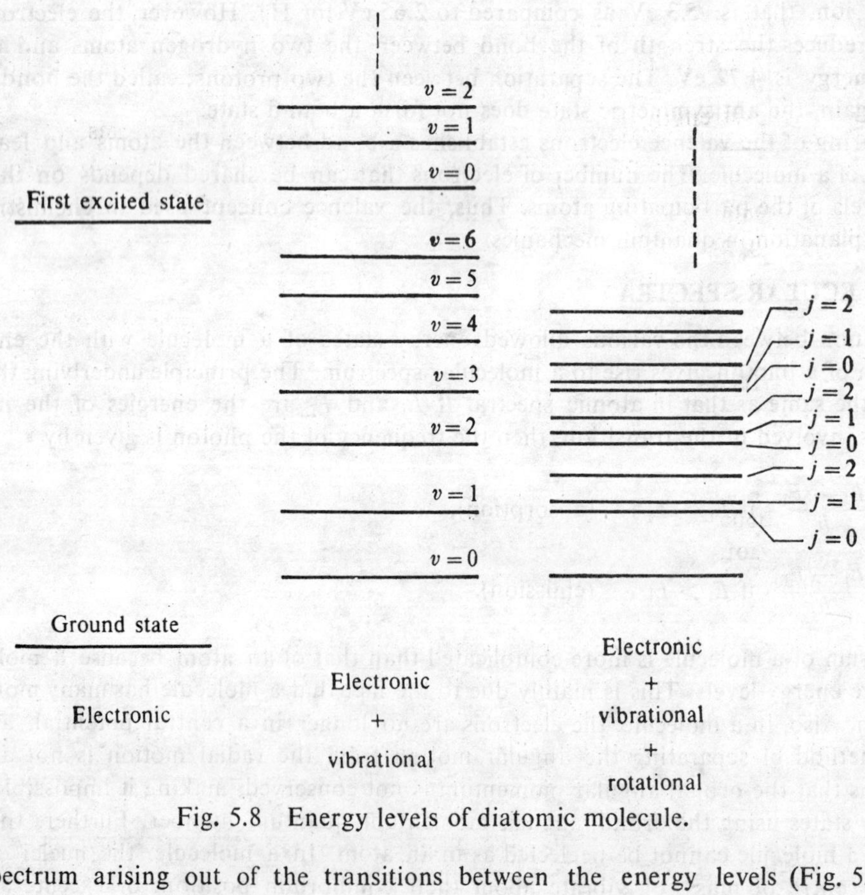

Fig. 5.8 Energy levels of diatomic molecule.

The spectrum arising out of the transitions between the energy levels (Fig. 5.8) can be categorized into one of the following three types:

(1) *Rotational spectrum* This corresponds to radiation absorbed or emitted when a transition between two rotational energy levels occurs while the vibrational state of the molecule and the electronic configuration of the atoms remain the same. Since the energy involved in such a transition is of the order of 0.001 eV, the emitted radiation is in the far-infrared or microwave region ($\sim 10^7$ Å).

(2) *Rotational-vibrational spectrum* This corresponds to radiation emitted or absorbed when a transition between two vibrational energy levels occurs while the electronic configuration of the atoms remains fixed. The rotational levels may or may not change. If the rotational levels do not change, we get a pure vibrational spectrum. Since the transition energy involved is of the order of 0.1 eV, the emitted radiation is in the near-infrared region ($\sim 10^5$ Å).

(3) *Electronic spectrum* This corresponds to radiation emitted or absorbed when a transition between two electronic energy levels of an atom in a molecule occurs. The radiation is in the visible and ultraviolet regions.

ROTATIONAL ENERGY LEVELS OF DIATOMIC MOLECULES

Consider two atoms of mass m_1 and m_2 forming a molecule. We shall assume that the atoms are point masses. If the two atoms rotate with an angular velocity ω about an axis perpendicular to the line joining the two and passing through the centre of mass (see Fig. 5.9), then the moment of inertia of the system about this axis is

$$I = m_1 r_1^2 + m_2 r_2^2. \tag{5.23}$$

Since, in the centre-of-mass coordinates, $m_1 r_1 = m_2 r_2$ and $R = r_1 + r_2$, we have

$$r_1 = m_2 R/(m_1 + m_2),$$
$$r_2 = m_1 R/(m_1 + m_2). \tag{5.24}$$

Substituting for r_1 and r_2 in Eq. (5.23), we get

$$I = \mu R^2, \tag{5.25}$$

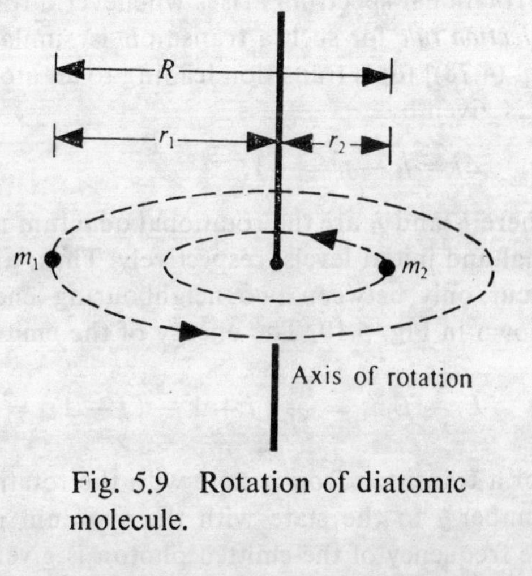

Fig. 5.9 Rotation of diatomic molecule.

where the reduced mass $\mu = m_1 m_2/(m_1 + m_2)$. The energy of the system is given by

$$E = \tfrac{1}{2} I\omega^2 = \frac{1}{2I}(I\omega)^2 = \frac{J^2}{2I}, \tag{5.26}$$

where J^2 is the angular momentum of the system.

If we wish to obtain the quantum mechanical version of the foregoing classical analysis, we have to replace J by its corresponding operator. The rotational energy levels are then given by the eigenvalues of J^2 [$j(j+1)\hbar^2$ with $j = 0, 1, 2, \ldots$]. Thus, the rotational energy levels are

$$E_j = \frac{j(j+1)\hbar^2}{2I} = Bj(j+1), \qquad j = 0, 1, 2, \ldots, \tag{5.27}$$

where $B = \hbar^2/(2I)$.

EXAMPLE 5.3 Calculate the moment of inertia of an HCl molecule about an axis passing through the centre of mass and perpendicular to the line joining the two nuclei H and Cl and obtain the rotational energy levels of HCl in eV. Given $R = 1.3$ Å.

From Eq. (5.25), we have

$$I = \frac{m_1 m_2 R^2}{m_1 + m_2} = \frac{m_H m_{Cl}}{m_H + m_{Cl}} \times (1.3 \times 10^{-10})^2 \sim \tfrac{35}{36} \times m_H \times (1.3 \times 10^{-10})^2$$

$$= \tfrac{35}{36} \times 1.0078 \text{ u} \times (1.3 \times 10^{-10})^2 = \tfrac{35}{36} \times 1.0078 \times 1.66 \times 10^{-27} \times (1.3 \times 10^{-10})^2$$

$$= 2.75 \times 10^{-47} \text{ kg-m}^2.$$

The rotational energy levels are

$$E_j = \frac{\hbar^2}{2I} j(j+1) = \frac{(1.05 \times 10^{-34})^2}{2 \times 2.75 \times 10^{-47}} j(j+1)$$

$$= 2 \times 10^{-22} j(j+1)j = 1.25 \times 10^{-3} j(j+1) \text{ eV}.$$

Rotational Spectra

A rotational spectrum arises whenever a transition occurs between rotational energy levels. The *selection rule* for such a transition is similar to that [see Eq. (4.78)] for a transition leading to an atomic spectrum. It is given by

$$\Delta j = j_f - j_i = \pm 1, \tag{5.28}$$

where j_f and j_i are the rotational quantum numbers of the final and initial levels, respectively. Thus, a transition can occur only between two neighbouring energy levels as shown in Fig. 5.10. The energy of the emitted photon is

$$E_j - E_{j-1} = \frac{\hbar^2}{2I} j[j + 1 - (j - 1)] = 2Bj. \tag{5.29}$$

For a transition from a state with the rotational quantum number j to the state with the quantum number $j - 1$, the frequency of the emitted photon is given by

$$\nu_j = \frac{2B}{h} j \quad (j = 1, 2, 3, \ldots). \tag{5.30}$$

Thus, the spectrum has the frequencies

$$\frac{2B}{h}, \frac{4B}{h}, \frac{6B}{h}, \ldots \tag{5.31}$$

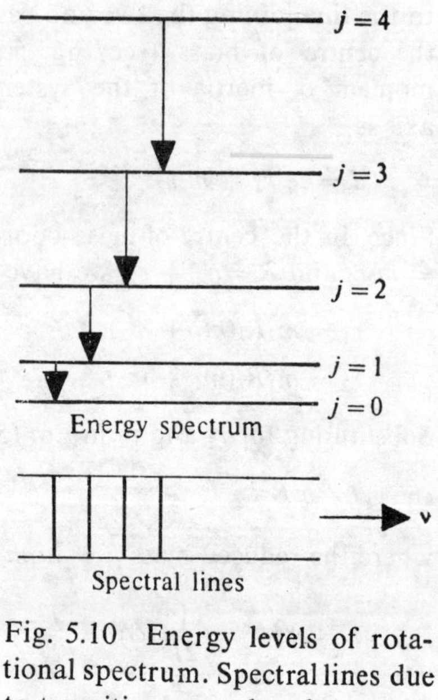

Fig. 5.10 Energy levels of rotational spectrum. Spectral lines due to transitions are also shown.

The experimental values of the frequencies can be used to determine the value of the constant B. The constant B can also be calculated in terms of the internuclear distance and the atomic masses.

EXAMPLE 5.4 The lines of the rotational spectrum for an HCl molecule appear at the wave numbers $(1/\lambda)$ 83.5 cm^{-1}, 104.1 cm^{-1}, 124.7 cm^{-1}, 145.4 cm^{-1}, 166.0 cm^{-1}. Find the value of the constant B. If the interatomic distance in the HCl molecule is 1.3 Å, calculate the value of B and compare it with the value of B obtained from the given wave numbers.

From Eq. (5.30), for a transition from j to $j-1$,

$$\left(\frac{1}{\lambda}\right)_j = \frac{\nu_j}{c} = \frac{2B}{hc}j \quad (j = 1, 2, 3, \ldots). \tag{5.32}$$

Since we do not know the value of j associated with the given wave numbers, we use the difference between two consecutive wave numbers to eliminate j. Using Eq. (5.32), we find this difference is given by

$$\left(\frac{1}{\lambda}\right)_{j+1} - \left(\frac{1}{\lambda}\right)_j = \frac{2B}{hc}. \tag{5.33}$$

Further, from the given wave numbers, we see that the difference between two consecutive wave numbers is almost 20.6 cm^{-1}. Thus,

$$2B/(hc) = 20.6 \text{ cm}^{-1} = 2060 \text{ m}^{-1}$$

or

$$B = \frac{6.63 \times 10^{-34} \times 3 \times 10^8}{2} \times 2.06 \times 10^3$$

$$\sim 2 \times 10^{-22} \text{ J} \sim 1.25 \times 10^{-3} \text{ eV}.$$

In Example 5.3, we found that the rotational energy for the HCl molecule with an internuclear separation of 1.3 Å is

$$E_j = Bj(j+1) = 1.25 \times 10^{-3} j(j+1) \text{ eV}.$$

From this, we see that the value of $B = 1.25 \times 10^{-3}$ eV. Thus, the two values of B are the same.

In the foregoing discussion, we have considered the rotation only about an axis perpendicular to the line joining the two nuclei. What about the rotation of the molecule about this line? The moment of inertia about this line is very small because the line passes through the nuclei where most of the mass is concentrated. In fact, the moment of inertia would be smaller by a factor $\sim 10^{-8}$ (radius of the nucleus/radius of the atom $= 10^{-4}$). Therefore, the energies associated with the rotational states (5.27) would be very large ($\sim 10^5$ eV). If such an energy is supplied to the molecule, it will dissociate, and thus we cannot hope to see the levels arising from the rotation of the molecule about the line joining the two nuclei.

Not all diatomic molecules exhibit rotational spectra. For example, an H_2 molecule does not show a rotational spectrum. Though the precise reason for such a behaviour can be explained in terms of quantum mechanics, we shall present a qualitative and classical version of the argument. When an electric dipole moment oscillates, radiation is emitted or absorbed. A molecule such as HCl has an electric dipole moment and when it rotates, the dipole moment oscillates, causing an emission or absorption of radiation. The reason an HCl molecule has a dipole moment is as follows. As we know, an electron is transferred to Cl from H and the molecule acquires the form H^+Cl^-. This creates the electric dipole moment. However, a symmetric molecule such as H_2 does not have an electric dipole moment about the line joining the nuclei, and hence it cannot emit or absorb radiation. This does not mean that the rotational levels do not exist in an H_2 molecule, but only that the transitions between the rotational levels do not normally occur.

QUESTION 5.2 Give an argument to explain the absence of the electric dipole moment in a hydrogen molecule.

We shall not discuss the rotational levels of polyatomic molecules. Their energy levels can be obtained using an analysis similar to that given for the diatomic molecules.

VIBRATIONAL ENERGY LEVELS OF DIATOMIC MOLECULES

We have seen that the atoms in a molecule are bound to each other with a definite distance (called bond length) between them. Figure 5.11 shows the potential energy between the atoms as a function of R, the distance between the atoms in a molecule. A molecule in equilibrium has a bond length R_0. When energy is supplied to a molecule, one of the ways in which it can absorb the energy is through the vibration of the atoms in it.

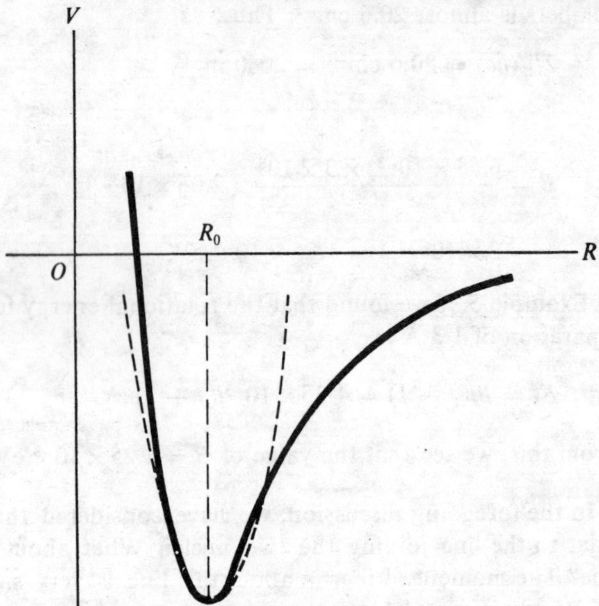

Fig. 5.11 Potential energy between atoms as function of R. Dashed curve represents harmonic oscillator approximation to potential.

We shall now consider the vibration of the atoms around R_0. If the vibration is small, we can expand the potential energy function $V(R)$ about R_0, using the Taylor series, as

$$V(R)=V(R_0)+\tfrac{1}{2}(R-R_0)^2 \times \left.\frac{d^2V(R)}{dR^2}\right|_{R=R_0}, \qquad (5.34)$$

remembering that the first differential of $V(R)$ is zero at $R=R_0$ as R_0 represents the minimum of the potential. Denoting $R-R_0=x$, we find the potential energy of vibration is

$$V(R_0)+\tfrac{1}{2}V''(R_0)x^2. \qquad (5.35)$$

In this expression, the second term represents the potential energy of a simple harmonic oscillator with a force constant $V''(R_0)$, the first term being a constant. Thus, the atoms undergo a simple harmonic oscillation about their mean position. Before we discuss the vibrational energy levels in a molecule, let us consider the motion of two atoms, say, A and B, in a molecule whose masses are m_A and m_B, respectively. We can think of the vibration in the molecule to be caused by a spring, with a spring constant $V''(R_0)$, between the two atoms (see Fig. 5.12). If x_A and x_B are the coordinates of A and B (considering one-dimensional motion), we then have

$$x_B - x_A = R_0 + x, \quad (5.36)$$

$$m_A\ddot{x}_A = V''(R_0)x, \quad (5.37)$$

$$m_B\ddot{x}_B = -V''(R_0)x. \quad (5.38)$$

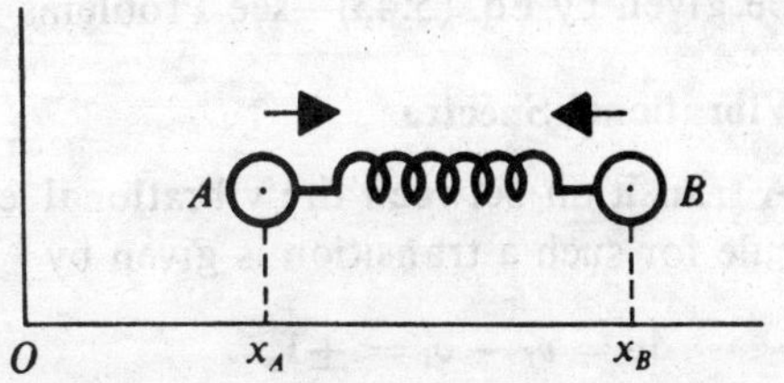

Fig. 5.12 Model for vibration of diatomic molecule.

Multiplying Eqs. (5.37) and (5.38) by m_B and m_A, respectively, and subtracting the results so obtained, we get

$$m_A m_B(\ddot{x}_B - \ddot{x}_A) = -(m_A + m_B)V''(R_0)x. \quad (5.39)$$

Differentiating Eq. (5.36) twice, we have

$$\ddot{x}_B - \ddot{x}_A = \ddot{x}. \quad (5.40)$$

From Eqs. (5.39) and (5.40), we obtain

$$\frac{m_A m_B}{m_A + m_B}\ddot{x} = \mu\ddot{x} = -V''(R_0)x, \quad (5.41)$$

where μ is the reduced mass. Thus, the frequency of the oscillator is

$$\nu_0 = \frac{1}{2\pi}\left(\frac{V''(R_0)}{\mu}\right)^{1/2}. \quad (5.42)$$

Quantum mechanically, the energy of such an oscillator is given by Eq. (3.6), i.e.,

$$E_v = (v + \tfrac{1}{2})h\nu_0, \quad (5.43)$$

where v (=0, 1, 2, . . .) is referred to as the *vibrational quantum number*. The lowest vibrational state has an energy $V(R_0) = h\nu_0/2$.

EXAMPLE 5.5 Calculate the vibrational energy levels of an HCl molecule, assuming the force constant to be 516 N-m^{-1}.

From Eq. (5.42), the natural frequency of an oscillator is

$$\nu_0 = \frac{1}{2\pi}\left(\frac{516}{\mu}\right)^{1/2}.$$

The reduced mass μ for the HCl molecule is

$$\mu = (35/36) \times 1.0078 \times 1.66 \times 10^{-27} \text{ kg}.$$

The energy levels are then given by

$$h\nu_0(v + 1/2) = 6.6 \times 10^{-34} \times \frac{1}{2\pi}\left(\frac{516 \times 36}{35 \times 1.0078 \times 1.66 \times 10^{-27}}\right)^{1/2}(v + 1/2) \text{ J}$$

$$\sim 5.9 \times 10^{-20}(v + 1/2) \text{ J}$$

$$\sim 0.37(v + 1/2) \text{ eV} \quad (v = 0, 1, 2, \ldots).$$

We have approximated the potential $V(R)$ by a simple harmonic oscillator. This is valid only for a small oscillation or for a small value of v. As v becomes large, the energy levels will not be given by Eq. (5.43)—see Problems 5.10 and 5.11.

Vibrational Spectra

A transition between the vibrational energy levels gives a vibrational spectrum. The selection rule for such a transition is given by

$$\Delta v = v_f - v_i = \pm 1, \tag{5.44}$$

where v_i and v_f refer to the vibrational quantum numbers of the initial and final states, respectively. The transition energy in the harmonic oscillator approximation is given by

$$\Delta E = E_{v+1} \quad E_v = h\nu_0. \tag{5.45}$$

This means that only one frequency is emitted in the vibrational spectrum. However, a vibrational-rotational spectrum is obtained when, in general, a vibrational transition mixes with a rotational transition.

Such homonuclear molecules as H_2, N_2 do not show a vibrational spectrum. This is because, as already pointed out, these molecules do not have an electric dipole moment. In such molecules, even though no transition between the vibrational energy levels occurs, the vibrational levels do exist. For example, they can be found when studying the Raman effect.

Vibrational-Rotational Spectra

We have so far considered the rotational motion and the vibrational motion of a molecule separately. A molecule can execute both these motions simultaneously.

We can treat the rotational motion and the vibrational motion independently, and therefore the energy of the molecule is the sum of the energies associated with the two types of motion. Thus, the allowed energy levels are specified by the two quantum numbers v and j. From Eqs. (5.27) and (5.43), the energy levels are

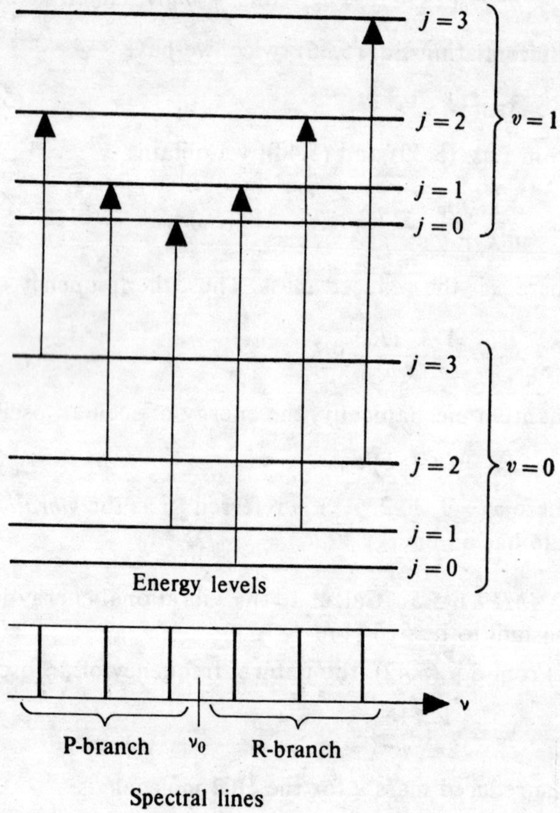

Fig. 5.13 Energy levels for $v = 0$, $v = 1$ for various values of j. P-branch and R-branch of rotational-vibrational spectrum are also shown.

$$E_{v,j} = (v + 1/2)h\nu_0 + j(j+1)\frac{\hbar^2}{2I} \quad (v = 0, 1, 2, \ldots, j = 0, 1, 2, \ldots). \tag{5.46}$$

The various energy levels for $v = 0$, $j = 0, 1, 2, \ldots$, and $v = 1$, $j = 0, 1, 2, \ldots$ are shown in Fig. 5.13. Depending on the quantum number associated with the transition, the spectrum can be divided into the following three branches:

(1) *R-branch* ($\Delta v = +1$, $\Delta j = +1$) The frequency of the photon here is

$$\nu_R = \frac{1}{h}[E_{1,\, j+1} - E_{0,\, j}] = \nu_0 + (\frac{j+1}{2\pi})\frac{\hbar}{I} \qquad (j = 0, 1, 2, \ldots). \tag{5.47}$$

Comparing these frequencies with the frequency ν_0 of a pure vibrational spectrum, we find the frequencies here are higher by $\hbar/(2\pi I)$, $\hbar/(\pi I)$, $\ldots$.

(2) *Q-branch* ($\Delta v = 1$, $\Delta j = 0$) The energy of this branch corresponds to the fundamental frequency ν_0. This transition is not allowed in a diatomic molecule.

(3) *P-branch* ($\Delta v = 1$, $\Delta j = -1$) Here, the frequency of the photon is

$$\nu_P = \nu_0 - \frac{(j+1)\hbar}{2\pi I} \qquad (j = 0, 1, 2, \ldots). \tag{5.48}$$

If we compare these frequencies with the frequency ν_0 of a pure vibrational spectrum, we find the frequencies here are lower by $\hbar/(2\pi I)$, $\hbar/(\pi I)$, $\ldots$.

5.4 ELECTRONIC SPECTRA

An excited state of a molecule can also arise through an excited state of the electrons in the molecule. The transition between these energy levels gives rise to an electronic spectrum. The spacing between these levels is of the order of electron volts, much greater than the spacing between the rotational or vibrational levels. Therefore, the radiation due to the transition is emitted or absorbed in the visible or ultraviolet region. When a transition occurs between the two electronic levels of a molecule, the initial and final vibrational states need not be the same. In other words, the vibrational quantum numbers associated with the two electronic levels are, in general, different. Figure 5.14 shows a typical transition between the electronic levels having different vibrational quantum numbers. It should be noted that the vibrational levels here are not quite equally spaced as expected from Eq. (5.43). This is because the potential is not a truly harmonic oscillator type. It has some anharmonicity. Each electronic level has its own vibrational levels. Similarly, each of the vibrational levels has its own set of rotational levels. We can write the total energy E of a molecule as the sum

$$E = E_e + E_v + E_j, \tag{5.49}$$

where E_e is the energy associated with the electronic excitation, and E_v and E_j are the vibrational and rotational energies given by Eqs. (5.43) and (5.27). In a transition, the total energy change is the sum

$$E_i - E_f = (E_{e_i} - E_{e_f}) + (E_{v_i} - E_{v_f}) + (E_{j_i} - E_{j_f})$$

or

$$\Delta E = \Delta E_e + \Delta E_v + \Delta E_j, \tag{5.50}$$

where ΔE_e, ΔE_v, and ΔE_j represent the changes in the electronic, vibrational, and rotational energies, respectively. These are of the order of

$$\Delta E_e \sim 1 \text{ eV}, \qquad \Delta E_v \sim 0.1 \text{ eV}, \qquad \Delta E_j \sim 0.001 \text{ eV}.$$

For a given electronic transition, the spectrum consists of many lines arising due to the energy

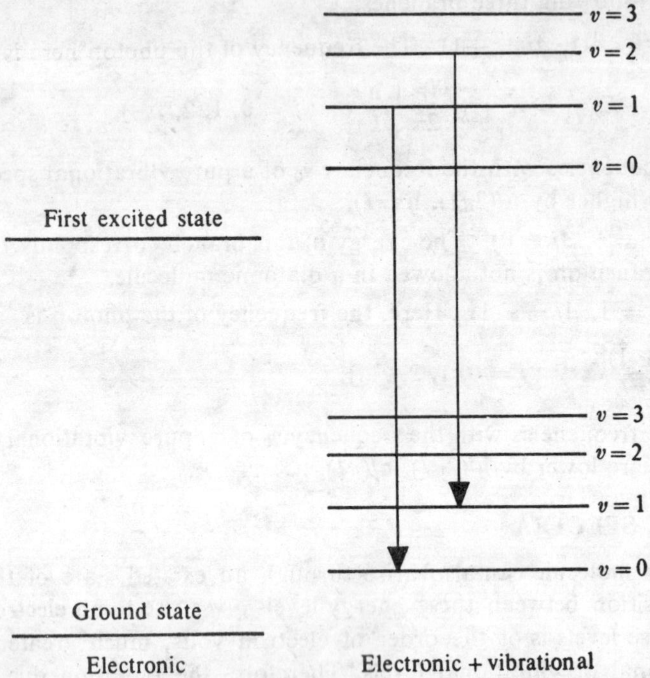

Fig. 5.14 Transitions between two electronic levels having different vibrational quantum numbers. Spacing between vibrational levels decreases as v increases due to anharmonicity of potential. Rotational levels are not shown.

changes ΔE_v and ΔE_j. These lines are so close to each other in frequency that the spectrum looks like a band.

5.5 THE RAMAN EFFECT

When a photon of frequency ν_1 falls on a molecule, the molecule emits a photon having a frequency ν_2 different from ν_1. The energy of the emitted photon can be greater or less than that of the incident photon. As a result, the state of the molecule undergoes a change. In particular, the vibrational or rotational state of the molecule changes. The scattering of light by a molecule, resulting in a frequency change is known as the Raman effect, named after its discoverer, C. V. Raman. The study of Raman effect gives information about the rotational and vibrational levels of a molecule.

ROTATIONAL RAMAN SPECTRA

The energy of a molecule in a given electronic and vibrational state is

$$E = E_e + E_v + Bj(j+1). \tag{5.51}$$

If a photon of frequency ν_1 is absorbed by the molecule when the latter was initially in the rotational state j_i, and the emitted photon has a frequency ν_2 with the molecule being in the rotational state j_f, then

$$h\nu_1 + E_e + E_v + Bj_i(j_i + 1) = h\nu_2 + E_e + E_v + Bj_f(j_f + 1)$$

or

$$h\nu_1 + Bj_i(j_i + 1) = h\nu_2 + Bj_f(j_f + 1). \tag{5.52}$$

The frequency of the emitted photon is

$$\nu_2 = \nu_1 + \frac{B}{h}[j_i(j_i + 1) - j_f(j_f + 1)]. \tag{5.53}$$

The selection rule associated with the Raman effect is

$$j_f - j_i = \pm 2. \tag{5.54}$$

Therefore, we have

$$\nu_2 = \nu_1 - \frac{2B}{h}(2j + 3) \quad \text{for } j_f = j_i + 2$$

$$= \nu_1 + \frac{2B}{h}(2j - 1) \quad \text{for } j_f = j_i - 2. \tag{5.55}$$

The emitted frequency spectrum has lines on both sides of ν_1.

VIBRATIONAL RAMAN SPECTRA

The basic principle in vibrational Raman spectra is essentially the same as in rotational Raman spectra, except that here the vibrational levels get excited.

The Raman spectra are very useful in studying the vibrational and rotational energy levels of homonuclear (e.g., N_2 and O_2) molecules.

PROBLEMS

5.1 Show that the mutual potential energy of the ions A^+ and B^- in the molecule A^+B^- is $\sim 14.4/R$ eV, where R is the separation (in angstroms) between A^+ and B^-. What would be the mutual potential energy of the two ions in the molecule $A^{++}B^{--}$? Assume R is larger than the atomic sizes of A and B.

5.2 The ionization potential for lithium is 5.4 eV, and the electron affinity of chlorine is 3.6 eV. Evaluate the internuclear distance beyond which LiCl becomes unstable.

5.3 Assume the potential energy curve between Na and Cl in a Na^+Cl^- molecule to be of the form

$$V(R) = \frac{C}{R^n} - \frac{e^2}{4\pi\epsilon_0 R},$$

where C and n are constants. Calculate C and n, given the internuclear distance to be 2.4 Å and the dissociation energy of NaCl to be 3.7 eV.

5.4 Show that Eq. (5.13) is the solution of the Schroedinger equation (5.12) with the boundary condition $\psi(a + b) = \psi(-a - b) = 0$.

5.5 The arrangement of a CO_2 molecule is a linear chain of the form O═C═O, where the carbon atom shares two electrons with each oxygen atom. Give an argument to show that CO_2 does not have a permanent electric dipole moment.

5.6 Calculate the energy levels of the rotational spectrum of a DCl molecule, assuming that the distance R is the same as in an HCl molecule (see Example 5.3).

5.7 The separation between the rotational levels $j = 1$ and $j = 0$ of an HI molecule results in a transition of wave number 13.10 cm^{-1}. Calculate the bond length of the molecule.

5.8 The wave numbers of the vibrational transitions occurring in the HF, HCl, and HI molecules are 4141.3 cm^{-1}, 2988.9 cm^{-1}, and 2309.5 cm^{-1}, respectively. Compare the force constants of the three molecules.

5.9 Calculate the zero point energy of an HCl molecule from the data in Problem 5.8.

5.10 The potential energy curve of a diatomic molecule is better described by the Morse potential $V(R) = D_e[1 - \exp\{-a(R - R_0)\}]^2$.

(a) Sketch the potential and show that its minimum occurs at $R = R_0$.

(b) Near $R = R_0$, the potential can be approximated by a simple harmonic potential. Show that the frequency associated with the simple harmonic potential is $\nu_0 = 2a(2\mu/D_e)^{-1/2}$, where μ is the reduced mass of the molecule.

(c) Show that the dissociation energy of the molecule is $D_e - \hbar\omega/2$, where $\omega = 2\pi\nu_0$.

5.11 The vibrational energy levels of a molecule described by the Morse potential (see Problem 5.10) are given by

$$E_v = (v + 1/2)\hbar\omega - (v + 1/2)^2\frac{\hbar^2 a^2}{\mu} \qquad (v = 0, 1, 2, \ldots).$$

(a) Plot the energy levels for $v = 0, 1, 2, 3, 4$.

(b) Obtain an expression for the frequency for the transition v to $v - 1$.

(c) The wave numbers for the transitions in an HCl molecule are as follows:

Transition	*Wave number*
$v = 1$ to $v = 0$	2782.1 cm^{-1}
$v = 2$ to $v = 1$	2678.9 cm^{-1}
$v = 3$ to $v = 2$	2576.2 cm^{-1}

Calculate D_e and the dissociation energy.

5.12 The values of the constants B and ν_0 for a KCl molecule are 1.43×10^{-5} eV and 8.4×10^{12} sec^{-1}, respectively. Find the number of rotational levels between the vibrational levels $v = 0$ and $v = 1$.

5.13 Calculate the first three lines of the R-branch of a KCl molecule. Use the data given in Problem 5.12.

6

Equilibrium Statistical Mechanics*

6.1 INTRODUCTION

Some of the major studies in physics deal with the behaviour of solids, liquids, and gases. For instance, Boyle's law and Charles' law were important developments in the description of the properties of gases. In general, such studies lead to a thermodynamical description. Let us try to understand this by considering an example. Suppose we take a flask containing oxygen at room temperature. To specify this system, we need to know the mass, volume, and temperature of oxygen. Once these variables are known, the other physical properties of the system get fixed and can be experimentally measured. The number of variables required to describe a system depends on the nature of the system. A simple system requires a few variables, whereas a complex one may need as many as ten or more variables. For example, in a complex system such as a mixture of the oxygen and nitrogen gases, we need to specify the total mass, volume, and temperature, in addition to the percentage of oxygen (or nitrogen).

> **QUESTION 6.1 For the oxygen gas in a flask, list three physical quantities which can be measured besides mass, volume, and temperature.**

In the foregoing discussion, we have assumed that the system is in a state of equilibrium, that is, the physical variables of the system are independent of time. A system can be made to achieve such a state by isolating it from the influence of its surroundings. In what follows, we shall consider only those systems that are under equilibrium conditions. A discussion on the non-equilibrium systems is beyond the scope of our study. The important point to realize is that the properties of a system (solid, liquid, or gas) can be specified by only a few variables. A detailed study of the dependence of various physical quantities of a system on these variables leads to the thermodynamic description of the system. Boyle's law and Charles' law for ideal gases are examples of such a description.

We can also examine a substance from the molecular point of view. Thus, the oxygen contained in a flask can be thought of as a collection of a large number of molecules. The exact number of molecules will depend on the mass of the gas. The number of molecules, if the mass of oxygen is 32 kg, is equal to the Avogadro number 6.023×10^{26}. If we take a litre of gas at normal temperature and pressure, the number of molecules is of the order of 10^{22}. Each molecule is moving around in the flask and occasionally colliding with the other molecules. We can study the motion of such molecules either classically, using Newton's law of motion, or quan-

*The essential results of this chapter have been summarized in Chapter 7. Therefore, a reader not interested in the derivation of these results can skip this chapter.

tum mechanically, using Schroedinger's equation. Such an attempt would mean that we would have to write of the order of 10^{22} equations, solve these equations, and then obtain the various properties of oxygen. Clearly, this is a superhuman task.

Our earlier thermodynamic discussion, however, points out that the number of variables required to specify a system is really very small, certainly much less than 10^{22}. Thus, there seems to be a contradiction in the two ways of describing the system. It turns out, however, that the thermodynamic behaviour of a system can in fact be understood from a microscopic point of view. This is the subject matter of statistical mechanics.

6.2 MICROSCOPIC AND MACROSCOPIC SYSTEMS

A microscopic system is one which has atomic dimensions (10^{-10} m). A macroscopic system, on the other hand, consists of a large number of microscopic systems ($\sim 10^{20}$ or more). An oxygen molecule is a microscopic system, whereas the oxygen gas in a flask is a macroscopic system.

We now summarize our discussion so far in terms of microscopic and macroscopic systems. When we consider a macroscopic system we can describe it:

(1) By specifying a few thermodynamic variables. These variables are called the macroscopic parameters.

(2) Classically by stating the position and velocity of each microscopic part, or quantum mechanically by stating the wave function of each microscopic part, of the macroscopic system. Such a description is called a microscopic description of the system and the variables needed are known as the microscopic variables. The number of microscopic variables required for this description is very large.

Using our earlier example of the oxygen gas in a flask, we see that the mass, volume, and temperature are the macroscopic variables, whereas the position and velocity of each molecule are the microscopic variables of the system. We shall see that a macroscopic system can be described by only a few macroscopic variables mainly because most such variables are insensitive to the microscopic variables of the system. Thus, for understanding the behaviour of a macroscopic system under different conditions, a detailed knowledge of the microscopic variables is not necessary. It is indeed fortunate that this is so, otherwise it would have been virtually impossible to understand the behaviour of many-particle systems.

EXAMPLE 6.1 List some of the macroscopic variables for describing the following systems:

(a) A mixture of the oxygen and hydrogen gases in a box.

(b) An iron bar in a magnetic field.

(a) In this system, the possibility is that water will be formed. Some of the macroscopic variables are pressure, volume, temperature, and relative proportion of oxygen, hydrogen, and water.

(b) Here, the iron bar will get magnetized. Some of the macroscopic variables are magnetic moment, temperature, and dimensions of the iron bar.

We illustrate through an example the point that a knowledge of most microscopic variables of a system is quite unnecessary for a macroscopic description of the system.

Consider a box of volume V containing N molecules of a gas. Each molecule is assumed to move in a straight line until it collides with another molecule or with the box wall. We also assume that the system is in equilibrium, that is, the macroscopic variables, e.g., pressure and density, are independent of time. We know from experience that the density of a gas under equilibrium conditions is constant. Let us see whether we can understand this from the microscopic point of view. We divide the box into two equal parts, each of volume $V/2$, and label them as the left-part and the right-part. Let n_r and n_l be the number of molecules in the right-part and left-part, respectively. Since the total number of molecules is N, we have

$$n_r + n_l = N. \tag{6.1}$$

Let us find the probability of such an occurrence, that is, the probability of having n_r molecules in the right-part and n_l molecules in the left-part. We assume that *none of the two parts is 'preferred' by the molecules.* In other words, the probability of finding a molecule in either part is equal, i.e., the probability of finding a molecule in the left-part is $\frac{1}{2}$ and that of finding it in the right-part is also $\frac{1}{2}$.

We should note that we are interested in having only n_r molecules in the right-part. It does not matter which of the N molecules form the set n_r. This is because the density depends on the number of molecules and not on any particular set of molecules. Therefore, let us choose any n_r molecules out of the N molecules. This can be done in ${}^N C_{n_r}$ ways. Of these, we take one set of n_r molecules and put it in the right-part, putting at the same time the remaining molecules in the left-part. The probability of such a distribution of molecules is

$$(\tfrac{1}{2})^{n_r} \cdot (\tfrac{1}{2})^{n_l} = (\tfrac{1}{2})^N. \tag{6.2}$$

Therefore, the probability of finding n_r molecules in the right-part is

$$p(n_r) = {}^N C_{n_r} (\tfrac{1}{2})^N. \tag{6.3}$$

Each of the $p(n_r)$ possible ways gives a microscopic description of the system. This is referred to as a *microstate* of the system. The probability of a microstate to occur is $(\frac{1}{2})^N$ and the number of microstates is ${}^N C_{n_r}$, each leading to the same macroscopic state with n_r molecules in the right-part and n_l molecules in the left-part.

Let us now consider two situations, namely, (1) when $N = 4$ and (2) when $N = 10$.

(1) $N = 4$ Using Eq. (6.3), we get

$$p(n_r = 0) = \tfrac{1}{16} = p(n_r = 4), \qquad p(n_r = 1) = \tfrac{1}{4} = p(n_r = 3), \qquad p(n_r = 2) = \tfrac{3}{8}.$$

Here, each microstate has the probability $\frac{1}{16}$. The macrostate ($n_r = 0$, $n_l = 4$) has only one microstate, the macrostate ($n_r = 1$, $n_l = 3$) has four microstates, the macrostate ($n_r = 2$, $n_l = 2$) has six microstates, and so on. The probability of observing a macrostate is given by the product of the probability of observing a microstate ($\frac{1}{16}$) and the number of microstates associated with the macrostate.

(2) $N = 10$ Again, using Eq. (6.3), we obtain

$$p(n_r = 0) = \tfrac{1}{1024} = p(n_r = 10), \qquad p(n_r = 1) = \tfrac{10}{1024} = p(n_r = 9),$$
$$p(n_r = 2) = \tfrac{45}{1024} = p(n_r = 8), \qquad p(n_r = 3) = \tfrac{120}{1024} = p(n_r = 7),$$
$$p(n_r = 4) = \tfrac{210}{1024} = p(n_r = 6), \qquad p(n_r = 5) = \tfrac{252}{1024}.$$

As can be noted, in both (1) and (2), the probability is maximum when $n_r = N/2 = n_l$. In (2), we see, in addition, that the probability falls rapidly as we move away from this value of n_r.

QUESTION 6.2 Verify the relation $\sum_{n_r=0}^{N} p(n_r) = 1$.

We now consider the situation when N is really large, say, of the order of 10^{23} or so. We then have, from Eq. (6.3),

$$p(n_r) = \frac{1}{2^N} {}^N C_{n_r} = \frac{1}{2^N} \frac{N!}{(n_r)!(N-n_r)!}. \tag{6.4}$$

In Appendix 7, we have proved that the probability $P(n_r)$, for a system having a very large number of particles, is given by

$$P(n_r) = \left(\frac{2}{\pi N}\right)^{1/2} \exp\left(-\frac{2x^2}{N}\right), \tag{6.5}$$

where $x = n_r - N/2$. From this expression, we note that

(1) the maximum of the probability function $P(n_r)$ occurs for $n_r = N/2$ and corresponds to $x = 0$, and

(2) the probability decreases as $|x|$ increases. The decrease is a very sensitive function of the number of particles, i.e., the larger the number of particles, the steeper the curve (see the example that follows).

EXAMPLE 6.2 Consider a box containing in all 10^{23} particles. What is the probability of finding in the right-half of the box (a) 5×10^{22} particles? (b) 6×10^{22} particles? Compare the two probabilities thus obtained.

(a) Here, $x = n_r - N/2 = 0$. Using Eq. (6.5), we get

$$P(5 \times 10^{22}) = \left(\frac{2}{\pi \times 10^{23}}\right)^{1/2} = \left(\frac{1}{5\pi}\right)^{1/2} \times 10^{-11}.$$

(b) In this situation, $x = n_r - N/2 = 10^{22}$. Using Eq. (6.5) once again, we obtain

$$P(6 \times 10^{22}) = \left(\frac{2}{\pi \times 10^{23}}\right)^{1/2} \times \exp(-2 \times 10^{21}).$$

The ratio of the two probabilities is

$$\frac{P(6 \times 10^{22})}{P(5 \times 10^{22})} = \exp(-2 \times 10^{21}) \sim 10^{-8.6 \times 10^{20}}.$$

As can be noted, the probability of finding $n_r = 5 \times 10^{22}$ is *small*, and that of finding $n_r = 6 \times 10^{22}$ is comparatively *exceedingly small*.

The usual way of describing the steepness of a curve is to specify the full width of the curve at half its maximum value which is known as *full width at half maximum* (FWHM). This is shown in Fig. 6.1. To calculate the FWHM, we have to find the value of x for which the probability becomes half the maximum value, $P(N/2)$. Thus,

$$P(n_r = N/2 + x) = \tfrac{1}{2} P(N/2). \tag{6.6}$$

From Eq. (6.5), we get, after cancelling $[2/(\pi N)]^{1/2}$ from both sides,

$$\exp\left(-\frac{2x^2}{N}\right) = \tfrac{1}{2} \quad \text{or} \quad x = \pm\left(\frac{N}{2}\ln 2\right)^{1/2}.$$

Therefore,

$$\text{FWHM} = \left(\frac{N}{2}\ln 2\right)^{1/2} = a(N)^{1/2}, \tag{6.7}$$

where $a\,[=(\ln 2/2)^{1/2}]$ is a numerical constant of the order of 1. This FWHM is very small as compared to N, and so the curve is very steep.

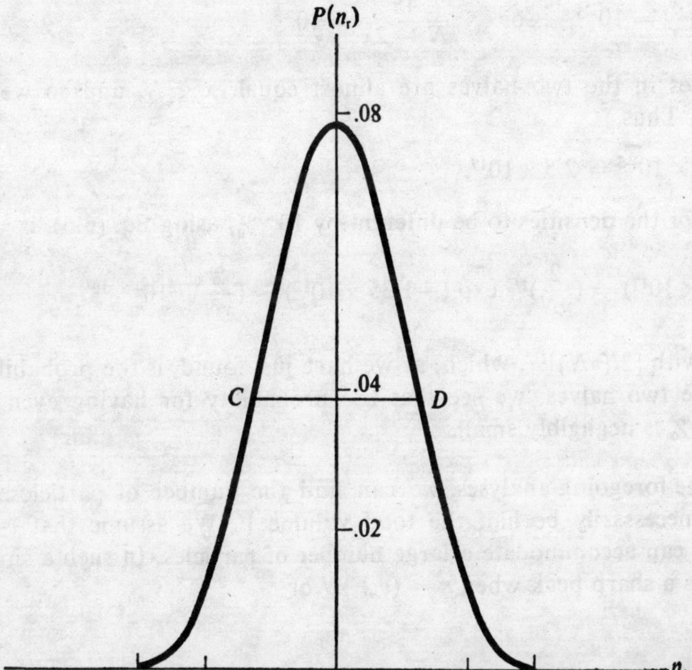

Fig. 6.1 Probability versus n_r for $N = 100$. CD is full width at half maximum (FWHM).

To get a physical picture of Eq. (6.5), let us carry out the foregoing analysis in the following way. The probability for the system to have an equal number of particles in the left-half and right-half is maximum. This corresponds to equal density in the two halves. The probability of having different densities in the two halves is negligibly small. We illustrate this through the example that follows.

EXAMPLE 6.3 Consider a system consisting of $N(=10^{23})$ particles in a box of unit volume. What is the probability for the particle density in the two halves of the box being (a) the same? (b) different by $10^{-3}\%$?

(a) Here, $n_r = n_l = N/2$. Using Eq. (6.5), we have

$$P(\frac{N}{2}) = (\frac{2}{\pi N})^{1/2}.$$

(b) Let the particle density in the right-half and the left-half be d_r and d_l, respectively. We then have

$$|1 - \frac{d_r}{d_l}| = 10^{-5} \quad \text{or} \quad |1 - \frac{n_r}{n_l}| = 10^{-5}.$$

Using $n_r = N/2 + x$, we get

$$|1 - \frac{N/2 + x}{N/2 - x}| = 10^{-5} \quad \text{or} \quad \frac{4x}{N - 2x} = 10^{-5}.$$

Since the densities in the two halves are almost equal, $x \ll N$, and so we can neglect x in comparison to N. Thus,

$$|x| = \tfrac{1}{4} \times N \times 10^{-5} = 2.5 \times 10^{17}.$$

The probability for the densities to be different by $10^{-3}\%$, using Eq. (6.5), is

$$P(x = 2.5 \times 10^{17}) = (\frac{2}{\pi N})^{1/2} \exp(-1.35 \times 10^{12}) \sim (\frac{2}{\pi N})^{1/2} 10^{-6 \times 10^{11}}$$

Comparing this with $[2/(\pi N)]^{1/2}$, which, as we have just found, is the probability for the density to be equal in the two halves, we see that the probability for having even a minute density difference of $10^{-3}\%$ is negligibly small.

Generalizing the foregoing analyses, we can find the number of particles n in a volume v, which may not necessarily be half the total volume V. We assume that v is a macroscopic volume, and so it can accommodate a large number of particles. In such a situation, the probability function has a sharp peak when $n = (v/V)N$ or

$$n/v = N/V, \tag{6.8}$$

i.e., the probability is maximum when the density is uniform. For any slight deviation from the density value N/V, the probability decreases rapidly.

In all our discussions, the only assumption we have made is that a particle has no particular preference and is equally likely to be found anywhere in the box. That is, the probability of finding a particle in a volume v is v/V. This assumption leads to the result *that the density is most likely uniform throughout the box.* The question on the location of each particle never entered our discussion. So we see that a detailed specification of the microscopic variables is unnecessary to understand the macroscopic property that the gas has a uniform density.

EXAMPLE 6.4 Consider a system of three distinguishable particles in a box of volume V. Calculate the probability of finding $n\,(=0, 1, 2, 3)$ particles in a volume $V/3$.

If n is the number of particles in the volume $V/3$, then $(3 - n)$ is the number of particles in the remaining volume. The probability of finding a given particle in the volume $V/3$ is $1/3$, and hence the probability of finding a given set of n particles in the volume $V/3$ is $(1/3)^n(2/3)^{3-n}$ since the remaining (i.e., $3 - n$) particles must not be in this volume. For the macrostate $(n, 3 - n)$, the a priori probability of occurrence of each microstate is $(1/3)^n(2/3)^{3-n}$. The number of microstates for this macrostate is ${}^3C_n$. The probability for the occurrence of a macrostate is the product of the number of microstates available to it and the a priori probability of occurrence of one such microstate. The probabilities are listed in Table 6.1. As can be noted from this

TABLE 6.1 Computation of macroscopic probability

Macrostate $(n, 3 - n)$	Number of microstates (Ω)	a priori probability per microstate (p)	Probability for macrostate $(P = p\Omega)$
(3, 0)	1	$(\frac{1}{3})^3(\frac{2}{3})^0$	$\frac{1}{27}$
(2, 1)	3	$(\frac{1}{3})^2(\frac{2}{3})^1$	$\frac{2}{9}$
(1, 2)	3	$(\frac{1}{3})^1(\frac{2}{3})^2$	$\frac{4}{9}$
(0, 3)	1	$(\frac{1}{3})^0(\frac{2}{3})^3$	$\frac{8}{27}$

table, the sum of all the probabilities of finding the different macrostates, that is, different types of distributions, is one.

6.3 MICROSCOPIC DESCRIPTION

CLASSICAL: PHASE SPACE

The microscopic state of a classical system can be conveniently described in terms of the phase space of the system. Consider a system of N particles. In order to completely specify it, we must specify, at some given time, the position and velocity of each particle. Instead of using the velocity, it is more convenient to use the momentum as a variable. There are $3N$ variables of the position (3 for each particle) and $3N$ variables of the momentum. Thus, we have to fix $6N$ quantities to completely specify the system. Let us imagine a $6N$-dimensional space in which the $3N$ coordinates $x_1, y_1, z_1, \ldots, x_N, y_N, z_N$ fix the positions of the particles, and the other $3N$ coordinates $p_{x1}, p_{y1}, p_{z1}, \ldots, p_{xN}, p_{yN}, p_{zN}$ specify the momenta of the particles. Such a $6N$-dimensional space is called the *phase space*. Of course, it is very difficult to imagine such a space when $N = 10^{23}$. Even when we have one particle, the phase space is six-dimensional. If we consider a simple situation where a single particle moves only in one dimen-

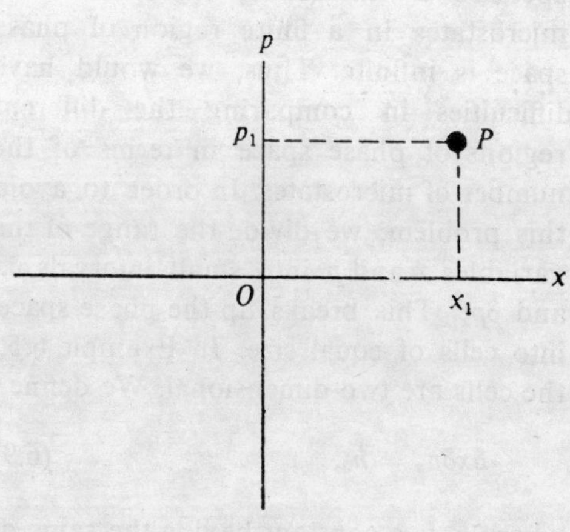

Fig. 6.2 Phase space for single particle in one dimension. Point P represents microstate.

sion, then we need only two coordinates, the position and the momentum, to specify the system. For this system, the phase space is quite simple and is as shown in Fig. 6.2. If the particle has a momentum p_1 and is at the position x_1 from the origin, then the point P (see the figure) represents the state of the system in the phase space. In general, a point in a phase space is specified by $6N$ coordinates and represents a microscopic state, i.e., microstate, of an N-particle system.

EXAMPLE 6.5 Consider a free particle inside a one-dimensional box of length L, having an energy between E_1 and E_2 $(E_2 > E_1)$. With the help of a diagram, show the allowed regions in the phase space of the system.

For one particle, the phase space is just the two-dimensional $(p_x\text{-}x)$-space. Let the box lie between $x = 0$ and $x = L$. Since the particle is assumed to be free inside the box, $p_x = (2mE)^{1/2}$, where E is the energy of the particle. If its energy has to be between E_1 and E_2, its momentum will then be restricted to

$$(2mE_1)^{1/2} \leqslant p_x \leqslant (2mE_2)^{1/2} \quad \text{or} \quad -(2mE_2)^{1/2} \leqslant p_x \leqslant -(2mE_1)^{1/2}.$$

The shaded portions in Fig. 6.3 are the allowed regions.

The foregoing example illustrates that a restriction on a macroscopic variable limits the region allowed in the phase space. Thus, confining the particle in a box and requiring the energy of the system to be between E_1 and E_2 are the two macroscopic restrictions.

Since each point in the phase space specifies a microstate, the number of microstates in a finite region of phase space is infinite. Thus, we would have difficulties in comparing the different regions of phase space in terms of the number of microstates. In order to avoid this problem, we divide the range of the variables x and p into small intervals δx and δp_x. This breaks up the phase space into cells of equal size. In Example 6.5, the cells are two-dimensional. We define

$$\delta x \delta p_x = h_0, \tag{6.9}$$

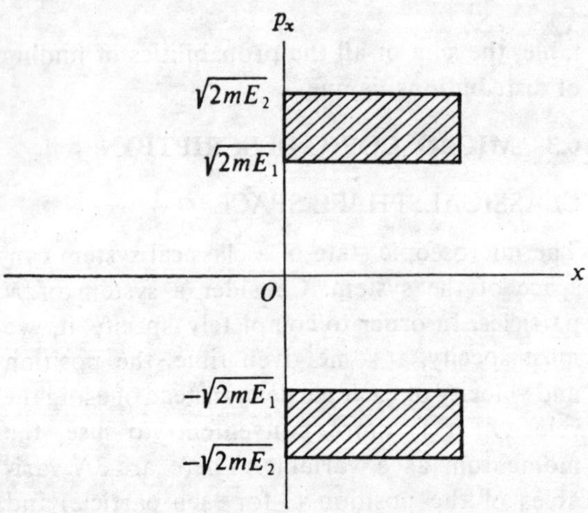

Fig. 6.3 Allowed regions of phase space for particle in one-dimensional box.

where h_0 is a constant having the same dimension as Planck's constant. A microscopic state of a system is specified by determining the cell it is in. The number of cells in a given region of phase space is the number of microstates in the region.

EXAMPLE 6.6 Calculate the number of microscopic states for Example 6.5.

The number of microstates Ω is the number of cells in the allowed region of phase space.

Thus,

$$\Omega = \frac{\text{area in phase space}}{h_0} = \frac{2[(2mE_2)^{1/2} - (2mE_1)^{1/2}]L}{h_0}.$$

If we have a particle in a cubic box, the phase space will be six-dimensional. The size of a cell in the phase space will be

$$(\delta x_1 \delta p_{x_1})(\delta y_1 \delta p_{y_1})(\delta z_1 \delta p_{z_1}) = h_0 \times h_0 \times h_0 = h_0^3. \tag{6.10}$$

We can generalize this to a system of N particles in three dimensions. The phase space in this case is $6N$-dimensional, and so the phase space is divided into $6N$-dimensional cells, each of size

$$(\delta x_1 \delta p_{x_1})(\delta y_1 \delta p_{y_1})(\delta z_1 \delta p_{z_1}) \ldots (\delta x_N \delta p_{x_N})(\delta y_N \delta p_{y_N})(\delta z_N \delta p_{z_N}) = h_0^{3N}. \tag{6.11}$$

Again, the number of cells in a given region of phase space is the number of microstates in the region.

EXAMPLE 6.7 Consider a single particle enclosed in a square of side L. Calculate the number of microstates available to this system if the total energy of the system is between U and $(U + \Delta U)$.

Let the position and momentum coordinates be x, y and p_x, p_y, respectively. Then, the energy of the system is

$$E = \frac{p_x^2 + p_y^2}{2m}, \tag{6.12}$$

where m is the mass of the particle. The position of the particle is restricted between 0 and L, that is, $0 \leqslant x \leqslant L$, $0 \leqslant y \leqslant L$.

The number of cells $\Omega(U)$ in the energy range $U \leqslant E \leqslant (U + \Delta U)$ is given by the ratio of the area of the allowed region of phase space and the area of a cell, i.e.,

$$\Omega(U) = \frac{1}{h_0^2} \iiiint\limits_{\substack{\text{region of}\\ \text{integration}=\Gamma}} dx\, dy\, dp_x\, dp_y, \tag{6.13}$$

where the region Γ is obtained by using the restrictions that the energy of the system is between U and $(U + \Delta U)$ and that the particle lies in the box, i.e.,

$$U \leqslant \frac{p_x^2 + p_y^2}{2m} \leqslant U + \Delta U \qquad (0 \leqslant x \leqslant L,\ 0 \leqslant y \leqslant L). \tag{6.14}$$

Since x, y can vary from 0 to L, we have

$$\int_0^L \int_0^L dx\, dy = L^2 = A,$$

and thus the number of microstates is

$$\Omega(U) = \frac{A}{h_0^2} \iint_{\Gamma_p} dp_x\, dp_y, \tag{6.15}$$

where Γ_p is the allowed region in the momentum space. The allowed values of p_x and p_y can be determined using

$$p_x^2 + p_y^2 = 2mU, \qquad p_x^2 + p_y^2 = 2m(U + \Delta U). \tag{6.16}$$

Equations (6.16) represent circles in the $(p_x\text{-}p_y)$-space as shown in Fig. 6.4. If we take a point inside circle 1 (excluding the circumference), then $p_x^2 + p_y^2 < 2mU$, and so the point cannot be a part of Γ_p. Similarly, a point outside circle 2 satisfies $p_x^2 + p_y^2 > 2m(U + \Delta U)$, and so the

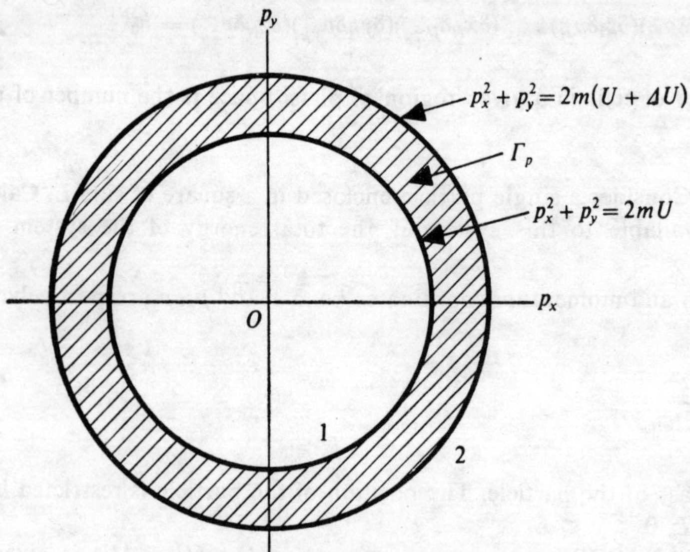

Fig. 6.4 Allowed region of integration Γ_p in momentum space.

point is excluded from Γ_p. On the other hand, any point in the shaded region (see Fig. 6.4), including the circumference of the circles, is a part of Γ_p. Let us assume that the cells are so tiny that h_0 is very small. Let us also assume that the area of the shaded region is much larger than h_0^2. This means that the number of microstates is far greater than one. In this approximation, we do not have to worry whether or not all the cells are completely inside the boundary of the shaded region because the number of cells close to the boundary is negligibly small. The number of cells in the shaded region is

$$\Omega(U) = \frac{A}{h_0^2} \iint_{\Gamma_p} dp_x\, dp_y = \frac{A(\text{area of shaded region})}{h_0^2}$$

$$= \pi A(R_2^2 - R_1^2)/h_0^2, \tag{6.17}$$

where the radii R_1 and R_2 of circles 1 and 2 are $(2mU)^{1/2}$ and $[2m(U+\Delta U)]^{1/2}$, respectively. Thus,

$$\pi(R_2^2 - R_1^2) = 2m\pi(U + \Delta U - U) = 2m\pi\Delta U. \tag{6.18}$$

Finally, the number of microstates is

$$\Omega(U) = 2m\pi A\Delta U/h_0^2. \tag{6.19}$$

The results obtained in the foregoing example can be generalized to a particle in a three-dimensional cube of side L, having an energy E lying between U and $(U+\Delta U)$. The number of cells $\Omega(U)$ in the energy range

$$U \leqslant E\left(=\frac{p_x^2+p_y^2+p_z^2}{2m}\right) \leqslant (U+\Delta U)$$

is given by the ratio of the volume of the allowed region of phase space and the volume of a cell, i.e.,

$$\Omega(U) = \frac{1}{h_0^3}\iiiint\!\!\iint_{\Gamma} dx\, dy\, dz\, dp_x\, dp_y\, dp_z. \tag{6.20}$$

The region Γ in phase space is specified by the restriction on the energy [i.e., $U \leqslant E \leqslant (U+\Delta U)$] and the restriction on the position of the particle (i.e., $0 < x, y, z < L$). Relation (6.14) is now generalized to

$$\int_0^L\int_0^L\int_0^L dx\, dy\, dz = L^3 = V, \tag{6.21}$$

where V is the volume of the cube. Thus, the number of microstates is

$$\Omega(U) = \frac{V}{h_0^3}\iiint_{\Gamma_p} dp_x\, dp_y\, dp_z. \tag{6.22}$$

The allowed region Γ_p lies between two concentric spheres of radii R_1 and R_2 given by $(2mU)^{1/2}$ and $[2m(U+\Delta U)]^{1/2}$, respectively. Thus, retaining only the first order terms in $\Delta U/U$, we get

$$\Omega(U) = \frac{V}{h_0^3}\frac{4\pi}{3}(R_2^3 - R_1^3) = \frac{4\pi V}{3h_0^3}(2m)^{3/2}[(U+\Delta U)^{3/2} - U^{3/2}] = \frac{2\pi V}{h_0^3}(2m)^{3/2}U^{1/2}\Delta U. \tag{6.23}$$

An alternative method to evaluate $\Omega(U)$ is to write the three components of momentum in Eq. (6.22) in spherical polar coordinates. Thus,

$$\Omega(U) = \frac{V}{h_0^3}\int_p^{p+\Delta p}\int_0^{\pi}\int_0^{2\pi} p^2 \sin\theta\, d\theta\, d\phi\, dp,$$

where $p^2/(2m) = U$ and $(p+\Delta p)^2/(2m) = U + \Delta U$. Evaluating this integral, we get

$$\Omega(U) = \frac{4\pi V}{h_0^3}\frac{p^3}{3}\bigg|_p^{p+\Delta p} = \frac{4\pi V}{h_0^3}\frac{(2m)^{3/2}}{3}[(U+\Delta U)^{3/2} - U^{3/2}] = \frac{2\pi V}{h_0^3}(2m)^{3/2}U^{1/2}\Delta U.$$

This is the same as Eq. (6.23).

QUANTUM MECHANICAL

Quantum mechanically, a microscopic state of a system is specified by the wave function $\psi(\boldsymbol{r}_1, \ldots, \boldsymbol{r}_N)$ of the system, where $\boldsymbol{r}_1, \ldots, \boldsymbol{r}_N$ refer to the position vector of the particles. Consider a particle confined in a cuboid of sides a, b, c. The wave function describing the system (see Problem 2.25) is

$$\psi(\boldsymbol{r}) = \left(\frac{8}{abc}\right)^{1/2} \sin\left(\frac{\pi n_x x}{a}\right) \sin\left(\frac{\pi n_y y}{b}\right) \sin\left(\frac{\pi n_z z}{c}\right), \tag{6.24}$$

where $\boldsymbol{r} = x\hat{i} + y\hat{j} + z\hat{k}$ and n_x, n_y, n_z take the integral values 1, 2, ... The energy associated with this system is

$$E = \frac{\hbar^2\pi^2}{2m}\left(\frac{n_x^2}{a^2} + \frac{n_y^2}{b^2} + \frac{n_z^2}{c^2}\right), \tag{6.25}$$

where m is the mass of the particle. A set (n_x, n_y, n_z) specifies one microstate of the system. If we now consider a system of N particles in the cuboid, all moving freely and independent of one another, then the wave function describing the system (generalizing Problem 2.27) is

$$\psi(\boldsymbol{r}_1, \ldots, \boldsymbol{r}_N) = \psi_1(\boldsymbol{r}_1) \ldots \psi_k(\boldsymbol{r}_k) \ldots \psi_n(\boldsymbol{r}_n). \tag{6.26}$$

Here, $\psi_k(\boldsymbol{r}_k)$ is the wave function of the k-th particle and is given by

$$\psi_k(\boldsymbol{r}_k) = \left(\frac{8}{abc}\right)^{1/2} \sin\left(\frac{\pi n_{x_k} x_k}{a}\right) \sin\left(\frac{\pi n_{y_k} y_k}{b}\right) \sin\left(\frac{\pi n_{z_k} z_k}{c}\right), \tag{6.27}$$

where $\boldsymbol{r}_k = x_k\hat{i} + y_k\hat{j} + z_k\hat{k}$ and $n_{x_k}, n_{y_k}, n_{z_k}$ take the integral values 1, 2, 3, ... The energy E_k of the k-th particle is

$$E_k = \frac{\hbar^2\pi^2}{2m_k}\left(\frac{n_{x_k}^2}{a^2} + \frac{n_{y_k}^2}{b^2} + \frac{n_{z_k}^2}{c^2}\right), \tag{6.28}$$

where m_k is the mass of the k-th particle. A microscopic state of the N-particle system is thus specified by $3N$ integers, namely,

$$n_{x_k}, \quad n_{y_k}, \quad n_{z_k} \qquad (k = 1, 2, \ldots, N). \tag{6.29}$$

EXAMPLE 6.8 Find the number of microstates for Example 6.7 quantum mechanically.

A particle moving in a two-dimensional box is described by the wave function

$$\psi = \frac{2}{L} \sin \frac{\pi n_x x}{L} \sin \frac{\pi n_y y}{L} \tag{6.30}$$

and its energy is given by

$$E = \frac{\hbar^2\pi^2}{2mL^2}(n_x^2 + n_y^2), \tag{6.31}$$

where the quantum numbers n_x and n_y take positive integral values. For each set (n_x, n_y), there is one microstate. The energy constraint equation reads

$$U \leqslant \frac{\hbar^2\pi^2}{2mL^2}(n_x^2 + n_y^2) \leqslant (U + \Delta U) \tag{6.32}$$

or

$$\frac{2mL^2}{\hbar^2\pi^2}U \leqslant (n_x^2 + n_y^2) \leqslant \frac{2mL^2}{\hbar^2\pi^2}(U + \Delta U). \tag{6.33}$$

The number of microstates is obtained by counting the number of sets (n_x, n_y) which satisfy inequalities (6.33). To do this, let us plot n_x on the x-axis and n_y on the y-axis (Fig. 6.5). Let us identify the points (1, 1), (1, 2), (2, 1), (2, 2), . . . , (n, m) by dots as done in the figure. Draw two circles of radii

$$R_1 = \left(\frac{2mL^2}{\hbar^2\pi^2}U\right)^{1/2}, \qquad R_2 = \left[\frac{2mL^2}{\hbar^2\pi^2}(U + \Delta U)\right]^{1/2}.$$

All the points inside circle 1 satisfy $n_x^2 + n_y^2 < R_1^2$ and all the points outside circle 2 satisfy $n_x^2 + n_y^2 > R_2^2$. We must therefore consider the set of points between the two circles which satisfy the conditions

$$R_1^2 \leqslant (n_x^2 + n_y^2) \leqslant R_2^2. \tag{6.34}$$

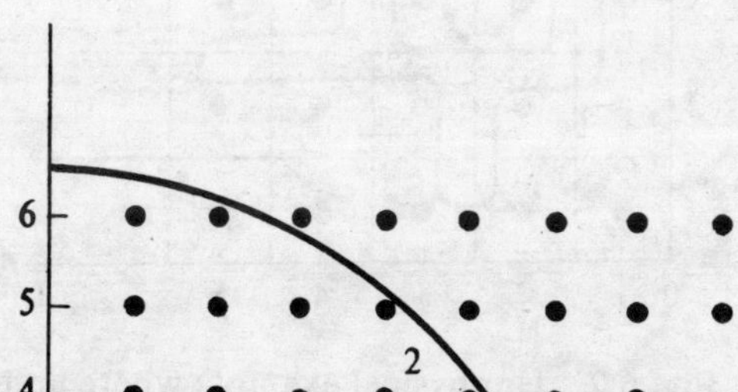

Fig. 6.5 Counting of microstates in quantum mechanical two-dimensional problem.

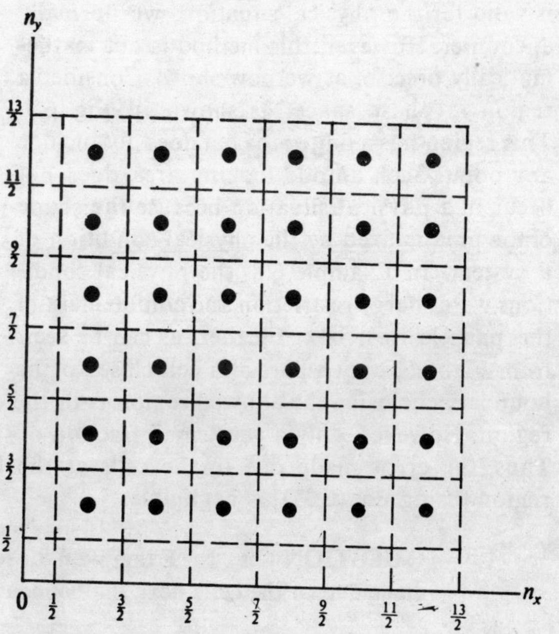

Fig. 6.6 Division of plane into cells to count number of microstates.

These inequalities are the same as those expressed by (6.33). To calculate the number of points in a given area, we divide the $(n_x$-$n_y)$-plane into cells as shown in Fig. 6.6. Each cell here has a unit area and contains one point. Therefore, the number of points in a given area A is equal to A. The number of microstates lying between U and $(U + \Delta U)$ is

$$\Omega_{QM}(U) = \frac{\pi}{4}(R_2^2 - R_1^2).$$

It should be noted that n_x and n_y take on only positive integral values. Therefore, we count the number of points in the first quadrant. It is because of this that the factor 4 appears in the denominator. Substituting for R_2^2 and R_1^2 in the foregoing equation, we get

$$\Omega_{QM}(U) = \frac{4}{\pi}\frac{2mL^2}{\hbar^2\pi^2}\Delta U = \frac{2\pi mL^2}{\hbar^2}\Delta U. \tag{6.35}$$

It should be noted that the values of n_x, n_y normally encountered are very large ($\sim 10^{30}$). This is because the energy constraints are of macroscopic dimensions, that is, of the order of joules. Thus, the number of microstates Ω is far greater than one and, as a result, the error involved in counting the points at the boundary will be negligible.

The procedure of counting the number of microstates as number of points in a given area is valid for the physical situations we normally encounter. However, this method is not mathematically precise, as we now show. Consider a region in phase space, as shown in Fig. 6.7. This region has a finite area but does not include any point. Such an odd looking area does not occur in a physical situation because the shape of the area is fixed by the physical condition of a system. In Example 6.8, the physical conditions were energy restriction and confinement of the particle in a box. Further, as can be seen from Fig. 6.5, only a part of a cell close to the boundary enclosing the allowed region is in the region. However, only a very small fraction of the total number of cells lies near the boundary. Thus, the error made due to the cells at the boundary in counting the cells in the allowed region [using Eq. (6.35)] is negligible.

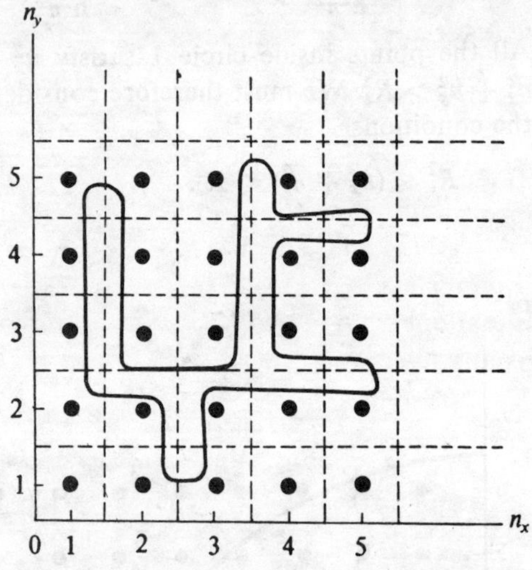

Fig. 6.7 Unphysical example where finite area does not include any point.

QUESTION 6.3 In Example 6.8, roughly estimate the fractional error made due to the cells near the boundary in counting the cells.

We note that the number of microstates obtained in Examples 6.7 and 6.8 is the same if we replace h_0 in Eq. (6.19) by Planck's constant h. This result is also valid for a three-dimensional system of a particle confined in a cube. Thus,

$$\Omega_{QM}(U) = \Omega(U) \tag{6.36}$$

if $h_0 = h$, the Planck constant.

The foregoing discussion has to be modified if we have a particle with spin s. For a particle having the wave function (6.24), the different spin states are $(2s + 1)$. The total number of microstates is then $(2s + 1)\Omega(U)$.

6.4 BASIC POSTULATE OF STATISTICAL MECHANICS

According to the basic postulate of statistical mechanics, the probability of *an isolated system in equilibrium being in any of the microscopic states available to it is equal if the microscopic states are consistent with the macroscopic description of the system.* If the system is not in equilibrium, all the microstates available to it need not then have equal likelihood of occurrence. We shall illustrate this through a simple example.

Consider a box divided into two parts by a partition. Let the left-half of the box contain some gas and the right-half be empty. Assume that the gas is under equilibrium in the left-half. Thus, all the microstates, say, Ω_i, available to the system are in the left-part. Then, according to the basic postulate, the probability of occurrence of each of the Ω_i microstates is equal.

Let the partition be now removed. At the instant it is removed, the gas molecules are in the left-half. Also, the states available get suddenly extended to include those that permit occupation of the right-half, and so the system is not in equilibrium. Thus, now the probability of occurrence of all states is not equal. After the partition has been removed, the gas expands rapidly till it occupies the entire box and the equilibrium condition is again reached. Now, the probability of occurrence of all the microscopic states becomes equal. On removing the constraint (in this case, the partition), the number of states available becomes $\Omega_f\ (>\Omega_i)$. After the equilibrium is reached, the probability of occurrence of all the microstates Ω_f is equal. In this situation, what is the probability of finding all the gas in the left-half of the box? This question is answered by the basic postulate that the probability of each microstate is equal. Therefore, the probability is

$$p_i = \Omega_i/\Omega_f. \tag{6.37}$$

EXAMPLE 6.9 Consider a box divided into two equal parts, with a removable partition. Initially, there are N particles, all in the left-half of the box. When the partition is removed, the system achieves equilibrium after some time. Calculate the probability of again finding all the N particles in the left-half of the box.

Let the number of microstates initially available to one particle be n. Since the number of microstates available to a system is directly proportional to the volume available to the system [see Eq. (6.23)], the number of available microstates will be $2n$ after the partition is removed. Thus, the probability p_1 of finding a single particle in the left-half of the box is

$$p_1 = \frac{n}{2n} = \tfrac{1}{2}.$$

The probability of finding all the particles in the left-half then is

$$p = p_1 p_2 \ldots p_N = \tfrac{1}{2}\tfrac{1}{2} \ldots \tfrac{1}{2} = (\tfrac{1}{2})^N.$$

It should be noted that normally $N \sim 10^{23}$, and the probability of finding all the particles in the left-half of the box after the partition is removed is $2^{-10^{23}}$ which is exceedingly small.

From the macroscopic or thermodynamic point of view, the removal of the partition is an irreversible process because we cannot go back to the original state without making simultaneous changes in the surrounding.

QUESTION 6.4 What simultaneous change in the surrounding would be required to restore the system to its original state, that is, to bring the entire gas back to the left-half and replace the partition?

In an irreversible process, the entropy of the system always increases. If S_i and S_f are the entropies of the initial and final states of the system, respectively, then $S_f > S_i$. From the microscopic point of view, we saw that $\Omega_f > \Omega_i$. Therefore, we expect some relation between thc entropy of the system and the number of microstates available to the system. It can be shown (see Appendix 8) that the entropy of a system is related to the number of microstates available to the system by the Boltzmann relation

$$\boxed{S = k \ln \Omega,} \tag{6.38}$$

where k is the Boltzmann constant ($=1.3 \times 10^{-23}$ J/K).

6.5 CLASSICAL STATISTICS: MAXWELL-BOLTZMANN DISTRIBUTION

We shall now use the basic ideas we have so far developed to discuss how energy is distributed among the various particles in a system when it is in equilibrium. Consider a system consisting of N identical particles, for example N molecules of the oxygen gas, in a container. Let the total energy of the system be E. We wish to determine the distribution $N(E)$, defined as the number of molecules having an energy E in a unit energy interval. In particular, the number of particles having an energy between E and $(E + dE)$ is $N(E)\, dE$.

We formulate the problem mathematically for an ideal gas. In other words, we assume that the N molecules move like the molecules in an ideal gas. We recall that the molecules in an ideal gas behave like free point particles moving independently. Let the gas be contained in a box of volume V in equilibrium at a temperature T. The total energy E is assumed to be distributed amongst the N particles as follows:

Energy	*Number of particles*
E_1	N_1
E_2	N_2
$\vdots$	$\vdots$
E_k	N_k

$E_1, E_2, \ldots, E_k$ and $N_1, N_2, \ldots, N_k$ satisfy the equations

$$N_1 + N_2 + \ldots + N_k + \ldots = \sum_k N_k = N, \tag{6.39}$$

$$N_1E_1 + N_2E_2 + \ldots + N_kE_k + \ldots = \sum_k N_kE_k = E. \tag{6.40}$$

The problem is to find $N_1, N_2, \ldots$.

According to the basic postulate of statistical mechanics, each microstate of a system in equilibrium has equal a priori probability. Thus, we have to find the total number of microstates for distributing the energy amongst the N particles under the macroscopic constraints (6.39) and (6.40). In Section 6.4, we obtained the number of microstates for a single particle with an energy E_k in a small energy interval [see Eqs. (6.23) and (6.36)]; we now denote it by g_k and write

$$g_k = \Omega(E_k) = \frac{2\pi V}{h^3}(2m)^{3/2}(E_k)^{1/2} \times [\text{energy interval}]. \tag{6.41}$$

The number of particles is N_k, each having an energy around E_k. Thus, we have to arrange these N_k particles in g_k microstates available to each of these particles.

In classical statistics, all particles, even when identical, are *distinguishable* from one another. This means that each of the N_k particles can be put in the g_k states in g_k different ways. Thus, the total number of different ways of arranging N_k particles in g_k states is

$$\underset{(N_k \text{ times})}{g_k g_k \ldots g_k} = g_k^{N_k}. \tag{6.42}$$

Each of these arrangements of N_k particles in g_k states constitutes a microstate for the N_k particles. Therefore, the number of microstates available for the N_k particles, each having an energy around E_k, is $g_k^{N_k}$ ($k = 1, 2, \ldots$). The number of microstates available for the configuration of the system having a given set of N_1 particles with energy E_1, another given set of N_2 particles with energy $E_2, \ldots$ is then the product

$$g_1^{N_1} g_2^{N_2} \ldots g_k^{N_k} = \prod_k g_k^{N_k}. \tag{6.43}$$

This is not yet the total number of microstates for the system since it is based on a given set of N_1 particles identifiable with energy E_1 and similar identifiable sets $N_2, \ldots, N_k$. However, a given partition of the N particles is not to be preferred to other possible partitions. The number of ways in which N_1 particles have energy E_1, N_2 particles have energy $E_2, \ldots$ is obtained by multiplying $\prod_k g_k^{N_k}$ by the number of ways the N particles can be divided into the groups $N_1, N_2, \ldots, N_k, \ldots$. The number of ways in which the N particles can be divided into the groups $N_1, N_2, \ldots, N_k, \ldots$ is

$$\frac{N!}{N_1! N_2! \ldots N_k! \ldots}, \tag{6.44}$$

and thus the total number of microstates for the configuration in which N_1 particles have energy E_1, N_2 particles have energy $E_2, \ldots$ is

$$\prod_k g_k^{N_k} \frac{N!}{N_1! N_2! \ldots N_k!} = \prod_k g_k^{N_k} \frac{N!}{\prod_s N_s!}. \tag{6.45}$$

There are clearly many possible values of $N_1, N_2, \ldots$ satisfying Eqs. (6.39) and (6.40). All such distributions will occur with a probability which is proportional to the number of microstates available to the system. The most probable distribution is the one which gives the maximum number of microstates. The most probable distribution obtained by maximizing Eq. (6.45) is the

Maxwell-Boltzmann distribution (see Appendix 9)

$$N(E)\,dE = \frac{2N}{(\pi)^{1/2}} \frac{E^{1/2}}{(kT)^{3/2}} \exp\left(-\frac{E}{kT}\right) dE. \tag{6.46}$$

QUESTION 6.5 At what stage in the derivation of Eq. (6.46) did we use the assumption that the gas is ideal?

6.6 QUANTUM STATISTICS: FERMI-DIRAC AND BOSE-EINSTEIN DISTRIBUTIONS

In quantum mechanics, a microscopic state of a system is specified by the wave function of the entire system. When we consider identical particles (for example, a collection of electrons), then the wave function is required to satisfy an important property. Consider two identical particles. Let $\boldsymbol{r}_1$ and $\boldsymbol{r}_2$ represent their coordinates and $\psi(\boldsymbol{r}_1, \boldsymbol{r}_2)$ be the wave function of the two-particle system. The requirement is that the wave function be symmetric or antisymmetric under the interchange of the two particles 1 and 2 (see Section 2.12). That is, either

$$\psi(\boldsymbol{r}_1, \boldsymbol{r}_2) = \psi(\boldsymbol{r}_2, \boldsymbol{r}_1) \tag{6.47}$$

or

$$\psi(\boldsymbol{r}_1, \boldsymbol{r}_2) = -\psi(\boldsymbol{r}_2, \boldsymbol{r}_1). \tag{6.48}$$

Equation (6.47) represents a symmetric wave function and Eq. (6.48) an antisymmetric wave function. If the wave function is symmetric, the particles are said to obey the *Bose-Einstein statistics*, and if it is antisymmetric, the particles are said to obey the *Fermi-Dirac statistics.* The particles obeying the Bose-Einstein statistics are called *bosons*, and those that obey the Fermi-Dirac statistics are called *fermions*.

If we have N identical bosons, the wave function $\psi(\boldsymbol{r}_1, \ldots, \boldsymbol{r}_N)$ is completely symmetric under the interchange of any two particles. This means

$$\psi(\boldsymbol{r}_1, \ldots, \boldsymbol{r}_i, \ldots, \boldsymbol{r}_j, \ldots, \boldsymbol{r}_N) = \psi(\boldsymbol{r}_1, \ldots, \boldsymbol{r}_j, \ldots, \boldsymbol{r}_i, \ldots, \boldsymbol{r}_N) \tag{6.49}$$

for all i and j from 1 to N. On the other hand, for N identical fermions, the wave function $\psi(\boldsymbol{r}_1, \ldots, \boldsymbol{r}_N)$ is completely antisymmetric under the interchange of any two particles. This means

$$\psi(\boldsymbol{r}_1, \ldots, \boldsymbol{r}_i, \ldots, \boldsymbol{r}_j, \ldots, \boldsymbol{r}_N) = -\psi(\boldsymbol{r}_1, \ldots, \boldsymbol{r}_j, \ldots, \boldsymbol{r}_i, \ldots, \boldsymbol{r}_N) \tag{6.50}$$

for all i and j taking values from 1 to N.

Let us consider two identical particles in a three-dimensional cubical box. The general wave function of this system is obtained by generalizing Example 2.23 [see Eq. (2.137)] and is

$$\psi(\boldsymbol{r}_1, \boldsymbol{r}_2) = A\psi_{\{n_1\}}(\boldsymbol{r}_1)\psi_{\{n_2\}}(\boldsymbol{r}_2) + B\psi_{\{n_2\}}(\boldsymbol{r}_1)\psi_{\{n_1\}}(\boldsymbol{r}_2). \tag{6.51}$$

Here, $\{n_1\}$ and $\{n_2\}$ represent the sets of quantum numbers (n_{1x}, n_{1y}, n_{1z}) and (n_{2x}, n_{2y}, n_{2z}), respectively. The wave function satisfies the condition

$$\psi(\boldsymbol{r}_1, \boldsymbol{r}_2) = \psi(\boldsymbol{r}_2, \boldsymbol{r}_1) \tag{6.52}$$

for bosons and

$$\psi(\boldsymbol{r}_1, \boldsymbol{r}_2) = -\psi(\boldsymbol{r}_2, \boldsymbol{r}_1) \tag{6.53}$$

for fermions. Thus, if the particles are bosons, the normalized wave function is

$$\psi(\boldsymbol{r}_1, \boldsymbol{r}_2) = a[\psi_{\{n_1\}}(\boldsymbol{r}_1)\psi_{\{n_2\}}(\boldsymbol{r}_2) + \psi_{\{n_2\}}(\boldsymbol{r}_1)\psi_{\{n_1\}}(\boldsymbol{r}_2)], \tag{6.54}$$

where $a = 1/2$ when $\{n_1\} = \{n_2\}$, and $a = 1/(2)^{1/2}$ when $n_1 \neq n_2$. For fermions, the normalized wave function is

$$\psi(\boldsymbol{r}_1, \boldsymbol{r}_2) = \frac{1}{(2)^{1/2}}[\psi_{\{n_1\}}(\boldsymbol{r}_1)\psi_{\{n_2\}}(\boldsymbol{r}_2) - \psi_{\{n_2\}}(\boldsymbol{r}_1)\psi_{\{n_1\}}(\boldsymbol{r}_2)]. \tag{6.55}$$

INDISTINGUISHABILITY OF IDENTICAL PARTICLES

There is one very important point which must be understood when determining the number of microstates of a quantum mechanical system. When counting the microstates available, the particles must be considered as *indistinguishable*. The wave function for bosons [i.e., Eq. (6.54)] represents two particles, one in the state n_1 and the other in the state n_2. Interchanging the two particles does not give a new wave function, and thus the two situations correspond to the same microstate. This means that the particles must be treated as indistinguishable. The same argument holds for the fermions. The fact that the particles are indistinguishable must be kept in mind when enumerating all the states. What matters is the number of particles in a given state and not a set of any specific particles in the state.

In the Fermi-Dirac statistics, *no more than one particle can be in one state.* This is because the wave function is antisymmetric under the interchange of any two particles. Therefore, when two particles occupy the same state, the wave function vanishes. This, in fact, is Pauli's principle, i.e., not all the quantum numbers of two identical fermions can be the same.

We should once again like to emphasize that in nature all particles obey either the Bose-Einstein statistics or Fermi-Dirac statistics. The reason we study the classical Maxwell-Boltzmann statistics is its simplicity and validity under certain approximations.

It has been found that a particle with an integral spin is a boson and a particle with a half-integral spin is a fermion.

EXAMPLE 6.10 Consider a two-particle system. Assume that three states a, b, and c, e.g., the first three energy levels of a particle in a one-dimensional box, are available to each particle. What are the possible states of the system when the particles are (a) bosons? (b) fermions? (c) classical?

(a) Each particle can be in any one of the three states. So, the system can be in any of the following six microstates:

State

a	b	c
××		
	××	
		××
×	×	
×		×
	×	×

(b) Each of the two particles can be in all the three states with the restriction that the two particles cannot be in the same state. Therefore, the system can be in any of the following three microstates:

State *a*	State *b*	State *c*
$\times$	$\times$	
	$\times$	$\times$
$\times$		$\times$

(c) Here, the two particles are distinguishable even though they are identical. Let us call one of them A and the other B. Then, the state

a	*b*	*c*
$\times^{A}$	$\times^{B}$	

is different from the state

a	*b*	*c*
$\times^{B}$	$\times^{A}$	

whereas, in the *quantum mechanical description, each of these states corresponds to the same state.* We then find the system can have the following nine microstates:

State *a*	State *b*	State *c*
$\times^{A}$	$\times^{B}$	
$\times^{B}$	$\times^{A}$	
$\times^{A}$		$\times^{B}$
$\times^{B}$		$\times^{A}$
	$\times^{A}$	$\times^{B}$
	$\times^{B}$	$\times^{A}$
$\times^{A}\times^{B}$		
	$\times^{A}\times^{B}$	
		$\times^{A}\times^{B}$

From this example, we have an interesting result, namely, the probability P of finding two particles in the same state divided by the probability of finding the particles in different states is

$$P = \tfrac{3}{3} = 1 \quad \text{for bosons}$$
$$= \tfrac{0}{3} = 0 \quad \text{for fermions}$$
$$= \tfrac{3}{6} = \tfrac{1}{2} \quad \text{for classical particles.}$$

This shows that the probability of two bosons occupying the same state is higher than that of the classical particles, whereas the probability of two fermions occupying the same state is zero. In other words, bosons 'attract' one another and fermions 'repel' one another.

We can now determine the distribution function for an N-particle system having a total energy E and obeying either the Bose-Einstein or Fermi-Dirac statistics. The procedure for obtaining the most probable distribution is identical with that used in the classical statistics [see Eq. (6.46)]. For the details on the method of computing the number of microstates and maximizing it, see Appendix 9. Here we give only the final results.

BOSE-EINSTEIN DISTRIBUTION

The number of particles $N(E)\,dE$ lying in the energy range E to $(E + dE)$ [see Eqs. (A9.25) and (6.41)] is

$$N(E)\,dE = \frac{CE^{1/2}\,dE}{\exp\,[\alpha + E/(kT)] - 1}, \tag{6.56}$$

where

$$C = \frac{2\pi(2m)^{3/2}V}{h^3}(2s + 1) \tag{6.57}$$

[the reason the factor $(2s + 1)$, s being the particle spin, occurs is explained following Eq. (6.36)], α is determined from the condition that the total number of particles should be N, i.e.,

$$\int_0^\infty N(E)\,dE = C\int_0^\infty \frac{E^{1/2}\,dE}{\exp\,[\alpha + E/(kT)] - 1} = N, \tag{6.58}$$

V is the volume occupied by the system, and T is the temperature at which the system is assumed to be in equilibrium. We might wonder why the upper limit of integration in Eq. (6.58) is infinity. This can be explained as follows. The maximum energy a particle can have should not exceed the total energy E of the system. However, in a practical situation, the total energy of a system is much more than kT, and so $N(E)$ is very small for an energy far greater than kT. Thus, it is a reasonable and convenient approximation to take the integration limit as infinity.

FERMI-DIRAC DISTRIBUTION

The number of particles $N(E)\,dE$ lying in the energy range E to $(E + dE)$ [see Eqs. (A9.25) and (6.41)] is

$$N(E)\,dE = \frac{CE^{1/2}\,dE}{\exp\,[\alpha + E/(kT)] + 1}, \tag{6.59}$$

where C is as given by Eq. (6.57) and α is determined from the condition

$$\int_0^\infty N(E)\,dE = \int_0^\infty \frac{CE^{1/2}\,dE}{\exp\,[\alpha + E/(kT)] + 1} = N. \tag{6.60}$$

MAXWELL-BOLTZMANN DISTRIBUTION

The number of particles $N(E)\,dE$ lying in the energy range E to $(E + dE)$ [see Eqs. (A9.25) and

(6.41)] is

$$N(E)\,dE = \frac{CE^{1/2}\,dE}{\exp\,[\alpha + E/(kT)]}, \tag{6.61}$$

where C is as given by Eq. (6.57) and α is determined from the condition

$$\int_0^\infty N(E)\,dE = N.$$

A convenient formula to represent all the three distributions together is

$$N(E)\,dE = \frac{CE^{1/2}\,dE}{\exp\,[\alpha + E/(kT)] + \delta}, \tag{6.62}$$

where

$$\begin{aligned} \delta &= +1 \quad \text{for Fermi-Dirac (F-D) distribution} \\ &= 0 \quad \text{for Maxwell-Boltzmann (M-B) distribution} \\ &= -1 \quad \text{for Bose-Einstein (B-E) distribution.} \end{aligned} \tag{6.63}$$

The likelihood of the occurrence of a distribution different from the most probable distribution, i.e., (6.62), is exceedingly small. Thus, it is a very good approximation to assume that the distribution is given by the most probable distribution. In each of the three distributions, the factor $C(E)^{1/2}$, i.e., the number of states in a unit energy interval available for a particle with energy E, is the same. The factor $1/[\exp\,\{\alpha + E/(kT)\} + \delta]$ distinguishes the three distributions, and we rewrite Eq. (6.62) as

$$N(E) = CE^{1/2}n(E),$$

where

$$n(E) = \frac{1}{\exp\,[\alpha + E/(kT)] + \delta} \tag{6.64}$$

is the probability of occupation of a single state at energy E and δ is $+1$ for F-D, 0 for M-B, and -1 for B-E statistics. We will refer to $n(E)$ as the *occupation probability* or the *distribution function*. The factor $CE^{1/2}$ is also referred to as the *density of states* and is denoted by $g(E)$. Thus,

$$N(E)\,dE = n(E)g(E)\,dE. \tag{6.65}$$

We recall that we derived (see Section 6.3) the relation $g(E) = CE^{1/2}$ for a particle with energy $E = p^2/(2m)$, i.e., for a free particle. Expression (6.65) is valid in general. It just states that the number of particles with an energy between E and $(E + dE)$ is equal to the product of the number of states in the energy interval and the occupation probability. If the particles are under the action of some common potential, then the density of states will be different from $CE^{1/2}$. In such a situation, the general expression (6.65) will still hold.

It should be mentioned that the occupation probability $n(E)$ is not quite the probability in the usual sense. It can be greater than one if the number of particles in a given state is more than one. Therefore, it is also referred to as distribution function or occupation index.

6.7 GENERAL PROPERTIES OF THE THREE DISTRIBUTIONS

MAXWELL-BOLTZMANN DISTRIBUTION

The Maxwell-Boltzmann distribution is given by

$$N(E)\,dE = CE^{1/2} \exp\left(-\alpha - \frac{E}{kT}\right) dE. \tag{6.66}$$

The distribution $N(E)$ and the occupation probability $n(E)$ are plotted in Fig. 6.8. The M-B

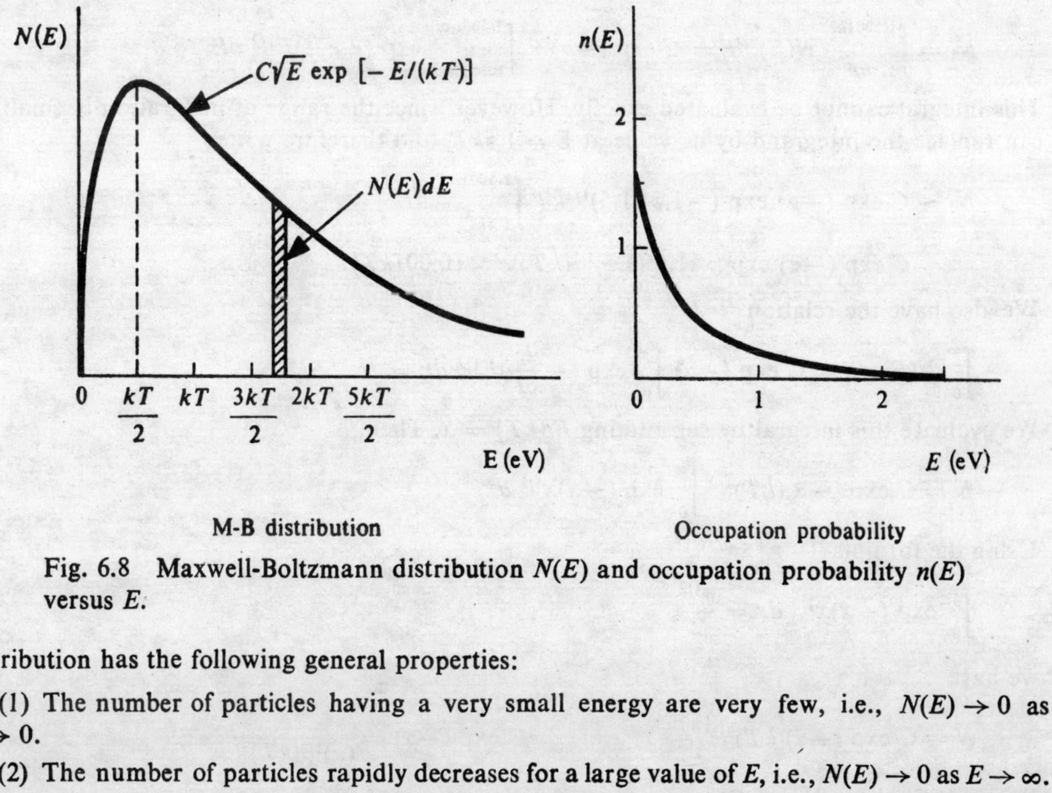

Fig. 6.8 Maxwell-Boltzmann distribution $N(E)$ and occupation probability $n(E)$ versus E.

distribution has the following general properties:

(1) The number of particles having a very small energy are very few, i.e., $N(E) \to 0$ as $E \to 0$.

(2) The number of particles rapidly decreases for a large value of E, i.e., $N(E) \to 0$ as $E \to \infty$.

(3) Most particles have an energy around kT. The maximum of the distribution is determined by

$$\frac{dN}{dE} = C \exp\left(-\alpha - \frac{E}{kT}\right)\left(\frac{1}{2}\,\frac{1}{E^{1/2}} - \frac{E^{1/2}}{kT}\right) = 0.$$

The energy at which this maximum occurs is

$$E = kT/2. \tag{6.67}$$

(4) The occupation probability has an exponential dependence on energy, i.e.,

$$n(E) = A \exp\left[-E/(kT)\right], \tag{6.68}$$

where A is a constant. The factor $\exp[-E/(kT)]$ is known as the *Boltzmann factor*.

EXAMPLE 6.11 Consider N molecules of hydrogen at room temperature (300 K). Of these, how many have an energy between $1.5kT$ and $1.5001kT$, assuming that they obey the M-B statistics?

The number of molecules with an energy E in an energy interval dE is

$$N(E)\,dE = C\exp(-\alpha)\exp(-\frac{E}{kT})E^{1/2}\,dE.$$

The number of particles N_1 in the given energy interval is obtained by integrating this equation. Thus,

$$N_1 = \int_{1.5kT}^{1.5001kT} N(E)\,dE = C\exp(-\alpha)\times\int_{1.5kT}^{1.5001kT}\exp(-\frac{E}{kT})E^{1/2}\,dE.$$

This integral cannot be evaluated exactly. However, since the range of integration is small, we can replace the integrand by its value at $E = 1.5kT$, and therefore write

$$N_1 \sim C\exp(-\alpha)\exp(-1.5)(1.5)^{1/2}kT\int_{1.5kT}^{1.5001kT} dE$$

$$\sim C\exp(-\alpha)\exp(-1.5)(1.5)^{1/2}(kT)^{1/2}\times(0.001kT).$$

We also have the relation

$$\int_0^\infty N(E)\,dE = C\exp(-\alpha)\int_0^\infty \exp(-\frac{E}{kT})E^{1/2}\,dE = N.$$

We evaluate this integral by substituting $E/(kT) = x$. Then,

$$N = C\exp(-\alpha)(kT)^{3/2}\int_0^\infty \exp(-x)x^{1/2}\,dx.$$

Using the formula

$$\int_0^\infty \exp(-x)x^{1/2}\,dx = \frac{\pi^{1/2}}{2},$$

we have

$$N = C\exp(-\alpha)(kT)^{3/2}\frac{(\pi)^{1/2}}{2} \quad \text{or} \quad C\exp(-\alpha) = \frac{2N}{(\pi)^{1/2}(kT)^{3/2}}.$$

Substituting the value of $C\exp(-\alpha)$ in the approximation for N_1, we get

$$N_1 = \frac{2N}{(\pi)^{1/2}(kT)^{3/2}}\exp(-1.5)\times(1.5)^{1/2}\times .001(kT)^{3/2}$$

$$= \frac{.002(1.5)^{1/2}\exp(-1.5)}{(\pi)^{1/2}}N = 3.07\times10^{-4}N.$$

FERMI-DIRAC DISTRIBUTION

We rewrite the occupation probability (6.64) as

$$n(E) = \frac{1}{\exp[(E-E_F)/(kT)]+1}, \tag{6.69}$$

where E_F, called the *Fermi energy*, is defined as

$$\alpha = -E_F/(kT). \tag{6.70}$$

We first consider the occupation probability in a low temperature limit. In this limit, the factor $(E - E_F)/(kT)$ is large and positive if $E > E_F$, and large but negative if $E < E_F$. Thus,

$$\begin{aligned} \exp[(E - E_F)/(kT)] &\gg 1 \quad \text{if } E > E_F \\ &\ll 1 \quad \text{if } E < E_F. \end{aligned} \tag{6.71}$$

At $E = E_F$,

$$\exp[(E - E_F)/(kT)] = 1. \tag{6.72}$$

When $E > E_F$, the factor $\exp[(E - E_F)/(kT)]$ is very much greater than 1 and the occupation probability is

$$n(E) = \frac{1}{\exp[(E - E_F)/(kT)] + 1} \sim \exp(-\frac{E - E_F}{kT}) \ll 1. \tag{6.73}$$

When $E = E_F$,

$$n(E) = \frac{1}{1+1} = \tfrac{1}{2}. \tag{6.74}$$

When $E < E_F$, we can neglect $\exp[(E - E_F)/(kT)]$ in comparison to one, and get

$$n(E) \sim 1. \tag{6.75}$$

In fact, at the lowest temperature ($T = 0$), the factor $\exp[-(E - E_F)/(kT)]$ is zero when $E > E_F$, and the factor $\exp[(E - E_F)/(kT)]$ is zero when $E < E_F$. Thus, at $T = 0$, we have

$$\begin{aligned} n(E) &= 0 \quad \text{for } E > E_F \\ &= \tfrac{1}{2} \quad \text{for } E = E_F \\ &= 1 \quad \text{for } E < E_F. \end{aligned} \tag{6.76}$$

The occupation probability takes a simple form as $T \to 0$, as shown in Fig. 6.9.

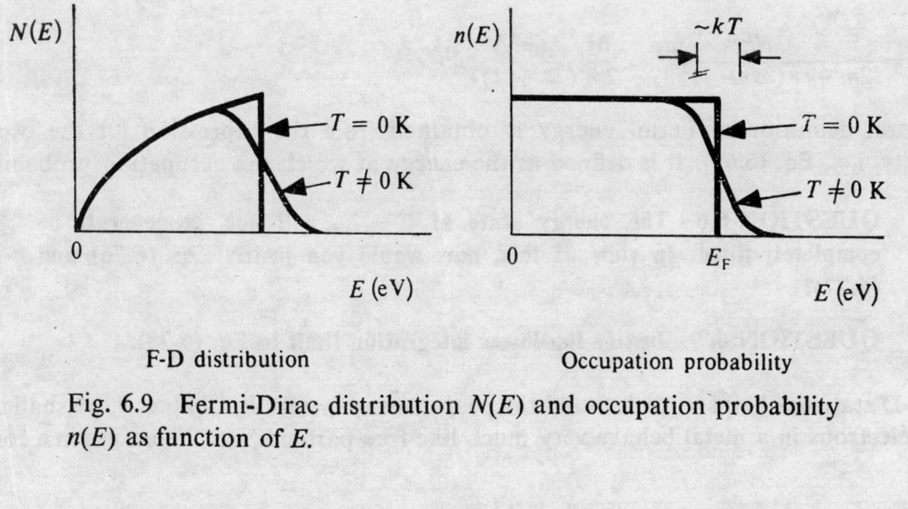

Fig. 6.9 Fermi-Dirac distribution $N(E)$ and occupation probability $n(E)$ as function of E.

The occupation probability for the F-D distribution at $T = 0$ can be understood physically if we realize that at $T = 0$ the particles tend to be in the lowest energy state. However, all the particles cannot occupy the lowest energy state because the fermions obey the Pauli exclusion principle. According to this principle, a given state can have at the most one particle. If the number of states corresponding to the lowest allowed energy (E_0) is $g(E_0)$, then we can accommodate at the most $g(E_0)$ particles in the lowest energy state. The number of states $g(E_0)$ with energy E_0 is the degeneracy of the state with energy E_0. If E_1 is the next higher allowed energy state with degeneracy $g(E_1)$, then it can accommodate $g(E_1)$ particles. The particles thus occupy the allowed energy states in the ascending order of energy. This means that, below a certain energy, all the states will be filled and, above this energy, all the states would be unoccupied. This is precisely the distribution represented by Eqs. (6.76).

The Fermi energy E_F at $T = 0$ has a definite physical significance. It is the maximum value of energy the fermions can have at $T = 0$. The Fermi energy is related to the total number of particles in a system. If a system has N particles, then

$$g(E_0) + g(E_1) + \ldots + g(E_F) = \sum_{E_i=E_0}^{E_i=E_F} g(E_i) = N. \tag{6.77}$$

For particles in a container of macroscopic dimensions, the energy levels are very closely spaced, and we can write the sum in Eq. (6.77) in an integral form as

$$\int_0^{E_F} g(E)\, dE = N. \tag{6.78}$$

From Eq. (6.41), we have

$$\int_0^{E_F} g(E) = (2s + 1)\frac{4\pi V}{h^3}(2m)^{3/2} \int_0^{E_F} E^{1/2}\, dE, \tag{6.79}$$

where s is the spin of a particle. Thus, we have

$$(2s + 1)\frac{4\pi V}{3h^3}(2m)^{3/2}(E_F)^{3/2} = N$$

or

$$E_F = \frac{1}{2m}\left[\frac{3Nh^3}{4\pi V(2s + 1)}\right]^{2/3} = \frac{\hbar^2}{2m}\left[\frac{6\pi^2 N}{(2s + 1)V}\right]^{2/3}. \tag{6.80}$$

The general definition of Fermi energy is obtained from the expression for the occupation probability, i.e., Eq. (6.69). It is defined as the energy at which the occupation probability is $\frac{1}{2}$.

> **QUESTION 6.6 The energy state at $E = E_F$ will not, in general, be completely filled. In view of this, how would you justify Eqs. (6.76) and (6.77)?**
>
> **QUESTION 6.7 Justify the lower integration limit in Eq. (6.78).**

The F-D statistics helps us understand the properties of metals. In Chapter 7, we shall discover that the electrons in a metal behave very much like free particles, i.e., like a gas in a container.

Since an electron has a half-integral spin, it obeys the Fermi-Dirac statistics. Therefore, we can treat the electrons in a metal as an ideal gas (also known as Fermi gas) obeying the Fermi-Dirac statistics.

EXAMPLE 6.12 Calculate the Fermi energy E_F at $T = 0$ for the electrons in copper (electron density $= 8.4 \times 10^{28}\ \text{m}^{-3}$).

From Eq. (6.80),

$$E_F = \frac{\hbar^2}{2m}\left(\frac{6\pi^2 N}{2V}\right)^{2/3} \qquad (s = 1/2 \text{ for an electron})$$

$$= \frac{(1.05 \times 10^{-34})^2}{2 \times 9.1 \times 10^{-31}}\left(\frac{6\pi^2 \times 8.4 \times 10^{28}}{2}\right)^{2/3} \text{J} \sim 11 \times 10^{-19}\ \text{J} \sim 6.9\ \text{eV}.$$

The Fermi energy of most metals is of the order of a few eV.

For finite temperatures, the occupation probability $n(E)$ is as shown in Fig. 6.9. For a small value of $E(\to 0)$, $(E - E_F) \sim -E_F$ and is of the order of a few eV. Thus, we can use the approximation

$$\exp\,[(E - E_F)/(kT)] \sim \exp\,[-E_F/(kT)]. \tag{6.81}$$

At room temperature, $kT \sim 0.025$ eV, and hence $E_F \gg kT$, which leads to a very small value for $\exp\,[(E - E_F)/(kT)]$. Therefore, for a small value of E, we have

$$n(E) = \frac{1}{\exp\,[(E - E_F)/(kT)] + 1} \sim 1. \tag{6.82}$$

If E is greater than E_F and $(E - E_F) \gg kT$, then $\exp\,[(E - E_F)/(kT)] \gg 1$, giving

$$n(E) = \frac{1}{\exp\,[(E - E_F)/(kT)] + 1} \sim \exp\left(-\frac{E - E_F}{kT}\right) \sim 0. \tag{6.83}$$

We see that the occupation probability tends to 1 for $(E - E_F) \ll -kT$ and to 0 for $(E - E_F) \gg kT$. Comparing this with the occupation probability [i.e., Eqs. (6.76)] at $T = 0$, we find that, for electrons with an energy different from the Fermi energy by an amount larger than kT, the occupation probability does not alter at a finite temperature. In other words, the occupation probabilities at $T = 0$ and at room temperature differ only in a small energy band of a width of a few kT around the Fermi energy E_F. This is also shown in Fig. 6.9.

The foregoing analysis of the occupation probability can also be explained in terms of the physical considerations based on the Pauli exclusion principle. As we know, when a gas is heated to a temperature T, the average energy acquired by a molecule of the gas is of the order of kT. At $T = 0$, all the energy levels up to E_F are occupied. Consider a fermion having an energy $E < E_F$. On heating, the fermion tends to absorb an energy kT, and therefore it moves to a level with an energy $(E + kT)$ provided the level is not already occupied. Thus, the chances for the fermion to absorb the energy are very small unless $(E + kT) > E_F$. In other words, only an electron lying in an energy band $(E_F - E) < kT$ can change its energy. As we shall see in Chapter 7, this provides a natural explanation for the observed variation in the specific heat of metals with temperature.

BOSE-EINSTEIN DISTRIBUTION

We can also analyze the Bose-Einstein distribution in a manner similar to that for the Maxwell-Boltzmann or Fermi-Dirac distribution. However, instead of doing this, we shall study the application of the B-E distribution to photons and obtain Planck's radiation formula. In fact, S. N. Bose discovered the now-called B-E statistics in his attempt to rederive this formula. Subsequently, Einstein realized the general nature of Bose's argument and suggested that it can be applied also to a system consisting of particles other than photons. Now, as we know, the B-E statistics is applicable to a system consisting of particles with an integral spin, i.e., bosons.

Black-Body Radiation

Consider a box containing electromagnetic radiation (i.e., photons) at a temperature T. These photons would be absorbed and re-emitted by the container walls. Under thermal equilibrium, the energy distribution can be determined by the Bose-Einstein statistics. However, there is an important difference between the photons and the other particles, namely, the photon number is not conserved. This is because the walls can absorb a photon of energy E and emit two photons of energy E_1 and E_2 such that $E_1 + E_2 = E$. Thus, we have only the energy restriction [i.e., Eq. (6.40)] and no restriction on the photon number. This means that the distribution $N(E)\,dE$, for the photons in a box at a temperature T, will (see Appendix 9) be

$$N(E)\,dE = g(E)\,dE \frac{1}{\exp{[E/(kT)]} - 1}, \tag{6.84}$$

where $g(E)\,dE$ is the number of states available to a photon between an energy E and $(E+dE)$. We note that the function here does not contain the term $\exp{(\alpha)}$. This is because, as already stated, there is no restriction on the photon number (see Appendix 9). Using the relation $E = h\nu$, we can write the distribution as

$$N(E)\,dE = g(E)\,dE \frac{1}{\exp{[h\nu/(kT)]} - 1}. \tag{6.85}$$

It should be noted that for calculating $g(E)$ we cannot use expression (6.23) or (6.41) because to derive each of these expressions we had employed the relation $p^2 = 2mE$ which is not valid for photons. For photons, we have

$$p = E/c. \tag{6.86}$$

To evaluate $\Omega_p(E)$, we again write Eq. (6.22), i.e.,

$$g(E)\,dE = \frac{2V}{h^3} \int_{F_p} dp_x\,dp_y\,dp_z. \tag{6.87}$$

The reason the factor 2 appears here is that a photon of definite momentum p can be in two possible states of polarization. This corresponds to the two transverse polarization modes of electromagnetic waves. The integral in Eq. (6.87) can be easily evaluated if we write it in terms of the spherical polar coordinates as

$$g(E)\,dE = \frac{2V}{h^3} \int_p^{p+dp} \int_0^{\pi} \int_0^{2\pi} p^2\,dp \sin\theta\,d\theta\,d\phi = \frac{2V}{h^3}\frac{4\pi}{3}[(p+dp)^3 - p^3]$$

$$= \frac{2V}{h^3}\frac{4\pi}{3}[(\frac{E+dE}{c})^3 - (\frac{E}{c})^3] = \frac{8\pi V}{h^3c^3}E^2\,dE = \frac{8\pi V}{c^3}\nu^2\,d\nu, \tag{6.88}$$

where the higher order terms are neglected. Thus, the distribution $N(E)$ is given by

$$N(E) = \frac{8\pi V}{h^3c^3} \frac{E^2}{\exp\,[h\nu/(kT)] - 1}. \tag{6.89}$$

We now define u_ν as the energy of photons per unit volume in a unit frequency range, that is,

$$u_\nu\, d\nu = \frac{EN(E)}{V}\, dE = \frac{8\pi}{h^3c^3} \frac{E^3\, dE}{\exp\,(\frac{h\nu}{kT}) - 1} = \frac{8\pi\nu^3}{c^3} \frac{h\, d\nu}{\exp\,(\frac{h\nu}{kT}) - 1}$$

or

$$u_\nu = \frac{8\pi\nu^3}{c^3} \frac{h}{\exp\,[h\nu/(kT)] - 1}. \tag{6.90}$$

The experimental results are generally expressed in terms of u_λ rather than u_ν, where $u_\lambda\, d\lambda$ is the energy density in the wavelength band λ to $(\lambda + d\lambda)$. Defining

$$u_\lambda = u_\nu \left|\frac{d\nu}{d\lambda}\right| = \frac{c}{\lambda^2} u_\nu, \tag{6.91}$$

we can write Eq. (6.90) in terms of u_λ as

$$u_\lambda = \frac{8\pi hc}{\lambda^5} \frac{1}{\exp\,[hc/(\lambda kT)] - 1}. \tag{6.92}$$

This is *Planck's law of radiation*. It gives us information about the pattern of energy density distribution. Further, it is applicable for electromagnetic radiation in equilibrium, with the walls enclosing the radiation at some temperature T. A system consisting of equilibrium radiation inside a furnace at a temperature T is an example where this law is applicable. In such a system, the spectral distribution of energy in the radiation can be analyzed by making a small hole in the furnace and examining the radiation emitted. This emitted radiation, it has been experimentally found, satisfies Planck's law. Such a system is referred to as a *black body*, and the radiation in equilibrium is known as the *black-body radiation*. The reason for calling the equilibrium radiation black-body radiation is historical, and is as follows. A body painted black and heated to some temperature T, it was discovered, emits radiation which depends only on the temperature and not on the material of the body. This is because a black body absorbs, but does not reflect, the thermal radiation incident on it. A nonblack body, on the other hand, partly reflects and partly absorbs the thermal radiation

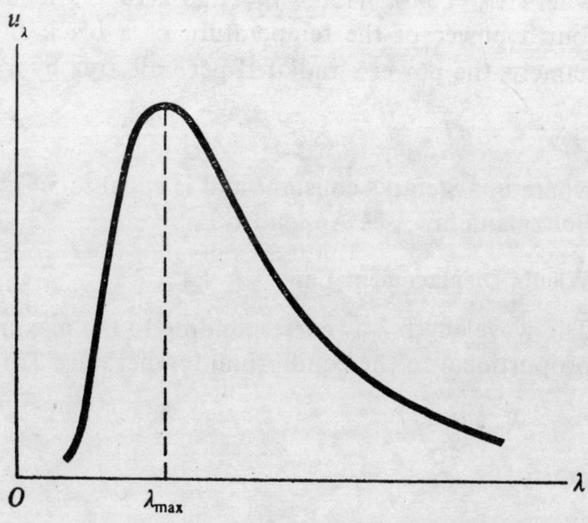

Fig. 6.10 u_λ as function of λ for black body at fixed temperature: Planck's law of radiation.

incident on it. Thus, the thermal radiation emitted by a black body is at equilibrium with the temperature of the body, but the radiation reflected by a nonblack body is not at equilibrium with the temperature of the body. A furnace with a small hole can be considered as a black body. This is because if a radiation is incident on the hole, it would be repeatedly reflected by the furnace walls. The chances of it being reflected outside the furnace through the hole are very little. Thus, all the radiation incident through the hole is effectively absorbed in the furnace. In other words, the furnace acts like a totally absorbing body, i.e., a black body. The observed wavelength distribution of the radiation from a black body is shown in Fig. 6.10. From this figure, we note that the energy density per unit wavelength u_λ first increases with the wavelength reaching a maximum at λ_{max} and then decreases for higher wavelengths.

The Stefan-Boltzmann Law

The total energy density U in electromagnetic radiation can be obtained by integrating u_λ over all wavelengths. Thus,

$$U = \int_0^\infty u_\lambda \, d\lambda = 8\pi hc \int_0^\infty \frac{d\lambda}{\lambda^5 [\exp \{hc/(\lambda kT)\} - 1]}.$$

Substituting $hc/(\lambda kT) = x$, we get, using $d\lambda = -[hc/(x^2 kT)]\, dx$,

$$U = -8\pi hc\left(\frac{kT}{hc}\right)^4 \int_\infty^0 \frac{x^3\, dx}{\exp (x) - 1} = \frac{8\pi k^4}{h^3 c^3} T^4 \int_0^\infty \frac{x^3\, dx}{\exp (x) - 1}.$$

It can be shown that the integral has the value $\pi^4/15$. Thus,

$$U = \frac{8\pi^5 k^4}{15 c^3 h^3} T^4 = U_0 T^4,$$

where $U_0 = 8\pi^5 k^4/(15 c^3 h^3) = 7.56 \times 10^{-16}$ J/m^3-K^4. The energy density is proportional to the fourth power of the temperature of a black body. This leads to the Stefan-Boltzmann law, namely, the power e radiated per unit area by a black body is proportional to T^4, that is,

$$e = \sigma T^4, \tag{6.93}$$

where σ is Stefan's constant and is equal to 5.67×10^{-8} W/m^2-K^4. For a derivation of the Stefan-Boltzmann law, see Appendix 9.

Wien's Displacement Law

The wavelength λ_{max} corresponding to the maximum of u_λ (see Fig. 6.10) is found to be inversely proportional to the equilibrium temperature T, i.e.,

$$\lambda_{max} \propto 1/T \tag{6.94}$$

or

$$\lambda_{max} T = \text{constant} = 2.89 \times 10^{-3} \text{ m-K}.$$

This is referred to as Wien's displacement law and can be derived from Planck's law of radiation (see Problem 6.21).

The Wien displacement law has been useful in determining the surface temperature of stars.

To do this, u_λ is measured as a function of λ for a given star and Eq. (6.92) is then used to fit the result so obtained. The value of T for the best fit gives the temperature of the emitting surface of the star.

Recently, an interesting use of Planck's law of radiation was made by R. H. Dicke, A. A. Penzias, and R. W. Wilson. While studying the long wavelength electromagnetic radiation impinging on the earth, they found that there was a background radiation of the same intensity from all directions. This background radiation has a distribution given by Planck's radiation law with a characteristic temperature 3 K. The presence of this radiation supports the big bang theory of the universe, which assumes that the universe started as a fireball some 20 billion years ago. This is discussed in detail in Chapter 12.

QUESTION 6.8 What is the wavelength at which the human body radiates maximum energy?

The electromagnetic radiation a human body emits can be detected by a detector. In fact, this radiation is used to discover the presence of an individual in the dark. Such a detector is known as an infrared sighter.

6.8 CLASSICAL LIMIT OF QUANTUM STATISTICS*

We have seen that classical mechanics is a very good approximation to quantum mechanics when the particles are macroscopic. In fact, we have learnt that a quantum mechanical description approaches the classical description when the quantum numbers are large. A similar result is obtained when we study the behaviour of a quantum statistical system. As we know, all particles obey either the Bose-Einstein or Fermi-Dirac statistics. We shall now see that in a high temperature, low density limit the two quantum statistics become identical with the classical statistics.

The occupation probability in quantum statistics is

$$n(E) = \frac{1}{\exp\,[\alpha + E/(kT)] \pm 1} \qquad (-\text{ for B-E, } + \text{ for F-D}), \tag{6.95}$$

whereas, in the classical statistics, it is

$$n(E) = \frac{1}{\exp\,[\alpha + E/(kT)]}.$$

Clearly, the B-E and F-D statistics go over to the M-B statistics if

$$\exp\,[\alpha + E/(kT)] \gg 1. \tag{6.96}$$

This relation should hold for all values of energy E. If $E = 0$, condition (6.96) takes the form

$$\exp \alpha \gg 1. \tag{6.97}$$

If $E > 0$, inequality (6.96) will certainly be satisfied because $\exp\,[E/(kT)] > 1$. When inequality (6.97) is satisfied, the distribution of particles as a function of energy, using Eqs. (6.63), is

*A student taking an introductory course may skip this discussion.

given by

$$N(E) = Cn(E)E^{1/2} = C \exp\left(-\alpha - \frac{E}{kT}\right)E^{1/2}. \tag{6.98}$$

The total number of particles N is then given by

$$N = \int_0^\infty N(E)\, dE = C \exp(-\alpha) \int_0^\infty \exp\left(-\frac{E}{kT}\right)E^{1/2}\, dE = C \exp(-\alpha)(kT)^{3/2}\frac{(\pi)^{1/2}}{2}, \tag{6.99}$$

where we have used the relation

$$\int_0^\infty \exp(-x)x^{1/2}\, dx = \pi^{1/2}/2.$$

Substituting in Eq. (6.99) the value of C from Eq. (6.57), we get

$$N = \frac{2\pi(2m)^{3/2}V}{h^3}(2s+1)\exp(-\alpha)(kT)^{3/2}\frac{(\pi)^{1/2}}{2}$$

or

$$\exp(\alpha) = \frac{(2\pi mkT)^{3/2}}{h^3}\frac{V}{N}(2s+1), \tag{6.100}$$

where s is the spin of a particle. Therefore, condition (6.97) now becomes

$$\frac{(2\pi mkT)^{3/2}}{h^3}\frac{V}{N}(2s+1) \gg 1. \tag{6.101}$$

This is the mathematical condition under which the F-D or B-E statistics yields the same result as the classical statistics. It has a simple physical interpretation. Since kT is approximately the average energy of a particle in a gas at a temperature T, the term $(2mkT)^{1/2}$ is approximately the average magnitude of momentum. The de Broglie wavelength associated with this momentum is

$$\lambda = \frac{h}{(2mkT)^{1/2}}, \tag{6.102}$$

and so the volume occupied by the particle (wave packet) in the quantum mechanical description is

$$\lambda^3 = \frac{h^3}{(2mkT)^{3/2}}. \tag{6.103}$$

On the other hand, (V/N) is the average volume available per particle. We expect the quantum mechanical description to go over to the classical description if the volume occupied by the wave packet associated with the particle is much smaller than the average volume available to the particle. Thus, physically, we expect the classical description to be valid when

$$\frac{V}{N} \gg \frac{h^3}{(2mkT)^{3/2}}$$

or

$$\frac{(2mkT)^{3/2}}{h^3}\frac{V}{N} \gg 1. \tag{6.104}$$

This is the same as the mathematical condition (6.101) except for the factor $\pi^{3/2}(2s+1)$ which is of the order of one. We note that condition (6.104) will be satisfied when T is large or V/N is large (i.e., density is low).

EXAMPLE 6.13 Can the classical statistics be applied to a system of (a) 10^{23} molecules of the hydrogen gas occupying a volume of 20 litres at NTP? (b) electrons in metals at room temperature (electron density $\sim 10^{29}$ m^{-3})?

We shall evaluate exp (α) for the two systems and see whether the result obtained is far greater than one.

(a) We have here

$$\frac{N}{V} = \frac{10^{23}}{20 \times 10^{-3}} = 5 \times 10^{24}/\text{m}^3.$$

The approximate de Broglie wavelength is

$$\frac{h}{(2\pi mkT)^{1/2}} = \frac{6.6 \times 10^{-34}}{(2 \times \pi \times 1.6 \times 10^{-27} \times 1.36 \times 10^{-23} \times 300)^{1/2}} = 1.03 \times 10^{-10} \text{ m}.$$

Then,

$$\exp(\alpha) = \frac{(2\pi mkT)^{3/2}}{h^3}\frac{V}{N} = \frac{1}{(1.03)^3 \times 10^{-30}}\frac{1}{5 \times 10^{24}} = 1.7 \times 10^5 \gg 1.$$

Thus, the classical statistics is applicable to the system.

(b) Here, $N/V \sim 10^{29}$, and the de Broglie wavelength is

$$\frac{h}{(2\pi mkT)^{1/2}} = \frac{6.6 \times 10^{-34}}{(2 \times \pi \times 9.1 \times 10^{-31} \times 1.36 \times 10^{-23} \times 300)^{1/2}} = 0.43 \times 10^{-8} \text{ m}.$$

Then,

$$\exp(\alpha) = \frac{(2\pi mkT)^{3/2}}{h^3}\frac{V}{N}(2s+1) = \frac{1}{(0.43)^3 \times 10^{-24} \times 10^{29}} \times 2 = 2.5 \times 10^{-4} \ll 1.$$

Therefore, we cannot apply the classical statistics to the system. In this system, the electrons have to be treated as a Fermi-Dirac gas.

6.9 THE LASER

Consider a system of N atoms in thermal equilibrium at a temperature T. Let N_1 and N_2 be the number of atoms in the energy levels E_1 and E_2, respectively. Assuming that the levels are non-degenerate, the ratio of the two numbers we have in the classical description is

$$\frac{N_2}{N_1} = \exp\left(-\frac{E_2 - E_1}{kT}\right). \tag{6.105}$$

If $E_2 > E_1$, the exponent is negative and $N_1 > N_2$. Thus, if we take a box containing the hydrogen gas at a temperature T, the number of atoms in any excited state will be less than the number of atoms in the ground state. Also, the atoms in the ground state can go over to an excited state through collisions and the atoms in an excited state can come down to the ground state through the emission of photons. In an equilibrium situation, the number of transitions upward is equal to the number of transitions downward. Thus, the system consisting of the photons and hydrogen atoms is in thermal equilibrium at the temperature T, assuming that the box holds within it both the atoms and the photons.

Let us now consider, for simplicity, only two levels of a hydrogen atom, the ground state and the first excited state. We also ignore for the time being the degeneracy of the first excited state. If $R_{1\to 2}$ and $R_{2\to 1}$ are the transition rates (i.e., the number of transitions per atom per second) from the ground state to the first excited state and from the first excited to the ground state, respectively, then, under equilibrium,

$$N_1 R_{1\to 2} = N_2 R_{2\to 1}. \tag{6.106}$$

The transition rate $R_{2\to 1}$ consists of two parts, namely, spontaneous emission and induced

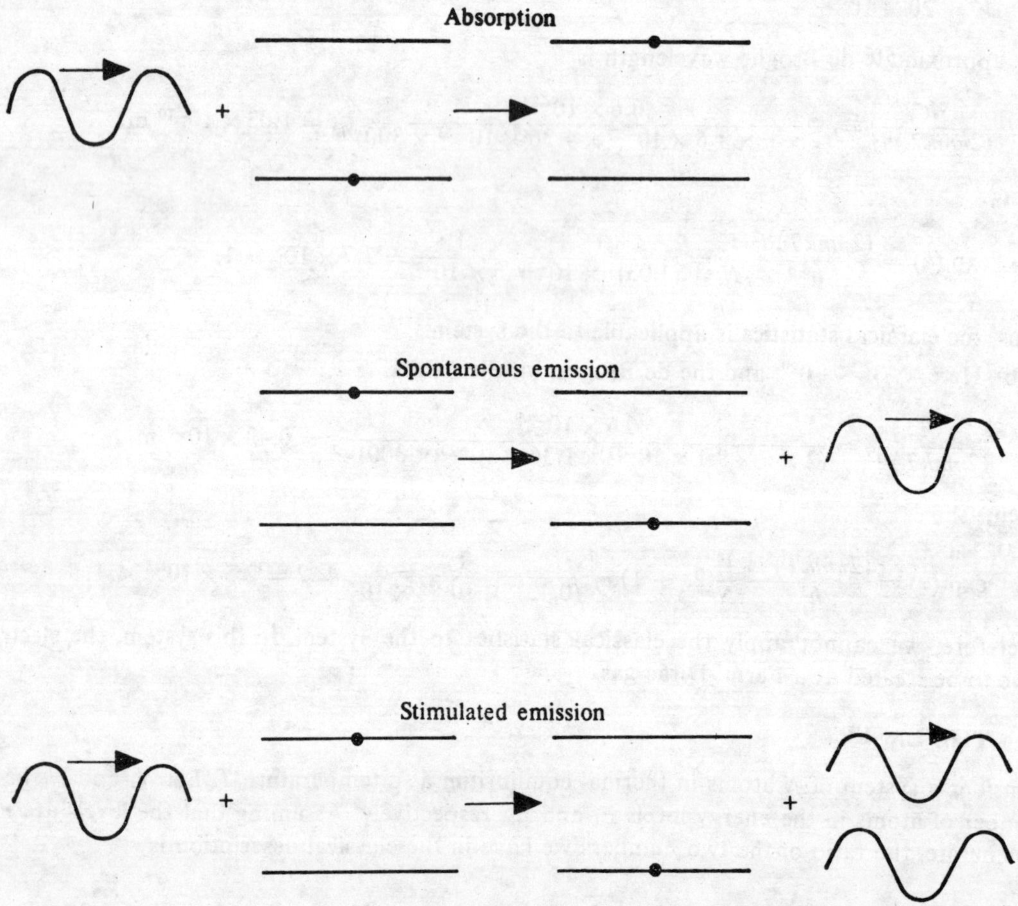

Fig. 6.11 Absorption, spontaneous emission, and stimulated emission processes.

emission. An atom in an excited state makes a transition to the ground state by itself when it is isolated. This is due to the spontaneous emission of a photon. An atom can also come to the ground state by interacting with a photon of frequency $\nu = (E_2 - E_1)/h$. This process is referred to as induced emission. Thus, an excited atom can reach the ground state through either of these two processes. Figure 6.11 shows the emission and absorption processes.

The amount of induced emission of photons depends on the number of excited atoms available. In a normal system, this is less than the number of atoms in the ground state by the Boltzmann factor, and hence the induced emission is small, whereas the absorption of the photons is large.

We now consider a system where the number of atoms in an excited state is larger than that in the ground state. If we send a beam of photons of frequency $\nu = (E_2 - E_1)/h$ through the system, then the number of photons emitted will be larger than the number of photons absorbed. Thus, the number of photons of frequency ν will increase. These photons lead to a further stimulation, causing the emission of more photons. In other words, the number of photons is amplified. This is the basic principle of the working of the laser, the acronym for *light amplification by stimulated emission of radiation*.

How do we obtain a system consisting of more atoms in an excited state than in the ground state? To answer this, let us consider a ruby laser. The relevant energy levels of a chromium atom are shown in Fig. 6.12. (A ruby crystal consists of Al_2O_3 and Cr_2O_3.) The energy level E_3

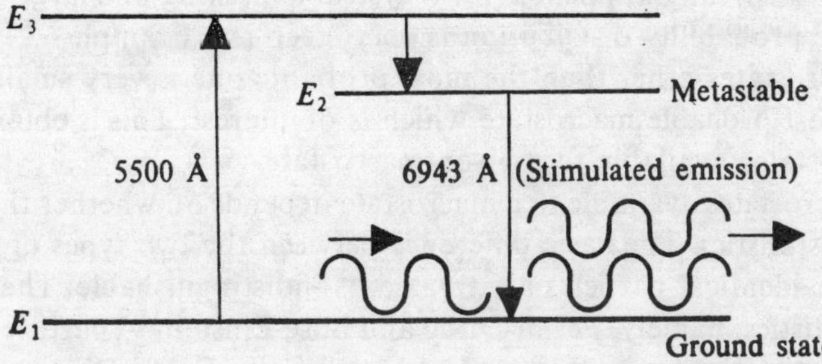

Fig. 6.12 Energy levels of ruby laser. 5500 Å is wavelength of photon with energy $E_3 - E_1$, and 6943 Å is wavelength of photon with energy $E_2 - E_1$.

is short-lived and decays into an intermediate, metastable state of energy E_2. A state, let us recall, is metastable if it slowly decays into the ground state. The lifetime of the energy level E_3 is about 10^{-8} sec, whereas that of the energy level E_2 is about 3×10^{-3} sec. Under equilibrium conditions, $N_3 < N_2 < N_1$, where N_1, N_2, N_3 are the number of atoms occupying the three energy levels. If a radiation of wavelength 5500 Å falls on the system, many atoms will absorb the radiation and get excited to the energy level E_3. Such atoms can also get de-excited to a lower energy level E_2 or E_1. Since the lifetime of the state with energy E_3 is very small, the de-excitation occurs almost immediately after the system reaches this state. The atoms that reach the energy level E_2 stay in this state for an appreciable period because it is a metastable state. Under suitable circumstances, it is possible to realize a condition of the system where the number of atoms in the first excited state is more than that in the ground state. If photons of wavelength 6943 Å (see Fig. 6.12) are incident on the system, then the transition from the state E_2 to the state E_1 causes stimulated emission, thus increasing the number of photons of wavelength 6943 Å.

A laser consists of a cylindrical rod with optically flat and partly reflecting ends. The emitted photons of wavelength 6943 Å reflect back and forth in the rod, producing more stimulated radiation. Thus, the photon build-up is rapid and the resulting radiation from the rod ends is of a large intensity. This photon beam is coherent. This is because the emitted electromagnetic radiation is in-phase with the wave stimulating the transition. The coherence and the large intensity of the outcoming radiation from a laser have played an important role in modern science and technology. For example, the laser has been used in radioastronomy, microwave spectroscopy, communications, and various industrial applications.

6.10 SUMMARY

The basic postulate of equilibrium statistical mechanics states that, for an isolated system in equilibrium, all microstates consistent with the macroscopic condition are equally probable. A microscopic state is described by a point in phase space for a classical system and by a wave function for a quantum system. For a given physical condition of a system, several macroscopic states are possible. Of these, the state with the maximum number of microstates is the most probable one. If the system consists of a small number of particles, the most probable distribution can be explicitly evaluated (see Example 6.4). However, the probabilities for the other macrostates are not small.

Statistical mechanics is really applicable for a system consisting of a large number of particles. In such a system, the probability distribution is very steep (see Example 6.2) and the probability of observing the macrostates other than the most probable one is very small. For all practical purposes, it is the most probable macrostate which is of interest. This is obtained by maximizing the number of microstates available to a given macrostate.

The number of microstates available to a macrostate depends on whether the particles obey the classical or quantum statistics. The basic difference between the two types of statistics is that, in quantum physics, the identical particles are treated as indistinguishable. There are two different kinds of quantum statistics, namely, Fermi-Dirac and Bose-Einstein. A particle with a half-integral spin (e.g., an electron, a proton, and a neutron) obeys the Fermi-Dirac statistics, whereas a particle with an integral spin (e.g., a photon and a pion) obeys the Bose-Einstein statistics. The basic difference between these two kinds of statistics is that a fermion obeys the Pauli exclusion principle, whereas there is no such restriction for a boson. The restriction due to the Pauli exclusion principle is that no two particles can have all their quantum numbers the same. In the limit of high temperature and low density, the two types of quantum statistics yield results similar to those obtained using the classical statistics. Thus, the Maxwell-Boltzmann statistics is a good approximation in the limit of high temperature and low density.

To obtain $N(E)\,dE$, i.e., the number of particles with an energy between E and $(E + dE)$, we have applied the principle of statistical mechanics. It is given by the product of the occupation probability $n(E)\,dE$ and the density of states $g(E)$.

The occupational probability $n(E)$ is expressed as

$$n(E) = \frac{1}{\exp\left[\alpha + E/(kT)\right] + \delta},$$

where δ is $+1$ for the F-D statistics, -1 for the B-E statistics, and 0 for the M-B statistics.

The density of states $g(E)$ depends on the system. For free classical particles, it is given by Eq. (6.23). For the free particles restricted to move in a box, $g(E)$ obtained quantum mechanically is the same as that evaluated classically, and is given by Eq. (6.36). In general, the density of states $g(E)$ for a quantum system is obtained from the Schroedinger equation.

PROBLEMS

6.1 Consider a box containing 5 particles a_1, a_2, a_3, a_4, and a_5. The box is divided into two equal parts, the right and the left. Write all the different ways in which we can have 3 particles in the left-part and 2 particles in the right-part of the box.

6.2 Calculate the probability of finding 0.499×10^{23} particles in the right-part of a box and 0.501×10^{23} particles in the left-part of the box (the total number of particles is 10^{23}), the volume in the right- and left-part being the same.

6.3 Consider a small volume v in a box of volume V. The box contains N identical but distinguishable particles. What is the probability that n particles are in the volume v and $(N - n)$ particles are in the remaining volume $(V - v)$? Maximize this probability and prove that $n = (v/V)N$ when the probability is maximum. Use the approximation $\ln N! = N \ln N - N$ to show that the probability distribution has a gaussian shape.

6.4 A box is separated by a partition which divides its volume in the ratio 3 : 1. The larger portion of the box contains 1000 molecules of the Ne gas and the smaller one has 100 molecules of the He gas. The partition is punctured to make a small hole and let the equilibrium be attained. Assuming that all the particles are distinguishable, find the

(a) mean number of molecules of each category on either side of the partition,

(b) probability of the larger portion having 1000 molecules of the Ne gas and the smaller portion 100 molecules of the He gas, that is, the same distribution as in the initial system.

6.5 Assume that a lecture hall contains 10^{28} air molecules and is occupied by 100 students. Let one-hundredth of the volume of the hall be available to each student. What is the probability that a given student will have no air molecules in his part of the volume?

6.6 Consider a one-dimensional harmonic oscillator of mass m whose total energy is $E = p^2/(2m) + \frac{1}{2}m\omega^2 x^2$. Find the volume in the phase space of the oscillator with the restriction that the energy lies between U and $(U + \Delta U)$.

6.7 Consider the simple harmonic oscillator in Problem 6.6. Use the basic postulate of statistical mechanics to calculate the probability $P(x)\,dx$ of finding the oscillator between x and $(x + dx)$.

6.8 Consider a system of four distinguishable particles, each with spin 1/2, at rest in a magnetic field of intensity B. Each particle has an energy $+\mu B$ (for $m_s = 1/2$) or $-\mu B$ (for $m_s = -1/2$), where μ is a constant. Calculate the number of microstates if the total energy of the system is (a) $-4\mu B$, (b) $-3\mu B$, (c) $-2\mu B$, (d) $-\mu B$, (e) 0, (f) μB, (g) $2\mu B$, (h) $3\mu B$, (i) $4\mu B$. Here, a microstate of the system can be specified in terms of the spin state of each particle.

6.9 Consider a system of N distinguishable particles, each with spin 1/2, in a magnetic field of intensity B. Each particle has two energy states, $+\mu B$ and $-\mu B$. If n is the number of particles in the spin-up (i.e., energy $= \mu B$) state:

(a) Find the energy of the system.

(b) Calculate the number of microstates in the system.

(c) Determine the entropy of the system.

[*Hint*: Use Stirling's formula $\ln N! = N \ln N - N$.]

(d) Use the relation $\partial S/\partial E = 1/T$ to show that

$$\frac{n}{N} = \frac{1}{1 + \exp[2\mu B/(kT)]}.$$

6.10 Four particles are to be distributed amongst five energy states *a, b, c, d,* and *e*. Enumerate all the possible ways for such a distribution if the particles are (a) fermions, (b) bosons, and (c) classical particles.

6.11 Find quantum mechanically the number of microstates for a one-dimensional simple harmonic oscillator having an energy between E and $(E + dE)$ $(dE \gg \hbar\omega)$.

6.12 Find quantum mechanically the number of microstates for a particle of mass m moving in a two-dimensional simple harmonic oscillator having a potential $\frac{1}{2}k(x^2 + y^2)$ and an energy between E and $(E + \Delta E)$. Assume $\hbar\omega \ll \Delta E < E$, where $\omega = (k/m)^{1/2}$.

6.13 Calculate the number of microstates for a particle having an energy between U and $(U + \Delta U)$ if the relation between the energy E and momentum p is given by $E = Cp^n$, where C and n are constants.

6.14 A vessel contains the hydrogen gas at a temperature 400 K. Use the classical statistics, not ignoring the degeneracy of the hydrogen atom energy levels, to calculate the ratio of the number of hydrogen atoms in the $(n = 2)$- and $(n = 1)$-state.

6.15 Show that the Maxwell-Boltzmann velocity distribution of classical particles, each of mass m, moving with a speed v in a system at a temperature T is

$$N(v) = \frac{4}{(\pi)^{1/2}}\left(\frac{m}{2kT}\right)^{3/2} v^2 \exp\left(-\frac{mv^2}{2kT}\right),$$

where $N(v) = N(E)\, dE/dv$.

6.16 Calculate the average energy of each particle at a temperature T for the Maxwell-Boltzmann distribution.

6.17 Consider a system of noninteracting fermions at a temperature T. Calculate the average energy of each particle in terms of the Fermi energy E_F of the system.

6.18 For the system in Problem 6.17, calculate the average velocity and average speed of each particle in terms of the Fermi energy E_F and the mass m of the particle.

6.19 Consider a cubical box of sides L. What is the lowest energy of a system consisting of 10 free particles in this box if the particles obey the (a) F-D statistics? (b) B-E statistics?

6.20 Consider an N-particle ideal Fermi gas in two dimensions. Show that the Fermi energy of this gas is given by $(N/A)[h^2/(2\pi m)]$, where A is the area to which the gas is confined and m is the mass of each particle.

6.21 Calculate the wavelength λ_{max} for which u_λ [see Eq. (6.92)] is a maximum and obtain the value of the constant in Wien's displacement law.

6.22 Given $\lambda_{max} = 5000$ Å for the radiation from the sun, calculate the surface temperature of the sun.

6.23 Assuming the radius of the sun to be 7×10^8 m and the temperature of its surface to be 6000 K, find the amount of energy radiated by the sun.

6.24 The intensity of radiation from the north star is maximum for $\lambda = 3500$ Å. Find the surface temperature of the star.

6.25 Two identical particles of mass m are moving in a one-dimensional simple harmonic oscillator having a potential $\frac{1}{2}kx^2$. The system is in its ground state. Write the normalized wave function of the system when the particles are (a) fermions, (b) bosons.

6.26 Consider a system of two independent identical particles of mass m in a one-dimensional box extending from $x = 0$ to $x = L$. The system is in its ground state. Calculate the average value of the square of the separation between the two particles if the particles obey the (a) F-D statistics, (b) B-E statistics.

6.27 Consider a system of two noninteracting identical particles of mass m confined in a two-dimensional square box of sides L. Find the wave function of the ground state of the system when the particles are (a) fermions, (b) bosons.

7

Solid State Physics*

7.1 INTRODUCTION

Matter occurs in many forms, e.g., gas, liquid, and solid. In a gas, the molecules are far apart. The behaviour of a gas can be explained by assuming that the molecules are more or less free. Such a study leads to the kinetic theory of gases. On the other hand, the molecules in a solid are closely packed and their mutual interaction determines the various properties of the solid. In a liquid, the situation is somewhere between that in a gas and in a solid. In this chapter, we shall study the properties of solids using the quantum mechanical ideas we have so far developed.

A solid occurs in two forms, namely, *crystalline* and *amorphous*. In a crystalline solid, the atoms or molecules are arranged in a regular fashion. This regular pattern repeats itself over the entire solid. Thus, a crystal has a long-range order. For example, rock salt consists of units of NaCl arranged in a face-centered cubic structure, and the chemically similar CaCl has its units arranged in a body-centered cubic structure.

A number of solids, e.g., glass and plastic, are not crystalline. These solids are called amorphous. When such a solid is heated, it gradually softens and becomes fluid. An amorphous solid has no long-range order, that is, the atoms in it are not arranged in a regular fashion over its entire structure. It is, however, possible that some ordering of the atoms exists over short distances (of the order of 50 Å) within the solid.

We shall be concerned only with crystalline solids. Some of them melt below room temperature (e.g., for H_2O, $T_{melt} = 273$ K and for Ne, $T_{melt} = 24.4$ K) and some at a very high temperature (e.g., for TaC, $T_{melt} = 4150$ K and for Be, $T_{melt} = 1556$ K). Some are metallic [e.g., copper and brass] and some are insulating [e.g., rock salt (NaCl), glass, and mica]. At a very low temperature, some metals completely lose their electrical resistance and become superconductors (e.g., Pb at $T \sim 7.2$ K). There is an equally wide range of optical, thermal, magnetic, and mechanical properties associated with solids. Of these, some are enormously sensitive to the presence of impurities or imperfections. For example, the room temperature resistivity of pure Ge decreases by a factor of 10^3 when only one part of As is added to 10^9 of Ge. We shall use the principles of quantum mechanics to understand the basic properties of solids.

We shall first discuss the metallic solids because there is a simple physical picture which helps us understand their properties. This is the free-electron theory of metals.

7.2 FREE-ELECTRON THEORY OF METALS

A metal is characterized by its ability to conduct electricity. In contrast, a nonmetal is usually a

*In Chapter 6, we derived the Fermi-Dirac and Bose-Einstein distributions. In this chapter, we shall only summarize the relevant results of Chapter 6 since their derivation is not required for understanding the content of this chapter.

nonconductor of electricity. Since the electron is the basic carrier of electricity, the electrons must be freer to move in a metal than in a nonmetal. In the free-electron theory, we assume that the electrons in a metal are completely free. However, they are restricted to be inside the metal since a piece of metal left to itself remains electrically neutral.

Our picture of a metal can be summarized as consisting of atoms arranged in a regular order, forming a crystalline solid. The electrons are free to move from one atom to another inside the metallic solid. Since the average electric field inside a conductor is zero, we expect that on the average the force acting on the electrons will be zero. In other words, the *electrons can be thought of as an ideal gas* moving freely inside the metal. Thus, the free-electron model of metals is identical with the model of noninteracting electrons confined in a box. The restriction that an electron is inside the metal means that it has a potential energy shown in Fig. 7.1. Since the potential is constant, say, $-V_0$, inside the metal, no force is acting on the electron. When the electron reaches the boundary of the metal, it cannot escape because of the sudden rise in its potential energy.

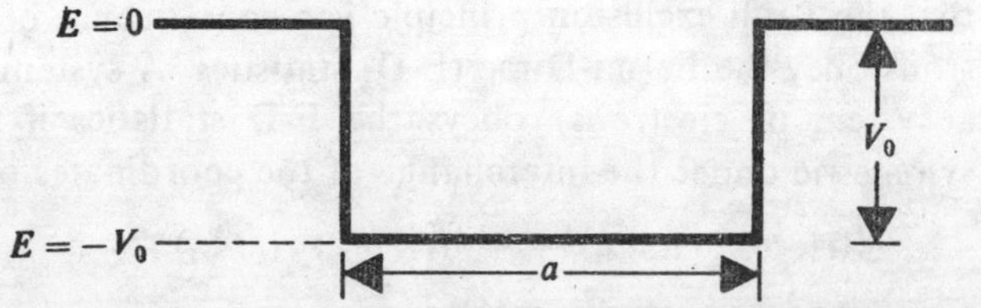

Fig. 7.1 Potential energy of electrons inside solid of dimension *a*.

QUESTION 7.1 Give a physical argument to support the fact that the potential at the boundary suddenly rises.

In Section 7.3, we shall see that only the valence electrons can be treated as free particles, each of the other electrons being bound to its atom. Thus, the free-electron theory of metals is concerned only with the valence electrons.

Let us recollect the important properties of an ideal gas. Consider N particles of an ideal gas under equilibrium at a temperature T confined to a volume V. Not all these particles will have the same speed or energy. Some particles may even momentarily be at rest and some others may have a large speed. Also, any single particle will change its energy when it collides with another particle. Thus, the particles are continuously exchanging energy with each other, even though the total energy of the N particles is constant. When the gas is in equilibrium, the distribution $N(E)\,dE$, that is, the number of particles having an energy between E and $(E+dE)$, can be defined. This distribution describes the manner in which the total energy of the system is divided amongst its various particles. A classical ideal gas obeys the Maxwell-Boltzmann distribution

$$N(E)\,dE = \frac{2N}{\pi^{1/2}(kT)^{3/2}} \exp\left(-\frac{E}{kT}\right)E^{1/2}\,dE, \tag{7.1}$$

where k is the Boltzmann constant. We can calculate several properties of an ideal gas using this distribution.

There are two important differences between an electron gas inside a metal and a classical ideal gas. These are:

(1) The electrons obey the Pauli exclusion principle, whereas the classical particles do not.

(2) The electrons are indistinguishable, but the classical particles are distinguishable.

Pauli's Exclusion Principle

In Chapter 4, we studied the Pauli exclusion principle and applied it to build up the periodic table. According to this principle, no two electrons can occupy a single quantum mechanical state. In other words, no two electrons can have all their quantum numbers the same. We recall that the Pauli exclusion principle is a consequence of a more general statement, namely, electrons obey the Fermi-Dirac (F-D) statistics. A system consisting of two identical particles (e.g., a system of electrons) obeys the F-D statistics if the wave function of the system is antisymmetric under the interchange of the coordinates of particle 1 and particle 2. That is,

$$\psi(\mathbf{r}_1, m_{s_1}; \mathbf{r}_2, m_{s_2}) = -\psi(\mathbf{r}_2, m_{s_2}; \mathbf{r}_1, m_{s_1}). \tag{7.2}$$

It should be noted that interchanging the two particles means the interchange of the position coordinates and the spin coordinates. The wave function vanishes if the two particles have the same quantum numbers. This is because the arguments $\mathbf{r}_1$, m_{s_1} and $\mathbf{r}_2$, m_{s_2} of ψ in Eq. (7.2) become equal and the wave function then is the negative of itself, implying that it is zero. The vanishing of the wave function implies that the probability of the system occupying the configuration $\mathbf{r}_1 = \mathbf{r}_2$, $m_{s_1} = m_{s_2}$ is zero. This statement is nothing but the Pauli exclusion principle, i.e., no two particles can have all their quantum numbers the same.

Electrons, protons, and neutrons are examples of the particles obeying the F-D statistics. In fact, all particles with a half-integral spin (namely, $\hbar/2$, $3\hbar/2$, ...) obey the F-D statistics. The particles obeying the F-D statistics are called the *fermions*.

QUESTION 7.2 If a particle obeys the Bose-Einstein statistics, will it also obey Pauli's exclusion principle?

Indistinguishability of Electrons

If we have two identical classical particles, they are distinguishable. We can keep track of each particle as it travels around. For example, we can find out the particle reaching a given point. On the other hand, in quantum mechanics, we can only describe a system of two identical particles by a wave function $\psi(x_1, x_2)$. The probability that one of the particles is in the interval x_1 and $(x_1 + dx_1)$ and the other is in the interval x_2 and $(x_2 + dx_2)$ is then given by the wave function to be $|\psi(x_1, x_2)|^2\, dx_1\, dx_2$. We cannot distinguish the particle in position 1 from the particle in position 2. This will become clear if we consider the wave function in Example 2.23, namely,

$$\psi(x_1, x_2) = A[\psi_{n_1}(x_1)\psi_{n_2}(x_2) \pm \psi_{n_2}(x_1)\psi_{n_1}(x_2)], \qquad + \text{ for bosons, } - \text{ for fermions,}$$

where $A = (1/2)^{1/2}$ when $n_1 \neq n_2$ and $A = 1/2$ when $n_1 = n_2$. This wave function describes the system with one of the two particles in the state n_1 and the other in n_2. We cannot, however, distinguish the particle in the state n_1 from the particle in the state n_2. In other words, identical particles are treated as indistinguishable from each other in quantum mechanics.

Since the electrons are indistinguishable and obey the Fermi-Dirac statistics, their energy distribution $N(E)$ is very different from the Maxwell-Boltzmann distribution. Consider N electrons in a box of volume V at an equilibrium temperature T. Let $N(E)\, dE$ be the number of electrons having an energy between E and $(E + dE)$. Then, if we ignore the Coulomb interaction between the electrons, we can show that

$$N(E)\,dE = \frac{4\pi(2m)^{3/2}}{h^3} V \frac{E^{1/2}\,dE}{\exp[(E - E_F)/(kT)] + 1} \tag{7.3a}$$

where E_F is the Fermi energy and k is the Boltzmann constant. We also define the occupation probability $n(E)$ as the probability that a given quantum state of energy E is occupied. It is given by

$$n(E) = \frac{1}{\exp[(E - E_F)/(kT)] + 1}. \tag{7.3b}$$

Both $N(E)$ and $n(E)$ are plotted in Fig. 7.2 for $T = 0$ K and $T \sim 300$ K (room temperature). For

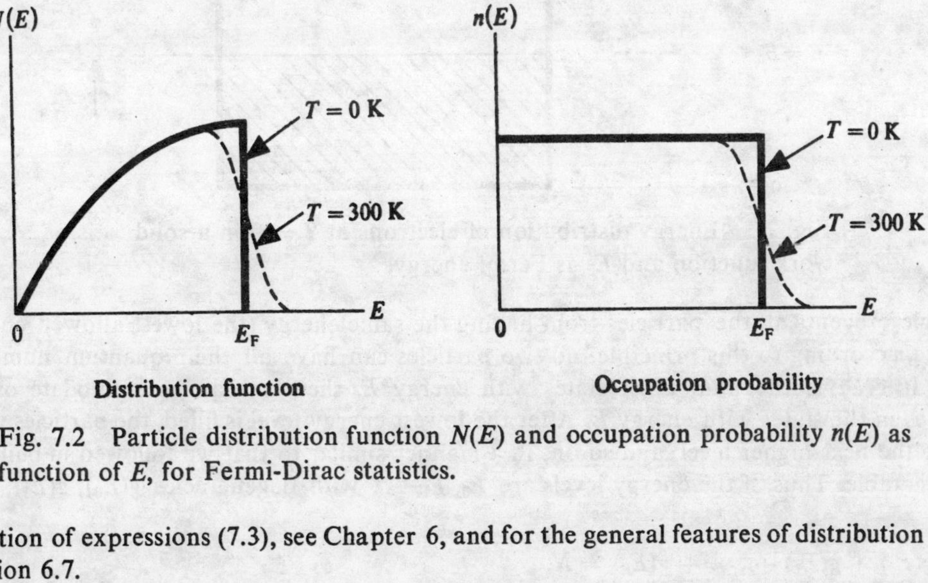

Fig. 7.2 Particle distribution function $N(E)$ and occupation probability $n(E)$ as function of E, for Fermi-Dirac statistics.

a derivation of expressions (7.3), see Chapter 6, and for the general features of distribution (7.3a), see Section 6.7.

We shall now summarize some of the important steps needed to understand the behaviour of metals.

The distribution $N(E)$ is related to the occupation probability as

$$N(E)\,dE = n(E)g(E)\,dE, \tag{7.4}$$

where $g(E)\,dE = [4\pi(8m^3E)^{1/2}V/h^3]\,dE$ is the number of energy levels lying in the energy interval between E and $(E + dE)$, with $g(E)$ being the *density of states*. In writing Eq. (7.4), we have treated the particles to be free inside the box of volume V.

At $T = 0$, the occupation probability distribution looks like a step function. This can be seen mathematically by noting that, if $E > E_F$,

$$(E - E_F)/(kT) \to \infty \quad \text{as} \quad T \to 0,$$

and hence

$$n(E) \to \frac{1}{\exp(\infty) + 1} = 0.$$

If $E < E_F$,

$$(E - E_F)/(kT) \to -\infty \quad \text{as} \quad T \to 0,$$

and hence

$$n(E) \to \frac{1}{\exp(-\infty) + 1} = 1.$$

The physical reason for the step function nature of $n(E)$ is clear if we realize that at $T = 0$ all the particles tend to have the lowest possible energy (see Fig. 7.3). However, the Pauli exclusion

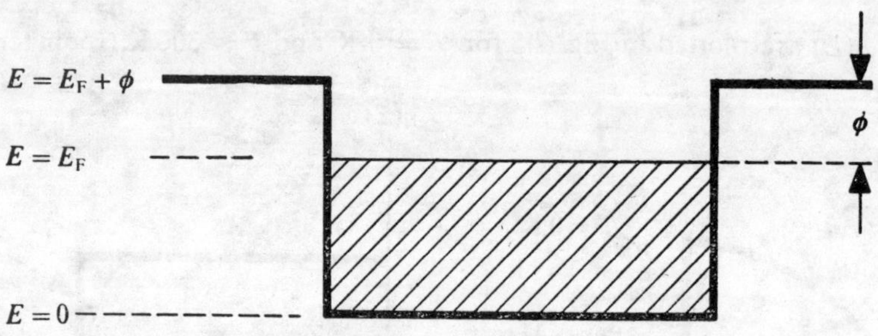

Fig. 7.3 Energy distribution of electrons at $T = 0$ in a solid. ϕ is work function and E_F is Fermi energy.

principle prevents all the particles from having the same energy (the lowest allowed energy E_0) because, according to this principle, no two particles can have all their quantum numbers the same. If $g(E)$ is the number of states with energy E, then we can accommodate only $g(E)$ particles in the states with energy E. After the lowest energy level is filled, the particles will have to take the next higher level, and so on, in a manner similar to that we followed in building the periodic table. Thus, if the energy levels are $E_0, E_1, \ldots$ with degeneracies $g(E_0), g(E_1), \ldots$, we have

$$g(E_0) + g(E_1) + \ldots + g(E_F) = N. \tag{7.5}$$

We see that at $T = 0$ all the levels below a certain energy E_F (Fermi energy) are occupied and all the levels above this energy are unoccupied. This means that the occupation probability of the levels having an energy $E < E_F$ is one, and an energy $E > E_F$ is zero. From Eq. (7.3b), we see that the occupation probability $n(E) = 1/2$ at $E = E_F$. This is the general definition of Fermi energy.

> **QUESTION 7.3 Show that ϕ (see Fig. 7.3) represents the work function of the metal.**

EXAMPLE 7.1 Calculate the Fermi energy of the electrons in a typical metal (electron density = $10^{28}/\text{m}^3$).

As already noted, the number of electrons having an energy between E and $(E + dE)$ is given by Eq. (7.3a), i.e.,

$$N(E)\,dE = \frac{4\pi(2m)^{3/2}}{h^3} V \frac{E^{1/2}\,dE}{\exp[(E - E_F)/(kT)] + 1}.$$

Also, as we know, at $T = 0$ K, $\exp[(E - E_F)/(kT)] \to 0$ for $E < E_F$ and $\to \infty$ for $E > E_F$. Thus,

$$N(E)\,dE = \frac{4\pi(2m)^{3/2}}{h^3} V E^{1/2}\,dE \quad \text{for } E < E_F$$

$$= 0 \qquad\qquad \text{for } E > E_F.$$

If N is the total number of electrons, then

$$N = \int_0^\infty N(E)\,dE = \int_0^{E_F} \frac{4\pi(2m)^{3/2}}{h^3} V E^{1/2}\,dE = \frac{4\pi(2m)^{3/2}}{h^3} V \frac{2E_F^{3/2}}{3}$$

or

$$E_F = \frac{h^2}{2m}\left(\frac{3N}{8\pi V}\right)^{2/3} = \frac{(6.6 \times 10^{-34})^2}{2 \times 9.1 \times 10^{-31}} \times \left(\frac{3}{8\pi} \times 10^{28}\right)^{2/3} \text{ J} = 1.7 \text{ eV}.$$

The distributions at finite temperatures are almost the same except in the region around the Fermi energy. The distribution at $T = 300$ K shown in Fig. 7.2 can be understood physically if we realize that the Fermi energy E_F is of the order of a few eV (see Example 7.1) and kT ($\sim 1/40$ eV at $T = 300$ K) is much smaller than the Fermi energy.

When we heat a gas from 0 K to T K, the average energy absorbed by the gas particles is $\sim kT$. A particle having an initial energy E will move to the energy state $(E + kT)$ provided this level is not occupied, the reason being that we are considering the particles obeying the Fermi-Dirac statistics. But, as we know, all the energy levels below E_F are occupied. Thus, only the electrons having an energy of the order of $(E_F - kT)$ can move to the unoccupied energy levels. In other words, the occupation probability can change appreciably only for those electrons that lie in the energy band of a width of a few kT around the energy E_F.

Specific Heat of Metals

The picture of a metal in the free-electron theory is that the atoms are fixed at definite centres and the valence electrons move freely between them. If we supply energy to a metal, say, by heating the metal, then some of the energy is absorbed by the atoms and they vibrate about their mean position, and some of the energy is taken up by the valence electrons.

The specific heat of a substance is the energy absorbed by a unit mass of the substance when the temperature increases by 1 degree. The specific heat is composed of two components, namely, specific heat of vibration and electronic specific heat. The energy absorbed by the atoms in their vibrational mode when the temperature of a solid increases by 1 degree is called the *specific heat of vibration*. The energy absorbed by the electrons for a unit increase in the temperature is called the *electronic specific heat*. The total specific heat of a solid is the sum of the specific heat of vibration and the electronic specific heat.

We shall first consider the specific heat of vibration. If we classically treat the vibrational motion of atoms in a solid, we can then show (see Appendix 10) that the average vibrational energy per atom is $3kT$, where T is the temperature of the solid. In one mole of a solid, the total number of atoms is N_0 (i.e., equal to Avogadro number), and hence the total vibrational energy of the solid is

$$E_{vib} = 3N_0kT = 3RT, \tag{7.6}$$

where R is the universal gas constant. The specific heat of one mole of the solid is then

$$C_{vib} = \frac{dE_{vib}}{dT} = 3R. \tag{7.7}$$

We now turn to the electronic specific heat. We shall use the physical arguments to obtain an approximate expression for the electronic specific heat. As we have already noted, if an electron gas is heated from 0 K to T K, most electrons that can absorb the energy lie in the energy band of width of a few kT around the energy $E = E_F$. The fraction of electrons that can absorb the energy is proportional to the ratio of the energy band of width kT to the Fermi energy. Thus, if the total number of electrons is N, then the number of electrons that can absorb the energy is

$$n \sim \frac{kT}{E_F}N. \tag{7.8}$$

The energy absorbed by the electrons when their temperature is raised from 0 to T is

$$E_{abs} \sim nkT \sim \frac{(kT)^2}{E_F}N. \tag{7.9}$$

The electronic specific heat C_{elec} is then

$$C_{elec} = \frac{dE_{abs}}{dt} \sim \frac{2kT}{E_F}Nk. \tag{7.10}$$

A precise mathematical treatment gives the result

$$C_{elec} = \frac{\pi^2}{2}z\frac{kT}{E_F}Nk, \tag{7.11}$$

where z is the number of valence electrons per atom.

The ratio of the two components of the specific heat is then

$$\frac{C_{elec}}{C_{vib}} \sim \frac{kT}{E_F}. \tag{7.12}$$

For $T \sim 300$ K, $kT \sim 10^{-2}$ eV, and hence the electronic specific heat is of the order of $(1/100)Nk$ or about 1% of the specific heat of vibration. Thus, to a good approximation, the specific heat of a metal is $3Nk$. That is, the specific heat of a metal is entirely due to the specific heat of vibration. This result has been experimentally verified and is referred to as the *Dulong-Petit law*.

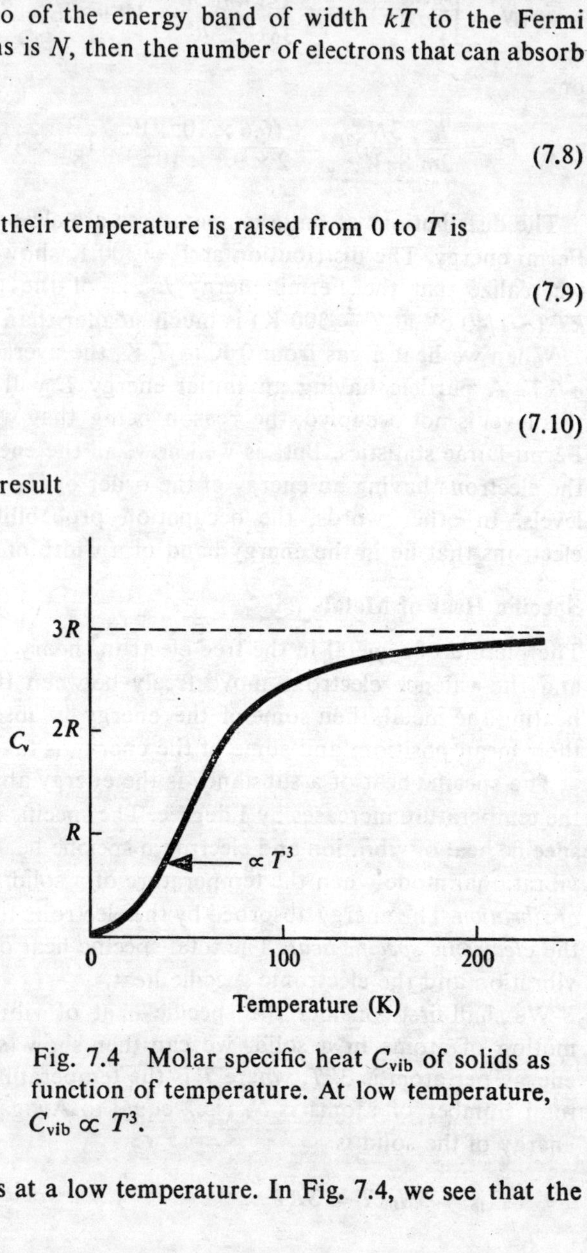

Fig. 7.4 Molar specific heat C_{vib} of solids as function of temperature. At low temperature, $C_{vib} \propto T^3$.

The Dulong-Petit law fails when a metal is at a low temperature. In Fig. 7.4, we see that the specific heat of vibration for one mole of a solid, as given by this law, is $\sim 3R$ for $T \sim 200$ K. However, for a temperature lower than 200 K, the specific heat of vibration decreases rapidly, and at a very low temperature ($\sim$20 K),

$$C_{vib} \propto T^3. \quad (7.13)$$

Such a behaviour of a metal at a low temperature can be explained in terms of quantum mechanics. The relation expressed by (7.13) is known as Debye's T^3 law.

At a high temperature, as already noted, C_{vib} dominates the electronic specific heat. At a low temperature, the two components, viz., C_{elec} and C_{vib}, become comparable (see Fig. 7.5) and the total specific heat is then given by

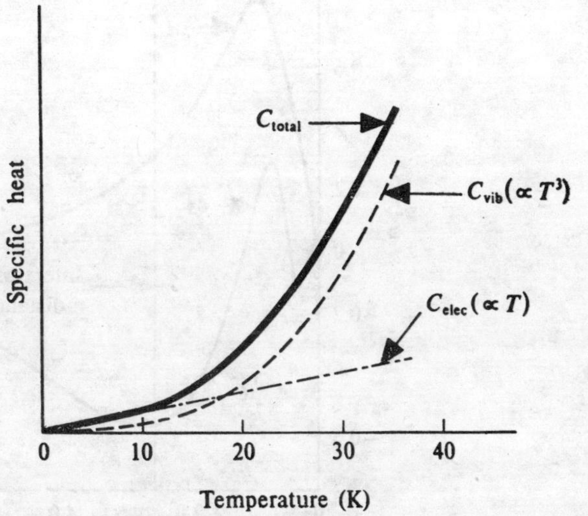

Fig. 7.5 Electronic and vibrational specific heat of solid at very low temperatures.

$$C = C_{elec} + C_{vib}$$
$$= AT + BT^3, \quad (7.14)$$

where A and B are constants.

7.3 BAND THEORY OF SOLIDS

A solid consists of a large number of atoms in close proximity to one another. For example, in a metal, the interatomic separation is about a few angstroms and the radius of an atom is also of the same order of magnitude. As we know, no definite radius is associated with the electron motion in an atom. Therefore, we can determine only the probability density and the average radius of an electron. An electron has a finite probability of extending well beyond the atomic radius. The wave function of an electron in an atom gives a better picture of the atom (see Fig. 7.6) than the Bohr model. The electrons in an atom are therefore influenced by the electrons of the neighbouring atoms. Thus, the atoms in a crystal cannot be considered as independent. The energy of a solid containing a large number of atoms is the sum of the kinetic energy of all the electrons, potential energy due to the mutual Coulomb interaction of the electrons, potential energy due to the Coulomb interaction of the electrons with the nuclei, and kinetic energy of the nuclei. The Schroedinger equation $H\psi = E\psi$ for the solid gives all the energy eigenfunctions and energy eigenvalues of the solid as a whole. However, the solution of this equation is quite complicated. Therefore, we need a simple model of the solid which is physically realistic and easier to deal with mathematically. In order to formulate such a model, we shall qualitatively examine the situation where a number of sodium atoms are brought together to form a solid. In so doing, we shall discover the reason why some solids are conductors whereas some others are insulators.

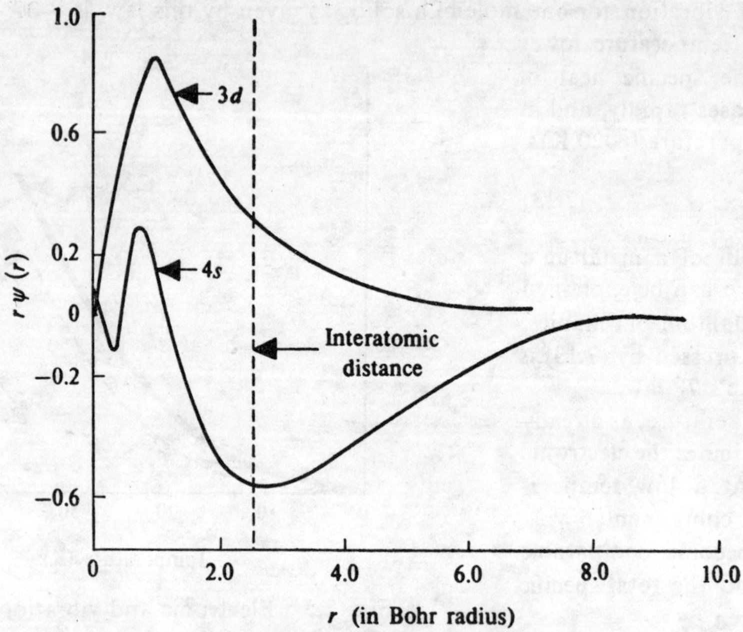

Fig. 7.6 $r\psi(r)$ of electrons in 4*s*- and 3*d*-state as function of *r*. Interatomic distance r_0 is also shown.

Metallic Sodium as Example

Consider first an isolated sodium atom. This atom has sharp discrete energy levels as shown in Fig. 7.7. The two electrons in the 1*s*-state are closest to the nucleus, the average radius r_{1s} being 0.1 Å. Therefore, these electrons experience nearly the full attractive potential of the nucleus (having $Z = 11$). Thus, 1*s* is the lowest energy level. In order of increasing energy, we have then the energy levels 2*s*, 2*p*, and 3*s*, with $r_{2s} = .32$ Å, $r_{2p} \sim .28$ Å, and $r_{3s} \sim 1.9$ Å (see Fig. 7.8). We know that the ($n = 1$)- and ($n = 2$)-shell are full and the 3*s*-state has one (valence) electron. As can be noticed, the difference in the radii, as in the energies, of the levels 1*s*, 2*s*, 2*p*, and 3*s* is large. The reason is as follows. Crudely speaking, the energy and size of the 1*s*-state correspond to those of the ($n = 1$)-state of a hydrogen-like atom with $Z = 11$. The electrons in the 2*s*-state 'see' the nucleus and the two inner electrons in the 1*s*-state, i.e., see a core with an effective charge $Z_{eff} = 9$. (A more careful treatment of the Na atom using the Schroedinger equation shows that $Z_{eff} = 9.3$.) The radius r_{2s} and the energy E_{2s} therefore correspond to those of the ($n = 2$)-state of a hydrogen atom with

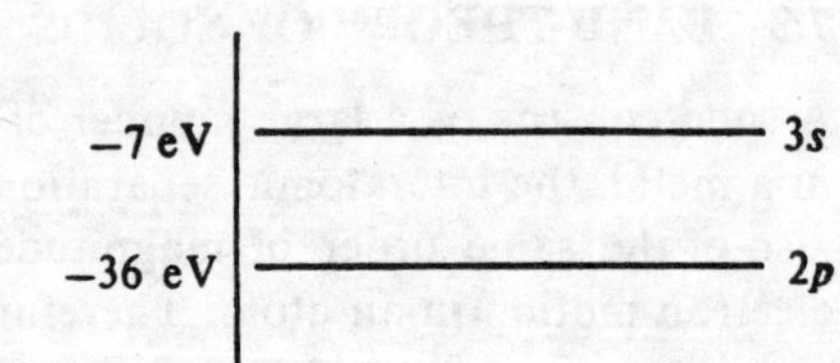

Fig. 7.7 Energy levels of sodium atom. Energy level 1*s* is very deep. Diagram not drawn to scale.

$Z_{eff} = 9.3$. Finally, the 3*s*-state is rather loosely bound, r_{3s} being approximately the radius of

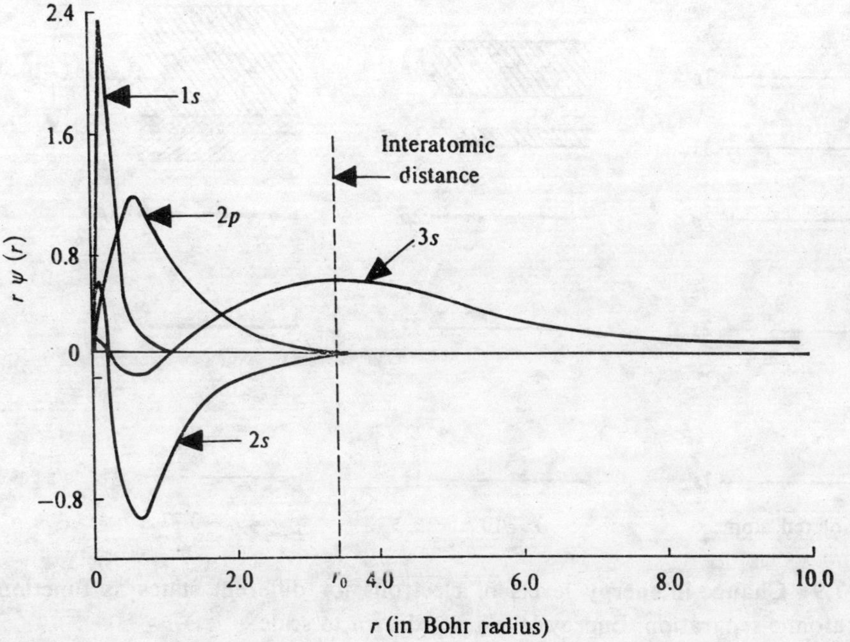

Fig. 7.8 $r\psi(r)$ of electrons in 1*s*-, 2*s*-, 2*p*-, and 3*s*-state in sodium atom. Interatomic distance r_0 is also shown.

a hydrogen atom with $n = 3$ and $Z_{eff} = 2.2$.

QUESTION 7.4 What will be the approximate value of Z_{eff} for the electrons in the 2*s*-state in potassium?

Consider now the situation where a number of sodium atoms are brought together to form a solid. When the atoms are very far apart, their effect on one another is negligible. If the number of sodium atoms is N, the system as a whole will have $2N$ states (the factor 2 being due to electron spin) with energy E_{1s}, $2N$ states with energy E_{2s}, $6N$ states with energy E_{2p}, and $2N$ states with energy E_{3s}. Accordingly, the energy levels of the solid (i.e., the N-atom system) are said to be $2N$-, $2N$-, $6N$-, and $2N$-fold degenerate. When the atoms are brought together, an electron in a Na atom will experience the potential due to the other sodium atoms. Therefore, the energy of this electron will change. Clearly, an exact calculation of the energy levels is very difficult. We can, however, obtain some idea about these levels from our earlier study of the H_2^+ ion (see Fig. 5.7). In Section 5.2, we found that the two levels corresponding to the symmetric and antisymmetric states are degenerate for a very large interatomic separation. Also, we noted that as the separation decreases, the energy difference between these two levels increases. Further, the total number of levels is two for all separations. Assuming that these general results (of Section 5.2) can be extended to the N-atom system, we can say, for example, that the $2N$-fold degenerate 3*s*-states will split into $2N$ nondegenerate levels. The maximum energy separation between the nondegenerate levels will be of the order of an eV when the interatomic separation is of the order of 10 Å (see Fig. 7.9). Since the number of levels within

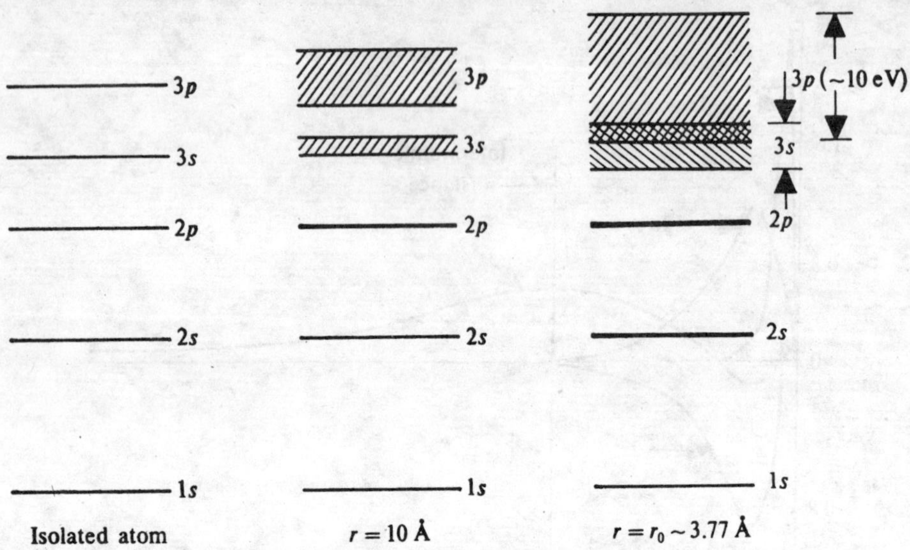

Fig. 7.9 Change in energy levels of electrons for different states as function of interatomic separation. Energy levels not drawn to scale.

this maximum separation is $2N$ ($\sim 10^{23}$), these levels will be very closely spaced and we can visualize them as forming a continuous band.

We note the following features of the energy band diagram of sodium (Fig. 7.9). For a given separation, the width of the band increases with the principal quantum number n. The $1s$-level has a very low energy. The $1s$-, $2s$-, and $2p$-level remain unaffected for $r \sim 4$ Å. The energy levels $3s$, $3p$, . . . , on the other hand, get substantially broadened (band width ~10–15 eV) for a typical interatomic separation. An electron in the $1s$-state of a given Na atom is very close to its nucleus and is farther from the other Na atoms than, say, an electron in the $2s$- or $2p$-state of the given atom. Thus, the effect of the proximity of the other atoms on the energy of the electron in the $1s$-state will be less than that on the energy of the electron in the $2s$- or $2p$-state. Let us now examine the state of affairs when the interatomic separation is r_0. As already noted, the band width of each of the states $1s$, $2s$, and $2p$ is negligibly small. The energy of the electrons in each of these states remains virtually unchanged, the reason being that each of the radii r_{1s}, r_{2s}, and r_{2p} is much less than r_0. Therefore, we can assume that the wave functions of the states $1s$, $2s$, and $2p$ are also unaffected. In an isolated Na atom, the $1s$-, $2s$-, and $2p$-state form a closed shell or an inert core. This seems to be also true in a solid. On the other hand, the $3s$-state of an atom is considerably broadened (band width ~10 eV) in the solid. Also, the band resulting from the energy level $3p$ considerably overlaps that resulting from the level $3s$, so that over an energy range the electronic state of the atom is neither pure s nor pure p. In a given Na atom, an electron in the $3s$-state is strongly influenced by the nearest Na atoms the electrons in the $3s$-states of which in turn are strongly influenced by their nearest Na atoms, and so on (ad infinitum). Thus, it is not accurate to consider the wave function of an electron in the $3s$-state as being confined to any specific atom. In view of this description of the outermost electron in sodium, it appears that a good approximation for a model of electron motion in a solid might be to altogether abandon the atomic viewpoint of the solid. Thus, we can say that the outermost electron has lost its atomic character (i.e., its location and definite angular momentum around a given atom) and moves freely from atom to atom, that is, throughout the crystal.

We are thus led to the following picture of solid Na. Each lattice site (see Fig. 7.10) has a core consisting of the Na nucleus and 10 electrons (1*s*, 2*s*, and 2*p*). The wave functions and

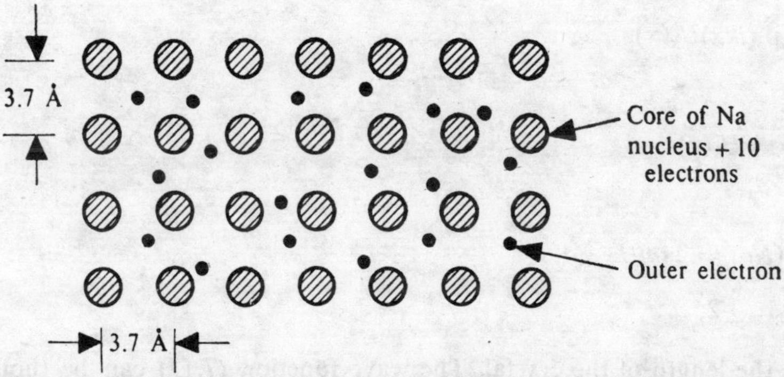

Fig. 7.10 Solid consisting of Na atoms. Each circle denotes lattice point consisting of Na nucleus and 10 electrons (1*s*, 2*s*, and 2*p*).

energies of the core electrons remain almost unchanged. Each ion core is identical and has a net charge of $+e$. Each outer electron moves in a spatially periodic potential due to the ion core lattices and the other $(N - 1)$ outer electrons. It is still difficult to obtain the energy levels in this picture since we have N outer electrons moving in the periodic potential. It turns out that, in most metals, the effect of the electron-electron interaction can be well approximated by a periodic potential representing the average effect on one outer electron of all the other outer electrons. This periodic potential is assumed to be the same for each outer electron. We thus have a model where each outer electron of a Na atom moves *independently* in the same periodic potential, this potential being due to the ion cores and the average effect of the other outer electrons.

The foregoing model of solid Na of simple, independent electrons in a periodic potential seems to be a good model for a large number of crystalline solids. The size and shape of the potential, its periodicity, depend on the crystalline solid, but the underlying features of the model are common to all crystalline solids. We shall now study the motion of an electron under the potential produced by the cores at the lattice sites, i.e., study the Schroedinger equation with a periodic potential as done by F. Bloch, and state the result in the form of a theorem known as *Bloch's theorem*. We shall confine our discussion to one dimension for simplicity. For a mathematical discussion of Bloch's theorem in one dimension, see Appendix 11. Consider a solid consisting of ion cores at the lattice sites $a, 2a, \ldots, Na$ and electrons moving independently in the potential produced by the ion cores. The potential is periodic with a period a, i.e.,

$$V(x + a) = V(x). \tag{7.15}$$

Bloch realized that it is natural to impose the boundary condition

$$\psi(x + Na) = \psi(x) \tag{7.16}$$

on the wave function of an electron. This condition is referred to as a *periodic boundary condition*. A wave function with a periodic boundary condition represents a particle anywhere in the lattice. The particle can be thought of as being bound to the crystal as a whole.

The general solution of the Schroedinger equation with a periodic potential and periodic boundary condition is (see Appendix 11) of the form

$$\psi(x) = \exp(ikx)U_k(x), \tag{7.17}$$

where

$$U_k(x + a) = U_k(x) \tag{7.18}$$

with

$$k = 2\pi m/(Na) = 2\pi m/L, \tag{7.19}$$

$$m = 0, \pm 1, \pm 2, \ldots, \tag{7.20}$$

and $L\,(=Na)$ is the length of the crystal. The wave function (7.17) can be thought of as equivalent to a plane wave modulated by a function $U_k(x)$ with the periodicity of a lattice [see Eq. (7.18)]. The wave function (7.17) is referred to as Bloch's function, and the result that $\psi(x)$ can be written in terms of exp (ikx) and a periodic function $U_k(x)$ is known as *Bloch's theorem*.

The Kronig-Penney Model

An application of Bloch's theorem can be illustrated by considering a simple form of the periodic potential (see Fig. 7.11), known as the Kronig-Penney model. The form of the Kronig-Penney potential is

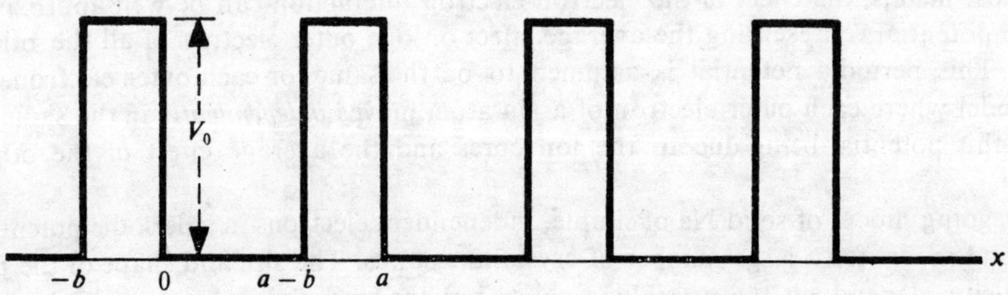

Fig. 7.11 Potential for electron moving in periodic lattice approximated as array of rectangular potential wells and barriers: Kronig-Penney model.

$$V(x) = 0 \qquad [0 \leqslant x \leqslant (a - b)]$$

$$= V_0 \qquad [-b \leqslant x \leqslant 0], \tag{7.21a}$$

$$V(x + a) = V(x). \tag{7.21b}$$

The procedure for obtaining the solution of the Schroedinger equation is straightforward. In Appendix 11, it has been shown that the solution is possible only if the relation

$$R\frac{\sin \alpha a}{\alpha a} + \cos \alpha a = \cos ka, \tag{7.22}$$

where $\alpha^2 = 2mE/\hbar^2$ and $k = 2\pi m/(Na)$, holds. [We have worked in the limit $V_0 \rightarrow \infty$, $b \rightarrow 0$ such that $R = (m/\hbar^2)V_0ab$ is finite.] To study the allowed values of αa which in turn give the allowed values of energy, we plot the left-hand side of Eq. (7.22) as a function of αa (see Fig. 7.12).

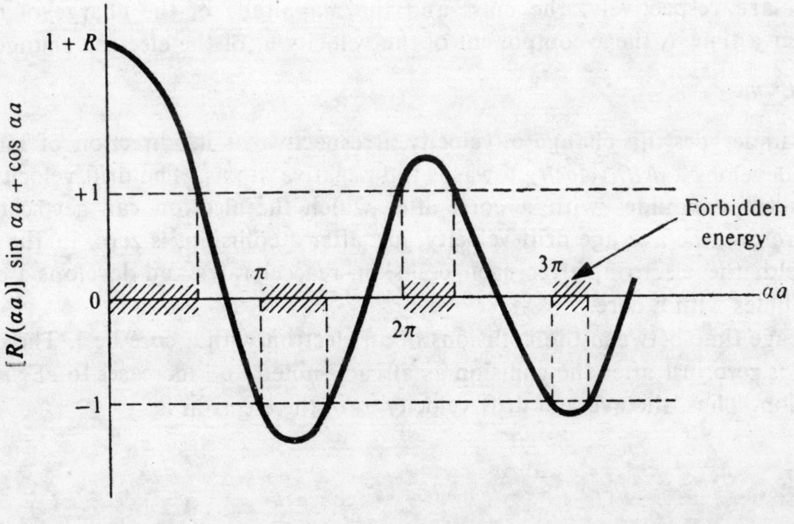

Fig. 7.12 $[R/(\alpha a)] \sin \alpha a + \cos \alpha a$ as function of αa to illustrate allowed energy regions. Shaded areas correspond to forbidden bands.

At $\alpha a = 0$, the function is $1 + R$ and then it decreases to $2R/\pi$ at $\alpha a = \pi/2$ and reaches -1 at $\alpha a = \pi$. Since $\cos ka$ must be between -1 and $+1$, condition (7.22) is satisfied whenever the value of αa is such that $\{[R/(\alpha a)] \sin \alpha a + \cos \alpha a\}$ lies between -1 and $+1$. There are regions of αa, where the value of $\{[R/(\alpha a)] \sin \alpha a + \cos \alpha a\}$ does not lie between -1 and $+1$. For these values of αa, and hence of the energy E, no solutions exist. In other words, there are regions of energy which are prohibited as we have no solution of the Schroedinger equation for these values of energy. Thus, we get a region of allowed energy levels followed by a forbidden region of energy. That is, we have alternating energy bands and energy gaps.

7.4 ELECTRICAL PROPERTIES OF SOLIDS

The band structure for the energy levels of electrons in a solid forms the basis for understanding the electrical properties of the solid. Let us first consider a simplified picture of how a solid, e.g., sodium crystal, conducts electricity. We recall that a sodium crystal consists of a core at each lattice site and the valence electrons. These electrons are free to move inside the crystal. On the other hand, each core vibrates about its mean position. The amplitude of vibration of the core depends on the temperature of the solid. The electrons occasionally collide with the core and scatter in different directions after the collision. The core, being much heavier than the electrons, does not leave the lattice site. However, its amplitude of vibration changes after the collision. Since the electrons are moving in all possible directions, there is no net transfer of charge in any single direction, and hence the net current in the solid is zero.

Let us now consider the solid in an electric field E pointing along the positive x-direction. The electrons will be accelerated along the negative x-axis because they are negatively charged. The magnitude of acceleration of an electron is given by

$$a = eE/m, \tag{7.23}$$

where m and e are, respectively, the mass and the magnitude of the charge of the electron. Therefore, after a time t, the x-component of the velocity v_x of the electron changes by

$$\Delta v_x = eEt/m. \tag{7.24}$$

Each electron undergoes this change of velocity, irrespective of its direction of motion. Thus, each electron develops a *drift velocity* towards the negative x-axis. The drift velocity of an electron increases till it collides with a core, after which the electron can get scattered in any direction. Therefore, the average drift velocity, just after a collision, is zero. In the presence of the electric field, the electron, after each collision, reaccelerates and develops a drift velocity till it again collides with a core.

Let the average time between the collisions of an electron with a core be τ. The drift velocity of the electron is zero just after the collision as already noted and increases to $eE\tau/m$ just before the next collision. Thus, the average drift velocity v_d of the electron is

$$v_d = \frac{eE}{2m}\tau. \tag{7.25}$$

If there are n valence electrons per unit volume, then the current density j, which is the flow of charge per unit time across a unit area, is

$$j = nev_d = \frac{ne^2}{2m}\tau E. \tag{7.26}$$

We should like to repeat that the direction of current is towards the positive x-axis.

EXAMPLE 7.2 A current of 1 ampere is flowing through a circular metallic wire of radius 10^{-4} m. Calculate the drift velocity of the electrons causing this current ($n = 10^{28}/\text{m}^3$).

The current density j is given by

$$j = \frac{I}{A} = \frac{1}{\pi \times 10^{-8}} = \frac{10^8}{\pi}\ \text{A/m}^2,$$

where $A(=\pi r^2)$ is the area of cross-section of the wire. From Eq. (7.26), we get

$$v_d = \frac{j}{ne} = \frac{10^8}{\pi \times 10^{28} \times 1.6 \times 10^{-19}} = \frac{1}{16\pi}\ \text{m/sec} = .02\ \text{m/sec}.$$

The energy of the valence electrons in a solid is of the order of a few electron volts (see Example 7.1). This corresponds to a speed of the order of 10^6 m/sec. The electrons move randomly inside the solid with this average speed. From Example 7.2, we see that the average drift velocity of an electron is very small as compared to the speed associated with its random motion. However, the drift velocity of each electron is in the same direction and since a large number of electrons participate in the drift motion, a sizeable current is produced in the solid.

Let us consider a solid of uniform cross-sectional area A and length l. If a voltage V is applied across this solid, the electric field will be given by V/l. Equation (7.26) then becomes

$$j = \frac{ne^2}{2m}\tau\frac{V}{l}. \tag{7.27}$$

If I is the total current flowing through the solid, then

$$I = jA = \left(\frac{ne^2\tau}{2m}\frac{A}{l}\right)V. \tag{7.28}$$

This can be written as

$$I = V/R, \tag{7.29}$$

where

$$R = \frac{2ml}{ne^2\tau A}. \tag{7.30}$$

This equation is nothing but Ohm's law and R is the resistance associated with the solid. If σ is the conductivity of the solid, then

$$R = \frac{1}{\sigma}\frac{l}{A} \tag{7.31}$$

and we have

$$\sigma = \frac{ne^2\tau}{2m} = ne\mu, \tag{7.32}$$

where μ $[=e\tau/(2m)]$ is called the *mobility*. From Eq. (7.25), it is clear that the mobility is the average drift velocity per unit electric field.

We see that the resistance of a solid depends on two factors, namely, the density of free electrons n, and the time between the collisions τ. It is inversely proportional to the density of free electrons and to the collision time. This is physically understandable because if the number of electrons is large, the carriers of electricity will increase, and thus the conductivity will increase. In other words, the resistance will decrease. Further, if the time between the collisions is long, then the electrons are freer to drift along the direction of the electric field for a long time, and hence the current will be large. This means that the resistance will decrease. In fact, if there were no collisions at all, the electrons would drift at very large speeds and the resistance would be negligible.

Let us now see how the picture of electrical conductivity in a solid we have described gets modified in quantum mechanics. As we have noted, an electron can occupy a definite allowed energy band. If an electric field is applied to an electron, it absorbs the energy (classically, it does so by accelerating) by moving to the next allowed energy level. The more the number of electrons in the electric field, the larger the current. This leads to a higher conductivity.

We now consider the wave function of the electrons in a solid. The Bloch function (7.17) gives the wave function of electrons in a periodic potential. However, we know that at a finite temperature the cores vibrate, and thus the periodicity of the potential is disturbed. If we consider the motion of electrons in a perfectly periodic potential (produced by fixed cores), we find that the value of k does not change. This means that the electrons do not undergo any collisions. However, in a solid, the potential is not perfectly periodic. The disturbance in the periodicity in a crystalline solid can be due to many reasons, for example, thermal vibration of the cores, presence of impurities, and defects in the structure. As a result, the electrons are scattered in different directions, leading to collisions. A detailed quantum mechanical description of collisions is beyond the scope of this text. However, for the general understanding of the electrical conductivity of a solid, it is sufficient to realize that (1) the electrons absorb energy from the electric field if the electrons have to conduct and (2) the conductivity is controlled by the collision of electrons with the cores. These collisions result from the factors affecting the periodicity of the solid.

QUESTION 7.5 The thermal vibration of a core contributes to the resistance of a solid. Does this contribution increase or decrease with temperature?

QUESTION 7.6 The presence of impurities in a solid contributes to the resistance of the solid. Why is this so? Explain why this contribution to resistance is independent of temperature.

EXAMPLE 7.3 Given the electrical conductivity of a metal to be of the order of 10^7/ohm-m and $n \sim 10^{28}/\text{m}^3$, calculate the average distance (mean free path) travelled by the electrons between collisions.

We have, from Eq. (7.32),

$$\tau = \frac{2m}{ne^2}\sigma \sim \frac{2 \times 9.1 \times 10^{-31}}{10^{28} \times (1.6)^2 \times 10^{-38}} \times 10^7.$$

The mean free path $\lambda \sim v\tau$, where v is the average speed of electrons in a metal. We have already seen that $v \sim 10^6$ m/sec. The mean free path is then given by

$$\lambda = \frac{2 \times 9.1}{(1.6)^2} \frac{10^{-31} \times 10^7}{10^{28} \times 10^{-38}} 10^6 \text{ m}$$

$$\sim 6 \times 10^{-8} \text{ m} = 600 \text{ Å}.$$

The lattice spacing is of the order of 5 Å. Thus, the electrons cross about 100 lattice sites before a collision occurs.

INSULATORS

We can understand the behaviour of an insulator using the band theory of solids. Consider a solid whose allowed energy bands are as shown in Fig. 7.13. The electrons in a solid occupy the lowest allowed levels. Suppose the number of electrons is such that the electrons completely fill only the lowest energy levels, i.e., the next allowed energy band is completely empty. Many

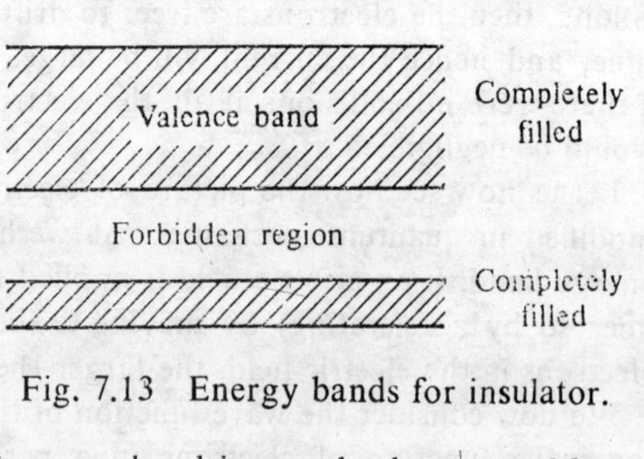

Fig. 7.13 Energy bands for insulator.

solids, e.g., diamond, silicon, and germanium, have this property. The topmost completely filled energy band is called the *valence band* and the next band of allowed energy is known as the *conduction band.* In an insulator, at a temperature close to zero, the valence band is completely filled and the conduction band is empty. The band gap, i.e., the energy difference between the bottom of the conduction band and the top of the valence band, is of the order of a few electron volts. For example, in diamond, it is 6 eV, whereas in silicon, it is 1.1 eV.

EXAMPLE 7.4 Estimate the amount of energy the electrons in a solid can absorb between collisions when the solid is subjected to an external electric field of strength 10^3 V/m.

When an electron travels a distance x in an electric field E, it absorbs an energy eEx. Since the electron travels a distance λ between collisions, the energy it absorbs is $W = \lambda eE$. In Example 7.3, we obtained $\lambda \sim 600$ Å. Thus,

$$W \sim 6 \times 10^{-8} \times 1.6 \times 10^{-19} \times 10^3 \text{ J} = 6 \times 10^{-5} \text{ eV}.$$

From the foregoing example, it is clear that in the presence of an electric field the electrons absorb an energy of about 10^{-4} eV and move to a level which is higher by an energy of about 10^{-4} eV provided this level is unoccupied. But, in an insulator, this is not possible, because all the levels in the valence band are occupied. Also, unless the electrons move to the conduction band, no transition can take place. Since the band gap is of the order of electron volts, the energy required for the transfer of an electron to the conduction band is also very large, of the order of electron volts. Such an energy can be generated only when the strength of the electric field is $\sim 10^8$ V/m, and the solid conducts electricity. In a solid in an electric field of strength less than 10^3 V/m, there are no electrons in the conduction band, and thus no conduction of electricity takes place.

A substance with a completely empty conduction band and a completely filled valence band is a perfect insulator. However, as we know, this is possible only when $T = 0$ K. At a finite temperature, the electrons in the valence band are thermally excited. Of these, some move to the conduction band and the substance has a small conductivity.

The transparency of diamond to visible radiation is due to the large band gap. This radiation has an energy of the order of 2 eV per photon, which is insufficient to excite the electrons from the valence band to the conduction band. Thus, light is not absorbed by diamond.

CONDUCTORS

Consider a solid in which the valence band is not completely filled. Such a situation is shown in Fig. 7.14a. For example, in solid sodium, one electron from each of the atoms belonging to the $3s$-state moves freely in the solid. If there are N such electrons, we then have $2N$ states (the factor 2 being due to spin). Thus, the valence band in solid sodium is only half-filled. The electrons in the valence band can be accelerated. This is because they can move over to the unfilled states in the valence band itself. Solid sodium is a good conductor and its conductivity depends on the collision time. The collision time, as we have seen, depends on the temperature of and the impurities in a crystal. The metals (e.g., Cu, Ag, and Au) in group I of the periodic table which have a single electron in the s-state in the outermost orbit have a partly filled valence band. They are all good conductors.

In some solids, the valence band and the conduction band overlap (see Fig. 7.14b). In such a solid, even when one band is completely filled, the energy can change in a continuous manner,

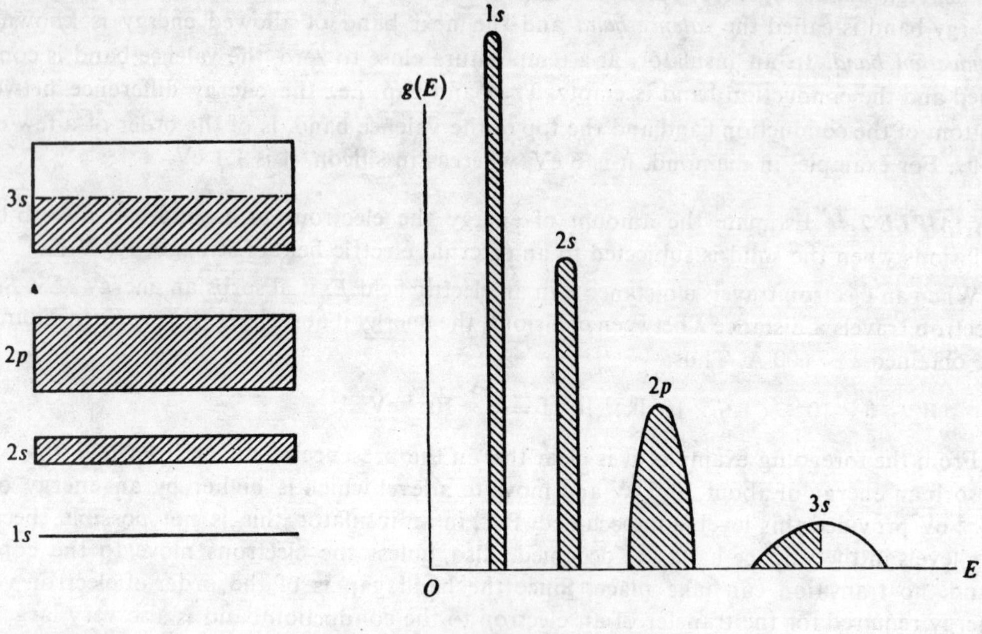

(a) Energy bands for group I conductor

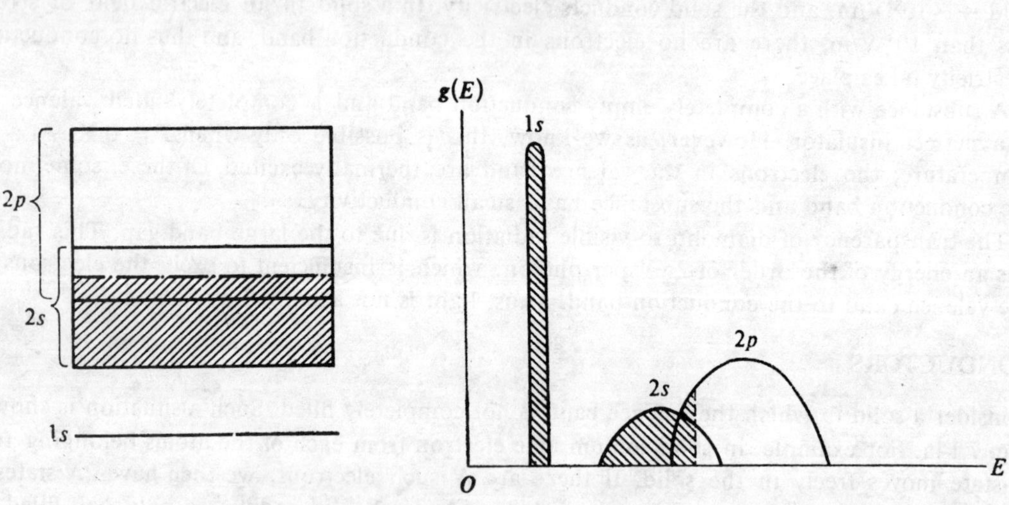

(b) Energy bands for group II conductor

Fig. 7.14 Energy levels, and density of states as function of energy. Shaded regions represent states filled by electrons.

and the solid can conduct electricity. The metals in group II of the periodic table belong to this category.

QUESTION 7.7 Solidified argon is an insulator. Why?

QUESTION 7.8 Why is a metallic solid opaque to visible radiation?

7.5 SEMICONDUCTORS

In a solid, if the energy gap between the bottom of the conduction band and the top of the valence band is about 1 eV, then the solid is said to be a semiconductor, e.g., pure germanium and silicon. Such a semiconductor is referred to as an *intrinsic semiconductor*. Semiconductors can also be made by adding a small number of impurity atoms to certain substances. Such a semiconductor is referred to as an *extrinsic semiconductor*.

INTRINSIC SEMICONDUCTORS

The pure materials most commonly used as semiconductors are germanium and silicon. The band gap in germanium is about 0.7 eV and in silicon, it is about 1.1 eV. The conduction takes place because some electrons from the valence band move over to the conduction band due to the thermal excitation. When an electric field is applied, the electrons in the conduction band are free to absorb energy as this band is almost empty.

Once an electron moves out from an energy state, say, E_1, of the valence band, this state is then unoccupied. Thus, an electron having an energy E_2 close to but less than E_1 can now absorb energy from the field (i.e., be accelerated by the field) and occupy the energy state E_1. The level with energy E_2 now becomes empty and can be occupied by another electron, and so on. Another way of looking at the absorption of energy in the valence band is to realize that when an electron moves out of the valence band, it leaves a *hole* in the valence band. This hole is positively charged as the electron has left the valence band and the original system of the electron and the core was initially electrically neutral. A neighbouring electron can now move into this hole, leaving a hole at the place from which it moves. Thus, the motion

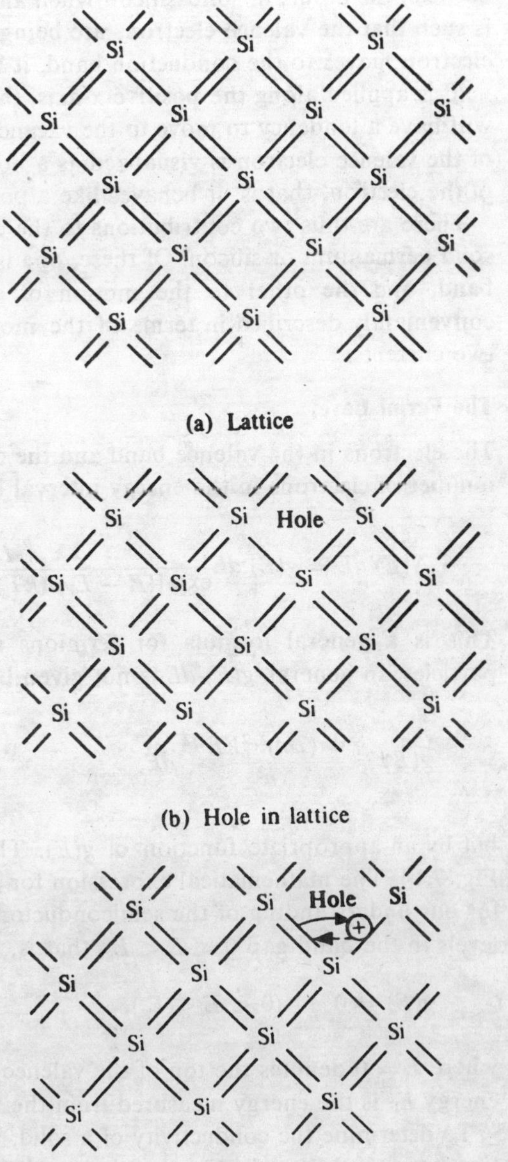

(a) Lattice

(b) Hole in lattice

(c) Motion of hole

Fig. 7.15 Motion of hole in valence band in presence of electric field. Dashed line with arrow denotes motion of electron.

of the electrons can be thought of as a hole moving in the direction opposite to that of the electrons.

The foregoing description can be visualized as follows. Figure 7.15a shows the bonding between the atoms in solid silicon when all the electrons are in the valence band. The bonding is such that the valence electrons are being shared by the neighbouring atoms. If one valence electron moves to the conduction band, it leaves a vacancy (see Fig. 7.15b). If now an electric field is applied along the positive x-axis, then the valence electrons to the right of this vacancy will have a tendency to move to the vacancy. Such a transfer is shown in Fig. 7.15c. The absence of the valence electron is visualized as a hole. The hole moves in the direction opposite to that of the electron, that is, it behaves like a positively charged particle.

There are thus two contributions to the electric current when an external field is applied to solid germanium or silicon. Of these, one is due to the motion of electrons in the conduction band, and the other to the motion of electrons in the valence band. The latter is more conveniently described in terms of the motion of holes. The total current is the sum of the two currents.

The Fermi Level

The electrons in the valence band and the conduction band behave like a Fermi-Dirac gas. The number of electrons in the energy interval between E and $(E + dE)$ [see Eq. (7.4)] is

$$N(E)\,dE = g(E)\,dE \frac{1}{\exp\,[(E - E_F)/(kT)] + 1}. \tag{7.33}$$

This is a general formula for fermions although in Chapter 6 we had derived it for free particles. In general, $g(E)\,dE$ is not given by

$$g(E) = \frac{4\pi(2m)^{3/2}VE^{1/2}}{h^3}\,dE,$$

but by an appropriate function of $g(E)$. The density of states for typical metals is shown in Fig. 7.14. The mathematical expression for $g(E)$ is, in general, complicated and is not needed for our understanding of the semiconductors. All we require to remember is that there are no levels in the band gap $0 < E < E_g$, that is,

$$g(E) = 0 \qquad (0 < E < E_g),$$

where $E = 0$ denotes the top of the valence band and E_g is the *energy* (or *band*) *gap*. The Fermi energy E_F is the energy measured from the top of the valence band.

To determine the conductivity of a solid, we must know the number of electrons in the conduction band. If the conduction band extends from E_g to E_c (see Fig. 7.16), then the number of electrons in the conduction band is

$$N_c = \int_{E=E_g}^{E=E_c} \frac{g(E)\,dE}{\exp\,[(E - E_F)/(kT)] + 1}.$$

In the conduction band, $(E - E_F)/(kT) \gg 1$ when $T \to 0$, and we can write the number of

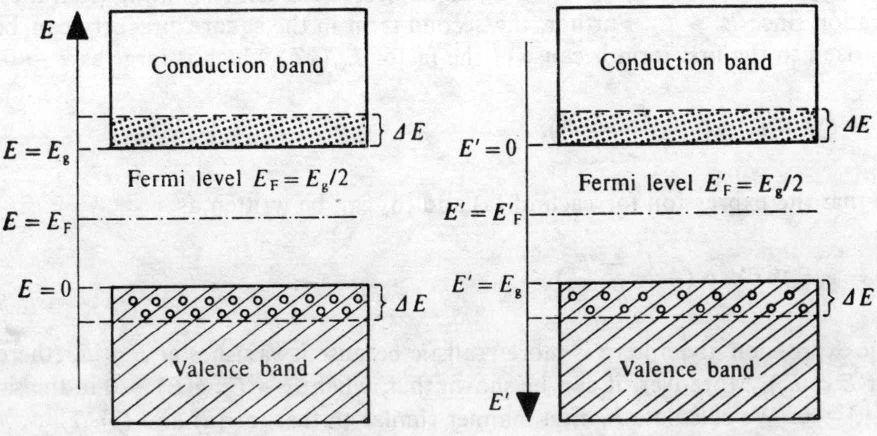

Fig. 7.16 Distribution of electrons (dots) in conduction band and holes (hollow circles) in valence band. Energy E is measured upwards and energy E' is measured downwards.

electrons in the conduction band as

$$N_c = \int_{E=E_g}^{E=E_c} g(E) \exp\left(-\frac{E - E_F}{kT}\right) dE. \tag{7.34}$$

This can be further simplified if we note that most of the contribution of the integral to N_c comes from the lower limit of the integral because the exponential factor rapidly decreases. In the example that follows, we shall illustrate this.

EXAMPLE 7.5 Evaluate N_c for (a) $g(E) = C$, (b) $g(E) = CE$, and (c) $g(E) = C(E - E_g)^{1/2}$, where C is a constant.

(a) From Eq. (7.34), we have

$$N_c = C \int_{E=E_g}^{E=E_c} \exp\left(-\frac{E - E_F}{kT}\right) dE = C(kT)\left[-\exp\left(-\frac{E_g - E_F}{kT}\right)\right]\Big|_{E_g}^{E_c}$$

$$= CkT\left[\exp\left(-\frac{E_g - E_F}{kT}\right) - \exp\left(-\frac{E_c - E_F}{kT}\right)\right].$$

Since $E_c \gg E_g$, we can neglect the term $\exp[-(E_c - E_F)/(kT)]$ in comparison to the term $\exp[-(E_g - E_F)/(kT)]$, and thus write the foregoing equation as

$$N_c = g(E_g)kT \exp\left(-\frac{E_g - E_F}{kT}\right).$$

(b) Proceeding as in (a), we have

$$N_c = C \int_{E=E_g}^{E=E_c} E \exp\left(-\frac{E - E_F}{kT}\right) dE = C(kT)^2\left[-\frac{E}{kT}\exp\left(-\frac{E - E_F}{kT}\right) - \exp\left(-\frac{E - E_F}{kT}\right)\right]\Big|_{E_g}^{E_c}$$

$$\sim C(kT)^2\left[\frac{E_g}{kT}\exp\left(-\frac{E_g - E_F}{kT}\right) + \exp\left(-\frac{E_g - E_F}{kT}\right)\right].$$

In obtaining this expression, we have again neglected the term resulting from the upper limit of integration since $E_c \gg E_g$. Further, the second term in the square brackets can be neglected in comparison to the first term because of the factor $E_g/(kT)$ which is large as $T \to 0$. Thus,

$$N_c = C(kT)E_g \exp\left(-\frac{E_g - E_F}{kT}\right).$$

We note that the expression for each of (a) and (b) can be written as

$$N_c = g(E_g)kT \exp\left(-\frac{E_g - E_F}{kT}\right).$$

(c) The expression given here is more realistic because it vanishes at $E = E_g$ (there being no states for $E < E_g$). Moreover, it can be shown that, when $E \to E_g$, $g(E) \to 0$ in the same way as $(E - E_g)^{1/2} \to 0$. We calculate N_c in a manner similar to that for (a) and (b). Thus,

$$N_c = C \int_{E_g}^{E_c} (E - E_g)^{1/2} \exp\left(-\frac{E - E_F}{kT}\right) dE$$

$$= C \exp\left(\frac{E_F - E_g}{kT}\right)(kT)^{3/2} \int_0^{(E_c - E_g)/(kT)} u^{1/2} \exp(-u)\, du,$$

where $u = (E - E_g)/(kT)$. Here, since $(E_c - E_g)/(kT) \gg 1$, the upper limit can be replaced by ∞ and the integration can be carried out. We then have

$$N_c = C(kT)^{3/2} \exp\left(\frac{E_F - E_g}{kT}\right)\frac{\pi^{1/2}}{2}.$$

Using the fact that

$$g(E_g + kT) = C(E_g + kT - E_g)^{1/2} = C(kT)^{1/2},$$

we get

$$N_c = \frac{\pi^{1/2}}{2} g(E_g + kT)kT \exp\left(\frac{E_F - E_g}{kT}\right) = \bar{g}(E_g)kT \exp\left(-\frac{E_g - E_F}{kT}\right),$$

where

$$\bar{g}(E_g) = \frac{\pi^{1/2}}{2}(E_g + kT)$$

near the bottom of the conduction band. In this expression for N_c, the exponential factor decreases rapidly with the increase in electron energy. Therefore, most of the electrons are near the bottom of the conduction band and are in the energy range E to $(E + kT)$ (see Fig. 7.16). In this energy range, $g(E)$, being a smooth function, is more or less constant, and we have represented this as $\bar{g}(E_g)$. The expression $\bar{g}(E_g)$ represents a more accurate definition of the density of states. However, for purposes of simplicity, we shall not be using it in our subsequent study.

At a normal temperature, $(E - E_F)/(kT)$ is greater than one, and therefore the approximation used in the foregoing example holds. Thus, the energy of the electrons lies between E_g and $(E_g + \Delta E)$, where ΔE is a small energy interval. Hence, the total number of electrons in the conduction band is

$$N(E_g)\Delta E = \frac{g(E_g)\Delta E}{\exp[(E_g - E_F)/(kT)] + 1} \sim g(E_g) \exp\left(-\frac{E_g - E_F}{kT}\right)\Delta E. \tag{7.35}$$

The vacancies left in the valence band allow the remaining electrons in the band to absorb energy and move to the vacant states. Thus, these electrons also contribute to the electrical conductivity. As we have already seen, the motion of electrons in the valence band can be treated in terms of the motion of holes. This provides an equivalent and simpler way to keep track of the electrons in the valence band.

We can analyze the holes in the valence band in the same manner as electrons in the conduction band. Most of the electrons that get excited to the conduction band are released from the top of the valence band. This is because higher energy is needed to make the electrons in the lower part of the valence band move over to the conduction band. The occupation probability $n_h(E)$ of a hole is given by

$$n_h(E) = 1 - n(E) = 1 - \frac{1}{\exp[(E - E_F)/(kT)] + 1}. \tag{7.36}$$

Thus,

$$n_h(E) = \frac{1}{\exp[-(E - E_F)/(kT)] + 1}. \tag{7.37}$$

Here, E is the energy measured upwards from the bottom of the valence band. This expression is in fact the Fermi distribution for holes when the energy is measured downwards from the bottom of the conduction band. In order to see this explicitly, we introduce a new energy variable, namely, E', given by

$$E' - E'_F = -(E - E_F), \tag{7.38}$$

where E' and E'_F are the energies of the hole and its Fermi level measured in the downward direction from the bottom of the conduction band as shown on the right-hand side of Fig. 7.16. Rewriting Eq. (7.37) in the new variable, we have

$$n_h(E') = \frac{1}{\exp[(E' - E'_F)/(kT)] + 1}. \tag{7.39}$$

We note that this expression is the same as that for the electron occupation probability $n(E)$. When $E = E_g$, we see, from Fig. 7.16, that $E' = 0$. Then, from Eq. (7.38), we get

$$E_F + E'_F = E_g. \tag{7.40}$$

If $N_h(E')$ is the distribution associated with the holes, then

$$N_h(E')\,dE' = \frac{g_h(E')\,dE'}{\exp[(E' - E'_F)/(kT)] + 1}, \tag{7.41}$$

where $g_h(E')$ is the density of states for the holes of energy E'. This can be interpreted to mean that the holes make a transition from the conduction band to the valence band.

We assume that the holes at the top of the valence band are in an energy band ΔE. Thus, the total number of holes is $N_h(E_g)\Delta E$ and can be written as

$$N_h(E_g)\Delta E = \frac{g_h(E_g)\Delta E}{\exp[(E_g - E_F')/(kT)] + 1} \sim g_h(E_g) \exp\left(-\frac{E_g - E_F'}{kT}\right)\Delta E. \tag{7.42}$$

In a pure semiconductor, the number of conduction electrons equals the number of holes in the valence band. This is because each electron, after it moves to the conduction band from the valence band, leaves behind a hole. Therefore,

$$N_h(E_g) = N(E_g) \tag{7.43}$$

or

$$\frac{g_h(E_g)}{\exp[(E_g - E_F')/(kT)] + 1} = \frac{g(E_g)}{\exp[(E_g - E_F)/(kT)] + 1}. \tag{7.44}$$

$(E_g - E_F')$ and $(E_g - E_F)$ are of the order of electron volts, whereas kT is about .025 eV at room temperature. Thus, the exponential factors in the denominators of Eq. (7.44) are much larger than 1, and it is a good approximation to write this equation as

$$g_h(E_g) \exp\left(-\frac{E_g - E_F'}{kT}\right) = g(E_g) \exp\left(-\frac{E_g - E_F}{kT}\right). \tag{7.45}$$

Taking the logarithm of both sides of Eq. (7.45), we get

$$kT \log g_h(E_g) - (E_g - E_F') = kT \log g(E_g) - (E_g - E_F). \tag{7.46}$$

At a normal temperature, we can neglect the terms proportional to kT on both sides of this equation. We then obtain

$$E_F = E_F'. \tag{7.47}$$

This is the exact result when $T \to 0$. Combining this with Eq. (7.40), we get

$$E_F = E_F' = E_g/2. \tag{7.48}$$

This equation shows that the Fermi level lies half-way between the conduction and valence bands (see Fig. 7.16). Substituting for E_F, we have

$$N_e = g(E_g) \exp\left(-\frac{E_g - E_F}{kT}\right)\Delta E = g(E_g) \exp\left(-\frac{E_g}{2kT}\right)\Delta E, \tag{7.49}$$

$$N_h = g_h(E) \exp\left(-\frac{E - E_F'}{kT}\right)\Delta E = g_h(E_g) \exp\left(-\frac{E_g}{2kT}\right)\Delta E, \tag{7.50}$$

where N_e is the number of electrons in the conduction band and N_h is the number of holes in the valence band.

Thus, the picture of the intrinsic semiconductor we have obtained from the band theory of solids is as follows. The conduction of electricity takes place by the motion of electrons in the conduction band and holes in the valence band. The number of electrons in the conduction band equals the number of holes in the valence band. Most of the electrons are close to the bottom of the conduction band and can be assumed to have an energy between E_g and $(E_g + \Delta E)$, where $\Delta E \leqslant E_g$. Similarly, most of the holes are near the top of the valence band. The explicit expressions for the numbers N_e and N_h can be obtained by taking the Fermi level to be midway in the gap between the conduction and valence bands. These expressions are given by (7.49) and (7.50).

As we have learnt, in a metal, the Fermi level at $T = 0$ is the energy below which all the states are occupied and above which all the states are empty. Such a simple definition of the Fermi energy is not consistent with expressions (7.49) and (7.50). The general definition of Fermi energy is obtained from the general formula (7.33). The result that the Fermi level lies midway in the band gap [see Eq. (7.48)] can be arrived at by using Eq. (7.33).

Variation of Conductivity with Temperature

As already noted, the expression for conductivity σ is (7.32). In an intrinsic semiconductor, both the electrons and the holes contribute to the conductivity. Thus, the total conductivity is given by

$$\sigma = N_e e\mu_e + N_h e\mu_h, \tag{7.51}$$

where μ_e and μ_h are the mobilities of the electrons and the holes, respectively. Using the expressions for N_e and N_h, namely, (7.49) and (7.50), we get

$$\sigma = \exp\left(-\frac{E_g}{2kT}\right)e[g(E_g)\mu_e + g_h(E_g)\mu_h]. \tag{7.52}$$

Here, the dependence of $g(E_g)$ is less than that of the exponential factor on the temperature. Thus, the conductivity increases exponentially as the temperature increases. On the other hand, the conductivity of a metal decreases as the temperature increases since the collision time decreases. This is because the lattice core vibrations increase as the temperature increases. In a semiconductor also, the collision time decreases, but the increase in the number of charge carriers as the temperature increases is such that it overwhelms the effect due to the lattice collisions. In a metal, on the other hand, there is no significant increase in the number of charge carriers with increase in the temperature.

Referring to Eq. (7.52), we see that a plot between ln σ and $1/(kT)$ gives a straight line with a slope $-E_g/2$. This is one of the methods of finding the band gap E_g.

QUESTION 7.9 The conductivity of a silicon crystal increases when light falls on it. Why?

EXTRINSIC SEMICONDUCTORS

n-Type

In a silicon crystal, let an arsenic atom replace a silicon atom. An arsenic atom has five valence electrons, whereas a silicon atom has four. The four valence electrons of the arsenic atom will form covalent bonds with the neighbouring silicon atoms, leaving behind one electron. This electron is bound by the arsenic core of charge $+e$. However, in a silicon crystal, the Coulomb attraction between the arsenic core and the electron is much less because of the high dielectric constant of silicon.

EXAMPLE 7.6 Use Bohr's model to calculate the binding energy and orbit radius of the extra electron going around the arsenic core (of charge $+e$) inside a silicon crystal. The dielectric constant of silicon is 10.

The Coulomb force between the core and the extra electron is $e^2/(4\pi\epsilon_0 kr)$, where k is the dielectric constant and r is the distance between the electron and the core. The binding energy and orbit radius of the electron can be obtained by replacing ϵ_0 by $\epsilon_0 k$ in Bohr's model (4.19)–(4.21). Then,

$$r_0 = \frac{(4\pi\epsilon_0 k)\hbar^2}{me^2}, \qquad E_0 = \frac{me^4}{2(4\pi\epsilon_0 k)^2\hbar^2}.$$

Substituting the value of the dielectric constant in these equations, we obtain

$$E = \frac{13.6}{k^2} = 0.1 \text{ eV}, \qquad r = 0.5 \times k \text{ Å} \sim 5 \text{ Å}.$$

This is a very rough estimate of the binding energy and the radius of the orbit. To obtain the precise values, the mass of the electron m has to be replaced by the effective mass of the electron, which is much less than m. As a result, the binding energy decreases and the radius of the Bohr orbit considerably increases.

From the foregoing example, we note that the fifth electron is loosely bound to the As atom (see Fig. 7.17a). If the binding energy is zero, the electron would be just free to move anywhere in the Si crystal. In fact, it would be at the bottom of the conduction band of Si. That is, the

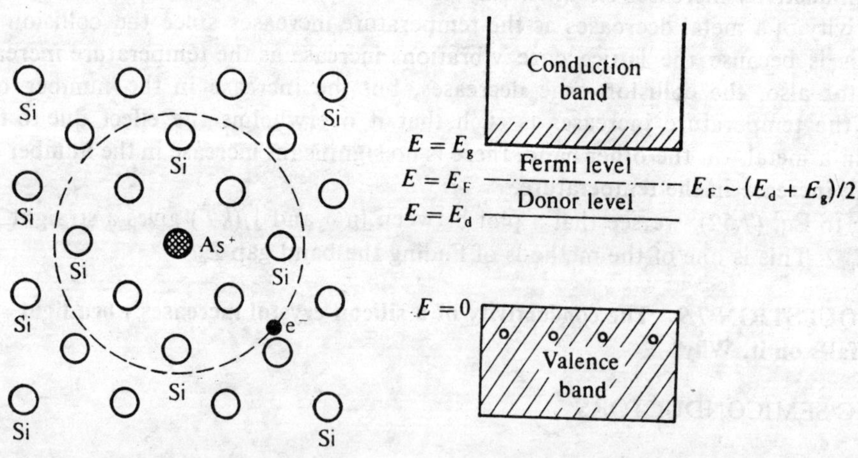

(a) Electron moving around As^+ ion in Si crystal

(b) Energy bands

Fig. 7.17 n-type semiconductor.

presence of the arsenic atom in silicon introduces an energy level very close to the conduction band (see Fig. 7.17b). The electron from this level can easily be excited as it is close to the conduction band. Such a level is known as the *donor level* and the impurity, i.e., arsenic, is known as the *donor*.

At a low temperature, all the electrons in the conduction band come from the donor levels. The number of electrons excited from the valence band to the conduction band is negligible. Therefore, the main charge carriers responsible for conduction are the electrons contributed by the donors. Since the excitation from the valence band is small, there are very few holes in this band. Thus, the contribution from the holes is small. The negatively charged electrons are the majority carriers and the holes are the minority carriers. Since the excitation takes place from a donor level, it can be shown (see Problem 7.22) that the Fermi level at a small value of T lies close to $(E_g + E_d)/2$, where E_d is the energy of the donor level (see Fig. 7.17).

A semiconductor formed by adding an element belonging to group V of the periodic table (e.g., As) as an impurity to an element of group IV (e.g., Si) is called an *n-type* semiconductor.

p-Type

If a trivalent impurity such as boron is introduced in silicon or germanium, we have one electron too few at the boron site to form four covalent bonds. Here, the valence band has a hole and the electrons from this band can move into this hole, creating new holes. We can treat this positively charged hole as a bound state of the B^- ion. This system, again, then is loosely bound (see Fig. 7.18a). The energy levels of the system are shown in Fig. 7.18b. Because boron

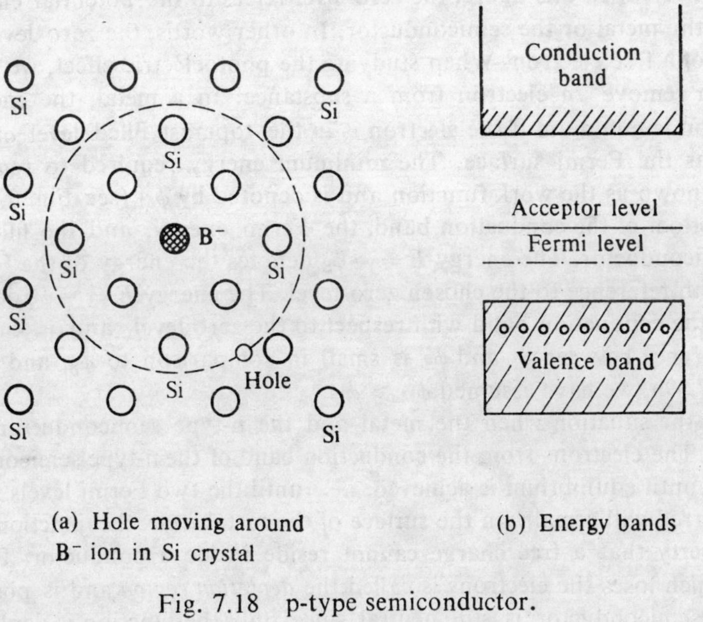

(a) Hole moving around B^- ion in Si crystal

(b) Energy bands

Fig. 7.18 p-type semiconductor.

has only three valence electrons, it accepts an electron. Subsequently, the energy of this electron becomes slightly more than the energy associated with the top of the valence band. Thus, an electron in the valence band moves to the *acceptor level* of boron, leaving behind a hole in the valence band. Therefore, the main carriers of electricity are the holes (majority carriers) and the electrons are the minority carriers. The Fermi level is half-way between the top of the valence band and the acceptor level, as shown in Fig. 7.18b.

A semiconductor formed by adding an element belonging to group III of the periodic table (e.g., boron) as an impurity to an element of group IV (e.g., Si) is called a *p-type* semiconductor.

QUESTION 7.10 What type of an extrinsic semiconductor will be formed when germanium is doped with aluminium?

7.6 SEMICONDUCTOR DEVICES

The electronic devices made with semiconductors have revolutionized the electronics technology. We shall discuss the physics of some of the simple electronic devices. In order to understand the working of such devices, we need the following basic result. If two solids are brought in electrical contact with each other such that the solids can exchange electrons, then, *after the equilibrium condition is achieved, the Fermi energies of the two solids become equal.* This follows from the thermodynamical result that the chemical potentials of two solids in electrical contact become the same in equilibrium condition. It can be shown that the chemical potential of the Fermi gas is in fact the Fermi energy of the gas.

METAL-SEMICONDUCTOR JUNCTION

The energy levels of a metal and an n-type semiconductor when the two are electrically separate are shown in Fig. 7.19a. In this figure, the zero level refers to the potential energy of an electron just outside the metal or the semiconductor. In other words, the zero level represents the potential energy of a free electron. When studying the photoelectric effect, we learnt that work has to be done to remove an electron from a substance. In a metal, the energy required to remove an electron is minimum if the electron is in the topmost filled level of the conduction band, that is, near the Fermi surface. The minimum energy required to remove an electron from a metal is known as the work function and is denoted by ϕ_M (see Fig. 7.19a). This figure also shows the bottom of the conduction band, the Fermi energy, and the filled valence band of the n-type semiconductor. The energy $E = -\phi_S$ denotes the energy of the Fermi level of the semiconductor with reference to the chosen zero level. The energy $E = -\phi'_S$ denotes the energy of the bottom of the conduction band with respect to the zero level, and is known as electron affinity. The difference between ϕ_S and ϕ'_S is small in comparison to ϕ_S, and can be ignored. In drawing Fig. 7.19a, we have assumed $\phi_M > \phi_S$.

Let us consider the situation when the metal and the n-type semiconductor are brought in electrical contact. The electrons from the conduction band of the n-type semiconductor will then flow to the metal until equilibrium is achieved, i.e., until the two Fermi levels match. The electrons thus transferred will remain on the surface of the metal near the junction because of the electrostatic property that a free charge cannot reside inside a conductor. The region of the semiconductor which loses the electrons is called the *depletion region* and is positively charged. The bulk of the semiconductor is still neutral since only the junction is depleted of electrons (see Fig. 7.19b). The electrons on the surface of the metal near the junction give rise to a potential barrier at the junction, preventing a further flow of the electrons from the junction to the metal. The potential barrier (i.e., the increase in potential energy of the electrons as we go from the semiconductor to the metal) is also shown in Fig. 7.19b.

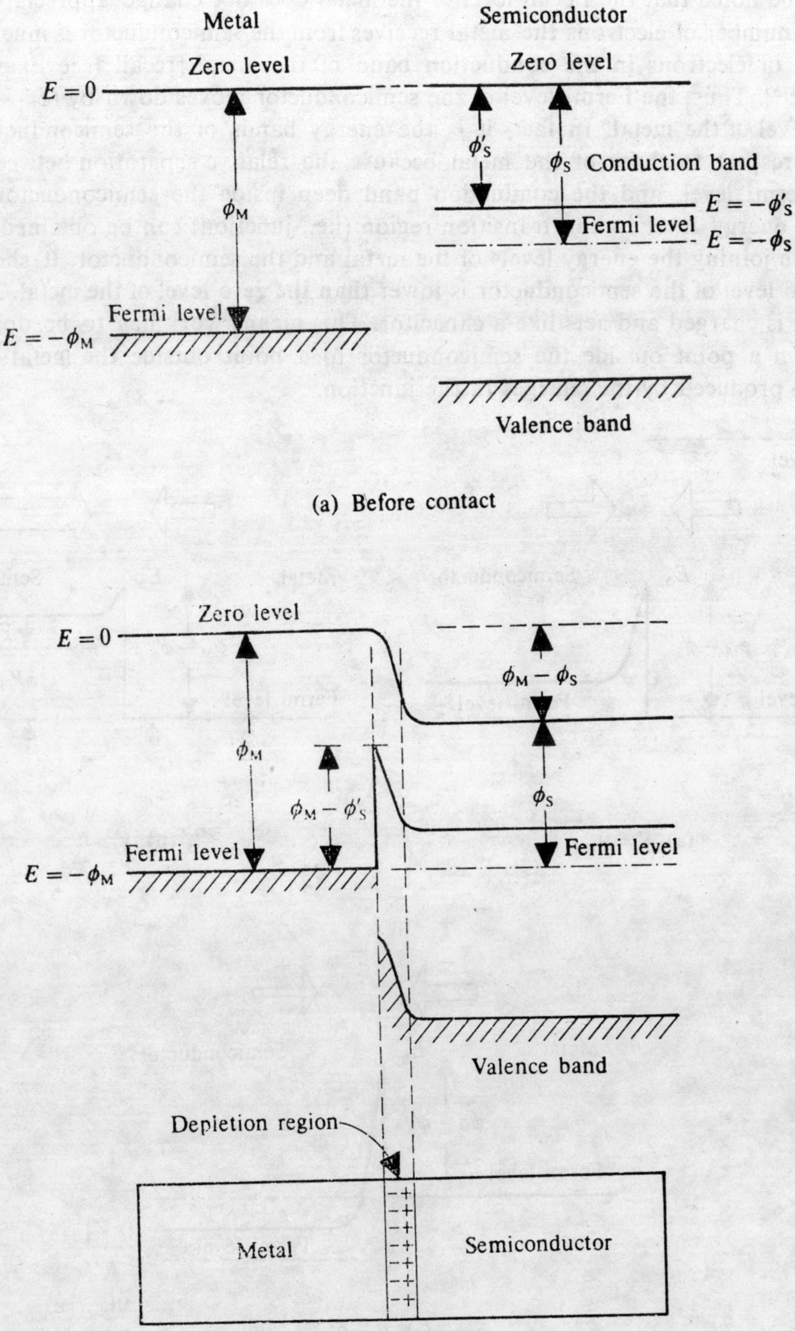

Fig. 7.19 Electron energy levels of metal and n-type semiconductor before and after contact.

It should be noted that the Fermi level of the metal does not change appreciably. This is so because the number of electrons the metal receives from the semiconductor is much smaller than the number of electrons in the conduction band of the metal [recall (see Example 7.1) that $E_F \propto (N/V)^{2/3}$]. Thus, the Fermi level of the semiconductor moves down by $(\phi_M - \phi_S)$ to match the Fermi level of the metal. In fact, it is the energy bands of the semiconductor that move down with respect to those of the metal because the relative separation between the valence band, the Fermi level, and the conduction band deep inside the semiconductor remains unaltered. An energy level in the transition region (i.e., junction) can be obtained by drawing a smooth curve joining the energy levels of the metal and the semiconductor. It should be noted that the zero level of the semiconductor is lower than the zero level of the metal. This is because the junction is charged and acts like a capacitor. This means work has to be done to take an electron from a point outside the semiconductor to a point outside the metal because of the electric field produced by the charges in the junction.

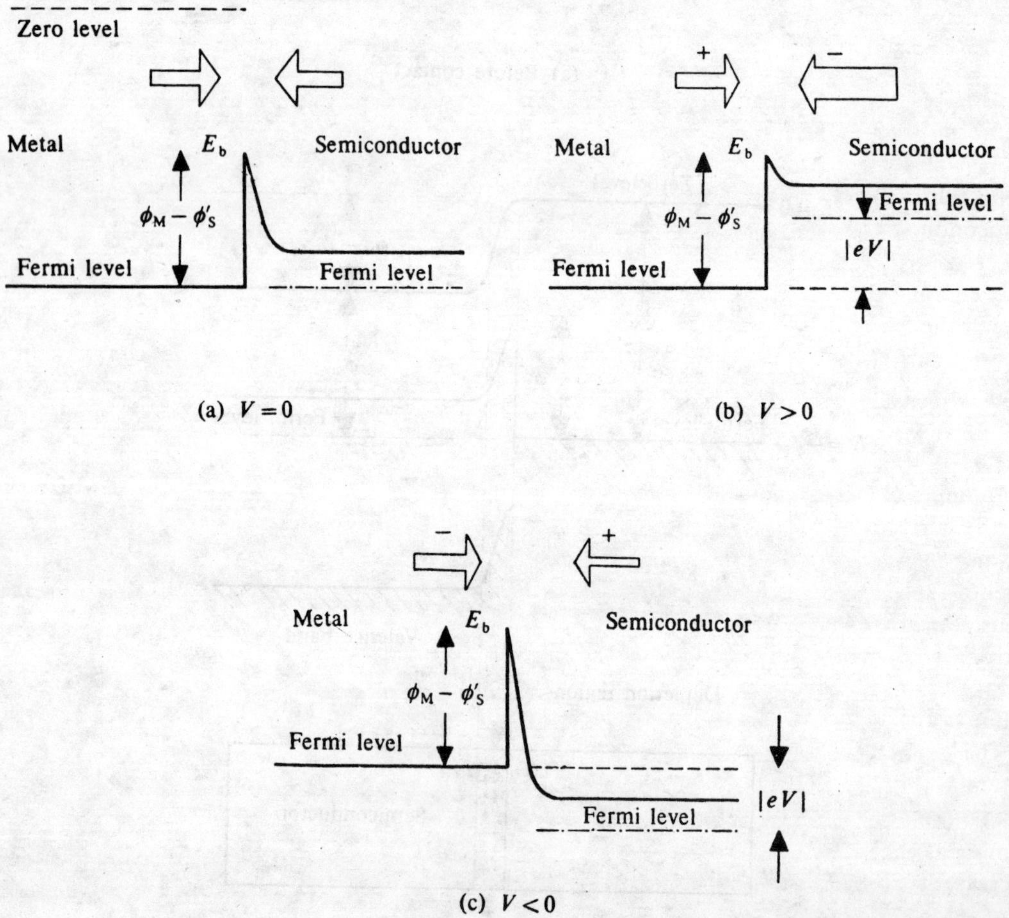

Fig.7 .20 Potential energy of an electron in metal-semiconductor junction. V is applied external potential; arrows denote direction of electron motion and their width indicates magnitude of current.

The potential barrier at the junction causes a potential difference ($\phi_M - \phi'_S$) between the two sides of the junction. The potential energy of an electron near the junction is shown in Fig. 7.20a. It should be noted that in the transition region the difference in energy of the conduction band and the Fermi level increases as we approach the metal. This means that the number of electrons in the conduction band has been reduced. This reduction is a function of the position of the electron. The electrons from the transition region have moved to the metal. This is the depletion region, and is typically of the order of 10^{-8} m in width.

We shall see that the metal-semiconductor junction device we have described acts as a rectifier. That is, it allows the current to flow from the metal to the semiconductor when the metal is kept at a positive potential with respect to the semiconductor. On the other hand, there is hardly any flow of current when the voltage is reversed.

In order to understand the working of the foregoing device, let us consider the potential energy of an electron as a function of its position. Figure 7.20a shows the potential energy diagram of a metal-semiconductor junction when there is no applied potential. The potential energy increases as we move closer to the surface of contact since the metal side of the contact is negatively charged. Under the equilibrium condition, the electrons with an energy greater than E_b can move to either side of the junction. The electrons with such an energy exist on both the sides as we are considering the metal and the semiconductor at room temperature. Since the Fermi levels on both the sides are equal, the number of electrons having an energy greater than E_b on both the sides is the same, and the current caused by the flow of electrons from the metal to the semiconductor is cancelled by the current caused by the flow of electrons from the semiconductor to the metal. These flows are indicated by arrows of equal thickness in Fig. 7.20a.

Let a positive potential $V(>0)$ be applied to the metal with respect to the semiconductor. The energy levels of the electrons in the semiconductor will then move up (see Fig. 7.20b) and so will the Fermi level of the semiconductor. The two Fermi levels will now not be equal, and therefore electrons will flow from the semiconductor to the metal. Since the Fermi level of the semiconductor has gone up by Ve, the number of electrons in the semiconductor having an energy greater than E_b increases. In fact, the number of such electrons increases by a factor $\sim\exp[eV/(kT)]$. On the other hand, the number of electrons in the metal having an energy greater than E_b has not increased because the Fermi level of the metal has not changed. Thus, a current proportional to $[\exp eV/(kT) - 1]$ flows from the metal to the semiconductor.

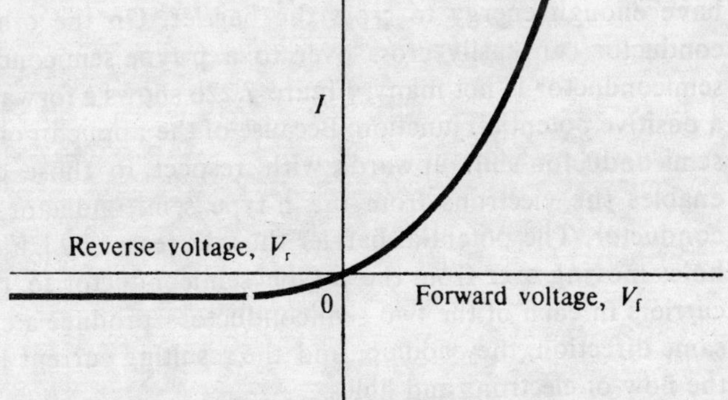

Fig. 7.21 *I V* characteristic of metal-semiconductor (n-type) junction.

If now the voltage is reversed, that is, the semiconductor is kept at a potential higher than that of the metal, then the energy of the electrons in the semiconductor will decrease. Also, the Fermi level of the semiconductor will move down by eV (see Fig. 7.20c). Further, the number of electrons crossing over to the metal from the semiconductor would decrease by a factor $\exp[-|eV|/(kT)]$, whereas the number of electrons crossing over to the semiconductor from the metal will remain the same. Hence, the current from the semiconductor to the metal is proportional to $[1 - \exp\{-|eV|/(kT)\}]$. A plot of $[\exp eV/(kT) - 1]$ when $V > 0$ and $[1 - \exp\{-|eV|/(kT)\}]$ when $V < 0$ is shown in Fig. 7.21. When the metal is at a potential higher than that of the semiconductor, the junction is said to be *forward-biased*, and a large current can be obtained. When the junction is *reverse-biased*, that is, the metal is at a potential lower than that of the semiconductor, very little current flows through the junction.

QUESTION 7.11 Why is the contribution of holes not significant in a metal-semiconductor (n-type) junction?

p-n JUNCTION

We can also get a rectifier when a p-type semiconductor is grown on the surface of an n-type semiconductor or vice versa. Such a combination is referred to as a p-n junction.

The energy level diagram for electrons and holes provides an understanding of how a p-n junction works. Figure 7.22a shows such a diagram for a p-n junction, and has been drawn following the result that the Fermi levels of the n- and p-type semiconductors in contact match after the equilibrium is achieved. When the two semiconductors are in contact, a charge layer is formed at the junction with the negative charges moving over to the p-type semiconductor. Under equilibrium, the number of electrons (which are the minority carriers in a p-type semiconductor) going over to the n-type semiconductor from the p-type semiconductor equals the number of electrons moving over to the p-type semiconductor from the n-type semiconductor. Even though an n-type semiconductor has a large number of electrons, it is only the electrons with an energy greater than $(E_0 + E_F)$ that can cross the potential barrier. An electron from a p-type semiconductor can easily cross over to an n-type semiconductor. In a p-type semiconductor, the number of electrons is very few.

As we shall now discover, the arguments applicable to electrons are similar to those valid for holes. The holes moving over to an n-type semiconductor from a p-type semiconductor have to cross the potential barrier (the energy of holes, it should be noted, is measured downwards). Although a p-type semiconductor has a large number of holes, only a very few of these holes have enough energy to cross the barrier. On the other hand, the holes from an n-type semiconductor can easily cross over to a p-type semiconductor. The number of holes in an n-type semicondnctor is not many. Figure 7.22b shows a forward-biased (i.e., a p-type semiconductor at a positive potential) junction. Because of the applied potential V, the energy levels of the n-type semiconductor shift upwards with respect to those of the p-type semiconductor. This shift enables the electrons from the n-type semiconductor to easily cross over to the p-type semiconductor. The potential barrier thus decreases by $|eV|$. A similar argument is applicable to the holes moving over from the p-type semiconductor to the n-type semiconductor. The majority carriers in each of the two semiconductors produce a current. Since these currents flow in the same direction, they add up, and the resulting current is large. The arrows in Fig. 7.22b indicate the flow of electrons and holes.

Figure 7.22c shows a reverse-biased junction. Here, the barriers associated with the electrons and holes become larger than those under the equilibrium condition. Hence, the majority carriers in the two types of semiconductors hardly produce any current, and so very little current flows. The p-n junction therefore acts as a rectifier.

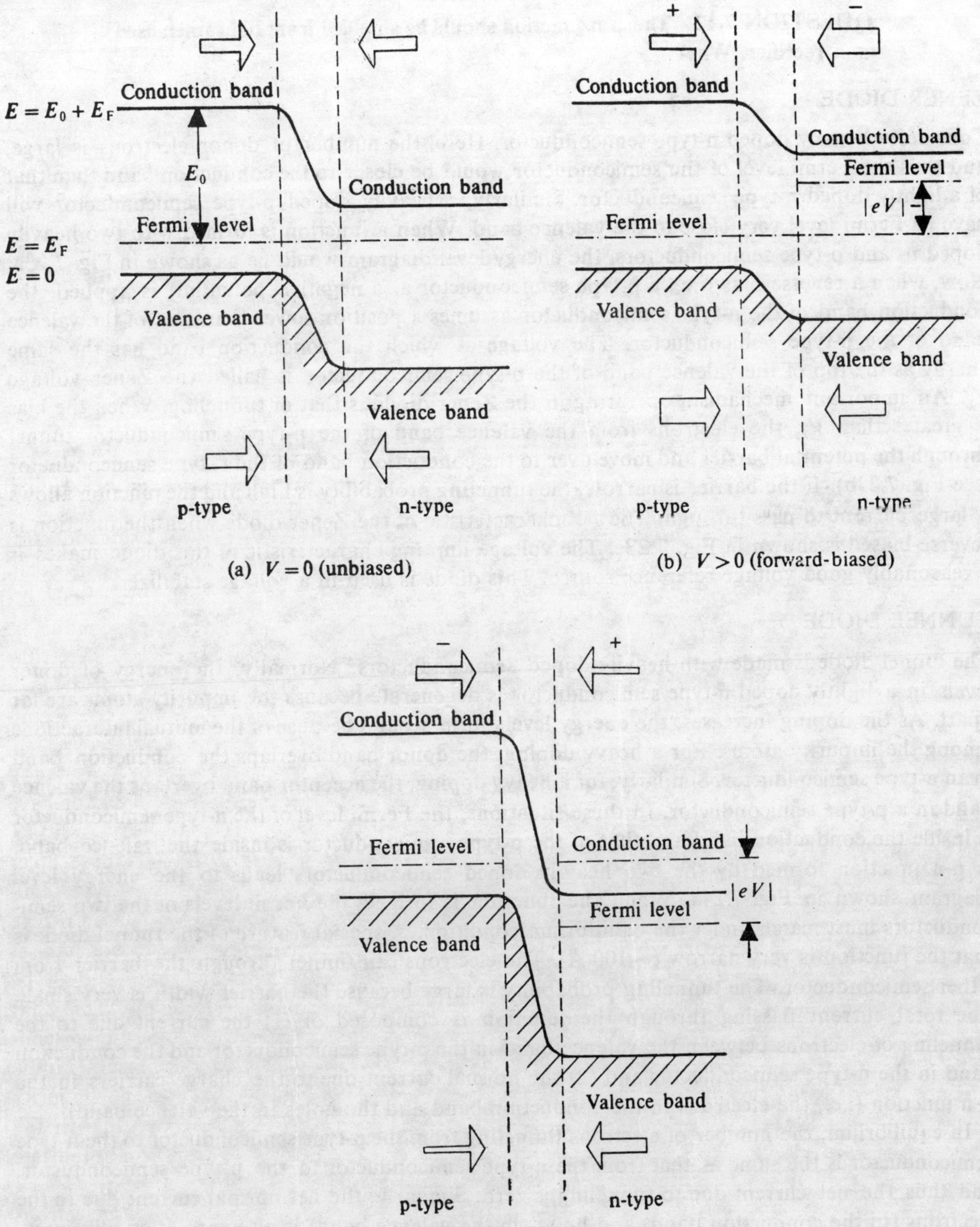

Fig. 7.22 Energy levels of p-n junction. V is applied potential; arrows in conduction band denote direction of electron motion and in valence band, direction of hole motion. Width of arrows indicates magnitude of current.

QUESTION 7.12 The p-n junction should be shielded from light when used as a rectifier. Why?

ZENER DIODE

Consider a heavily doped n-type semiconductor. Here, the number of donor electrons is large, and thus the Fermi level of the semiconductor would be closer to the conduction band than that of a lightly doped n-type semiconductor. Similarly, a heavily doped p-type semiconductor will have its Fermi level very close to the valence band. When a junction is formed with two heavily doped n- and p-type semiconductors, the energy level diagram would be as shown in Fig. 7.23a. Now, when a reverse bias (i.e., a p-type semiconductor at a negative potential) is applied, the conduction band of the n-type semiconductor assumes a position lower than that of the valence band of the p-type semiconductor. The voltage at which the conduction band has the same energy as the top of the valence band of the p-type semiconductor is called the Zener voltage V_z. An important mechanism operating in the Zener diode is that of tunneling. When the bias is greater than V_z, the electrons from the valence band of the p-type semiconductor tunnel through the potential barrier and move over to the conduction band of the n-type semiconductor (see Fig. 7.23b). If the barrier is narrow, the tunneling probability is high and the junction allows a large current to pass through. The *I-V* characteristic of the Zener diode when the junction is reverse-biased is shown in Fig. 7.23c. The voltage limiting characteristic of this diode makes it a reasonably good voltage reference source. This diode is used in a voltage stabilizer.

TUNNEL DIODE

The tunnel diode is made with heavily doped semiconductors. Normally, the energy of donor levels in a lightly doped n-type semiconductor is degenerate because the impurity atoms are far apart. As the doping increases, the energy levels form a band because of the mutual interactions among the impurity atoms. For a heavy doping, the donor band overlaps the conduction band in an n-type semiconductor. Similarly, for a heavy doping, the acceptor band overlaps the valence band in a p-type semiconductor. In these situations, the Fermi level of the n-type semiconductor is inside the conduction band and that of the p-type semiconductor is inside the valence band. A p-n junction formed by the two heavily doped semiconductors leads to the energy level diagram shown in Fig. 7.24a. When the junction is formed, the Fermi levels of the two semiconductors must match under the equilibrium condition. A special feature of the tunnel diode is that the junction is very narrow ($\sim$100 Å). The electrons can tunnel through the barrier from either semiconductor. The tunneling probability is large because the barrier width is very small. The total current passing through the junction is composed of (1) the current due to the tunneling of electrons between the valence band in the p-type semiconductor and the conduction band in the n-type semiconductor and (2) the normal current due to the charge carriers in the p-n junction (i.e., the electrons in the conduction band and the holes in the valence band).

In equilibrium, the number of electrons tunneling from the p-type semiconductor to the n-type semiconductor is the same as that from the n-type semiconductor to the p-type semiconductor, and thus the net current due to tunneling is zero. Similarly, the net normal current due to the electrons (in the conduction band) and holes (in the valence band) is also zero at equilibrium.

When a forward bias is applied, the energy levels in the n-type semiconductor will move up with respect to those in the p-type semiconductor. For a small forward bias, the number of electrons tunneling through from the n-type semiconductor to the p-type semiconductor is more than that in the opposite direction. This is because the electrons in the filled states in the

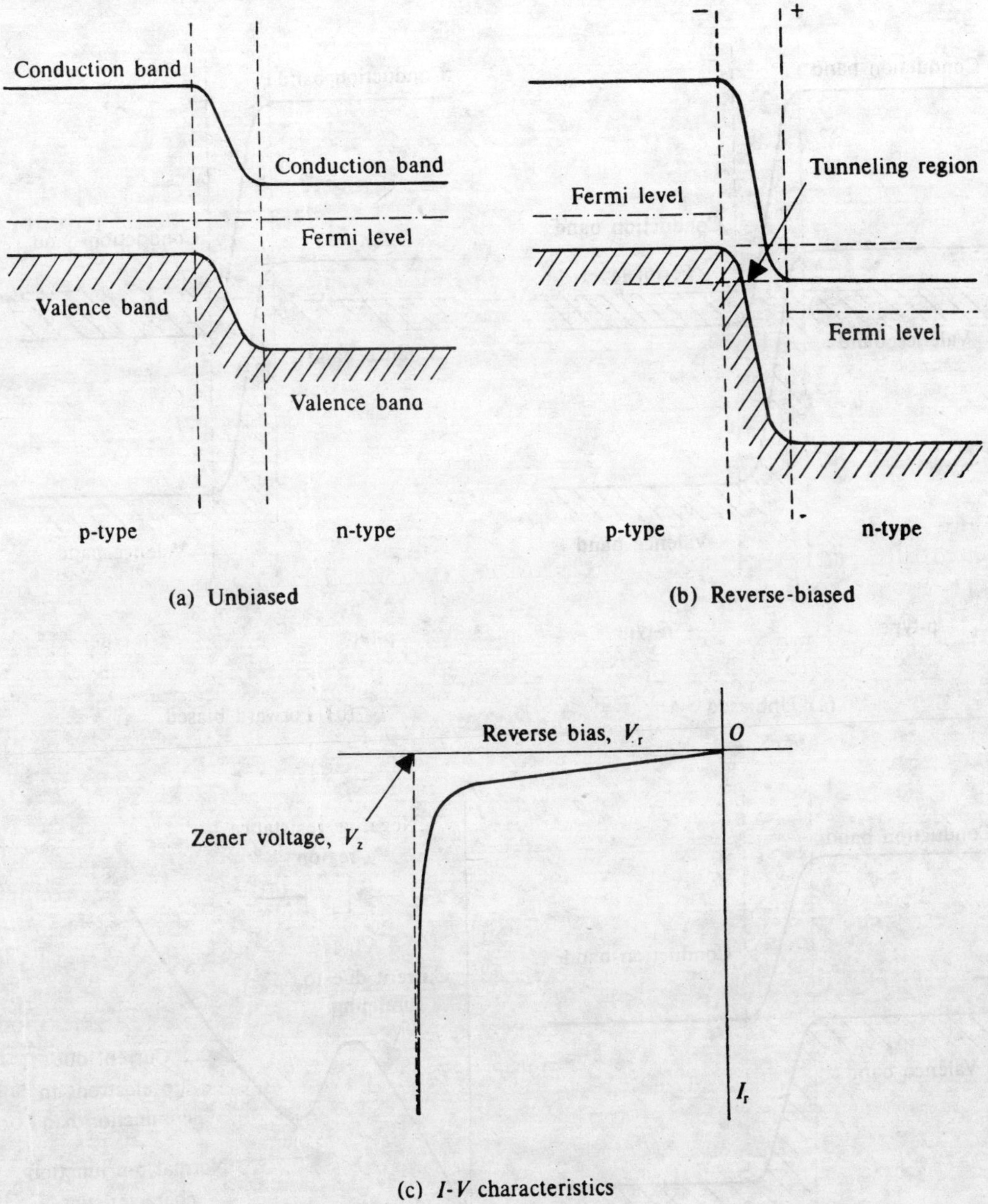

Fig. 7.23 Energy levels of Zener diode and its *I-V* characteristic.

conduction band of the n-type semiconductor move to the unfilled states in the valence band of the p-type semiconductor. This has been indicated by an arrow in Fig. 7.24b. Therefore, in the total current passing through the junction, the component of current due to tunneling is large as compared to the normal current component. As the forward bias increases, the energy of the bottom of the conduction band in the n-type semiconductor rises, thus increasing the tunneling

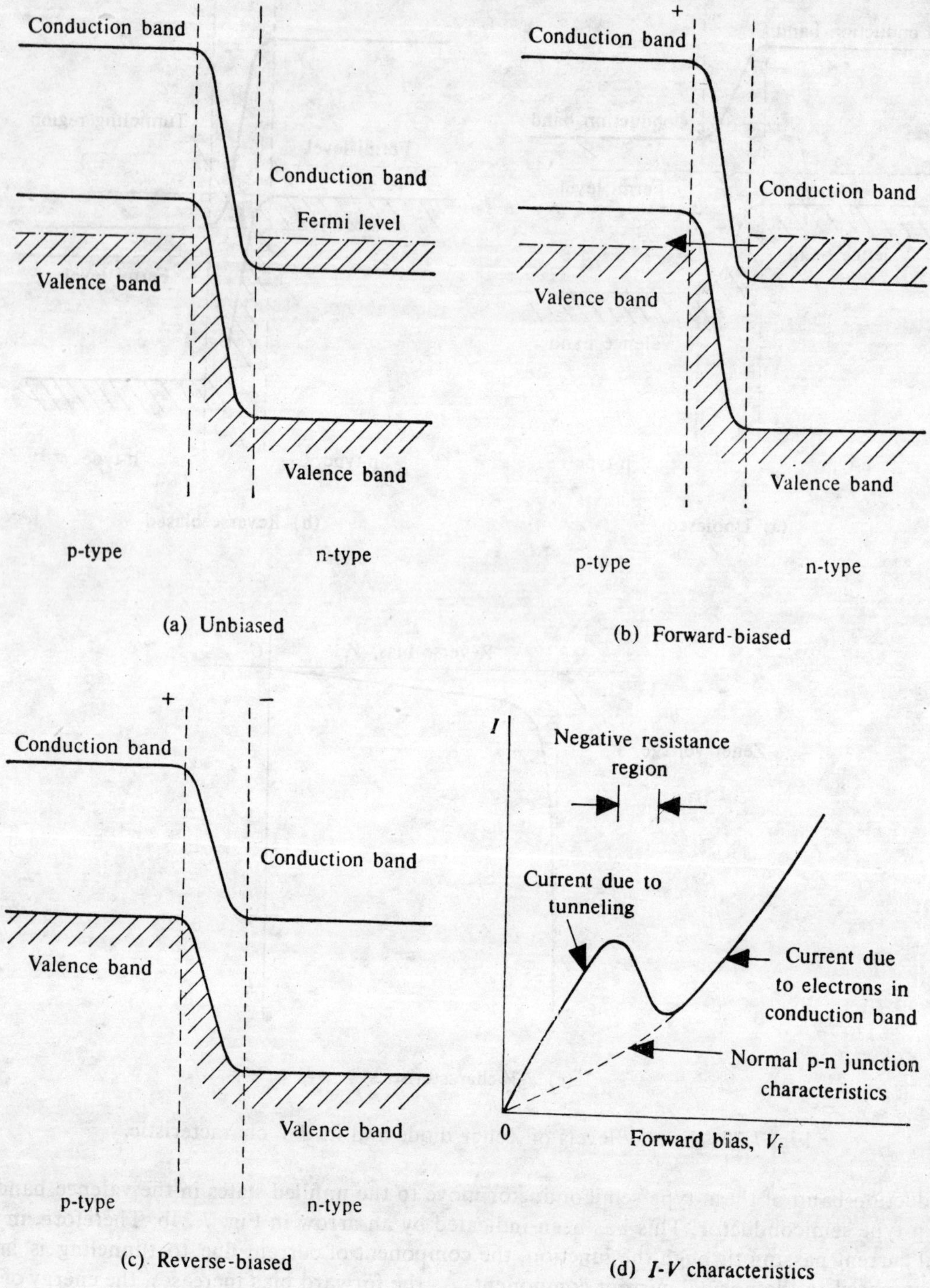

Fig. 7.24 Energy levels of tunnel diode and its *I-V* characteristic.

current. Further, when this energy approaches the energy of the top of the valence band in the p-type semiconductor, the number of available states in the valence band gradually decreases. Moreover, when this energy exceeds the energy of the top of the valence band, the tunneling is no longer possible because, at this energy, no levels are available in the p-type semiconductor (see Fig. 7.24c). The total current now is the normal current due to the p-n junction charge carriers. In other words, the *I-V* characteristic becomes the same as that in a normal p-n junction, as depicted in Fig. 7.24d. The *I-V* characteristic in this figure shows a negative resistance in a certain region, implying that the current decreases as the voltage increases. This characteristic is used in making an oscillator.

THE SOLAR CELL

Consider a p-n junction. If a beam of light falls on this junction, then an electron from the valence band of the p-type semiconductor can move to the conduction band of the n-type semiconductor (the photons of sufficiently small energy can cause this transition), and thus the two sides of the junction are oppositely charged. This is because the p-type semiconductor has a positively charged hole in its valence band, whereas the n-type semiconductor has an additional electron in its conduction band, making it negatively charged. This leads to a potential difference which can be used to give current to an external circuit. Thus, solar energy can be converted into electrical energy using a p-n junction. This device is known as the solar cell.

PROBLEMS

7.1 Each atom in silver metal has one free electron. Calculate the Fermi energy of the metal. (The density of silver is 10.5×10^3 kg-m^{-3} and its atomic weight is 108.)

7.2 Assume that the conduction electrons in a metal constitute a system of ideal Fermi gas. Show that the fractional number of conduction electrons having an energy $E > E_F$ at a temperature T is

$$\frac{N_1}{N} = \frac{\displaystyle\int_0^\infty \frac{(y + x_F)^{1/2}}{\exp(y) + 1}\,dy}{\displaystyle\int_{-x_F}^\infty \frac{(y + x_F)^{1/2}}{\exp(y) + 1}\,dy},$$

where $x_F = E_F/(kT)$ and $y = (E - E_F)/(kT)$. (We can use this expression to show that $N_1/N = kT/E_F$ if $kT \ll E_F$.)

7.3 Show that the occupation probability $n(E)$ satisfies the relation

$$n(E) = 1 - n(2E_F - E).$$

(This expression is useful in showing that the holes also follow the Fermi-Dirac distribution.)

7.4 Find the occupation probability $n(E)$ for $(E - E_F)/(kT) = 1, 3, 10$. If we approximate the F-D distribution by using the M-B distribution, what will be the fractional error in the probabilities obtained?

7.5 For lithium, the Fermi energy at $T = 0$ K is 4.7 eV and the density of electrons is 4.6×10^{28} per m^3. Find the electron density for a metal with Fermi energy 2.35 eV at $T = 0$ K.

7.6 Consider a system of ideal Fermi gas. Show that the number of gas particles per unit volume having a momentum between p and $(p + dp)$ is

$$n(p_x, p_y, p_z)\, dp_x\, dp_y\, dp_z = \frac{2dp_x\, dp_y\, dp_z}{h^3} \frac{1}{\exp[(E - E_F)/(kT)] + 1}.$$

Assuming that $\exp[(E - E_F)/(kT)] \gg 1$, show that the number of particles between p_x and $(p_x + dp_x)$ is

$$n(p_x)\, dp_x = \exp\left(\frac{E_F}{kT}\right)\frac{4\pi m_e kT}{h^3} \exp\left(-\frac{p_x^2}{2m_e kT}\right) dp_x.$$

7.7 Use the notation of Problem 7.6 to show that the number of electrons, with a momentum between p_x and $(p_x + dp_x)$, crossing per second a surface area A normal to the x-axis is $A(p_x/m_e)n(p_x)\, dp_x$. If these electrons were inside a metal, show that the current emitted by the area A would then have been $CT^2 \exp[-\phi/(kT)]$, where ϕ is the work function of the metal and $C = 4\pi k^2 m_e Ae/h^3$ with e being the electronic charge. This equation is known as the Richardson equation for thermionic emission.

7.8 Consider an alloy made of zinc and copper. Assume that the alloy has the same interatomic spacing as copper, which is possible only when the percentage of zinc in the alloy is small. Calculate the Fermi energy of the alloy at $T = 0$ as a function of the percentage of zinc. The Fermi energy of copper is 7.1 eV, and the valency of Zn is 2 and of Cu it is 1.

7.9 Calculate the molar electronic specific heat for silver at $T = 1000$ K, given the Fermi energy of silver to be 5.5 eV. Also calculate the percentage contribution of the result so obtained to the total specific heat of Ag. Assume that this total specific heat is given by the Dulong-Petit law.

7.10 The energy gaps for Si, Ge, and Ag are 1.1 eV, 0.7 eV, and 0 eV, respectively. Find the wavelength range of the electromagnetic waves to which these solids are opaque.

7.11 What would be the mobility of electrons in copper if there are 9×10^{28} valence electrons per m^3 and the conductivity of Cu is 6×10^7 mho/m?

7.12 A pure semiconductor having an energy gap of 1 eV is placed in a furnace of temperature 1000 K. Calculate the density of electrons in the conduction band of the semiconductor if the density of electrons in the valence band of the semiconductor is 10^{30} per m^3. Assume that the density of states in the conduction and valence bands is the same and is independent of temperature.

7.13 Find the temperature at which the number of electrons in the conduction band of a semiconductor increases by a factor of 10 over the number of electrons in it at the room temperature. The band gap for the semiconductor is 0.67 eV. (At room temperature, $kT \sim 1/40$ eV.)

7.14 A semiconductor has 6×10^{19} electrons and 7×10^{20} holes per cubic metre. If the mobilities of electrons and holes are 0.10 m^2/V-sec and 0.06 m^2/V-sec, respectively, calculate the conductivity of the semiconductor.

7.15 Silicon is doped with antimony. The concentration of Sb in Si is 10^{20} atoms/m^3. Calculate the difference between the energy of the bottom of the conduction band of Si and that of the donor level of Sb. Assume that the dielectric constant of Si is 12.

7.16 In Problem 7.15, estimate the number of silicon atoms that will be within the radius of the electron orbit of the impurity atom in its ground state. The separation between the silicon atoms is 3 Å.

7.17 Consider a doped semiconductor. In it, the concentration of electrons and holes is n_e and n_h, respectively. Show that $n_e n_h = n_i^2$, where n_i is the concentration of electrons (=holes) in the semiconductor before it is doped.

7.18 The conductivity of intrinsic Ge at a given temperature is doubled if the donor atoms with a concentration of 1.2×10^{20} per m^3 are introduced. Assume that all the donor atoms are ionized and the hole mobility in Ge is one-half the electron mobility in it. Calculate the concentration of each charge carrier in the intrinsic Ge and the concentration of holes in the doped Ge.

7.19 An intrinsic semiconductor has in its conduction band an electron density equal to 10^{20} per m^3. On doping with donor impurity, the conductor conductivity increases by a factor of 3. Calculate the number of conduction electrons and holes in the doped semiconductor. Assume that the mobility of electrons and holes in the semiconductors is the same.

7.20 An intrinsic semiconductor has a carrier density of 10^{20} per m^3 at room temperature. It is so doped with arsenic atoms that there are 10^{21} arsenic atoms per m^3. Calculate the density of hole carriers at room temperature, assuming that all the donor atoms are ionized.

7.21 Consider a model of an intrinsic semiconductor with the energy of its valence band lying in the range $-\infty < E < 0$ and that of its conduction band in the range $\infty > E > E_g$. The density of states in the two bands is

$$g(E) = A(E - E_g)^3, \qquad \infty > E > E_g,$$

$$g_h(E') = B(E' - E_g)^4, \qquad \infty > E' > E_g,$$

where $E' = E_g - E$ and A and B are positive constants having the same magnitude when the energies of the two bands are expressed in eV. Calculate the difference $E_F - E_g/2$.

7.22 Consider an n-type semiconductor with a band gap E_g and Fermi level E_F (recall that energy is measured from the top of the valence band). Let the energy of the donor level be E_d. The number of electrons N_e in the conduction band of the semiconductor is

$$N_e = g_e \Delta E \frac{1}{\exp[(E_g - E_F)/(kT)] + 1}$$

and the number of electrons N_d in the donor level is

$$N_d = L_d \frac{1}{\exp[(E_d - E_F)/(kT)] + 1},$$

where L_d is the total number of donor levels. Assuming that all the electrons in the conduction band originate from the donor levels, show that

$$E_F = \tfrac{1}{2}(E_g + E_d) + \frac{kT}{2} \ln\left(\frac{L_d}{g_e \Delta E}\right)$$

when each of $\exp[(E_g - E_F)/(kT)]$ and $\exp[(E_d - E_F)/(kT)]$ is very much greater than 1.

8

Special Theory of Relativity

8.1 INTRODUCTION

One of the most exciting events in the twentieth century is the development of the special and general theories of relativity by Einstein. In this chapter, we shall discuss only the special theory of relativity.

All physical laws relate some physically measurable quantities to one another. A physical quantity is measured in some specific reference frame. Thus, all the laws of physics can be verified only in a reference frame in which the measurements are made. For example, let us consider Newton's first law of motion. According to this law, a body is at rest or has a uniform velocity unless it is acted upon by some force. In order to verify this law, we consider a particle at rest at a given instant, say, $t = 0$, in some reference frame. If no force acts on this particle, then the particle continues to be at rest in the chosen reference frame. We now consider another frame of reference which is accelerating with respect to the first frame. We assume that the velocity of the new frame at $t = 0$, with respect to the first frame, is zero. Will now the particle viewed from the second frame obey Newton's first law? The answer is clearly no as this frame is accelerating with respect to the first frame in which the particle is at rest. Instead, the particle will appear to be accelerating. Thus, Newton's first law is applicable only in the first frame in our example. Newton postulated the presence of *absolute space* and defined frames known as inertial frames. The inertial frames are at rest or in uniform motion with respect to the absolute space. Newton's laws were then assumed to be valid only in these inertial frames.

Consider two inertial frames S and S', the latter moving with a velocity v along the positive x-axis with respect to S. The origins of the two frames coincide at the time $t = 0$. An event occurring at a point P can be observed from either of the frames S and S'. Let this event occur at a position (x, y, z) at the time t for an observer in the frame S and at (x', y', z') at a time t' for an observer in the frame S'. The transformations relating the two inertial frames S and S' can be mathematically represented as

$$x' = x - vt, \qquad y' = y, \qquad z' = z, \qquad t' = t. \tag{8.1}$$

Newton's laws, if they are valid in the frame S, can be shown to be also valid in the frame S'. This is clear because if a body is moving with a uniform velocity in one frame, then it will also move with a uniform (but different) velocity in the other frame. Further, the accelerations in both the frames are the same. Transformations (8.1) are known as the Galilean transformations.

QUESTION 8.1 Assume that a frame is rigidly attached to the surface of the earth. Is such a frame inertial?

Let us consider, as another example, Maxwell's equations of electromagnetism. As we know, these equations describe the electromagnetic waves in vacuum. In which frames are these equations valid? Maxwell assumed that light travels in a medium called ether, and the electromagnetic equations were held to be valid in a frame which is at rest with respect to ether. Ether was assumed to be at rest everywhere and it was reasonable to believe that this is the same frame in which Newton's laws are valid. Since the earth rotates about its axis and moves around the sun in an elliptic orbit, the surface of the earth must be moving with respect to ether. It was presumed that if the velocity of light could be measured, it would then be possible to determine the velocity of earth with respect to ether. This then would have made it possible to find the absolute frame in which the laws of mechanics and electrodynamics would be valid. Many experiments were performed to measure the velocity of earth through ether. However, all these experiments have been unsuccessful. Of these, we shall describe here the basic principle of one experiment, namely, the Michelson-Morley experiment.

8.2 THE MICHELSON-MORLEY EXPERIMENT

A schematic diagram of the experimental set-up A. A. Michelson and E. W. Morley used is shown in Fig. 8.1. In this experiment, a light beam from an extended source is made to fall on a partially silvered glass plate M. A part of the light beam then gets reflected to the plane mirror M_1 and the rest of the beam falls on the plane mirror M_2. The plate M is kept at 45° to the original beam so that the two beams 1 and 2 are at right angles to each other. Both M_1 and M_2 are kept almost normal to beams 1 and 2, respectively, so that these two beams are reflected back on almost the same path. A part of beam 1 reflected from M_1 is transmitted through the glass plate M to the telescope T. Similarly, a part of beam 2 reflected back from M_2 is reflected by the glass plate M onto the telescope. These two beams interfere with each other and the interference pattern can be seen through the telescope. The interference fringes are formed because of the phase difference between the two beams. The phase difference depends on the path length and velocity of light, and thus the observation of fringes should indicate the precise velocity of light and the velocity of earth through ether.

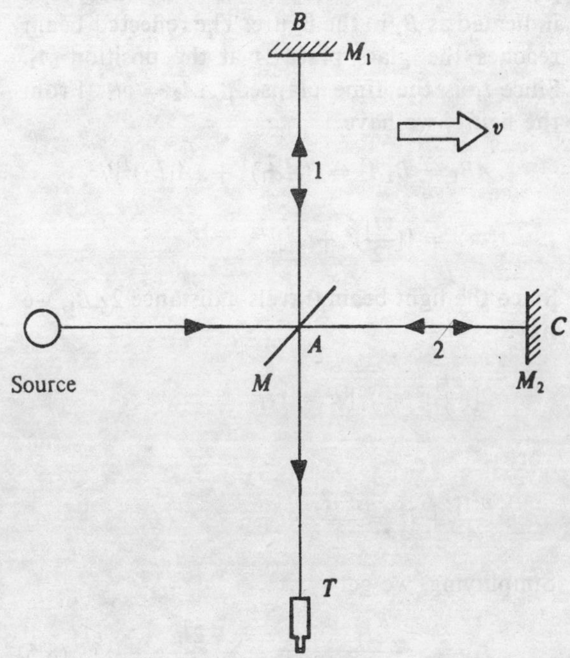

Fig. 8.1 Michelson-Morley experiment.

Let us assume that the apparatus is moving with a velocity v with respect to ether (Fig. 8.1). Let the length of AB and AC be l_1 and l_2, respectively. Since light is assumed to move with a velocity c in ether, let us look at the apparatus from the ether frame in which the apparatus has a velocity v along AC. If the light beam left A at $t = 0$ and struck the mirror M_2 at time $t = T_2$, then the mirror must have moved a distance vT_2. Thus, the light beam travels a distance

$l_2 + vT_2$ and not just l_2. Since c is the velocity of light and it has travelled for a time T_2, we have $l_2 + vT_2 = cT_2$ or

$$T_2 = l_2/(c - v). \tag{8.2}$$

If T_2' is the time needed for the light to travel from the mirror M_2 to the glass plate, then the distance travelled by light would be $l_2 - vT_2'$ since in time T_2' the glass plate would have moved forward a distance vT_2'. Therefore, we have $l_2 - vT_2' = cT_2'$ or

$$T_2' = l_2/(c + v). \tag{8.3}$$

Thus, t_2, the total time taken for the light beam to travel from A to C and back, is

$$t_2 = T_2 + T_2' = l_2\left(\frac{1}{c - v} + \frac{1}{c + v}\right) = l_2\frac{2c}{c^2 - v^2}. \tag{8.4}$$

In order to calculate the time t_1 taken by the light beam to travel from A to B and back, we use Fig 8.2. This figure has been drawn in the manner it would be viewed from the ether frame. The beam leaving A at the time $t = 0$ strikes the mirror B at a time $t_1/2$ after it has shifted by a distance $vt_1/2$ to the right. The position of the mirror at this time has been indicated as B_1 in the figure. The reflected beam reaches the glass plate A at the position A_2. Since t_1 is the time elapsed, $AA_2 = vt_1$. From the figure, we have

$$AB_1 = B_1A_2 = [(AA_1)^2 + (A_1B_1)^2]^{1/2}$$

$$= \left[\left(\frac{vt_1}{2}\right)^2 + l_1^2\right]^{1/2}.$$

Since the light beam travels a distance $2AB_1$, we have

$$2\left[\left(\frac{vt_1}{2}\right)^2 + l_1^2\right]^{1/2} = ct_1$$

or

$$\frac{v^2t_1^2}{4} + l_1^2 = \frac{c^2t_1^2}{4}.$$

Simplifying, we get

$$t_1 = \frac{2l_1}{(c^2 - v^2)^{1/2}} = \frac{2l_1}{c(1 - v^2/c^2)^{1/2}}. \tag{8.5}$$

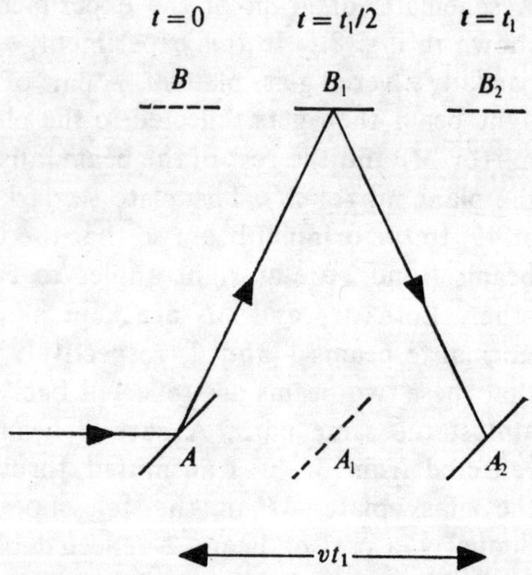

Fig. 8.2 Path of beam 1 as seen from ether frame.

The time difference between the two beams at the point A is

$$\Delta t = (t_1 - t_2) = \frac{2}{c}\left[\frac{l_1}{(1 - v^2/c^2)^{1/2}} - \frac{l_2}{1 - v^2/c^2}\right]. \tag{8.6}$$

If the apparatus is rotated by 90°, then the roles of the two paths are reversed. That is, the path CA (see Fig. 8.1) would be perpendicular whereas the path AB would be parallel to the velocity of the earth through ether. In this situation, the time difference $\Delta t'$ would be

$$\Delta t' = \frac{2}{c}\left[\frac{l_1}{1 - v^2/c^2} - \frac{l_2}{(1 - v^2/c^2)^{1/2}}\right]. \tag{8.7}$$

Since the two time differences (8.6) and (8.7) are not the same, we expect the interference pattern to change as we rotate the apparatus by 90°. In particular, the fringes will shift. The shift in the fringes is related to the time differences as

$$\Delta t' - \Delta t = \frac{2}{c}(l_1 + l_2)\left[\frac{1}{1 - v^2/c^2} - \frac{1}{(1 - v^2/c^2)^{1/2}}\right]. \tag{8.8}$$

If λ is the wavelength of the light used, then the time period of the wave is λ/c. The difference of one period in time between the two beams would cause a path difference of λ and there would be a fringe shift of one unit. Thus, when the apparatus is rotated by 90°, we expect a fringe shift $\Delta\lambda$ given by

$$\frac{\Delta\lambda}{\lambda} = \frac{2}{c}(l_1 + l_2)\left[\frac{1}{1 - v^2/c^2} - \frac{1}{(1 - v^2/c^2)^{1/2}}\right]\frac{c}{\lambda} = 2\frac{l_1 + l_2}{\lambda}\left[\frac{1}{1 - v^2/c^2} - \frac{1}{(1 - v^2/c^2)^{1/2}}\right]. \tag{8.9}$$

In the experiments Michelson and Morley conducted, the parameters were $l_1 = l_2 = 10$ m and $\lambda = 5.6 \times 10^{-7}$ m. If we assume that the speed of earth with respect to the sun is of the order of magnitude of the velocity of the earth through ether, then $v/c \sim 10^{-4}$. The fringe shift $\Delta\lambda$ will then be given by

$$\frac{\Delta\lambda}{\lambda} = \frac{2 \times 20}{5.6 \times 10^{-7}}\left[\frac{1}{1 - 10^{-8}} - \frac{1}{(1 - 10^{-8})^{1/2}}\right] \sim \frac{20 \times 10^{-8}}{5.6 \times 10^{-7}} \sim 0.4 \text{ fringe.}$$

The experiment was performed very carefully and was capable of detecting a shift of one-hundredth of a fringe. However, no such shift was detected. It was carried out during day and night and also at different times of the year to ensure that there was no accidental reason for obtaining a null result. The repetitive performance of the experiment clearly confirmed that there was no observable shift. The experiments that have more recently been carried out using microwaves have yielded an accuracy ten times better than that of the earlier measurements. However, the result of such experiments is the same, namely, there is no detectable velocity of the earth with respect to ether.

There have been several attempts to modify the ether theory but the various experimental results have not supported it. We shall not go into the details of these attempts but turn to Einstein's theory.

8.3 EINSTEIN'S THEORY OF RELATIVITY

In 1905, Einstein proposed his theory of relativity which completely solved the problem of measuring the velocity of earth through ether. The basic postulates of this theory are:

(1) The laws of physics are the same in all inertial frames.

(2) The speed of light c in free space is the same in all inertial frames.

The first postulate implies the absence of any preferred frame of reference. In other words, any inertial reference frame is as good as any other. If this postulate were not valid, then the physical laws would have been frame-dependent and it would have been possible to determine the objects at rest in space.

The second postulate can be understood from the first postulate as Maxwell's equations are valid in all inertial frames. The constant (i.e., $\mu_0\epsilon_0 = 1/c^2$) in Maxwell's equations is then independent of the reference frame. However, this constant is related to the velocity of light (see Appendix 1), and hence the velocity of light is the same in all inertial frames.

The result of the Michelson-Morley experiment can be easily appreciated once we assume that the velocity of light does not depend on the reference frame. As already noted, the time taken by the light beam to travel the two paths will be the same for any position of the apparatus. Also, there is no difference in the time lag when the apparatus is rotated, and thus there is no fringe shift.

The second postulate contradicts our concept of the velocity change when we move from one inertial frame to another. As we shall see, this postulate radically alters the concept of space and time. As an example, we shall show that, in a given inertial frame, two events occurring simultaneously at two different places need not be simultaneous when viewed from another inertial frame.

Consider a train AB moving with a speed v. There are two sets of observers. Of these, one set (P) is stationed in the train and the other set (Q) is stationed outside. The velocity v of the train is with respect to the set of observers Q. Let O be the centre of the train. One of the observers (of the set P) in the train sends a light signal from O at the time $t = 0$, which spreads in all directions (event one). The observers in the train, who are at the two ends of the train, will receive the light signal simultaneously (event two) because the light travels equal distances ($OA = OB$) in equal times (see Fig. 8.3). Let us consider the two events as viewed by the observers Q. Let Q' be an observer in the frame O' which coincides with O at the time $t = 0$ when the signal is flashed. Since the velocity of light in the frame O' is the same as in any other frame, the observers in this frame will see that the front end B of the train receives the light beam later than the rear end A of the train. This is because, while the light beam travels, the front end also moves, and hence the light beam has to cover a longer distance. This is illustrated in Fig. 8.4.

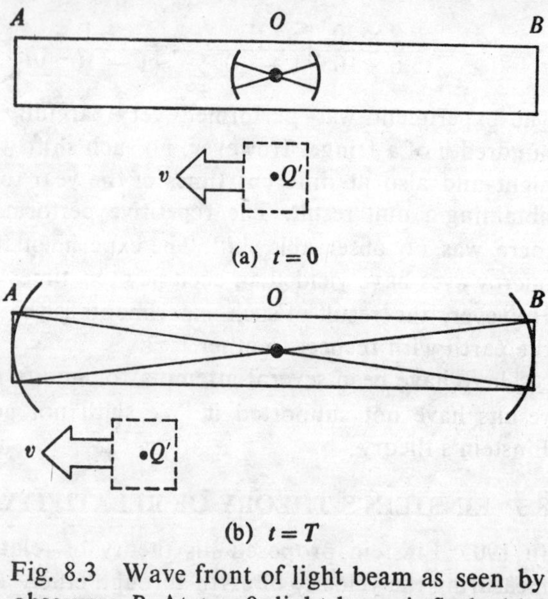

Fig. 8.3 Wave front of light beam as seen by observers P. At $t = 0$, light beam is flashed by observer at O. Observer Q' appears to move with velocity v in opposite direction. At $t = T$, light beam reaches end points of train. Observer Q' moves towards A.

The foregoing example leads to the radical conclusion that the events occurring simultaneously in one inertial frame are not simultaneous in another inertial frame. Simultaneity is not an absolute concept and depends on the inertial frame from which an event is observed. It has many implications. For example, the measurement of the length of any object is frame-dependent. This is because, in any measurement of length, say, of a rod, we have to simultaneously measure the position coordinates of its front end and rear end. As we shall soon discover, the measurement of time interval is also frame-dependent.

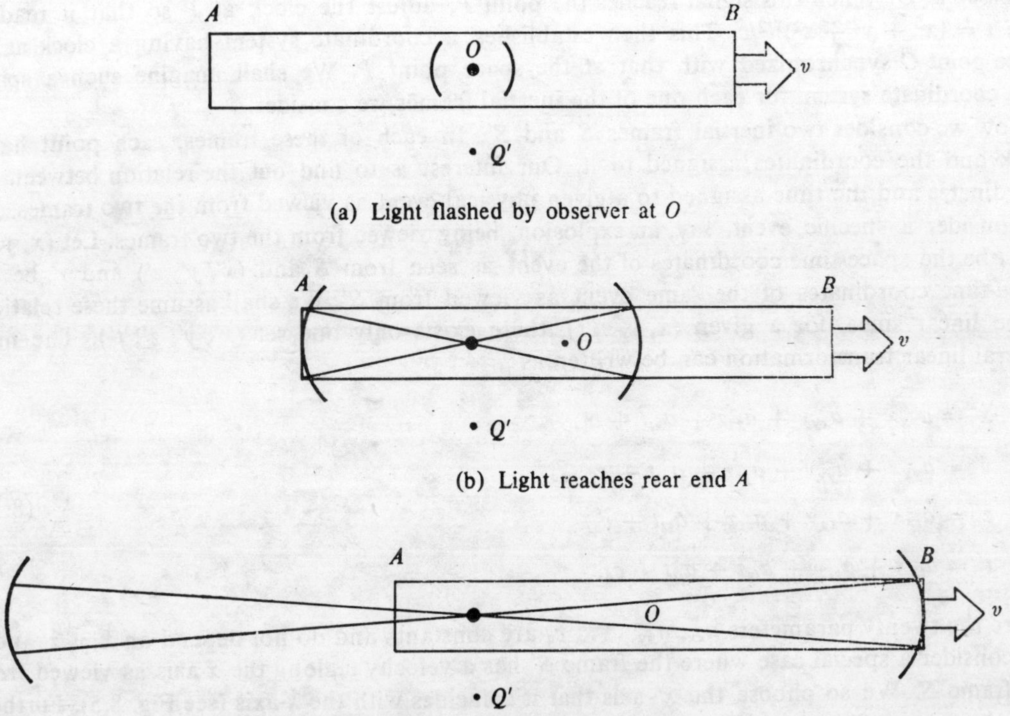

(a) Light flashed by observer at O

(b) Light reaches rear end A

(c) Light reaches front end and crosses rear end

Fig. 8.4 Wave front as seen by observer Q.

QUESTION 8.2 Consider two posts x_1 and x_2 equidistant from Q' (see Fig. 8.4). Will the light beam flashed by P at O reach x_1 and x_2 simultaneously (a) as seen by P? (b) as seen by Q?

8.4 THE LORENTZ TRANSFORMATION

The Lorentz transformation relates the space-time coordinates of an event measured from different inertial frames. First, we shall see how a set of coordinates can be established in an inertial frame. Let us choose a point, say, O, in space to be the origin and assume that there exists a standard clock at O and a right-handed coordinate system associated with O. We also assume that it is possible to determine the spatial coordinates (x, y, z) of any point P in space. One way to do this is to send a light beam from O to P and make it reflect back to O by placing a mirror at P. By measuring the time t taken by the light beam to go and come back, we can, as we know, find out the distance $OP[=t/(2c)]$. The coordinates (x, y, z) of P can now be obtained by measuring the angles between the line OP and the chosen set of x-, y-, and z-axis. Also, by keeping a free particle at rest with respect to the frame at (x, y, z), we can check whether we are in an inertial frame. If the frame is inertial, the particle will remain at rest. To obtain the time of occurrence of any event at (x, y, z), we need a clock at (x, y, z). This clock should be identical with the one kept at O. We can synchronize the clock at (x, y, z) with the one at the origin as follows. Let a light signal be sent from O at the time $t = 0$ as measured by the clock at O. When this signal reaches the point P, adjust the clock at P so that it reads a time $t = (x^2 + y^2 + z^2)^{1/2}/c$. This then establishes a coordinate system having a clock at the space point O synchronized with that at the space point P. We shall imagine such a space-time coordinate system for each one of the inertial frames we consider.

Now we consider two inertial frames S and S'. In each of these frames, each point has a clock and the coordinates assigned to it. Our interest is to find out the relation between the coordinates and the time assigned to a given physical event as viewed from the two frames. Let us consider a specific event, say, an explosion, being viewed from the two frames. Let (x, y, z) and t be the space-time coordinates of the event as seen from S and (x', y', z') and t' be the space-time coordinates of the same event as viewed from S'. We shall assume these relations to be linear since, for a given (x, y, z, t), there exists only one set (x', y', z', t'). The most general linear transformation can be written as

$$\begin{aligned} x' &= a_{xx}x + a_{xy}y + a_{xz}z + a_{xt}t + c_1, \\ y' &= a_{yx}x + a_{yy}y + a_{yz}z + a_{yt}t + c_2, \\ z' &= a_{zx}x + a_{zy}y + a_{zz}z + a_{zt}t + c_3, \\ t' &= a_{tx}x + a_{ty}y + a_{tz}z + a_{tt}t + c_4, \end{aligned} \tag{8.10}$$

where the twenty parameters $a_{xx}, a_{xy}, \ldots, c_4$ are constants and do not depend on x, y, z, and t. We consider a special case where the frame S' has a velocity v along the x-axis as viewed from the frame S. We so choose the x'-axis that it coincides with the x-axis (see Fig. 8.5). Further, the planes $y = 0$ and $y' = 0$ coincide for all times. Similarly, the planes $z = 0$ and $z' = 0$ coincide for all times. The clocks at O and O' will be assumed to start when the two origins coincide. Thus, the point $x = y = z = 0$ coincides with $x' = y' = z' = 0$ at the time $t = t' = 0$. This implies

$$c_1 = c_2 = c_3 = c_4 = 0. \tag{8.11}$$

Since $y' = 0$ and $y = 0$ coincide for all t, x, and z, we have

$$a_{yx}x + a_{yz}z + a_{yt}t = 0,$$

implying

$$a_{yx} = a_{yz} = a_{yt} = 0. \tag{8.12}$$

Therefore, from the second equation in (8.10), we get

$$y' = a_{yy}y, \tag{8.13}$$

where a_{yy} is a constant and can depend only on the relative velocity v between the two frames. Suppose there is another frame S'' which is moving with a velocity v along the negative x-axis (velocity is $-v$ along this axis) with respect to S. Then, from symmetry

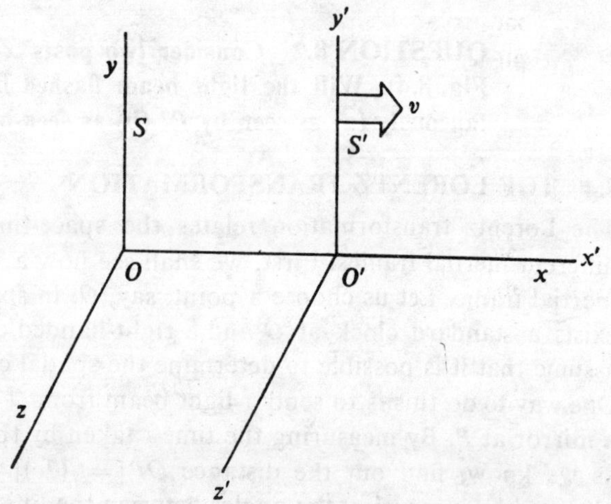

Fig. 8.5 Two inertial frames S and S'. S' is moving with velocity v with respect to S.

considerations, the coordinate y'' in S'' is, again, given by $y'' = a_{yy}y$. The transformation from S to S'' is identical with the transformation from S' to S because S appears to have a velocity $-v$ with respect to S'. Thus, we have

$$y = a_{yy}y'. \tag{8.14}$$

Combining Eq. (8.13) and Eq. (8.14), we get

$$y = a_{yy}^2 y \tag{8.15}$$

or $a_{yy}^2 = 1$. Therefore, we have

$$a_{yy} = \pm 1. \tag{8.16}$$

The coefficient a_{yy} cannot be -1 because this would lead to $y' = -y$ which is clearly unphysical since we expect the transformation to be continuously dependent on the relative velocity v. If the relative velocity v is gradually reduced to zero, we expect $y' = y$. Thus, the only solution is $a_{yy} = +1$. This gives

$$y' = y. \tag{8.17}$$

Similarly,

$$z' = z. \tag{8.18}$$

Thus, transformation (8.10) reduces to

$$\begin{aligned} x' &= a_{xx}x + a_{xy}y + a_{xz}z + a_{xt}t, \\ y' &= y, \\ z' &= z, \\ t' &= a_{tx}x + a_{ty}y + a_{tz}z + a_{tt}t. \end{aligned} \tag{8.19}$$

Consider now a particle at rest in S' at $(x' = 0, y', z')$. When viewed from the frame S, it will have a velocity v and its coordinates in S are $(x = vt, y, z)$ at a time t. Substituting $x = vt$ in transformation (8.19), we get

$$0 = a_{xx}vt + a_{xy}y + a_{xz}z + a_{xt}t. \tag{8.20}$$

This equation implies

$$a_{xy} = a_{xz} = 0, \qquad a_{xx}v = -a_{xt}. \tag{8.21}$$

Using Eqs. (8.21), we find the first equation in transformation (8.19) can be written as

$$x' = a_{xx}x - a_{xx}vt = a_{xx}(x - vt). \tag{8.22}$$

Let us again consider the transformation between S and S'', the latter having a velocity $-v$ with respect to S. Repeating the argument that led us to Eq. (8.22), we get

$$x'' = a_{xx}(x + vt). \tag{8.23}$$

The coefficient a_{xx} occurring in Eqs. (8.22) and (8.23) is the same since the only difference between the two equations is the direction of velocity. Using the fact that the relation between S and S'' is identical with that between S' and S, we obtain

$$x = a_{xx}(x' + vt'). \tag{8.24}$$

Applying Einstein's second postulate that the velocity of light is the same in all inertial frames, we get, for $x = ct$, $x' = ct'$ and vice versa. Substituting $x = ct$ and $x' = ct'$ in Eqs. (8.22) and (8.24), we have

$$ct' = a_{xx}(c - v)t, \tag{8.25}$$

$$ct = a_{xx}(c + v)t'. \tag{8.26}$$

Simplifying, we get

$$t' = a_{xx}^2 \frac{(c^2 - v^2)t'}{c^2} \tag{8.27}$$

or

$$a_{xx} = \frac{1}{(1 - v^2/c^2)^{1/2}}. \tag{8.28}$$

QUESTION 8.3 **Justify the reason for ignoring the solution** $a_{xx} = -1/(1 - v^2/c^2)^{1/2}$.

Substituting for a_{xx} from Eq. (8.28) in Eqs. (8.22) and (8.24), we get

$$x' = \frac{1}{(1 - v^2/c^2)^{1/2}}(x - vt), \tag{8.29}$$

$$x = \frac{1}{(1 - v^2/c^2)^{1/2}}(x' + vt'). \tag{8.30}$$

Eliminating x' from both these equations, we obtain

$$t' = \frac{1}{(1 - v^2/c^2)^{1/2}}\left(t - \frac{vx}{c^2}\right). \tag{8.31}$$

Collecting transformations (8.17), (8.18), (8.29), and (8.31), we have

$$x' = \gamma(x - vt), \qquad y' = y, \qquad z' = z, \qquad t' = \gamma(t - vx/c^2), \tag{8.32}$$

where $\gamma = 1/(1 - v^2/c^2)^{1/2}$. This transformation connecting the two frames S and S' (S' moving with a velocity v along the x-axis with respect to S) is referred to as the *standard Lorentz transformation.* It can be generalized to the situation where the velocity of S' is in an arbitrary direction. However, we shall be discussing only the standard transformation (8.32).

8.5 PROPERTIES OF THE LORENTZ TRANSFORMATION

We shall now take up the important properties of the Lorentz transformation. These properties help us understand many physical effects.

The Lorentz Transformation Is Symmetric between S and S'

It is possible to invert the Lorentz transformation (8.32) and express the coordinates (x, y, z, t) in terms of (x', y', z', t') (see Example 8.1) as

$$x = \gamma(x' + vt'), \qquad y = y', \qquad z = z', \qquad t = \gamma(t' + vx'/c^2), \tag{8.33}$$

where $\gamma = 1/(1 - v^2/c^2)^{1/2}$. It should be noted that transformation (8.33) is identical with transformation (8.32), except that, in the former, v has been replaced by $-v$. This is because the relative velocity of S as seen from S' is $-v$. Thus, the Lorentz transformation is completely symmetric between the frames S and S'.

EXAMPLE 8.1 Show that Eqs. (8.33) represent the inverse Lorentz transformation.

From transformation (8.32), we have

$$\frac{x'}{\gamma} = x - vt, \qquad \frac{t'}{\gamma} = t - \frac{vx}{c^2}.$$

Multiplying the second equation by v and adding the result to the first equation, we get

$$\frac{x' + vt'}{\gamma} = x - vt + vt - \frac{v^2 x}{c^2} = \frac{x}{\gamma^2} \qquad \text{or} \qquad x = \gamma(x' + vt').$$

Substituting for x in $x'/\gamma = x - vt$, we obtain

$$x'/\gamma = \gamma(x' + vt') - vt$$

or

$$vt = (\gamma - \frac{1}{\gamma})x' + v\gamma t',$$

$$\gamma - \frac{1}{\gamma} = \frac{\gamma^2 - 1}{\gamma} = \frac{1/(1 - v^2/c^2) - 1}{\gamma} = \frac{v^2/c^2}{\gamma(1 - v^2/c^2)} = \frac{v^2/c^2}{\gamma}\gamma^2 = \frac{v^2}{c^2}\gamma.$$

Thus,

$$vt = \frac{\gamma v^2 x'}{c^2} + v\gamma t' \qquad \text{or} \qquad t = \gamma(t' + \frac{vx'}{c^2}).$$

An Invariant Interval Exists

Let us consider two events P_1 and P_2 occurring at (x_1, y_1, z_1, t_1) and (x_2, y_2, z_2, t_2) in the inertial frame S and at (x_1', y_1', z_1', t_1') and (x_2', y_2', z_2', t_2') in the inertial frame S', respectively. Then, it can be shown that

$$c^2(t_1-t_2)^2-(x_1-x_2)^2-(y_1-y_2)^2-(z_1-z_2)^2=c^2(t_1'-t_2')^2-(x_1'-x_2')^2-(y_1'-y_2')^2-(z_1'-z_2')^2. \tag{8.34}$$

EXAMPLE 8.2 Prove relation (8.34).

Using Eqs. (8.32) for the events P_1 and P_2, we have

$$x_1' = \gamma(x_1 - vt_1), \qquad x_2' = \gamma(x_2 - vt_2),$$

$$y_1' = y_1, \qquad y_2' = y_2,$$

$$z_1' = z_1, \qquad z_2' = z_2,$$

$$t_1' = \gamma(t_1 - \frac{vx_1}{c^2}), \qquad t_2' = \gamma(t_2 - \frac{vx_2}{c^2}).$$

From these equations, we get

$$x_1' - x_2' = \gamma[x_1 - x_2 - v(t_1 - t_2)],$$

$$y_1' - y_2' = y_1 - y_2,$$

$$z_1' - z_2' = z_1 - z_2,$$

$$t_1' - t_2' = \gamma[t_1 - t_2 - \frac{v(x_1 - x_2)}{c^2}].$$

Substituting these equations in the right-hand side of Eq. (8.34), we obtain

$$c^2\gamma^2[(t_1 - t_2) - \frac{v}{c^2}(x_1 - x_2)]^2 - \gamma^2[(x_1 - x_2) - v(t_1 - t_2)]^2 - (y_1 - y_2)^2 - (z_1 - z_2)^2$$

$$= \gamma^2(t_1 - t_2)^2(c^2 - v^2) + \gamma^2(x_1 - x_2)^2(\frac{v^2}{c^2} - 1) - 2\gamma^2(t_1 - t_2)(x_1 - x_2)$$

$$\times (v - v) - (y_1 - y_2)^2 - (z_1 - z_2)^2$$

$$= (t_1 - t_2)^2 \frac{c^2 - v^2}{1 - v^2/c^2} + (x_1 - x_2)^2 \frac{v^2/c^2 - 1}{1 - v^2/c^2} - (y_1 - y_2)^2 - (z_1 - z_2)^2$$

$$= c^2(t_1 - t_2)^2 - (x_1 - x_2)^2 - (y_1 - y_2)^2 - (z_1 - z_2)^2$$

which is the same as the left-hand side of Eq. (8.34).

Consider now a light signal connecting the two events P_1 and P_2. That is, in the frame S, the light beam starts from (x_1, y_1, z_1) at a time t_1 and reaches (x_2, y_2, z_2) at a time t_2. Then,

$$(x_2 - x_1)^2 + (y_2 - y_1)^2 + (z_2 - z_1)^2 = c^2(t_2 - t_1)^2.$$

Similarly, in the frame S',

$$(x_2' - x_1')^2 + (y_2' - y_1')^2 + (z_2' - z_1')^2 = c^2(t_2' - t_1')^2.$$

Thus, when a light signal connects any two events, say, P_1 and P_2, then both the sides of Eq. (8.34) are zero. This is equivalent to saying that the velocity of light is the same in the two frames. The quantity

$$[c^2(t_2 - t_1)^2 - (x_2 - x_1)^2 - (y_2 - y_1)^2 - (z_2 - z_1)^2]^{1/2}$$

is called the *space-time interval* between the events P_1 and P_2. Equation (8.34) states that the space-time interval between any two events is the same in any inertial frame. We shall realize the significance of this statement in Section 8.6.

Frames Cannot Have Velocity Greater Than That of Light

The factor $\gamma = 1/(1 - v^2/c^2)^{1/2}$, as we shall now see, plays a crucial role in the Lorentz transformation. The value of γ is 1 when $v = 0$, and ∞ when $v = c$. If $v > c$, then $v^2/c^2 > 1$ and γ is imaginary. Thus, if the coordinates (x, t) are real, the coordinates (x', t') become imaginary. Since the Lorentz transformation connects two inertial frames specifying the same event, the relative velocity between the two frames must be less than the speed of light. Later in this chapter, we shall see that no particle or physical signal can travel with a speed greater than that of light.

One consequence of the fact that no physical signal can move with a velocity greater than that of light is that the concept of a rigid body is no longer valid. The concept of a rigid body implies that if we move a point in the body, then all the other points in it will move instantaneously in order to preserve the shape of the body. This can happen only if there is a signal which travels with an infinite velocity, and such signals do not exist.

The Lorentz Transformation Can Be Expressed in Hyperbolic Form

Let us define a parameter θ by the equation

$$\tanh \theta = v/c. \tag{8.35}$$

Then,

$$\sinh \theta = \frac{\tanh \theta}{(1 - \tanh^2 \theta)^{1/2}} = \frac{v}{c}\gamma, \tag{8.36}$$

$$\cosh \theta = (1 + \sinh^2 \theta)^{1/2} = \gamma. \tag{8.37}$$

Thus, the Lorentz transformation can be written in the form

$$\begin{aligned} x' &= x \cosh \theta - ct \sinh \theta, \\ y' &= y, \\ z' &= z, \\ ct' &= -x \sinh \theta + ct \cosh \theta. \end{aligned} \tag{8.38}$$

This form of the Lorentz transformation is used for many calculations. The corresponding inverse transformation is

$$\begin{aligned} x &= x' \cosh \theta + ct' \sinh \theta, \\ y &= y', \\ z &= z', \\ ct &= x \sinh \theta + ct' \cosh \theta. \end{aligned} \tag{8.39}$$

8.6 PHYSICAL EFFECTS OF THE LORENTZ TRANSFORMATION

Simultaneity Is Frame-Dependent

Suppose two events P_1 and P_2 occur simultaneously as seen by an observer in the frame S. Let the coordinates of P_1 and P_2 be (x_1, t_0) and (x_2, t_0), respectively (we shall consider only one space dimension). Then, the events, as seen by an observer in the frame S', will occur at times t_1' and t_2' given by the Lorentz transformations

$$t_1' = \gamma(t_0 - vx_1/c^2), \tag{8.40}$$

$$t_2' = \gamma(t_0 - vx_2/c^2), \tag{8.41}$$

where v is the relative velocity between the two frames. From transformations (8.40) and (8.41), it is clear that if $x_1 \neq x_2$, $t_1' \neq t_2'$. This shows that the two events do not occur simultaneously in the frame S'. Thus, the simultaneity concept is not absolute but frame-dependent. This is clearly a far-reaching conclusion and plays an important role in understanding how a physical phenomenon may look quite different in different inertial frames. As we go along, we shall also learn that a careless use of simultaneity leads to paradoxes which can be resolved only when we realize that simultaneous events in one frame are not, in general, simultaneous in another frame.

QUESTION 8.4 Assume that two events occur at the same place and at the same time for an observer. Will the occurrence of the two events be simultaneous in all frames? Explain.

Let an observer O measure the events P_1 and P_2 at (x_1, t_0) and (x_2, t_0) in the manner we have already described. In particular, we visualize the observer O having a set of clocks at rest in the frame S over all points in space and synchronized with one another. This set of clocks serves to measure the time coordinate in S. Similarly, there is a set of clocks synchronized with one another in the frame S'. The clocks of S are not synchronized with those in S' and vice versa. More specifically, when the events P_1 and P_2 occur, the clocks of S will show a time t, whereas those of S' will record t_1' and t_2'.

In order to picture the foregoing ideas, we consider the example of a rod lying at rest in the x-y plane in a frame S, with its two ends having the coordinates (x_1, y_1) and (x_2, y_2). The x- and y-component of the displacement represented by the rod are $(x_2 - x_1)$ and $(y_2 - y_1)$. If we choose a different frame S' making an angle with respect to S (S' being at rest with respect to S), then the coordinates (x_1', y_1') and (x_2', y_2') of the ends of the rod in S' will, in general, be different from those in S. Also, the x-component of the length of the rod $(x_2' - x_1')$ in S' will be different from $(x_2 - x_1)$. The same is true for the y-component of the length of the rod. However, as we know, the length of the rod is the same in both the frames. That is,

$$(x_2 - x_1)^2 + (y_2 - y_1)^2 = (x_2' - x_1')^2 + (y_2' - y_1')^2.$$

We now extend the foregoing concept of the length being invariant to the Lorentz transformation. As already noted, the time interval $(t_2 - t_1)$, just as the space interval $(x_2 - x_1)$, between any two events is frame-dependent. In the Lorentz transformation, the quantity that remains the same in the two frames is the space-time interval given by Eq. (8.34). Thus, the Lorentz transformation can be thought of as some generalized rotation in the $(x$-$t)$-space in which the invariant quantity is the space-time interval. It represents a generalized rotation in four dimensions (x, y, z, t). However, we have so far restricted our discussion mainly to the standard Lorentz transformation which involves one space coordinate x and the time coordinate t.

The graphical representation of the standard Lorentz transformation is done through a space-time diagram. We shall illustrate the frame-dependence of simultaneity in terms of such a diagram. Figure 8.6 shows the coordinates (x, ct) of an event as seen by an observer in a frame S. We have taken the time coordinate as ct rather than t in order to keep the dimension of the time coordinate the same as that of x. Consider now the event as viewed by an observer in a frame S'. The coordinates of the event can be obtained using the Lorentz transformation. Also, the ct'-axis is given by the equation $x' = 0$, and the x'-axis by the equation $ct' = 0$. Thus, we have, from Eq. (8.32),

$$x - \frac{v}{c}(ct) = 0 \qquad (ct'\text{-axis}),$$
$$ct - \frac{v}{c}x = 0 \qquad (x'\text{-axis}). \tag{8.42}$$

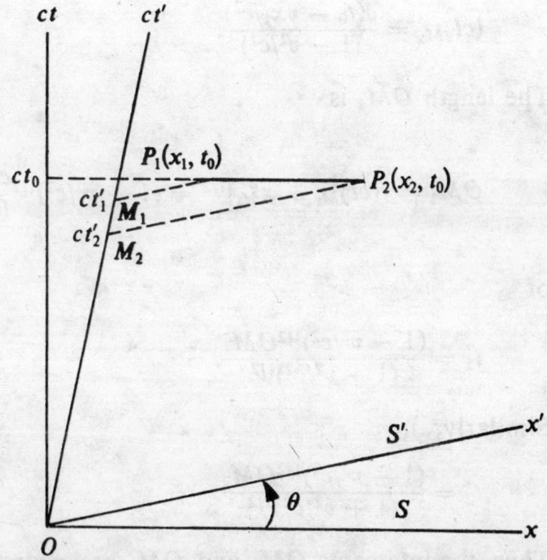

Fig. 8.6 Space-time diagram.

These two lines are also shown in Fig. 8.6. The angle θ is given by $\tan\theta = v/c$.

Let two events P_1 and P_2 occur at the same time when seen from the frame S. Thus, the coordinate ct of P_1 and P_2 is the same. To get the time coordinate in S', we draw two lines, each parallel to the x'-axis, one passing through P_1 and the other through P_2, meeting the ct'-axis at M_1 and M_2. The equation for the line P_1M_1 in terms of the coordinates (x, ct) can be obtained by noting that it is parallel to the line ox' which has a slope $v/c(=\tan\theta)$. This equation thus is

$$ct = \frac{v}{c}x + K, \tag{8.43}$$

where K is a constant (i.e., the intercept). We can determine this constant by observing that the line passes through the point $P_1(x_1, ct_0)$. Therefore,

$$ct_0 = \frac{v}{c}x_1 + K \qquad \text{or} \qquad K = ct_0 - \frac{v}{c}x_1.$$

This gives the equation for the line P_1M_1 as

$$ct = \frac{v}{c}x + ct_0 - \frac{v}{c}x_1 \qquad \text{or} \qquad ct - \frac{vx}{c} = ct_0 - \frac{vx_1}{c}.$$

This line meets the ct'-axis at the point M_1. To determine the point M_1, we substitute for x in this equation from Eq. (8.42). We thus have

$$ct(1 - \frac{v^2}{c^2}) = ct_0 - \frac{x_1}{c}.$$

Therefore, the coordinates of M_1, i.e., x_{M_1} and $(cT)_{M_1}$, are

$$x_{M_1} = \frac{v}{c}(ct)_{M_1},$$

$$(ct)_{M_1} = \frac{c(t_0 - vx_1/c^2)}{(1 - v^2/c^2)}. \tag{8.44}$$

The length OM_1 is

$$OM_1 = [(ct)^2_{M_1} + x^2_{M_1}]^{1/2} = (1 + v^2/c^2)^{1/2}\frac{c(t_0 - vx_1/c^2)}{(1 - v^2/c^2)^{1/2}} = \frac{(1 + \frac{v^2}{c^2})^{1/2}t_1'}{(1 - \frac{v^2}{c^2})^{1/2}} \tag{8.45}$$

or

$$t_1' = \frac{(1 - v^2/c^2)^{1/2}OM_1}{(1 + v^2/c^2)^{1/2}}. \tag{8.46}$$

Similarly,

$$t_2' = \frac{(1 - v^2/c^2)^{1/2}OM_2}{(1 + v^2/c^2)^{1/2}}. \tag{8.47}$$

Thus, the intercepts OM_1 and OM_2 are proportional to t_1' and t_2', respectively. The fact that $OM_1 \neq OM_2$ indicates that the two events occur at different times when seen from the frame S'.

Moving Rod Contracts (Length or Lorentz-Fitzgerald Contraction)

Consider a rod at rest in the frame S', with its ends E_1 and E_2 at $x'(E_1) = 0$ and $x'(E_2) = L$. Therefore, the value of x' for the two ends will be the same for all times t'. When the length of the rod is measured in the frame S, the rod appears to be moving with a velocity v with respect to an observer in the frame S. The positions of the ends E_1 and E_2 must be measured simultaneously as seen from S, say, at a time t_0. Then, in the frame S, the spatial coordinate of E_1 is

$$x(E_1) = \gamma[x'(E_1) + vt'], \tag{8.48}$$

where

$$t' = \gamma[t_0 - \frac{vx(E_1)}{c^2}] \tag{8.49}$$

or

$$x(E_1) = \gamma[x'(E_1) + v\gamma\{t_0 - \frac{vx(E_1)}{c^2}\}] \tag{8.50}$$

or

$$(1 + \frac{\gamma^2 v^2}{c^2})x(E_1) = \gamma[x'(E_1) + v\gamma t_0]. \tag{8.51}$$

Using the identity

$$1 + \gamma^2 v^2/c^2 = \gamma^2, \tag{8.52}$$

which can be derived by substituting $\gamma^2 = 1/(1 - v^2/c^2)$, we get

$$x(E_1) = \frac{1}{\gamma}[x'(E_1) + v\gamma t_0]. \tag{8.53}$$

Similarly,

$$x(E_2) = \frac{1}{\gamma}[x'(E_2) + v\gamma t_0]. \tag{8.54}$$

The length as seen from the frame S is

$$x(E_2) - x(E_1) = \frac{1}{\gamma}[x'(E_2) - x'(E_1)] = \frac{1}{\gamma}L. \tag{8.55}$$

Since $\gamma > 1$, the length of the rod as seen from S is less than L. In other words, there is a contraction in length in the direction of motion by a factor $(1 - v^2/c^2)^{1/2}$.

EXAMPLE 8.3 Consider a rod of length L at rest in the frame S. Find its length as measured by an observer in the frame S'.

Let the coordinates of the ends of the rod in S be $x(E_1) = 0$ and $x(E_2) = L$. If the rod is measured in S' at the time $t' = 0$, then the coordinates of the ends of the rod can be obtained by using the transformation

$$t' = 0 = \gamma(t - vx/c^2) \qquad \text{or} \qquad t = vx/c^2.$$

Thus, in S, the t-coordinate for the end E_1 is 0, whereas the t-coordinate for the end E_2 is vL/c^2. Using this information, we have

$$x'(E_1) = \gamma[x(E_1) - vt(E_1)] = 0,$$

$$x'(E_2) = \gamma[x(E_2) - vt(E_2)] = \gamma(L - \frac{vvL}{c^2}) = \gamma L(1 - v^2/c^2) = \gamma L\frac{1}{\gamma^2} = \frac{L}{\gamma}.$$

The length of the rod as measured by an observer in S' is L/γ.

We note that an observer in S sees the rod of S' contract, whereas an observer in S' sees the rod of S contract. Thus, the contraction is symmetric in the two frames.

QUESTION 8.5 The length of a rod measured in different frames is 10 cm, 10.1 cm, 9.5 cm, 7.8 cm. The rod is at rest in one of the frames. Which one?

EXAMPLE 8.4 Consider two bombs, kept at two points x_1 and x_2 separated by a distance of 1 km, exploding simultaneously as seen by an observer on the ground. Further, assume that an aeroplane is flying at a velocity of 1080 km/hr along the direction x_1x_2. What will be the difference in time between the two events (i.e., explosion of the bombs) for an observer in the frame of the plane?

Let us assume that there are clocks, at rest with respect to the ground, at the points x_1 and x_2 available for use by the observer on the ground for his measurements. Similarly, we also assume that there are clocks, at rest with respect to the plane, available for use by the observer in the frame of the plane. Then, from Eqs. (8.40) and (8.41), we have

$$t_1' - t_2' = \gamma\left(t_0 - \frac{vx_1}{c^2}\right) - \gamma\left(t_0 - \frac{vx_2}{c^2}\right) = \frac{\gamma v(x_2 - x_1)}{c^2} = \frac{1000\gamma v}{c^2} \text{ sec.}$$

Thus,

$$\frac{v}{c} = \frac{1080 \times 10^3}{3600 \times 3 \times 10^8} = 10^{-6},$$

$$\gamma = \frac{1}{(1 - v^2/c^2)^{1/2}} \sim 1 + \frac{v^2}{2c^2} \sim 1 + \tfrac{1}{2} \times 10^{-12},$$

$$t_1' - t_2' \sim \frac{10^{-3}}{3 \times 10^8} \sim 3 \times 10^{-12} \text{ sec.}$$

As we can see, the time difference is small.

EXAMPLE 8.5 A metre stick is observed from a frame in which it is moving with a speed of 1080 km/hr. What is the length of the stick?

The length of the metre stick, using the value of γ obtained in Example 8.4, is

$$L = \frac{1 \text{ m}}{\gamma} = (1 - \tfrac{1}{2} \times 10^{-12}) \text{ m.}$$

Moving Clock Goes Slow (Time Dilation)

Consider a time interval $\Delta t'$ measured by a clock stationed at the origin in a frame S'. As seen from an inertial frame S, the clock would move a distance $\Delta x = v\Delta t$, where Δt is the time interval as seen from S. From the invariance of the space-time interval [given by Eq. (8.34)], we have

$$c^2\Delta t^2 - \Delta x^2 = c^2\Delta t'^2 - \Delta x'^2.$$

Since the clock is at rest in S', $\Delta x' = 0$ and we have

$$c^2\Delta t^2 - v^2\Delta t^2 = c^2\Delta t'^2 \quad \text{or} \quad \Delta t^2 = \frac{c^2\Delta t'^2}{c^2 - v^2} = \frac{\Delta t'^2}{1 - v^2/c^2}$$

or

$$\Delta t = \frac{\Delta t'}{(1 - v^2/c^2)^{1/2}} = \gamma\Delta t'. \tag{8.56}$$

Thus, a one-second interval in S' (i.e., $\Delta t' = 1$) appears as γ seconds in S. Since $\gamma > 1$, the time interval appears dilated in S. In other words, the clock appears to go slow. The effect of time dilation is also symmetric. That is, a clock at rest in S when viewed from S' will also appear to go slow.

Time dilation has been experimentally studied and formula (8.56) has been verified. Certain particles, e.g., μ-mesons (see Chapter 11), have a lifetime of the order of 10^{-6} sec when at rest. However, when such particles decay in flight, the decay time is more due to time dilation. The dilation factor is the same as given by Eq. (8.56). We shall illustrate this in the example that follows.

EXAMPLE 8.6 The half-life of a π-meson at rest is 1.8×10^{-8} sec. (That is, to start with, if the number of π-mesons is N, only $N/2$ remain after 1.8×10^{-8} sec.) What will be the half-life of the π-meson when it travels with a speed $c/2$?

The factor γ is

$$\gamma = \frac{1}{(1 - v^2/c^2)^{1/2}} = \frac{1}{(1 - 1/4)^{1/2}} \sim \frac{2}{1.73}.$$

From Eq. (8.56), we note that the half-life will get dilated by the factor γ. Thus,

$$\text{half-life} = \frac{2}{1.73} \times 1.8 \times 10^{-8} \sim 2.08 \times 10^{-8} \text{ sec.}$$

No Signal Can Travel with Velocity Greater than c

Consider two events P_1 and P_2 connected by a signal which travels with a velocity $w > c$. Let the signal leave the point x_1, y_1, z_1 at a time t_1 (event P_1) and reach the point x_2, y_2, z_2 at a time t_2 (event P_2) as seen from a frame S. We must have $t_2 > t_1$ since the signal cannot reach P_2 before it leaves P_1. Let us view the two events from a frame S' moving with a velocity $v\,(<c)$ with respect to S. We have, from the Lorentz transformation,

$$t_1' = \gamma(t_1 - vx_1/c^2), \qquad t_2' = \gamma(t_2 - vx_2/c^2).$$

Subtracting the two, we get

$$\Delta t' = \gamma(\Delta t - v\Delta x/c^2),$$

where

$$\Delta t' = t_2' - t_1', \qquad \Delta t = t_2 - t_1, \qquad \Delta x = x_2 - x_1.$$

Since the signal travelled with a velocity w in the frame S, $\Delta x = w\Delta t$. Thus,

$$\Delta t' = \gamma(\Delta t - vw\Delta t/c^2) = \gamma\Delta t(1 - vw/c^2).$$

Since $w > c$, we can always find a velocity $v\,(<c)$ such that $vw > c^2$, and thus $\Delta t' < 0$. This implies that the signal reaches the destination even before it has left its starting point. This is clearly unphysical, and therefore we conclude that there exists no signal that can travel faster than light.

Are there particles with a velocity greater than that of light? This question has been and is being examined and its implications worked out.

> **QUESTION 8.6 If an event P_1 precedes an event P_2 in the frame S, will P_1 precede P_2 in all other inertial frames?**

8.7 VELOCITY TRANSFORMATION

Consider a particle whose coordinates in a frame S are given by (x, y, z, t). Then, the coordinates (x', y', z', t') of the particle in a frame S' can be obtained by using the Lorentz transformation (8.32), i.e.,

$$x' = \frac{x - vt}{(1 - v^2/c^2)^{1/2}}, \qquad y' = y, \qquad z' = z, \qquad t' = \frac{t - vx/c^2}{(1 - v^2/c^2)^{1/2}}.$$

Let us assume that the particle has a velocity u as seen from the frame S. The three components of the velocity are

$$u_x = dx/dt, \qquad u_y = dy/dt, \qquad u_z = dz/dt. \tag{8.57}$$

If the velocity of the particle as seen from the frame S' is u', then

$$u'_x = dx'/dt', \qquad u'_y = dy'/dt', \qquad u'_z = dz'/dt', \tag{8.58}$$

where u'_x, u'_y, and u'_z are the x-, y-, and z-component of the velocity as seen from S'. The calculation of the relation between the components of u' and u, using the Lorentz transformation, is straightforward. In order to do this, consider the Lorentz transformation, written in the infinitesimal form

$$dx' = \frac{dx - v\,dt}{(1 - v^2/c^2)^{1/2}}, \qquad dy' = dy, \qquad dz' = dz, \qquad dt' = \frac{dt - (v/c^2)\,dx}{(1 - v^2/c^2)^{1/2}}. \tag{8.59}$$

We have

$$u'_x = \frac{dx'}{dt'} = \frac{(dx - v\,dt)/(1 - v^2/c^2)^{1/2}}{[dt - (v/c^2)\,dx]/(1 - v^2/c^2)^{1/2}} = \frac{dx - v\,dt}{dt - v\,dx/c^2}$$

$$= \frac{dx/dt - v}{1 - (v/c^2)(dx/dt)}$$

or

$$u'_x = \frac{u_x - v}{1 - vu_x/c^2}. \tag{8.60}$$

Similarly,

$$u'_y = \frac{dy'}{dt'} = \frac{dy(1 - v^2/c^2)^{1/2}}{dt - v\,dx/c^2} = \frac{u_y(1 - v^2/c^2)^{1/2}}{1 - vu_x/c^2}, \tag{8.61}$$

$$u'_z = \frac{u_z(1 - v^2/c^2)^{1/2}}{1 - vu_x/c^2}. \tag{8.62}$$

We note the following points:

(1) The normal transformation of a velocity using the Galilean transformation $u'_x = u_x - v$, $u'_y = u_y$, $u'_z = u_z$ is not valid. However, if $v \ll c$ and $u_x \ll c$, we can neglect vu_x/c^2 and v^2/c^2 as compared to unity and the velocity transformations (8.61) and (8.62) become identical with those obtained using the Galilean transformation.

(2) The velocity transformations (8.61) and (8.62) are symmetrical between S and S'. In particular, we can invert these transformations (see Problem 8.10) and get

$$u_x = \frac{u'_x + v}{1 + vu'_x/c^2}, \tag{8.63}$$

$$u_y = \frac{u_y'(1 - v^2/c^2)^{1/2}}{1 + vu_x'/c^2}, \tag{8.64}$$

$$u_z = \frac{u_z'(1 - v^2/c^2)^{1/2}}{1 + vu_x'/c^2}. \tag{8.65}$$

(3) If the velocity of a particle is equal to c in a frame S, then it is also c in a frame S'. This can be easily seen by putting, for example, $u_x = c$, $u_y = u_z = 0$ in Eq. (8.60). Then,

$$u_x' = \frac{c - v}{1 - v/c} = c, \qquad u_y' = u_z' = 0.$$

This result should not surprise us as the entire relativity theory is based on the postulate that the velocity of light c is the same in all inertial frames.

(4) A particle as viewed from S can be thought of as being given a velocity $\mathbf{v} = (v, 0, 0)$ (to bring it at rest in S') followed by a velocity $u' = (u_x', u_y', u_z')$ with respect to S'. When the two velocities u' and v are added, the final velocity is not given by $(u' + v)$ but by the relativistic addition formulae (8.63) (8.65).

EXAMPLE 8.7 Consider two rockets 1 and 2 moving with a velocity $w = (w_x, 0, 0)$ and $-w = (-w_x, 0, 0)$, respectively, as measured from the frame S. Find the velocity of rocket 2 as measured from the frame of rocket 1.

The frame of rocket 1 (at rest with respect to rocket 1) is moving with a velocity $(w_x, 0, 0)$ with respect to S. Thus, we can consider the frame of rocket 1 as the frame S' moving with a velocity $v = w_x$. The given velocity of rocket 2 in the frame S is $(-w_x, 0, 0)$. Then, using Eq. (8.60) with $u_x = -w_x$, $v = w_x$, we get

$$u_x' = \frac{-w_x - w_x}{1 - (-w_x)w_x/c^2} = \frac{-2w_x}{1 + w_x^2/c^2}, \qquad u_y' = u_z' = 0.$$

Thus, the velocity of rocket 2 as calculated from the frame of rocket 1 is $[-2w_x/(1 + w_x^2/c^2), 0, 0]$.

Even if the velocity w_x of rocket 1 is c, the relative velocity of rocket 2 as seen from rocket 1 would be $-c$, that is, the two rockets (as seen from rocket 1) would approach each other with a velocity c.

EXAMPLE 8.8 An aeroplane moving at a speed of 1000 m/sec with respect to the ground shoots a rocket in the same direction as its own with a speed of 1000 m/sec as measured from the rocket frame. What will be the speed of the rocket as observed from the ground?

Here, we can assume the plane to be the frame S' and the ground to be the frame S. Then, $v = 1000$ m/sec and u_x' (the velocity of the rocket as measured from S') = 1000 m/sec. Therefore, using Eq. (8.63), we get the velocity of the rocket as observed from the ground as

$$u_x = \frac{1000 + 1000}{1 + \dfrac{(1000)(1000)}{9 \times 10^{16}}} = \frac{2000}{1 + \frac{1}{9} \times 10^{-10}} \sim (2000 - \tfrac{2}{9} \times 10^{-7}) \text{ m/sec.}$$

If we were to use the Galilean laws, the relative velocity u_x we would have obtained would have been 2000 m/sec, very close to the value we have worked out.

8.8 RELATIVISTIC MASS TRANSFORMATION

The motion of a particle under the action of a force is given by Newton's second law. According to this law, the rate of change of momentum of the particle is equal to the force applied. This law needs a modification because the calculation of the rate, which involves differentiation with respect to time, is frame-dependent. This modification can be made and Newton's laws can be written in a manner such that they are of the same form in all inertial frames as required by Einstein's first postulate. We shall not go through the details of doing this, but use the laws of conservation of momentum and energy to carry out the required modification.

From Newton's laws, we know that the net linear momentum of an isolated system is conserved. We assume that the law of conservation of momentum is valid also in the special theory of relativity. Therefore, it must be valid in all inertial frames.

Consider the collision of two identical spheres Q and Q'. Let Q be at rest initially in a frame S, and Q' be at rest initially in a frame S'. Let us imagine that in S the sphere Q' has an initial velocity v along the x-axis and, after the collision, Q develops a velocity $w(w_x, w_y)$. We shall assume that the collision occurs in the x-y plane. When viewed from S', we shall find Q approaching Q' with a velocity $-v$ along the x-axis and, after the collision, Q' developing a velocity $-w(-w_x, -w_y)$. This is because (the spheres being identical) there is a symmetry between the motion of Q in S and the motion of Q' in S'. Let us now consider the conservation of the y-component of momentum as seen from S. Initially, there is no y-component of momentum, and thus the total final y-component of momentum should be zero. The y-component of momentum of Q is Mw_y where M is the mass of Q and the Newtonian definition of momentum (namely, momentum = product of mass and velocity) is used. To calculate the y-component of momentum of Q' in S, we must know the y-component of velocity of Q' in S'. Since we know the velocity Q' in S', we can obtain its velocity in S by using Eq. (8.64). Thus, the y-component of velocity of Q' in S, after the collision, is

$$\frac{-w_y}{1 - vw_x/c^2}\left(1 - \frac{v^2}{c^2}\right)^{1/2},$$

and hence the momentum of Q' would be

$$-\frac{Mw_y}{1 - vw_x/c^2}\left(1 - \frac{v^2}{c^2}\right)^{1/2}.$$

We find that the momenta of Q' and Q after the collision when added do not result in zero. Since the principle of conservation of momentum is a universal law, we cannot assume that the mass of Q and Q' is the same in S. Let the mass of Q be M and of Q' be M' as seen by an observer in S. Then, we have, from the principle of conservation of momentum,

$$Mw_y - \frac{M'w_y(1 - v^2/c^2)^{1/2}}{1 - w_x v/c^2} = 0,$$

$$M = \frac{M'(1 - v^2/c^2)^{1/2}}{1 - w_x v/c^2}. \tag{8.66}$$

Let us now consider a glancing collision between Q and Q'. Then, as viewed from S, Q hardly moves and Q' continues to be in motion with its velocity v along the x-axis. In the limit $w_x = 0$, $w_y = 0$, we find $M = m_0$ is the mass of Q when it is at rest and $M' = m(v)$ is the mass of Q' which is moving with a velocity v in S'. Thus, from Eq. (8.66), we get

$$m_0 = m(v)(1 - v^2/c^2)^{1/2}$$

or

$$m(v) = \frac{m_0}{(1 - v^2/c^2)^{1/2}}. \tag{8.67}$$

Since we have assumed that the momentum is conserved, it follows that the mass of a particle depends on its velocity as given by Eq. (8.67). With this definition, it can be shown that the momentum conservation law is valid in all frames. If we carry out the analysis in S' in a manner similar to that in S, we shall reach the conclusion that the mass of Q is velocity-dependent and is given by Eq. (8.67).

We see that, when $v = 0$, the mass of a particle is the least and is equal to m_0. This is referred to as the rest mass of the particle. It is the mass of the particle measured in a frame in which the particle is at rest.

The dependence of mass on velocity has been experimentally tested and verified. The mass of an electron has been measured as a function of its speed and the result so obtained has confirmed the relativistic equation (8.67).

QUESTION 8.7 How fast must a particle move so that its mass becomes twice its rest mass?

8.9 MASS-ENERGY RELATION

The dependence of mass on velocity has far-reaching consequences. Consider a particle moving along the x-axis and acted upon by a force F in the same direction. If it moves a distance Δx, then the increase in its kinetic energy is $E = F\Delta x$. From Newton's second law, we know $F = d(mv)/dt$. Now m is no longer a constant but depends on the velocity. Thus,

$$\Delta E = \Delta x \frac{d(mv)}{dt}. \tag{8.68}$$

If the energy increases in time Δt, then $\Delta x = v\Delta t$, where v is the velocity of the particle. Therefore,

$$\Delta E = v\Delta t \frac{d(mv)}{dt} = v\Delta(mv) = v^2\Delta m + mv\Delta v. \tag{8.69}$$

Using $m = \gamma m_0$ and $\Delta m = \gamma^3 m_0 v\Delta v/c^2$, we have

$$\Delta E = \frac{m_0 v\Delta v}{(1 - v^2/c^2)^{1/2}}\left(\frac{v^2/c^2}{1 - v^2/c^2} + 1\right)$$

or

$$\Delta E = \frac{m_0 v dv}{(1 - v^2/c^2)^{3/2}} \tag{8.70}$$

or

$$\Delta E = \Delta mc^2. \tag{8.71}$$

Thus, the increase in kinetic energy is c^2 times the increase in the mass of the particle because of the change in particle velocity.

Consider now a particle at rest being acted upon by a force till its velocity becomes v. The kinetic energy is then

$$\int_{E=0}^{E_k} dE = E_k = c^2 \int_{m_0}^{m(v)} dm = c^2[m(v) - m(0)] = c^2\left[\frac{m_0}{(1 - v^2/c^2)^{1/2}} - m_0\right]. \tag{8.72}$$

This is the relativistic formula for the kinetic energy. If $v \ll c$, then, from the binomial expansion

$$\frac{1}{(1 - v^2/c^2)^{1/2}} = 1 + \frac{v^2}{2c^2} + \frac{3}{8}\frac{v^4}{c^4} + \ldots,$$

the kinetic energy of a particle with a velocity v is

$$\frac{m_0 v^2}{2} + \frac{3}{8}\frac{m_0 v^4}{c^2} + \ldots. \tag{8.73}$$

The first term here is the nonrelativistic expression for kinetic energy. This term is a good approximation for the kinetic energy when $v \ll c$.

QUESTION 8.8 What is the speed of a particle when its kinetic energy equals its rest mass energy?

The relation between the kinetic energy and the increase in mass suggests a more general interpretation of mass. Just as we have other forms of energy, e.g., thermal energy, internal energy, and potential energy, so too we have mass energy. The Joule constant relates thermal energy to mechanical energy. The factor c^2 relates mass energy to mechanical energy. Also, energy can be stored in the form of mass in the same way as it can be stored in the form of heat or chemical energy. In particular, energy and mass are interconvertible. Thus, it is possible to have a reaction in which the mass of the final constituents is less than that of the initial ones, the balance mass being converted into mechanical energy. In fact, as we shall learn in Chapters 9 and 10, a substantial amount of energy is released in a nuclear reaction, with the mass energy being converted into the kinetic energy.

To summarize, the total energy of a particle is given by

$$E = mc^2 \tag{8.74}$$

or

$$E = \frac{m_0 c^2}{(1 - v^2/c^2)^{1/2}} \tag{8.75}$$

and energy and mass are equivalent.

In a normal collision between objects when $v \ll c$, the change in mass is negligible and the mass can be assumed to be conserved. Besides, the energy is conserved. In an elastic collision (i.e., when kinetic energy is not transformed into the other forms of energy), the kinetic energy is conserved. However, in a relativistic collision, only the law of conservation of energy, including the mass energy, is applicable.

EXAMPLE 8.9 Calculate the energy release in the reaction $K^+ \rightarrow \pi^+ + \pi^0$, where the rest masses of the particles are $m_0(K^+) = 8.9 \times 10^{-28}$ kg, $m_0(\pi^+) \sim m_0(\pi^0) = 2.5 \times 10^{-28}$ kg.

The energy release is

$$\Delta E = \Delta mc^2 = [m_0(K^+) - \{m_0(\pi^+) + m_0(\pi^0)\}]c^2$$

$$= (8.9 - 2 \times 2.5) \times 10^{-28} \times 9 \times 10^{16} \text{ J} = 35.1 \times 10^{-12} \text{ J}$$

$$= \frac{35.1 \times 10^{-12}}{1.6 \times 10^{-19}} \text{ eV} \sim 2.2 \times 10^8 \text{ eV} = 220 \text{ MeV}.$$

We can obtain a relation between the energy E and the momentum p of a particle by using the equations

$$E = \frac{m_0 c^2}{(1 - v^2/c^2)^{1/2}},$$

$$p = \frac{m_0 v}{(1 - v^2/c^2)^{1/2}} \tag{8.76}$$

and eliminating v from the two equations. We then have

$$E^2 - c^2p^2 = \frac{m_0^2c^4 - m_0^2c^2v^2}{1 - v^2/c^2} = \frac{m_0^2c^4(1 - v^2/c^2)}{1 - v^2/c^2} = m_0^2c^4$$

or $E^2 = c^2p^2 + m_0^2c^4$. This gives the relativistic expression of energy as

$$E = (c^2p^2 + m_0^2c^4)^{1/2}. \tag{8.77}$$

Here we have taken only the positive square root as energy is a positive quantity.

Zero-Mass Particles

There are particles in nature, e.g., photons and neutrinos, with rest mass zero, that is, $m_0 = 0$. From Eq. (8.75), we see that the energy of such a particle will be zero unless $v = c$. If $v = c$, the equation gives an indeterminate value. Thus, a particle having a finite energy and rest mass zero travels with the velocity of light. For such a particle, the physical quantities that can vary are the energy and the momentum. Accordingly, Eq. (8.77) reduces to

$$E = cp. \tag{8.78}$$

We have already learnt that this relation is valid for a photon associated with an electromagnetic wave. Here we see that this is valid for any particle with zero rest mass.

A particularly interesting situation is the scattering of photons by electrons called Compton scattering (discussed as an example of the applications of relativistic mechanics in Appendix 12).

8.10 PARADOXES IN SPECIAL THEORY OF RELATIVITY

The special theory of relativity, as we have noted, has several concepts which seem to go against common sense. Simultaneity is one such concept since we treat it as an absolute concept. Careless thinking which ignores the frame-dependence of a concept can lead to an erroneous result. In order to clarify this, we shall discuss a few interesting paradoxes. A resolution of these paradoxes gives us a deeper insight into the various concepts of the special theory of relativity.

Paradox 1: Rocket Warfare

Consider two identical rockets A and B approaching each other, as shown in Fig. 8.7. Their lengths in their respective frames are l_0. When seen from the frame of A, the rocket B will have a length l_0/γ. Similarly, the rocket A will have a length l_0/γ when viewed from the frame of B. Consider now the situation, as seen from the frame of B, when a bullet is fired at the rocket A from the tail of the rocket B at the time the tip of B coincides with the tail of A (see Fig. 8.7a). In this situation, the bullet will not hit the rocket A. Now let us analyze the situation as viewed from the frame of A. Then, we would see the rocket B to be contracted (l_0/γ) and the rocket A having the length l_0 (see Fig. 8.7b). Thus, it would appear that the bullet fired from the rocket B will hit the rocket A. Does the bullet hit the rocket A or not?

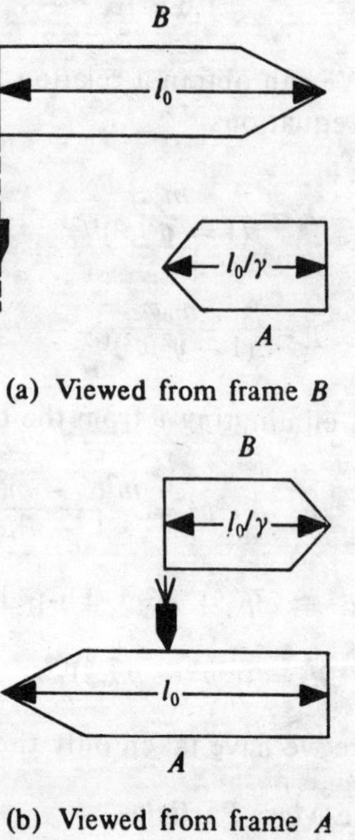

(a) Viewed from frame B

(b) Viewed from frame A

Fig. 8.7 Rocket warfare paradox.

Paradox 2: Car and Ditch

A car of length 5 m in its rest frame is travelling with a velocity $v = \sqrt{3}c/2$ (i.e., $\gamma = 2$) on a road. It meets a ditch of length 4 m (in the rest frame of the road). From the rest frame of the car, the ditch would appear contracted to a length 2 m and it would appear to the driver that he can pass over the ditch. On the other hand, an observer at rest with respect to the road would come to the opposite conclusion. He would see a car of length 2.5 m trying to cross a ditch of length 4 m. Therefore, he would argue that the car will fall into the ditch. Who is correct—the driver or the observer?

Paradox 3: Stick and Plane

A stick of length l (in its rest frame) is moving along the x-axis in the x-y plane with a velocity $v = \sqrt{3}c/2$ (i.e., $\gamma = 2$) with respect to an observer O in a frame S. There is a rectangular plane parallel to the x-y plane moving with a velocity u along the z-axis (see Fig. 8.8). This plane has a circular hole of diameter l (in the frame S) and the observer O sees the stick pass through the hole. To this observer, it appears that the stick has the length $l/2$ and the hole is of diameter l. To an observer moving with the stick, the hole appears to have the diameter $l/2$ and the stick seems to be longer (length l) than the hole. He would conclude that the stick would collide with the plane. What is the fallacy in this argument?

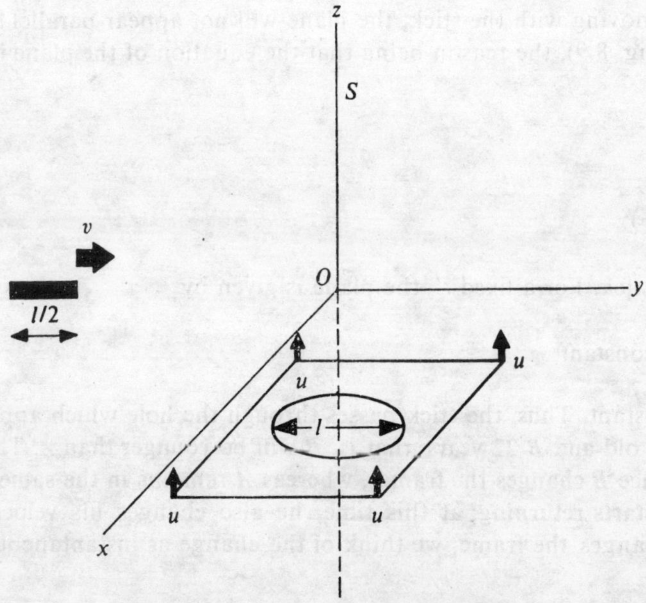

Fig. 8.8 Stick-and-plane paradox.

Paradox 4: Twin Paradox

A and B are two twins, each twenty years of age. A remains at rest at the origin O of a frame S, whereas B travels with a velocity $v = \sqrt{3}c/2$ (i.e., $\gamma = 2$) for 1 yr (his time) along the positive x-axis. He then turns back and travels with the same speed and reaches the starting point after the end of another year of his time. Then, B's age, according to B, would be $20 + 2 = 22$ yr, whereas the age of B, according to A, would be $20 + 2 \times \gamma = 20 + 2 \times 2 = 24$ yr. On the other hand, A's age, according to A, would be 24 yr, whereas A's age, according to B (A appears travelling in the direction opposite to that of B), would be 22 yr. Which of the two statements is correct? Will one of the two twins appear younger than the other? If so, why?

Resolution of Paradoxes

In the rocket warfare paradox, we used the concept of simultaneity when we said: "... when a bullet is fired at the rocket A from the tail of the rocket B at the time the tip of B coincides with the tail of A (see Fig. 8.7a)." The two events, namely, the tip of B coinciding with the tail of A and the firing of the bullet, are simultaneous only in the frame of B and, in particular, this is not so for the frame of A. In the frame of A, we can, using the Lorentz transformation (see Problem 8.15), show that the bullet is fired before the tip of B coincides with the tail of A, thus implying that the bullet indeed misses the rocket A. This means that Fig. 8.7b is wrongly drawn.

The car will fall into the ditch. The fallacy of the argument, as seen from the driver's frame, is as follows. The car cannot be considered as a rigid body (see Section 8.5), and hence as the front end of the car enters the ditch, the car will start falling. It will take a finite time for the elastic restoring forces in the car to react. As the front end bends and hits the ditch (the front end is below the road level now), it stops but the rear end of the car continues to be in motion since it does not 'know' that the front end has stopped. Therefore, the car falls into the ditch.

To the observer moving with the stick, the plane will not appear parallel to the x-y plane but slightly tilted (see Fig. 8.9), the reason being that the equation of the plane in the frame S, i.e.,

$$z = ut, \tag{8.79}$$

gets converted to

$$z' = u(t' + \frac{vx'}{c^2})\gamma \tag{8.80}$$

in the frame of the stick. For a fixed t', the plane is given by

$$z' = \frac{uvx'\gamma}{c^2} + \text{constant} \tag{8.81}$$

and not by $z' = $ constant. Thus, the stick passes through the hole which appears tilted.

A will be 24 years old and B 22 years, that is, B will be younger than A. There is no symmetry between B and A since B changes the frames, whereas A remains in the same frame. (B changes the frame when he starts returning; at this time, he also changes his velocity from v to $-v$.) When we say B changes the frame, we think of the change as instantaneous. As viewed by A, this is not so.

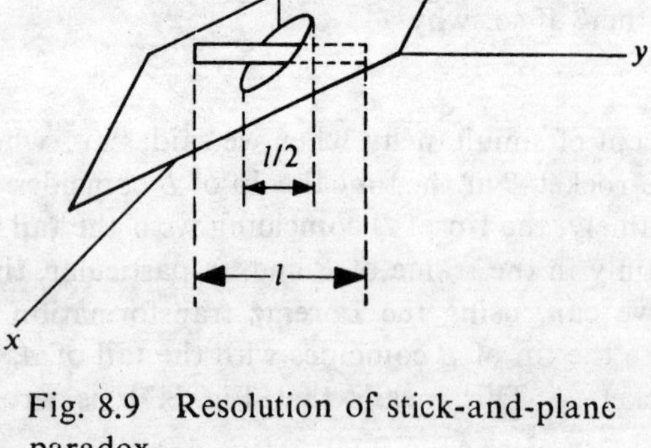

Fig. 8.9 Resolution of stick-and-plane paradox.

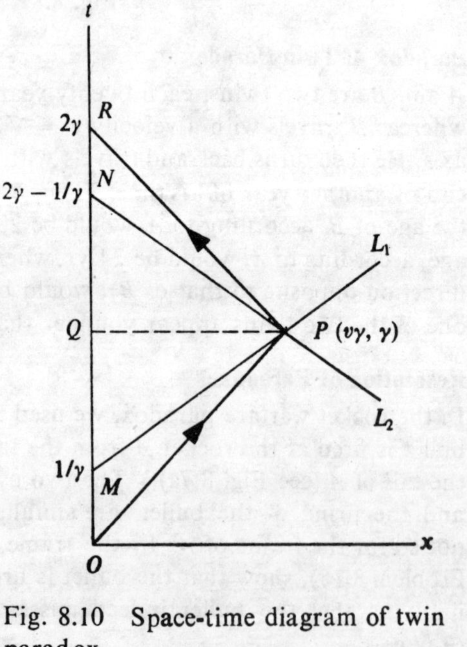

Fig. 8.10 Space-time diagram of twin paradox.

In order to see that the change of frame at P (see Fig. 8.10) is not instantaneous for the observer at O, let us look at the lines of simultaneity given by Eq. (8.32) with $t' = $ constant.

The line of simultaneity in the x-t plane for the outward journey is shown as L_1M (Fig. 8.10). Similarly, for the return journey, the line of simultaneity is L_2N. According to this figure, what appears to be simultaneous at P is not so for the observer at O. The time indicated by MN in the figure will be the time difference observed at O. It can be shown (see Problem 8.16) that $MN = 2\gamma - 2/\gamma$ and $OM = NR = 1/\gamma$. Thus, $OR = 2\gamma$ which is the time measured by A. On the other hand, the time measured by B is exactly two years.

PROBLEMS

8.1 When the origins O and O' of two frames S and S' coincide, suppose the time $t = 0$ and $t' = \alpha$. Write down the Lorentz transformation connecting the frames.

8.2 Suppose a frame S' has a velocity $\boldsymbol{v} = v_x\hat{i} + v_y\hat{j}$ with respect to a frame S. Write down the Lorentz transformation connecting the two frames.

8.3 The space-time coordinates of two events P_1 and P_2 as viewed in a frame S are, respectively, ($x = 3$ m, $y = 7$ m, $z = 13$ m, $t = 2 \times 10^{-8}$ sec) and ($x = 4$ m, $y = 7$ m, $z = 13$ m, $t = 10^{-8}$ sec). Further, the space-time coordinate of the event P_1 as viewed from a frame S' is ($x' = 2$ m, $y' - 7$ m, $z' = 13$ m, $t' = 2 \times 10^{-8}$ sec). Find the x-coordinate of the event P_2 in S' if its time coordinate is zero. Assume that S' is moving along the x direction.

8.4 Suppose the space-time coordinate in a frame S of an event is ($x = 200$ km, $y = 20$ km, $z = 7$ km, $t = 10^{-5}$ sec). Find the time coordinate of the event as viewed in a frame S' moving along the x-axis with a velocity of magnitude $0.8c$. The origins of the two frames coincide at $t' = t = 0$.

8.5 A metre stick moving with a speed v in a frame S appears to have a length 75 cm. Find the speed v. The direction of motion is along the length of the metre stick.

8.6 A solid has a density ρ in a frame in which it is at rest. What will be its density in a frame in which it moves with a speed v?

8.7 A μ-meson has a lifetime of 2×10^{-6} sec when it is at rest. How far will it travel if it decays while moving with a speed $v = 0.9c$?

8.8 Consider a light beam travelling in the x-y plane and making an angle θ with the x-axis in a frame S. Calculate (a) the x'- and y'-component of the beam velocity and (b) the angle θ' in a frame S'. Further, verify that the speed of light beam in S' remains unchanged.

8.9 A rocket is travelling with a speed of 900 m/sec away from earth. After how much time will a clock in the rocket differ by 1 sec from a clock on the ground?

8.10 Derive the inverse velocity transformation (8.63)–(8.65).

8.11 Two particles are travelling with the same speed v in directions perpendicular to each other. What will be the relative speed of one with respect to the other?

8.12 A rod of length 60 cm (in its rest frame) is travelling along its length with a speed $0.6c$ in a frame S. A particle moving in the direction opposite to that of the rod with a speed $0.6c$ in S passes the rod. How much time will the particle take to cross the rod (a) in the frame S? (b) in the rest frame of the particle?

8.13 A π^0-meson decays into two photons of equal energy. What will be the energy and wavelength of each photon? [The rest mass $M_0(\pi^0) = 2.5 \times 10^{-28}$ kg.]

8.14 A proton and a photon have the same energy, i.e., 30 GeV (1 GeV $= 10^3$ MeV). Calculate the difference in their momentum.

8.15 In paradox 1 (see Section 8.10), show that the bullet as seen from the frame of A is fired before the tip of B coincides with the tail of A. Further, show that the bullet as seen from the frame of A does not hit A.

8.16 Show that the line of simultaneity L_2N (see Fig. 8.10) is represented by the equation

$$\gamma\left(\frac{t + vx}{c^2}\right) - \frac{2v^2/c^2}{1 - v^2/c^2} = 1.$$

Further, show that (a) $OM = NR = 1/\gamma$ and (b) $MN = 2(\gamma - 1/\gamma)$.

9

The Nucleus

9.1 INTRODUCTION

The alpha-particle scattering experiments E. Rutherford and his collaborators conducted (see Chapter 4) suggested that an atom has a nucleus in which the positive charge of the atom is concentrated. These investigations further established that, for a distance greater than 10^{-14} m, the nucleus interacts with the alpha-particles by the Coulomb force. It was found, however, that when the distance between these and the nucleus is less than 10^{-14} m, the distribution of the scattered alpha-particles could not be understood in terms of the pure Coulomb force. This implied that a new type of force, the nuclear force, acts at a distance less than 10^{-14} m. It also indicated that the *nuclear force has a short range* and the radius of the nucleus of an atom is approximately 10^{-14} m, whereas the radius of an atom is about 10^{-10} m. Thus, the nucleus occupies an extremely small portion (10^{-12}) of the volume of an atom. In this chapter, we shall study some of the important properties of the nucleus.

9.2 NUCLEAR MASS

The mass of an atomic nucleus is very small when expressed in ordinary units (about 10^{-25} kg). It is more convenient to express it in terms of another unit, namely, the *atomic mass unit* (u), defined as 1/12 the mass of the carbon atom (1 u $= 1.66 \times 10^{-27}$ kg). For the masses of some atoms, see Table 9.1. It should be noted that a mass table usually gives the mass of the neutral atoms and not that of nuclei. To obtain a nuclear mass, the mass of the electron(s) has to be subtracted from the atomic mass, simultaneously taking into account the correction due to the electronic binding energy.

> **QUESTION 9.1 What is the mass of the $^{197}_{79}$Au nucleus, given the mass of an electron to be 5.3×10^{-4} u? Neglect the binding energy of the electrons. Use Table 9.1.**

To experimentally determine the exact mass of an atom, a mass spectrometer is used. This instrument can measure an atomic mass with an accuracy of a few parts in 10^7. A typical example of such an instrument is the one designed by K. T. Bainbridge (see Fig. 9.1). In this spectrograph, an ion source produces ions by removing one or more electrons from a neutral atom. The ions so generated are accelerated in the source itself by applying a suitable electric field. This ion beam then enters region 1 where both the electric and magnetic fields are present. The electric field is along the y-direction and the magnetic field is along the z-direction (coming out). The ion beam experiences a force qvB_1 due to the magnetic field and an opposite force qE_1 due to the electric field. The combination of the forces resulting from the two fields allows

TABLE 9.1 Atomic masses of some isotopes and their abundance

Symbol	Z	A	Mass (u)	Relative abundance (per cent)
n	0	1	1.008665	
H	1	1	1.007825	99.985
	1	2	2.014102	0.015
		3	3.016050	
He	2	3	3.016030	0.00013
		4	4.002603	100
C	6	12	12.000000	98.89
		13	13.003354	1.11
		14	14.003242	
O	8	16	15.994915	99.759
		17	16.999133	0.037
		18	17.999160	0.204
Ne	10	20	19.99244	90.92
		22	21.991385	8.82
Cl	17	35	34.96885	75.77
		37	36.96593	24.23
Ag	47	107	106.905096	51.82
		108	107.905949	
		109	108.904756	48.18
Au	79	197	196.966541	100
U	92	233	233.039522	
		235	235.043915	0.72
		238	238.050770	99.27

only the ions with a specific velocity v to enter region 2 through the slit S_3. These ions are those that are undeflected, that is, the magnitude of each of the two forces acting on these ions is the same. Thus, $qE_1 = qvB_1$ or

$$v = E_1/B_1. \qquad (9.1)$$

Region 2 has a uniform magnetic field B_2 along the z-direction. The ions now move in a circular orbit and strike the photographic plate. The distance from the slit S_3 to the image on this plate is $2R$, where R is the radius of the circular orbit. The centripetal force qvB_2 is provided by the

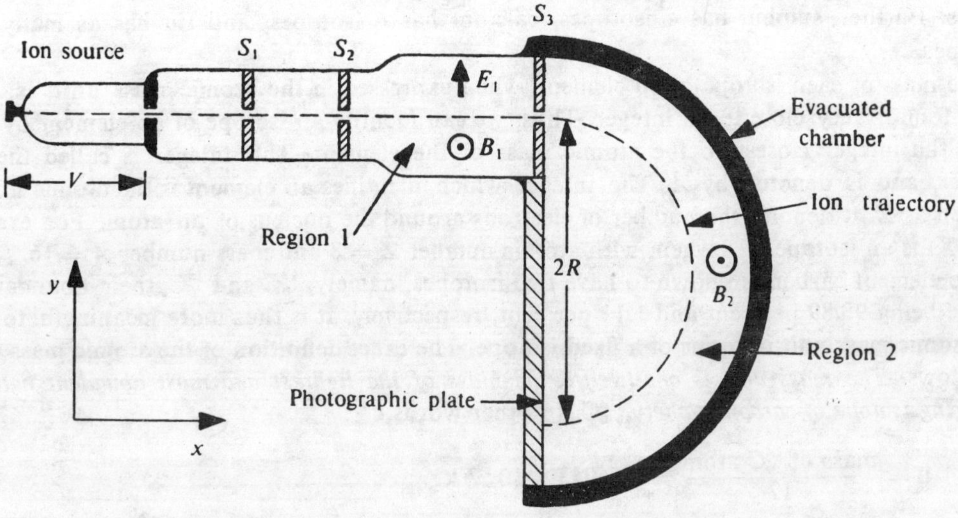

Fig. 9.1 Schematic diagram of Bainbridge-type mass spectrograph.

magnetic field B_2. Recalling that the centripetal force is also given by mv^2/R, we get

$$qB_2v = mv^2/R \tag{9.2}$$

or

$$\frac{q}{m} = \frac{v}{B_2R} = \frac{E_1}{B_1B_2R}. \tag{9.3}$$

Thus, a measurement of R, E_1, B_1, and B_2 determines q/m of the ions. If we know the charge of an ion, we can find out the mass of the ion. The charge of an ion is the number of electrons removed times the electronic charge. Once we know the mass of an ion, we can calculate the mass of the neutral atom by taking into account the mass of the electrons in the atom.

The study of ions by means of a mass spectrometer has shown that all atoms in an element do not have the same mass. For example, when the ratio q/m for the neon ions is measured, several different values are obtained, some of which are 4.81×10^6 C-kg^{-1}, 9.62×10^6 C-kg^{-1}, 4.56×10^6 C-kg^{-1}, and 4.35×10^6 C-kg^{-1}. These values have been obtained after making absolutely sure that no element other than neon is present in the ion source. As can be noted, the second value is exactly twice the first. This is so because some atoms lose two electrons instead of one, and thus have the ratio q/m twice that of an atom losing one electron. However, this reasoning cannot be applied to explain the other two values. This clearly indicates that atoms of varying mass are present in neon. Such atoms of an element are called *isotopes*. In chemical properties, however, the three types of neon atoms (whose q/m values we have already given) are identical because each of these atoms has the same number of electrons. Consequently, they cannot be distinguished by chemical methods of identification.

The number of isotopes found in nature varies from one element to another. The elements beryllium, fluorine, sodium, aluminium, and many others have only a single isotope each, that is, in these elements, every atom of the same element has the same mass. On the other hand, the number of isotopes found in nature is 2 for hydrogen, helium, boron, and many others. Further, sulphur has 4 isotopes, calcium has 6 isotopes, and tin has as many as 10 isotopes.

The mass of each isotope of an element, when expressed in the atomic mass unit, is, it has been found, very close to an integer. Thus, we can identify an isotope of an element by specifying the integer closest to the atomic mass of the element. This integer is called the *mass number*, and is denoted by A. The integer which identifies an element is the atomic number, written as Z. It denotes the number of electrons around the nucleus of an atom. For example, ${}^{16}_{8}\mathrm{O}$ $({}^{A}_{Z}\mathrm{X})$ is an isotope of oxygen, with atomic number $Z = 8$ and mass number $A = 16$.

The element carbon is known to have two isotopes, namely, ${}^{12}_{6}\mathrm{C}$ and ${}^{13}_{6}\mathrm{C}$, their abundance in nature being 98.89 per cent and 1.11 per cent, respectively. It is thus more meaningful to define the atomic mass unit in terms of a fixed isotope. The exact definition of the atomic mass unit is as follows. *The mass unit is one-twelfth the mass of the lightest and most abundant naturally-occurring isotope of carbon, namely,* ${}^{12}_{6}\mathrm{C}$. In other words,

$$1\ \mathrm{u} = \frac{\text{mass of } {}^{12}_{6}\mathrm{C} \text{ atom}}{12} = 1.66043 \times 10^{-27}\ \mathrm{kg}. \tag{9.4}$$

If we determine the atomic mass of an element without making any distinction between its different isotopes, we would discover that the mass we so obtain would be the same as that calculated by averaging the masses of the different isotopes of the element by taking into account the abundance of each such isotope.

EXAMPLE 9.1 What would be the average atomic mass of chlorine if no distinction is made between its different isotopes?

The element chlorine is a mixture of 75.77% ${}^{35}_{17}\mathrm{Cl}$ and 24.23% ${}^{37}_{17}\mathrm{Cl}$. Thus, its average atomic mass will be

$$\frac{75.77}{100} \times 34.96885 + \frac{24.23}{100} \times 36.96593 = 35.453\ \mathrm{u}.$$

The atomic mass of chlorine obtained chemically is the same as that arrived at in Example 9.1. This is so since a chemist when calculating the atomic mass of an element doesn't distinguish between its different isotopes. We must therefore remember that the *chemical* atomic weight (35.453 for chlorine) of an element is the *average* of its fundamental isotopic masses, but no atom with such an average mass exists. The chemically-determined atomic weight of an element we see in the periodic table is such an average mass.

9.3 NUCLEAR SIZE

The scattering experiments Rutherford conducted provided the first evidence that the nuclear volume is indeed a very small part of the total volume of an atom and indicated that the nuclear radius is of the order of 10^{-14} m. Since then, many measurements of the nuclear size and shape have been made using different methods. These attempts have given a consistent picture of the nuclear dimensions. The nucleus has been found to be approximately spherical in shape. Also, the radius R of a nucleus with $A > 4$ has been found to vary as

$$R = R_0 A^{1/3}. \tag{9.5}$$

Here, A is the mass number of the nucleus and the constant $R_0 = 1.2$ F, where $1\ \mathrm{F} = 10^{-15}$ m, the unit F being due to E. Fermi. From relation (9.5), we note the volume of the nucleus is proportional to A because $V \propto R^3 \propto A$. Thus, the volume divided by the mass number A is constant. In other words, the density of nuclear matter (i.e., the number of nucleons per unit volume) is independent of the number of nucleons in the nucleus. This is an important characteristic of nuclei.

QUESTION 9.2 For $^{198}_{79}$Au, what is the (a) nuclear radius and (b) density in kg/m^3 of nuclear matter?

9.4 ANGULAR MOMENTUM OF THE NUCLEUS

A nucleus, like an electron, has an intrinsic angular momentum, commonly known as *spin.* A spin operator $\boldsymbol{I}$ is associated with the nuclear angular momentum. As in the spin angular momentum of an electron, so too in the angular momentum of a nucleus, we can measure only the magnitude of the vector $\boldsymbol{I}$ and one of its components (usually the z-component). The eigenvalues of I^2 are

$$I(I+1)\hbar^2 \qquad (I = 0, \tfrac{1}{2}, 1, \tfrac{3}{2}, \ldots) \tag{9.6}$$

and the eigenvalues of I_z are

$$M_I\hbar \qquad (m_I = -I, -I+1, \ldots, I-1, I). \tag{9.7}$$

The magnitude of the spin is $[I(I+1)\hbar^2]^{1/2}$, where I is an integer or a half-integer. The maximum value this component can have is, as in the atomic angular momentum, $I\hbar$. A nucleus, like an atom, can exist in excited quantum states. In these states, the spin as well as the other properties of the nucleus may differ from those in the ground state of the nucleus. The nuclear angular momentum vector, like an atomic angular momentum vector, exhibits space quantization. A nucleus with a spin I, for example, can take only $2I+1$ orientations in an external field.

The spin of a nucleus can be experimentally found in several ways. The results so obtained show an important connection between the spin and the mass number, namely, a nucleus with an odd mass number has a half-integral spin, and that with an even mass number has an integral spin.

QUESTION 9.3 Is the $^7_3$Li nucleus a boson or a fermion?

9.5 CONSTITUENTS OF THE NUCLEUS

The simplest nucleus is that of hydrogen; it consists of just one proton. Next, according to the mass, is the deuteron nucleus which has a mass nearly equal to that of two protons and a unit positive charge. We might therefore imagine that the deuteron nucleus consists of two protons and one electron since this explains the mass (electron mass being very small as compared to the proton mass) and the charge of this nucleus. In fact, it was generally assumed (before 1932) that the nucleus of an atom of a mass number A and an atomic number Z is composed of A protons and $(A-Z)$ electrons. For example, the sodium atom has the mass number 23 and nuclear charge or atomic number 11. Therefore, it was believed that the nucleus of this atom has 23 protons and 12 electrons, thus giving the nucleus a net positive charge of 11 and a mass number 23 (since the mass of the electrons is negligible). However, the assumption that the electrons are contained in the nucleus presented a series of difficulties.

In order for an electron to be inside a nucleus, it must be confined in a nucleus of diameter$\sim 10^{-14}$ m. We can evaluate the momentum of this electron using the uncertainty principle. The uncertainty in position of the electron is $\Delta x \sim 10^{-14}$ m. By using the relativistic expression $E = cp$, we can show that the kinetic energy of the electron must be $\sim$20 MeV (see Problem 9.3). (However, the typical kinetic energy of an electron ejected from a nucleus is $\sim$1 MeV.) If an electron of such a high energy exists inside a nucleus, then the electron can be expected to have a finite probability of coming out of the nucleus with the same energy.

QUESTION 9.4 The uncertainty principle indicates that the electrons are not a part of the nucleus. Is this argument applicable also to protons?

As already noted, the nuclear spin is integral or half-integral depending on whether the mass number is even or odd. Both protons and electrons have a spin equal to $\frac{1}{2}$. If we consider a nucleus with odd A and even Z, then, according to the proton-electron model of the nucleus, it would have A (odd number) protons and $A - Z$ (odd number) electrons. These, when combined, would give an integral spin, but, as we know, a nucleus with odd A has a half-integral spin.

EXAMPLE 9.2 Determine the spin of the deuteron $^2_1$H nucleus, using the proton-electron model.

The given nucleus has the atomic number $Z = 1$ and the mass number $A = 2$. Thus, according to the proton-electron model,

$$\text{number of protons} = A = 2,$$

$$\text{number of electrons} = A - Z = 1.$$

The total angular momentum (spin) S of a nucleus is the vector sum of the spin angular momenta of the two protons and one electron, each having the value $s = \frac{1}{2}$ (in units of $\hbar$). Thus, S is given by

$$\mathbf{S} = \mathbf{S}_1 + \mathbf{S}_2 + \mathbf{S}_3.$$

The addition of the three spin angular momenta can be done in two stages. First, the spin angular momenta of the two protons are added, i.e.,

$$\mathbf{S}_{12} = \mathbf{S}_1 + \mathbf{S}_2.$$

In Section 4.16, we learnt that the vector addition for s_{12}, the spin quantum number of $\mathbf{S}_{12}$, gives the possible values $|\frac{1}{2} + \frac{1}{2}|$ and $|\frac{1}{2} - \frac{1}{2}|$, i.e., 1 and 0, in units of $\hbar$. Second, the spin angular momentum of the electron $\mathbf{S}_3$ is added to $\mathbf{S}_{12}$ to get $\mathbf{S}$, i.e.,

$$\mathbf{S} = \mathbf{S}_{12} + \mathbf{S}_3.$$

When $s_{12} = 0$, $s = s_3$, and hence the total spin quantum number would be $\frac{1}{2}$. Further, when s_{12} is 1, s will take the values $(1 + \frac{1}{2})$ and $(1 - \frac{1}{2})$, giving the total spin quantum number as 3/2 or 1/2. Thus, according to the proton-electron model, the spin angular momentum will be 1/2 or 3/2 in units of $\hbar$. The experimental value of the spin of the $^2_1$H nucleus $= \hbar$.

As we now realize, the experimental results, namely, (1) the typical energy of an electron emitted by a nucleus ~1 MeV and (2) a nucleus with odd A has a half-integral spin, cannot be understood on the basis of the electron-proton model of the nucleus. According to Rutherford, these results could be explained if the nucleus could be thought of as consisting of protons and another kind of particles identical with the protons in mass and spin but having zero charge. These particles are known as *neutrons*. Clearly, the problem of not understanding results (1) and (2) disappears if the nucleus is assumed to be composed of protons and neutrons. For example, in the neutron-proton model, the deuteron nucleus consists of one proton and one neutron. Each of these particles has a spin equal to $\frac{1}{2}$ and the total spin quantum number can be 1 or 0. (The experimentally observed spin of the deuteron is 1.) The neutron was discovered experimentally by J. Chadwick in 1932.

9.6 NEUTRON AS BUILDING BLOCK

With the discovery of neutron, it became clear that the neutron forms the second building block of an atomic nucleus, the first being the proton. In summary, an atom is composed of electrons, neutrons, and protons. These particles are called the *elementary atomic particles.* Their basic properties are summarized in Table 9.2. The region outside the nucleus, comprising practically

TABLE 9.2 Elementary atomic particles

	Electron	Nucleon	
		Proton	Neutron
Mass (kg)	9.109×10^{-31}	1.6725×10^{-27}	1.6748×10^{-27}
Mass (u)	.0005486	1.007825	1.008665
Rest energy (mc^2)	.511 MeV	938.3 MeV	939.6 MeV
Charge (C)	-1.602×10^{-19}	1.602×10^{-19}	0
Spin ($\hbar$)	$\frac{1}{2}$	$\frac{1}{2}$	$\frac{1}{2}$

the whole volume of the atom, contains only the negatively charged electrons. The nucleus, comprising practically the whole mass of the atom, contains protons and neutrons. Since each proton carries a charge of $+1$ (in terms of the magnitude of the electronic charge), the charge of a nucleus is equal to the number of protons in the nucleus, irrespective of the number of neutrons. The number of protons in a nucleus is referred to as the *atomic number*. The *mass number* A is equal to the total number of protons and neutrons. The term *nucleon* is often used for both protons and neutrons. The composition of a nucleus may then be summarized as follows:

$A =$ number of nucleons,

$Z =$ number of protons,

$A - Z =$ number of neutrons.

An atomic number and a mass number are by convention denoted by a subscript and a superscript, respectively, to a chemical symbol. For example, $^{13}_{6}C$ denotes a nucleus containing 6 protons. The superscript 13 is A, the number of nucleons (i.e., neutrons plus protons). The number of neutrons $(A - Z)$ is 7. As the mass of each proton and each neutron is approximately 1 u, the mass of a nucleus (in atomic mass units) is approximately equal to its mass number A.

Isotopes

In an element, each atom has the same number of electrons and, likewise, the same number of protons. However, the atoms in an element may differ in the number of neutrons. Each isotope of an element has the same atomic number (Z) but a neutron number (N), and hence the mass number (A) different from that of the other isotopes of the element. For example, $^{15}_{8}O$, $^{16}_{8}O$, $^{17}_{8}O$, and $^{18}_{8}O$ are the isotopes of oxygen.

Isobars

The nuclei with the same number of nucleons A, but different N and Z are called isobars. For example, $^{14}_{6}C$, $^{14}_{7}N$, and $^{14}_{8}O$ are the isobars corresponding to $A = 14$.

We can have nuclei with the same A, N, and Z, but different internal energies. Such a nucleus is similar to an excited atom. In the transition from this state to a lower excited state or to the ground state, as we know, photons are emitted. In the same way, when an excited nucleus assumes the ground state, high energy photons, known as gamma-rays, are emitted. However, while an atom cannot exist in an excited state for any appreciable time, some excited nuclei have a lifetime of several hours. A nucleus X in an excited state is denoted by X*, and a long-lived excited state is known as an *isomer*.

9.7 BINDING ENERGY

Let us consider a nucleus containing Z protons and N neutrons. If we attempt to separate this nucleus into individual protons and neutrons, a certain amount of work has to be done because these particles are held together by attractive forces. In other words, energy must be supplied to the nucleus to separate it into its individual constituents and move them to infinity with each constituent being at rest. The energy necessary to separate all the particles is called the *binding energy* of the nucleus. After the separation, if the nucleons interact and coalesce to form a nucleus, their energy must be less than what it was when they were separated. According to the Einstein mass-energy relation, their combined mass in the coalesced state must be less than the sum of their masses when they are far away from one another. If we denote the loss in mass by Δm, sometimes called the mass defect, then the energy emitted is Δmc^2 which is the binding energy, i.e., the binding energy (BE) is

$$\mathrm{BE} = \Delta mc^2. \tag{9.8}$$

If M, M_n, and M_p are the masses of a nucleus ($^{A}_{Z}X$), a neutron, and a proton, respectively, then the mass defect Δm is

$$\Delta m = NM_n + ZM_p - M \tag{9.9}$$

and the binding energy of the nucleus is

$$\mathrm{BE} = (NM_n + ZM_p - M)c^2. \tag{9.10}$$

The energy of the state of infinite separation of nucleons at rest is taken to be zero.

It is more convenient to work with the mass of an atom than with that of the nucleus. To do this, we add to and subtract from Eq. (9.10) the mass of Z electrons. Thus,

$$\mathrm{BE} = (NM_n + ZM_p + ZM_e - M - ZM_e)c^2.$$

Now, $ZM_p + ZM_e = Z(M_p + M_e)$. The quantity $(M_p + M_e)$ is just the mass of one atom of hydrogen (M_H) since the binding energy of an electron is negligible as compared to $(M_p+M_e)c^2$. Similarly, $(M({}^A_Z\mathrm{X}) + ZM_e)$ is the mass of the atom ${}^A_Z\mathrm{X}$ (M_A). Thus, the binding energy in terms of the atomic masses is

$$\mathrm{BE} = (NM_n + ZM_H - M_A)c^2. \tag{9.11}$$

EXAMPLE 9.3 Calculate the average binding energy per nucleon in the ${}^4_2\mathrm{He}$ nucleus, using the neutron mass and atomic masses of H and He, and neglecting the BE of the electrons in the He atom. Also, compute the error in the total BE of the ${}^4_2\mathrm{He}$ nucleus since the electronic BE is being neglected. (The electronic BE for He ~80 eV.)

The He atom contains two protons and two neutrons in its nucleus, and two electrons. The combined mass of these particles, i.e., $NM_n + ZM_p + ZM_e\,(=NM_n + ZM_H)$, is given by

$$2M_n + 2M_H = 2(1.008665 + 1.007825) = 4.032980.$$

The experimentally measured atomic mass of He is $M_A(\mathrm{He}) = 4.00260$. Thus, the mass defect is

$$\Delta m = 4.03298 - 4.00260 = .03038 \text{ u}.$$

Since $1 \text{ u} = 1.66 \times 10^{-27}$ kg, the energy equivalent to this mass, using Einstein's mass-energy relation, is

$$1 \text{ u} = 1.66 \times 10^{-27} \times (3 \times 10^8)^2 = 14.94 \times 10^{-11} \text{ J} \sim 931 \text{ MeV}.$$

The binding energy of the ${}^4_2\mathrm{He}$ nucleus is then

$$\mathrm{BE} = .03038 \times 931 = 28 \text{ MeV}.$$

The ${}^4_2\mathrm{He}$ nucleus has four nucleons. So, the *average* binding energy per nucleon, denoted by $\bar{B}$, would be 7 MeV. Thus, the mean energy required to remove a nucleon from the nucleus is 7 MeV.

The total BE of ${}^4_2\mathrm{He}$ is 28 MeV, whereas the atomic BE of the electrons is 80 eV or 80×10^{-6} MeV. Therefore, the error in the total BE is

$$\frac{80 \times 10^{-6}}{28} \times 100\% \sim .0003\%.$$

The error is very small as compared to the accuracy of measurement, and thus, when calculating the binding energy of a nucleus, the atomic binding energy is generally neglected.

The binding energy and mass defect are not peculiar to a nucleus. An electron in an atom, an atom in a molecule, and a molecule in a crystal each has a binding energy, but it is usually so small that its mass equivalent can be neglected. The mass equivalent of the binding energy of a nucleus is not negligible in comparison to the total mass of the nucleus because the nuclear forces are very strong.

The expression for the average binding energy per nucleon $\bar{B}$ is

$$\bar{B} = \frac{\text{BE}}{A} = \left(\frac{ZM_H + (A - Z)M_n - M}{A}\right)c^2, \tag{9.12}$$

where M is the mass of the atom. When $\bar{B}$ is plotted for all values of A for the known nuclei, we get the curve shown in Fig. 9.2. Here, except for the nuclei ${}^4_2$He, ${}^{12}_6$C, ${}^{16}_8$O, and ${}^{19}_9$F, the values of $\bar{B}$ lie on the curve. The value of $\bar{B}$ rises as the mass number increases until it reaches a maximum around $A = 56$ and then it slowly decreases.

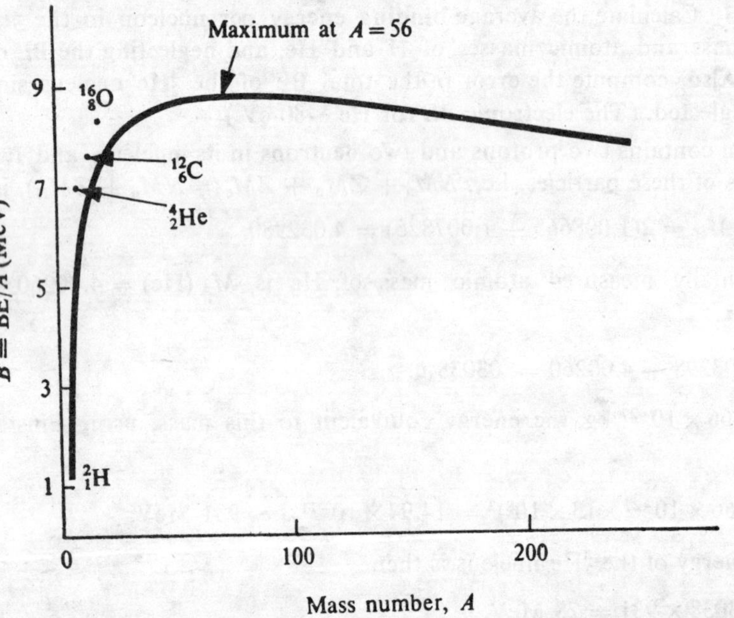

Fig. 9.2 Average binding energy per nucleon (B) for stable nuclei.

The two main features of this curve are the near constant value of $\bar{B}$ in the flat region (i.e., $A = 30$ to $A = 170$), and the low value of $\bar{B}$ in the regions $A < 30$ and $A > 170$. From this curve, we can readily deduce some general results. Consider two light ($A < 30$) nuclei combining to form a medium heavy nucleus. From Fig. 9.2, it is clear that the binding energy per nucleon is more for the final nucleus than for the initial nuclei. Thus, if light elements are combined together to produce a nucleus of medium A, a large amount of energy will be released. This is referred to as fusion and is the principle on which the hydrogen bomb is based. Similarly, if a heavy element is split up (fission) to form nuclei of medium A, the energy released would, again, be large. The atom bomb is based on this principle. The flat region (i.e., $30 < A < 170$) in Fig. 9.2 shows that a nucleus of intermediate mass is stable.

EXAMPLE 9.4 Find the amount of energy that will be released when two deuterium nuclei are fused together to form a $_2$He nucleus. The binding energy per nucleon for the $^2_1$H nucleus is 1.1 MeV and that for the $^4_2$He nucleus is 7 MeV.

The fusion reaction is $^2_1\text{H} + ^2_1\text{H} \rightarrow ^4_2\text{He}$. From the given data,

$$\text{energy of } ^2_1\text{H} = -(\text{BE of } ^2_1\text{H}) = -2 \times 1.1 = -2.2 \text{ MeV},$$

$$\text{energy of } ^2_1\text{H} + ^2_1\text{H} = -2 \times 2.2 = -4.4 \text{ MeV},$$

$$\text{energy of } ^4_2\text{He} = -(\text{BE of } ^4_2\text{He}) = -4 \times 7 = -28 \text{ MeV}.$$

Applying the energy conservation principle to the fusion reaction, the energy release we obtain is 28 MeV – 4.4 MeV = 23.6 MeV.

If we assume that the force between the nucleons exhibits the saturation property, then it is clear that the value of $\bar{B}$ in the range $30 < A < 170$ will be near constant. This means that a single nucleon interacts only with the few nucleons close to it and not with all the nucleons A in the nucleus. This in turn implies that the binding energy of the nucleon will depend only on these few nucleons, that is, on those nucleons that are at a distance less than the interaction distance from it (see Fig. 9.3). Therefore, the binding energy of a single nucleon depends on the number of nucleons inside the interaction volume and not on the total number of nucleons in the nucleus. As we have already noted, the density of nucleons in a nucleus is independent of A. Thus, the binding energy per nucleon is essentially constant.

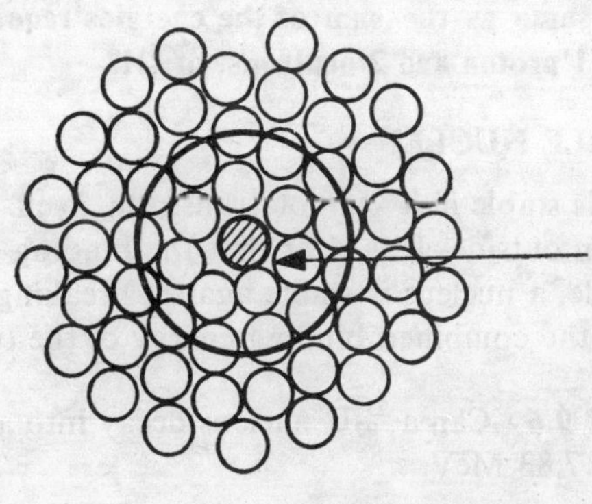

Fig. 9.3 Nucleus showing interaction volume.

The finite interaction distance between the nucleons results from the fact that the nuclear forces have a short range. This is in contrast with the behaviour of the electrostatic force. Charges interact with one another even when they are far apart, and hence the Coulomb force does not exhibit the saturation property.

QUESTION 9.5 Is the average binding energy of the electrons in an atom independent of *Z*?

EXAMPLE 9.5 Find the energy required to separate a tritium nucleus ($^3_1$H) from a $^{41}_{19}$K nucleus. Given $M(^3_1\text{H}) = 3.01605$ u, $M(^{41}_{19}\text{K}) = 40.96184$ u, and $M(^{38}_{18}\text{Ar}) = 37.96272$ u.

We are here required to calculate the separation energy $S_b(X)$, i.e., the energy needed to remove to infinity a part of the nucleus, say, b, from a nncleus X leaving a residual nucleus Y, both X and Y being in their ground states. From the energy conservation law,

$$S_b(X) = (M_b + M_Y - M_X)c^2, \tag{9.13}$$

where $X = ^{41}_{19}\text{K}$, $b = ^3_1\text{H}$, and the residual nucleus $Y = ^{38}_{18}\text{Ar}$. From the given masses, we get

$$M_b + M_Y - M_X = 0.01693 \text{ u}.$$

Thus,

$$S_b(X) = .01693 \times 931 = 15.8 \text{ MeV}.$$

It should be noted that Eq. (9.13) can be rewritten in terms of the binding energies of the three nuclei instead of in terms of the masses of the nuclei. To do this, we use Eq. (9.11) and obtain

$$S_b(X) = B_X - B_Y - B_b. \tag{9.14}$$

Thus, if we know the binding energies of the nuclei X, Y, b, we can, using Eq. (9.14), obtain the separation energy.

QUESTION 9.6 Is the separation energy calculated in Example 9.5 the same as the sum of the energies required to separate the constituents, i.e., 1 proton and 2 neutrons, of ^3_1H?

9.8 STABLE NUCLEI

A nucleus is stable if it cannot transform itself into another nucleus without the addition of energy from outside. It is clear that the binding energy is a measure of the stability of a nucleus For example, a nucleus is stable against breaking up into two fragments if its binding energy is more than the combined binding energy of the two fragments.

EXAMPLE 9.6 Can a $^{238}_{92}U$ nucleus decay into a $^{206}_{82}Pb$ nucleus? Given $\bar{B}(^{238}_{92}U) = 7.57$ MeV and $\bar{B}(^{206}_{82}Pb) = 7.83$ MeV.

For the $^{238}_{92}U$ nucleus,

$$\text{total BE} = 238 \times 7.57 = 1801.66 \text{ MeV}.$$

Further, for the $^{206}_{82}Pb$ nucleus,

$$\text{total BE} = 206 \times 7.83 = 1612.98 \text{ MeV}.$$

The difference in the two binding energies is equal to 188.68 MeV. If we consider the decay

$$^{238}_{92}U \rightarrow {}^{206}_{82}Pb + 10 \text{ protons} + 22 \text{ neutrons},$$

we find it cannot take place. However, this decay is possible if we supply 188.68 MeV energy from outside. On the other hand, if we consider the reaction

$$^{238}_{92}U \rightarrow {}^{206}_{82}Pb + 8\,^4_2He + 6 \text{ electrons},$$

the decay is possible. This is because, in this reaction, the binding energy in the final state (with the production of tightly bound ^4_2He nuclei) is larger than that of $^{238}_{92}U$, as we shall now explicitly show. The binding energy of the 8 α-particles is $8 \times 28.3 = 226.4$ MeV. So, the total binding energy of $^{206}_{82}Pb + 8\,^4_2He$ is $1612.98 + 226.4 = 1839.38$ MeV. This is larger than the binding energy of $^{238}_{92}U$ by 37.72 MeV.

The foregoing reaction takes place in stages, by a successive emission of α-, β-, and γ-particles as we shall see in Chapter 10.

In Fig. 9.4, all stable nuclei found in nature have been plotted on a neutron-proton diagram. The striking feature here is the tendency of all nuclei to group within a narrow band. A line

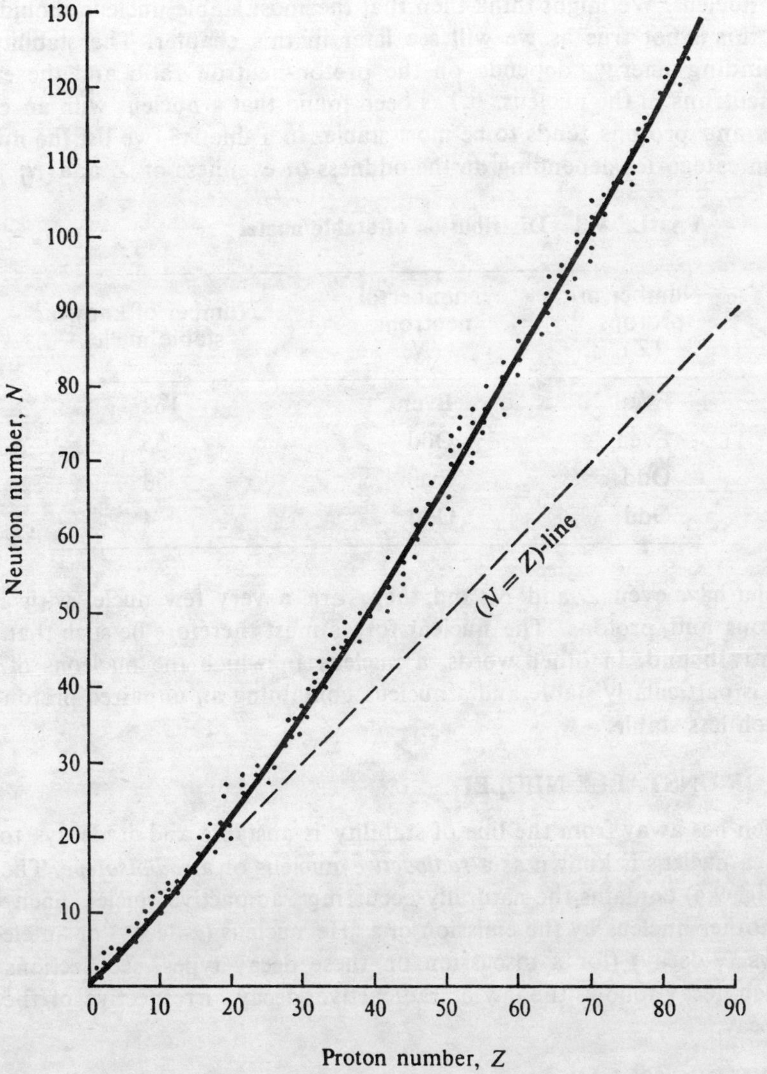

Fig. 9.4 Neutron-proton chart of stable nuclei.

drawn through this band is known as the *line of stability* or the stability curve. It indicates the general region in which most stable nuclei fall. As can be seen, for small values of N and Z, the stable nuclei lie close to the $(N = Z)$-line. The slope of this curve, it should be noted, is no longer 45° for $N, Z > 20$. At higher values of Z and N, the stable nuclei show an excess of neutrons. For example, the stable isotope of gold $(Z = 79)$ has a mass number equal to 197. It contains 118 neutrons and 79 protons. The neutron excess is $N - Z(= A - 2Z) = 39$. The reason the number of neutrons in a nucleus is more than the number of protons is as follows. A neutron interacts with the other nucleons only through the attractive nuclear forces. On the other hand, a proton, since it is positively charged, experiences the Coulomb repulsive forces from the other protons, besides the attractive nuclear forces. The repulsive forces increase the energy of the nucleus. We might think then that the most stable nucleus should consist only of neutrons. But this is not true as we will see later in this chapter. The stability of a nucleus, that is, its binding energy, depends on the proton-neutron ratio and the exact number of protons and neutrons in the nucleus. It has been found that a nucleus with an even number of both neutrons and protons tends to be most stable. In Table 9.3, we list the number of known stable nuclei in categories depending on the oddness or evenness of Z and N. More than half

TABLE 9.3 Distribution of stable nuclei

Number of protons (Z)	Number of neutrons (N)	Number of known stable nuclei
Even	Even	162
Even	Odd	55
Odd	Even	53
Odd	Odd	4

the stable nuclei have even Z and N, and there are a very few nuclei with an odd number of both neutrons and protons. The nuclear forces must therefore be such that an odd particle cannot be tightly bound. In other words, a nucleus in which the nucleons of the same kind occur in pairs is particularly stable and a nucleus containing an unpaired proton or an unpaired neutron is much less stable.

9.9 DECAY OF UNSTABLE NUCLEI

A nucleus which lies away from the line of stability is unstable and it decays to form a stable nucleus. Such a nucleus is known as a *radioactive* nucleus or a *radioisotope*. The region beyond $Z = 83$ (see Fig. 9.4) contains the naturally-occurring radioactive nuclei. Each of these nuclei decays into another nucleus by the emission of a ${}_2^4\text{He}$ nucleus (α-decay) or an electron (β-decay) or gamma-rays (γ-decay) (for a discussion on these decay types, see Sections 9.10, 9.11, and 9.12). Each such decay follows the law of radioactive decay, irrespective of the actual mechanism of the decay.

RADIOACTIVE DECAY LAW

The theory governing the rate of decay of an unstable nucleus (parent) into another nucleus (known as the daughter) is based on the statistical nature of the decay. In other words, the decay of an unstable nucleus is a random process. It is not possible to predict when an unstable nucleus will decay since the decay process is subjected to statistical laws. The probability that it will decay in the time interval between t and $(t + dt)$ is directly proportional to the time interval dt. Thus, the probability that a nucleus will decay in time dt is equal to $\lambda\, dt$, where λ is the *decay constant* and it varies from one nucleus to another. If N nuclei, all of the same type (say, ${}_{92}^{235}\text{U}$) are present at a time t, the change in the number of nuclei in a time interval dt will then be

$$dN = -\lambda N\, dt.$$

The negative sign here denotes that the number of nuclei has reduced. This equation can be rewritten as

$$dN/N = -\lambda\, dt. \tag{9.15}$$

On integration, we get

$$N_t = N_0 \exp(-\lambda t), \tag{9.16}$$

where N_0 is the number of nuclei originally present (say, at $t = 0$) and N_t is the number of undecayed nuclei present at the time t. This exponential decay law holds for all unstable systems as long as the decay is governed by a chance process.

If we start with a sample containing N radioactive nuclei or atoms, the number will go on decreasing exponentially with time (see Fig. 9.5). The exponential decay means that the time taken for all the radioactive atoms to decay will be infinite. Thus, it is difficult to quantitatively answer this question about the time the sample of N radioactive atoms will take to decay. However, we can certainly calculate the time the sample will take to reduce to some fraction of its initial amount.

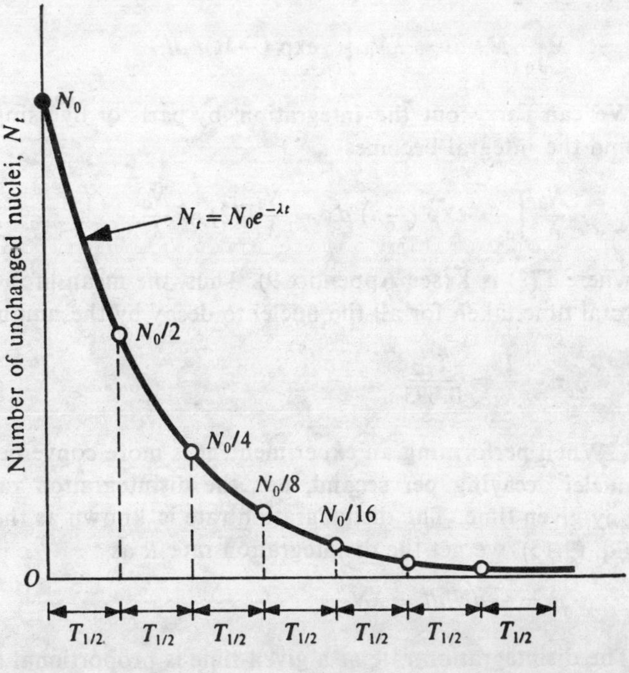

Fig. 9.5 Exponential decay curve.

We define the half-life $T_{1/2}$ as the time required for the number of atoms initially present to reduce to one-half the initial number. Let N_0 be the number initially present. Then, at $t = T_{1/2}$, we have

$$N_{t=T/2} = N_0/2 = N_0 \exp(-\lambda T_{1/2}). \tag{9.17}$$

Taking the logarithm of both the sides, we get $\lambda T_{1/2} = \ln 2$ or

$$T_{1/2} = \frac{\ln 2}{\lambda} = \frac{0.693}{\lambda}. \tag{9.18}$$

Thus, the characteristic of a decay of any radioactive nucleus can be denoted either by the decay constant λ or the half-life $T_{1/2}$ or any other appropriately defined constant. The half-life varies from one radioactive nucleus to another. Some radioactive nuclei are known to have a half-life as long as 10^{14} yr. At the other extreme are the nuclei with too short a half-life to be measured experimentally ($T_{1/2} \sim 10^{-14}$ sec), and thus such nuclei appear to decay instantaneously.

We also define the *mean life* τ of a nucleus as the average time the nucleus survives before it decays. It is obvious from the definition that we can determine τ by calculating the total lifetime of all the nuclei initially present (N_0) and dividing it by N_0. To do this, let $|dN|$ nuclei decay in a time interval t to $(t + dt)$ and let the lifetime of these nuclei be t. Then, the total lifetime of these nuclei is $t|dN|$. The total lifetime of all the nuclei is

$$\int_{N=N_0}^{N=0} t|dN|.$$

Using Eq. (9.15), we can convert the integral here into a time integral and get the total time for all the nuclei to decay as

$$\lambda \int_0^\infty N_t t\, dt = \lambda N_0 \int_0^\infty \exp(-\lambda t) t\, dt.$$

We can carry out the integration by parts or by using the Γ-function. Substituting $x = \lambda t$, we find the integral becomes

$$\frac{N_0}{\lambda} \int_0^\infty x \exp(-x)\, dx = \frac{N_0}{\lambda}\Gamma(2) = \frac{N_0}{\lambda},$$

where $\Gamma(2)$ is 1 (see Appendix 9). Thus, the mean life τ of the nuclei is obtained by dividing the total time taken for all the nuclei to decay by the number of these nuclei. Thus,

$$\tau = \frac{1}{\lambda} = \frac{T_{1/2}}{0.693}. \tag{9.19}$$

When performing an experiment, it is more convenient to measure the number of radioactive nuclei decaying per second, i.e., the disintegration rate, than the number of nuclei present at any given time. The disintegration rate is known as the *activity* of the radioactive nuclei. From Eq. (9.15), we get the disintegration rate R as

$$R = -dN/dt = \lambda N. \tag{9.20}$$

The disintegration rate at a given time is proportional to the number of nuclei present at that time. The number of nuclei decreases with time, and thus the negative sign in Eq. (9.20) makes R a positive quantity. We can easily see that the activity of a sample containing radioactive nuclei decreases exponentially with time. Multiplying both sides of Eq. (9.17) by λ, we get

$$-N\lambda = -N_0\lambda \exp(-\lambda t) \tag{9.21}$$

or

$$R_t = R_0 \exp(-\lambda t), \tag{9.22}$$

where R_0 is the initial activity at $t = 0$ and R_t is the activity at a time t.

The unit used to measure the activity is the *curie* (Ci), defined as the amount of radioactive material that produces 3.7×10^{10} disintegrations per second, i.e.,

$$1 \text{ Ci} = 3.70 \times 10^{10} \text{ disintegrations/sec.} \tag{9.23}$$

This definition originates from the fact that one gram of $^{226}_{88}$Ra gives 3.7×10^{10} disintegrations per second. Thus, if we want to express the activity of a substance in curies, we measure the number of disintegrations per second and divide it by 3.7×10^{10}.

EXAMPLE 9.7 A sample of 1 g of $^{209}_{83}Bi$ with a half-life of 2.7×10^7 yr decays into a stable isotope of thallium by emitting α-particles. What would be the activity of the sample? If a detector having 100% efficiency is placed in front of this sample in such a way that 30% of the α-particles emitted enter the detector, what then would be the rate at which these particles will be recorded by the detector? If the same measurement is carried out after one year, what would be the change in the rate of the α-particles recorded?

The activity is given by Eq. (9.20), i.e., $R = -dN/dt = \lambda N$. The decay constant of $^{209}_{83}Bi$ is

$$\lambda = \frac{0.693}{T_{1/2}} = \frac{0.693}{2.7 \times 10^7 \text{ yr} \times 3.16 \times 10^7 \text{ sec/yr}} = 8.12 \times 10^{-16} \text{ sec}^{-1}.$$

The number of nuclei in 1 g of $^{209}_{83}Bi$ is

$$\frac{6.02 \times 10^{23}}{209} \sim 2.88 \times 10^{21}.$$

Thus, the activity would be

$$R = 8.12 \times 10^{-15} \times 2.88 \times 10^{21} \sim 2.33 \times 10^6 \text{ disintegrations/sec}$$

or

$$R_t = \frac{2.33 \times 10^6}{3.7 \times 10^{10}} \text{ Ci} = 0.63 \times 10^{-4} \text{ Ci} = 0.063 \text{ mCi}.$$

The decay of $^{209}_{83}Bi$ is represented by the reaction

$$^{209}_{83}Bi \rightarrow {}^{205}_{81}Tl + {}^{4}_{2}He.$$

For each disintegrating $^{209}_{83}Bi$ nucleus here, one α-particle is emitted. Thus, the activity we have calculated directly gives the rate of emission of the alpha-particles. The number N_α of α-particles reaching the detector per second is

$$N_\alpha = \frac{2.33 \times 10^6 \times 30}{100} = 6.99 \times 10^5 \text{ sec}^{-1}.$$

Since the detector efficiency is 100%, all the particles reaching the detector are recorded.

The change in activity R_t with time is given by the exponential decay law

$$R_t = R_0 \exp(-\lambda t)$$
$$= R_0 \exp(-8.12 \times 10^{-16} \times 3.16 \times 10^7) = R_0 \exp(-2.56 \times 10^{-8}).$$

As can be seen, the exponential factor is almost 1. So, $R_t \sim R_0$ and therefore, after one year, there will be no noticeable change in the rate of α-particles being recorded.

We note that although approximately 0.7 million nuclei are disintegrating per second, yet there is no perceptible change in the disintegration rate. The activity reduces by half only after 27 million years (because $T_{1/2} = 2.7 \times 10^7$ yr).

CHAIN DISINTEGRATION

In some nuclei, the parent nucleus decays into a daughter nucleus which in turn decays into a granddaughter, and so on until, finally, a stable end product results. As an example, a chain disintegration starting from $^{232}_{90}Th$ and ending in $^{208}_{82}Pb$ is shown in Fig. 9.6. Thus, if we start with a sample containing pure $^{228}_{90}Th$, then, after some time, the sample would contain a mixture

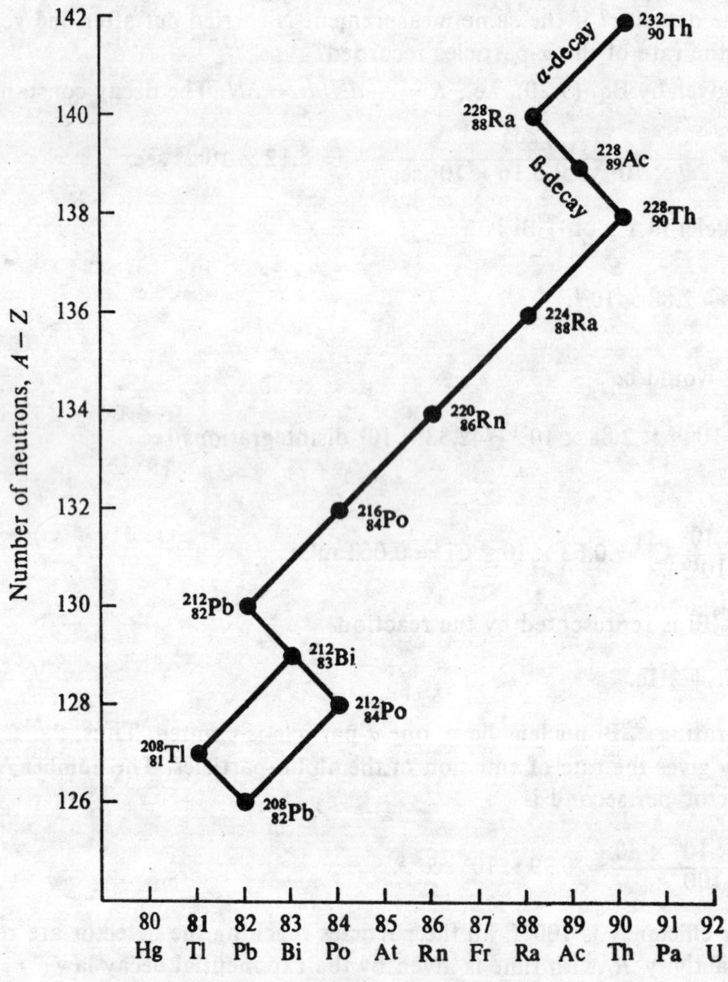

Fig. 9.6 Thorium (4*n*) series.

of all the nuclei given in this figure. The relative concentration of the nuclei in this *chain disintegration* would depend on the disintegration constants (λ's) of the nuclei. This dependence will clearly require solving a system of differential equations (see Problem 9.13).

One interesting implication of chain disintegration is that short-lived radioisotopes, with half-lives very small as compared to the age of the universe, exist in nature. The reason is simply that a short-lived radioisotope is in equilibrium with a long-lived parent and is being continuously formed in a decay chain. Such a chain is called a radioactive series. For example, $^{234}_{90}$Th has a half-life of only 24 days and yet it is found to occur in nature. This is because it is being continuously produced by $^{238}_{92}$U having a half-life of 4.5×10^9 yr which is comparable with the age of the earth.

APPLICATIONS OF RADIOACTIVITY

Dating

One of the earliest applications of radioactivity was in determining the age of a rock from the ratio of uranium and lead present in it (see Problem 9.12). This general technique is known as *dating*. There are numerous other dating techniques that have been developed. Here, we shall discuss only the technique used for determining the time intervals of the order of thousands of years.

The half-life of $^{14}_{6}$C is 5730 yr, and accordingly no $^{14}_{6}$C should have been left in nature since the time the element C was formed. But this is not true because $^{14}_{6}$C is continually being created in the atmosphere by high energy particles from outer space, called cosmic rays, through the reaction

$$^1n + {}^{14}_{7}N \rightarrow {}^{14}_{6}C + {}^{1}_{1}H.$$

Clearly, there is a kind of radioactive equilibrium between the production of $^{14}_{6}$C and its decay. The $^{14}_{6}$C thus produced exists as a small part of the total CO_2 in the atmosphere. This gas diffuses throughout the atmosphere and is picked up by any living organism utilizing carbon in its metabolic cycle. After the organism dies, its intake of $^{14}_{6}$C stops and the $^{14}_{6}$C already present in it starts decaying with its characteristic half-life. The age of the organism may therefore be determined by measuring the amount of $^{14}_{6}$C it contains and comparing this amount with the concentration of $^{14}_{6}$C in a live organism. This method is suitable for the dating of cultural history (time $\sim$5000 yr) in the same way as the technique using uranium is applicable for the dating of geological events (see Problems 9.11 and 9.12).

Tracer Technique

As we know, the chemical properties of a radioisotope of an element are identical with those of a stable isotope of the element. If the two, i.e., the stable isotope and the radioisotope [e.g., iodine $^{130}_{53}$I ($T_{1/2} = 12.6$ hr)], are mixed and introduced into some system, then both will undergo the same chemical, biological, and physical processes. If a chemical analysis of the mixture is carried out, it will not be possible to distinguish between the two constituents. However, since $^{130}_{53}$I decays via an electron emission, its progress through the system can be followed by means of an electron detector placed outside the container wall. The sensitivity of this detector is such that the detector will register the decay of even a single atom.

The element phosphorus is an essential constituent of a fertilizer. If a small amount of radioactive phosphorus $^{32}_{15}$P is added to a fertilizer, known as *labelling* or *tagging* the fertilizer, and the mixture spread on the soil, then, using an electron detector, the diffusion of the fertilizer through the soil can be traced. Such investigations have shown that it is unnecessary to spread a fertilizer uniformly on the soil.

The tracer technique is also frequently used in medical diagnostics. It has been found that a cancerous tissue can retain a larger amount of bismuth than a healthy tissue. By adding a small amount of radioactive bismuth $^{210}_{83}Bi$ to ordinary bismuth and administering the mixture to the body, we can trace the fate of bismuth by testing the radioactivity of the various parts of the body, and thus locate the cancerous tissue.

The tracer technique is being increasingly applied in industries. For example, we can make out the wear of a piston ring by incorporating in it a radioisotope and then measuring the accumulation of activity in the lubricating oil. Similarly, the wear of a car tyre can be investigated by incorporating $^{32}_{15}P$ into its outer tread and observing the decrease in counting rate as a function of time.

Activation Analysis

The activation analysis is a technique somewhat different from the tracer method. In this technique, the sample to be analyzed is bombarded with neutrons. The addition of a neutron to the nucleus of an element in the sample makes the element radioactive. By analyzing the characteristic radiation thus emitted, we can identify the element. This technique is particularly useful in determining the elements occurring in traces. For example, if we have a sample of a material containing traces of gold ($^{197}_{79}Au$), we can determine the concentration of gold by bombarding the sample with neutrons and finding the activity of $^{198}_{79}Au$. We can repeat this process with a sample containing a known amount of $^{197}_{79}Au$. Then, using this sample as standard, we can calculate the amount of $^{197}_{79}Au$ in the sample with unknown quantity of gold. This method can also be applied to evaluate the content of an element in a patient's body.

Therapeutic Effects of Radioactivity

We all know that the radiation emitted by a radioisotope can destroy cells and an excessive exposure to it can even cause death. However, this effect is beneficial in treating cancer. It has been found that by a controlled administration of radiation, cancer can be contained or cured. Similarly, by applying $^{131}_{53}I$, an enlarged thyroid can be reduced to its normal size.

Since radiation kills microorganisms, it can be used for food preservation. For example, irradiated potatoes are known to remain free for a long time from the sprouting disease.

9.10 ALPHA-DECAY

The nuclei with $Z > 82$ occur in nature but are basically unstable and tend to decay into more stable configurations. There are also a few naturally-occurring radioisotopes of a low atomic number, for example, $^{40}_{19}K$. Now we can produce artificial radioisotopes of all the elements by using a nuclear reactor or an accelerator. As we have already noted (see Fig. 9.4), each atomic number has a neutron-proton ratio in a certain range for which the nuclei are stable. A nucleus lying outside this range is radioactive and decays in such a way that it brings back the neutron-proton ratio within its stability range. A naturally-occurring or an artificially produced radioisotope spontaneously emits a ^4_2He nucleus (α-particle), or an electron or a positron (beta-particle), or a photon (gamma-ray), thereby reducing its energy and achieving a more stable configuration. Starting with the alpha-decay, we shall now study these three modes of decay of an unstable nucleus.

The emission of an α-particle from a parent nucleus leads to a daughter nucleus whose charge is reduced by 2 and atomic mass number is reduced by 4. The α-decay is represented by the equation

$$^A_ZX = {}^{A-4}_{Z-2}Y + {}^4_2He + Q. \tag{9.24}$$

The quantity Q is the energy available in the decay and is given by

$$Q = [M_x - (M_y + M_\alpha)]c^2, \tag{9.25}$$

where M_x, M_y, and M_α are the masses of the parent nucleus, daughter nucleus, and α-particle, respectively.

QUESTION 9.7 Why is an α-particle, rather than neutrons and protons, emitted from an unstable nucleus?

The kinetic energy of the α-particle T_α can be determined by applying the laws of conservation of energy and momentum. The law of conservation of energy specifies

$$(M_x - M_y + M_\alpha)c^2 = \tfrac{1}{2}M_\alpha v_\alpha^2 + \tfrac{1}{2}M_y v_y^2, \tag{9.26}$$

where v_α and v_y are the velocities of the α- and y-particle, respectively. Since the parent nucleus is initially at rest, its initial momentum is zero. Then, from the law of conservation of momentum, we have

$$M_\alpha v_\alpha = M_y v_y. \tag{9.27}$$

Combining Eqs. (9.26) and (9.27), we get

$$T_\alpha = \frac{1}{1 + M_\alpha/M_y} Q = \frac{M_y}{M_y + M_\alpha} Q. \tag{9.28}$$

Since the mass of the daughter nucleus (in u) is approximately $A - 4$ and the mass of the α-particle is approximately 4, we have

$$T_\alpha \sim \frac{A-4}{A} Q. \tag{9.29}$$

As can be seen, the kinetic energy of the emitted α-particle will be very close to the total disintegration energy Q available to a naturally-occurring radionucleus (with $Z > 83$). For example, in the α-decay of $^{26}_{88}Ra$, i.e.,

$$^{226}_{88}\text{Ra} = {}^{222}_{86}\text{Rn} + {}^{4}_{2}\text{He} + Q, \tag{9.30}$$

the disintegration energy $Q = 4.87$ MeV and the emitted α-particle is observed to have an energy = 4.78 MeV. In the decay represented by Eq. (9.30), the final system consists of just the α-particle and the residual (daughter) nucleus $^{222}_{86}Rn$, with the α-particle having a well defined energy. This is because both the parent and daughter nuclei are in a definite energy state. In certain situations, the α-particle spectrum is complex and includes many α-particles with different energies. This is because the daughter nucleus can be in its various possible excited states.

The half-life of alpha-decaying radioisotopes is found to vary over a very wide range, from 3×10^{-5} sec to 10^{17} yr. On the other hand, the lowest alpha-particle energy is about 4 MeV and the highest about 8.8 MeV. Table 9.4 lists some α-particle emitters with tremendously varying half-life but narrowly varying α-particle energy.

In their study of the known α-particle emitters, H. Geiger and J. M. Nuttal found the empiri-

TABLE 9.4 Some α-particle emitters

Nucleus	α-particle energy	Half-life ($T_{1/2}$)
${}^{212}_{84}Po$	8.8 MeV	3×10^{-7} sec
${}^{214}_{84}Po$	7.7 MeV	1.6×10^{-4} sec
${}^{210}_{84}Po$	5.3 MeV	1.38×10^{3} yr
${}^{226}_{88}Ra$	4.7 MeV	1.62×10^{3} yr
${}^{238}_{92}U$	4.1 MeV	4.5×10^{9} yr

cal relation between the decay constant λ and E_α to be

$$\log_{10} \lambda = C - \frac{D}{(E_\alpha)^{1/2}}, \tag{9.31}$$

where C and D are the slowly varying functions of Z but are independent of the neutron number N.

Theory of α-Decay

The α-decay characteristics could not be understood on the basis of classical physics but found a straightforward explanation in terms of quantum mechanics. The explanation, as given by G. Gamow, is as follows. An α-particle formed inside a nucleus (when two neutrons and two protons come close together) encounters a potential barrier. The alpha-decay occurs when the α-particle tunnels through this barrier.

In order to understand the presence of the barrier the α-particle encounters, let us examine the inverse of the α-decay reaction, namely,

$${}^{226}_{88}Ra \rightarrow {}^{222}_{86}Rn + \alpha(T_\alpha \sim 4.8 \text{ MeV}).$$

When this α-particle with a kinetic energy 4.8 MeV and charge $Z = 2$ approaches the daughter nucleus ${}^{222}_{86}Rn$ having a radius R, it experiences a Coulomb force. At a distance of the order of R from the daughter nucleus, the α-particle is also subjected to a strong attractive nuclear force. Therefore, the potential energy of the α-particle outside ${}^{222}_{86}Rn$ is purely electrostatic, i.e., $[2Ze^2/(4\pi\epsilon_0 r)]$, and inside ${}^{222}_{86}Rn$ it is the sum of the electrostatic and nuclear potential energies. The resultant potential energy curve is shown in Fig. 9.7a.

When the α-particle is very far from the daughter nucleus ($r \rightarrow \infty$), the potential energy is zero. In this situation, the kinetic energy of the α-particle (4.8 MeV) is also its total energy (shown by the dashed line in Fig. 9.7a). In the region between A and B, the α-particle is inside the nucleus and has a finite but undefined kinetic energy. Its total energy, i.e., the sum of the potential energy (negative) and the kinetic energy, then is 4.8 MeV. In the region between B and C, the kinetic energy of the α-particle is negative. Classically, it is a forbidden region. In order to take the α-particle out of the nucleus (i.e., from B to C) or to the nucleus from outside (i.e., from C to B), an energy ΔE has to be supplied to enable the particle to get over the potential energy hump. Thus, classically, no α-particle should be emitted. However, quantum mechanically, the α-particle has a finite probability of tunneling through the barrier. This is because the

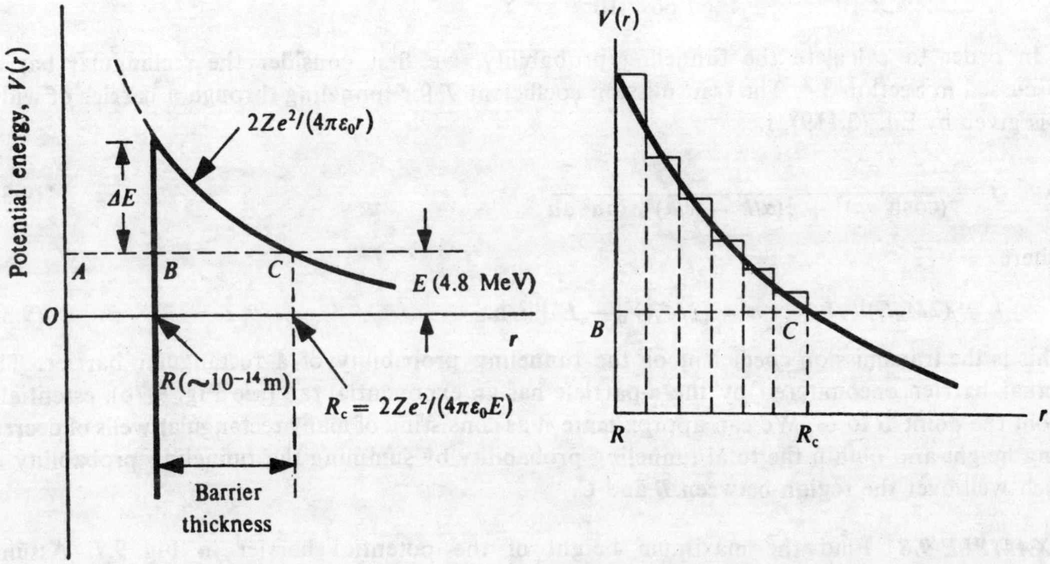

(a) Potential energy curve

(b) Barrier approximated by series of rectangular barriers

Fig. 9.7 Potential energy for $^{222}_{86}$Rn + α-particle system.

α-particle, when it is inside the nucleus, has a wave associated with it which tunnels through the barrier. This wave then emerges at C with a decreased amplitude.

We can express the decay probability as the product of two terms, i.e.,

$$\lambda = \nu T, \tag{9.32}$$

where ν is the number of times per second the α-particle strikes the potential wall at B and T is the transmission coefficient of the potential barrier.

Let us estimate the collision frequency ν. If the velocity of the α-particle inside the nucleus is v_α, the time taken by it to strike the wall is $2R/v_\alpha$, where R is the radius of the nucleus. Thus, the frequency with which it strikes the wall is $\nu \sim v_\alpha/(2R)$. Since the potential inside the nucleus is approximately constant (V_N), the total energy E is

$$E = \tfrac{1}{2}M_\alpha v_\alpha^2 + V_N \quad \text{or} \quad v_\alpha = \left[\frac{2(E - V_N)}{M_\alpha}\right]^{1/2}.$$

Thus,

$$\nu = \frac{1}{2R}\left[\frac{2(E - V_N)}{M_\alpha}\right]^{1/2}. \tag{9.33}$$

The radius of the nucleus ($^{222}_{86}$Rn), from Eq. (9.5), is approximately 8×10^{-15} m. The depth of the potential well (as estimated in Example 9.14) is ~ -50 MeV. Substituting these numbers in

Eq. (9.33), we get

$$\nu = \frac{1}{2\times 8\times 10^{-15}}[\frac{2(4.8+50)\times 1.6\times 10^{-13}}{4\times 1.66\times 10^{-27}}]^{1/2} \sim 3\times 10^{21}\ \text{sec}^{-1}.$$

In order to calculate the tunneling probability, we first consider the rectangular barrier discussed in Section 3.4. The transmission coefficient T for tunneling through a barrier of width a is given by Eq. (3.119), i.e.,

$$T = \frac{1}{(\cosh \alpha a)^2 + \frac{1}{4}(\alpha/k - k/\alpha)^2 \sinh^2 \alpha a}, \tag{9.34}$$

where

$$k = (2M_\alpha E)^{1/2}/\hbar, \qquad \alpha = [2M_\alpha(V_0 - E)]^{1/2}/\hbar. \tag{9.35}$$

This is the transmission coefficient or the tunneling probability of a rectangular barrier. The actual barrier encountered by the α-particle has an exponential tail (see Fig. 9.7b), essentially from the point B to ∞. We can approximate it as consisting of many rectangular wells of decreasing height and obtain the total tunneling probability by summing the tunneling probability of each well over the region between B and C.

EXAMPLE 9.8 Find the maximum height of the potential barrier in Fig. 9.7. Assume $R = 8\times 10^{-15}$ m, and treat the nuclear potential as an attractive square well potential of depth $V_N = 50$ MeV and the helium nucleus as a point particle. The charge is uniformly distributed inside the residual radon nucleus.

The Coulomb potential energy for the α-particle inside the radon nucleus can be calculated as follows. The electric field at a distance $r < R$ from the centre of the nucleus can be derived using Gauss's law and is given by

$$E_1 = \frac{2Ze^2}{4\pi\epsilon_0}\frac{r}{R^3}.$$

It should be noted that, for $r > R$, the radon nucleus can be considered as a point particle with the charge concentrated at its centre. The electric field in this situation is given by

$$E_2 = \frac{2Ze^2}{4\pi\epsilon_0}\frac{1}{r^2}.$$

The Coulomb potential energy at the point $r\,(>R)$ is

$$V_c(r) = -\int_\infty^r E_2\,dr = \frac{2Ze^2}{4\pi\epsilon_0}\frac{1}{r},$$

where $2e$ is the charge on the α-particle and Ze is the charge on the daughter nucleus, and that at the point $r\,(<R)$ is

$$V_c(r) = -\int_\infty^R E_2\,dr - \int_R^r E_1\,dr = \frac{2Ze^2}{4\pi\epsilon_0}\frac{1}{R}[1 + \frac{R^2 - r^2}{2R^2}].$$

The total potential can be obtained by adding the nuclear potential to the Coulomb potential.

The nuclear potential V_N is given by

$$V_N = 0 \qquad \text{for } r > R$$
$$= -50 \text{ MeV} \quad \text{for } r < R.$$

Thus,

$$V = V_c = \frac{2Ze^2}{4\pi\epsilon_0 r} \qquad (r > R)$$
$$= V_c - 50 \text{ MeV} \qquad (r < R)$$
$$= \frac{2Ze^2}{4\pi\epsilon_0}\frac{1}{R}[1 + \frac{R^2 - r^2}{2R^2}] - 50 \text{ MeV} \qquad (r < R).$$

It is clear that the potential is maximum at $r = R$ when

$$V = \frac{2Ze^2}{4\pi\epsilon_0 R} \sim 30 \text{ MeV}.$$

In Fig. 9.7a, the potential for $r < R$ has been shown to be constant. This is because in this figure we have neglected the Coulomb potential.

In most α-decays, the kinetic energy of the α-particle is a few MeV, much smaller than the barrier height (~30 MeV). Thus, $\alpha > k$ and

$$(\frac{\alpha}{k} - \frac{k}{\alpha}) \sim \frac{\alpha}{k}. \qquad (9.36)$$

We can assume $\alpha a \gg 1$, where a is of the order of BC, the width of the barrier (see Fig. 9.7a). We can estimate the value of αa from Example 9.8 as

$$\alpha a \sim (\frac{2M_\alpha V_0}{\hbar^2})^{1/2} BC.$$

We can also estimate that the width BC is of the order of 10^{-13} m since we know that the energy of the α-particle is 4.8 MeV. Thus,

$$\alpha a \sim (\frac{2 \times 4 \times 1.6 \times 10^{-27} \times 30 \times 1.6 \times 10^{-13}}{10^{-68}})^{1/2} \times 10^{-13} \sim 100.$$

We can now approximate cosh αa and sinh αa in Eq. (9.34) by exp ($\alpha a/2$). Therefore, we get

$$T = \frac{1}{\frac{\exp(2\alpha a)}{4} + \frac{1}{4}\frac{\alpha^2}{k^2}\frac{\exp(2\alpha a)}{4}} = [(\frac{16}{4 + \frac{\alpha^2}{k^2}})] \exp(-2\alpha a). \qquad (9.37)$$

When $\alpha a \gg 1$, the most important factor in this equation is the exponential. The factor in front of the exponential, which usually is of the order of unity, is not significant since its variation with V and E_0 is negligible as compared to the variation in the exponential itself. So, we can write

$$\ln T \sim -2\alpha a + \text{a term which varies slowly with energy.} \qquad (9.38)$$

For a rectangular well of thickness dr, we can write

$$\ln T = -2\alpha\, dr, \qquad (9.39)$$

where

$$\alpha^2 = \frac{2m}{\hbar^2}[V(r) - E] = \frac{2m}{\hbar^2}\left(\frac{2Ze^2}{4\pi\epsilon_0 r} - E\right), \tag{9.40}$$

making α a function of r. We have neglected the second term in Eq. (9.38). Summing $\ln T$ over all the rectangular wells, we can write

$$\ln T = -2\int_R^{R_c} \alpha(r)\, dr. \tag{9.41}$$

Here, R is the radius of the nucleus and R_c is the distance from the centre of the nucleus to the point C, where $V(r) = 4.8$ MeV, that is, $R_c = 2Ze^2/(4\pi\epsilon_0 E)$. Integrating Eq. (9.41), we get

$$\ln T = 2.97Z^{1/2}R^{1/2} - 3.95ZE^{-1/2}, \tag{9.42}$$

where R is in fermis and E is in MeV. Thus, the decay constant, using Eqs. (9.28) and (9.29), we get is

$$\lambda = \nu T = vT/(2R) = 10^{-21}T \tag{9.43}$$

or

$$\ln \lambda = \ln 10^{-21} + 2.97Z^{1/2}R^{1/2} - 3.95ZE^{-1/2}. \tag{9.44}$$

In the foregoing derivation, we have not considered the fact that the motion of the α-particle is three-dimensional. Instead, we have used the transmission coefficient of a one-dimensional barrier. Also, the approximation $\alpha/k > 1$ is not really valid near the region $r \sim R_c$. The value of λ obtained without this approximation and treating the α-particle motion in three dimensions is essentially the same as that given by Eq. (9.44). This expression is in very good agreement with the experimental observations. As can be noted, the form of Eq. (9.44) is the same as that of the Geiger-Nuttal law (9.31) with the Z-dependence of the coefficients C and D explicitly given by the former.

9.11 BETA-DECAY

The beta-decay is a process in which the atomic number of a nucleus changes by one unit but the mass number remains unchanged. In other words, the charge of the nucleus changes but the number of nucleons remains the same. In this process, an electron (β^--particle) or a positron (β^+-particle) is ejected from the nucleus or an orbital electron is absorbed by the nucleus. Since a nucleus contains only the neutrons and protons, how then do we understand the emission of an electron or a positron (beta-particles)? This can be explained by assuming that these particles are produced during the emission process in the same way a photon is produced during an emission from an excited atom. The three types of beta-decay we have mentioned correspond to the following basic reactions inside a nucleus:

$$\begin{aligned} &n \rightarrow p + e^- + \nu_e \qquad [\beta^-\text{-decay}],\\ &p \rightarrow n + e^+ + \nu_e \qquad [\beta^+\text{-decay}],\\ &p + e^- \rightarrow n + \nu_e \qquad [\text{electron capture (EC)}],\end{aligned} \tag{9.45}$$

where ν_e and $\bar{\nu}_e$ are the neutrino and antineutrino, respectively. Later in this chapter, we shall discuss the role of these particles in β-decay. The binding energy of a nucleus determines which, if any, of these three decay processes is energetically possible. For example, in the decay of the

$^{60}_{27}Co$ nucleus into the $^{60}_{28}Ni$ nucleus, a β⁻-particle is emitted, i.e.,

$$^{60}_{27}\text{Co} \rightarrow {}^{60}_{28}\text{Ni} + e^- + \bar{\nu}_e, \tag{9.46a}$$

releasing an energy of 1.56 MeV in the form of the kinetic energies of the three products. An example of the β^+-particle emission is the decay of $^{22}_{11}Na$, i.e.,

$$^{22}_{11}\text{Na} \rightarrow {}^{22}_{10}\text{Ne} + e^+ + \nu_e + 0.55 \text{ MeV}. \tag{9.46b}$$

Further, an example of electron capture is the decay of the $^{37}_{18}Ar$ nucleus into the $^{37}_{17}Cl$ nucleus after an atomic electron is absorbed, i.e.,

$$^{37}_{18}\text{Ar} + e^- \rightarrow {}^{37}_{17}\text{Cl} + \nu_e. \tag{9.46c}$$

The energy released in this reaction is 0.0186 MeV.

A nucleus lying above the region of stability, i.e., region of neutron excess (see Fig. 9.4), becomes stable by reducing the number of neutrons in it. In other words, such a nucleus disintegrates in a manner that its N/Z ratio gets reduced. This is achieved through the β^--decay where a neutron inside the nucleus changes into a proton, simultaneously creating and emitting an electron and a neutrino.

Similarly, a nucleus lying below the stability region becomes stable by increasing its N/Z ratio. Thus, such a nucleus decays by emitting a β^+-particle or by electron capture. In the β^+-decay, a proton inside the nucleus changes into a neutron, simultaneously creating and ejecting a positron and an antineutrino. In electron capture, an atomic orbital electron of a nucleus combines with a proton of the nucleus to change the proton into a neutron and a neutrino.

QUESTION 9.8 The electrons in the *K*-shell of an atom have the highest probability of being captured by the nucleus. Why?

To study the energetics of the beta-decay process, we write the mass-energy equations for this process in terms of the atomic masses:

$$^{A}_{Z}M = {}_{Z+1}^{A}M + Q/c^2 \qquad (\beta^-\text{-decay}), \tag{9.47}$$

$$^{A}_{Z}M = {}_{Z-1}^{A}M + 2m_e + Q/c^2 \qquad (\beta^+\text{-decay}), \tag{9.48}$$

$$^{A}_{Z}M = {}_{Z-1}^{A}M^* + Q/c^2 \qquad (\text{electron capture}). \tag{9.49}$$

In these equations, the neutrino mass does not appear because, as we shall learn later in this section, a neutrino (and also an antineutrino) is massless. Further, Q is the energy available in the decay. As can be noted, the electron mass does not appear on the right-hand side of Eq. (9.47). This is because an electron is needed for the atom $_{Z+1}^{A}M$ to become neutral. On the right-hand side of the β^+-decay equation, there are two electron masses. Of these, one is due to the positron the nucleus emits and the other is due to the extra electron (in the atom $_{Z-1}^{A}M$) the atom releases to make itself neutral. In electron capture [Eq. (9.49)], an orbital electron is captured by the nucleus, and thus the atom $_{Z-1}^{A}M$ produced is neutral but in an excited state, indicated by the superscript asterisk.

EXAMPLE 9.9 Show that in the β^--decay of a nucleus $^{A}_{Z}X$ the available energy in terms of the neutral atomic masses is given by $(^{A}_{Z}M - {}_{Z+1}^{A}M)c^2$. Neglect the binding energy of the electrons.

Assume that the parent nucleus $^{A}_{Z}X$ decays by electron emission. Let M'_P be the mass of the parent nucleus and M'_D the mass of the daughter nucleus. The energy conservation law implies

$$M'_P c^2 = M'_D c^2 + m_e c^2 + Q.$$

The rest mass of $\bar{\nu}_e$ is assumed to be zero (see the discussion on the neutrino later in this section). Writing the foregoing equation in terms of the atomic masses ${}^A_Z M$ and ${}_{Z+1}^{A}M$, we get

$${}^A_Z M - Zm_e = {}_{Z+1}^{A}M - (Z+1)m_e + m_e + Q/c^2.$$

Since ${}^A_Z M$ is the mass of the neutral atom, $({}^A_Z M - Zm_e)$ is the mass of the parent nucleus if we neglect the binding energy of the electrons. Thus, the available energy Q is

$$Q = [({}^A_Z M - Zm_e) - {}_{Z+1}^{A}M - (Z+1)m_e + m_e]c^2 = ({}^A_Z M - {}_{Z+1}^{A}M)c^2.$$

Beta-Particle Spectrum

In the alpha-decay, we noted that an α-particle has a discrete energy because its emission corresponds to a transition from a well defined energy of the parent nucleus to a well defined energy of the daughter nucleus. A β-particle spectrum, on the other hand, does not have this discrete character, but it does show a continuous energy distribution of the β-particles (see Fig. 9.8). This continuous nature of the energy distribution of β-particles at one time posed quite a problem in understanding the beta-decay. This problem was resolved with the prediction of the existence of the neutrino which was subsequently discovered.

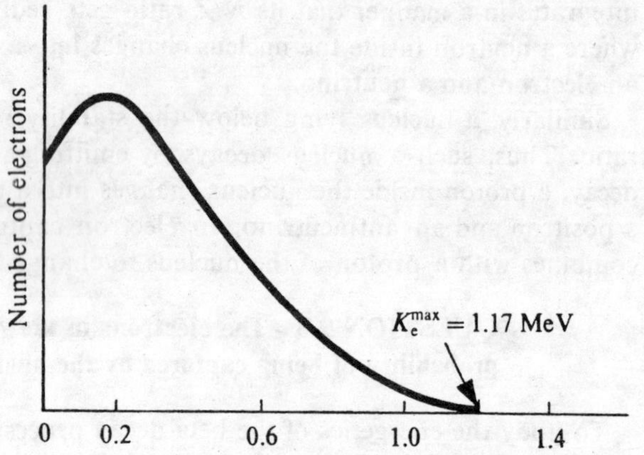

Fig. 9.8 Electron spectrum of β^--decay of ${}^{210}_{83}$Bi nucleus.

The Neutrino

Let us assume that in the β-decay of a nucleus ${}^A_Z$X at rest, no neutrino is emitted, i.e.,

$${}^A_Z X \rightarrow {}_{Z+1}^{A}Y + e^-.$$

This then represents a two-body decay like an alpha-decay. As already noted, the laws of conservation of momentum and energy, when applied to the alpha-decay, lead us to conclude that the emitted alpha-particle has a definite energy [see Eqs. (9.25) and (9.28)]. This energy depends only on the masses of the parent and daughter nuclei and alpha-particle. If the beta-decay were also a two-body process, the electrons would then have had a well defined energy. However, experimentally, the electrons were not found to be monoenergetic. Also, their maximum kinetic energy K_e^{max} so observed was consistent with the energy obtained assuming the beta-decay reaction to be a two-body type. But most electrons were noted to have an energy less than K_e^{max} distributed over the entire range from 0 to K_e^{max}. This distribution of energy could not be explained in terms of a two-body reaction, and led the physicists to question the validity of the laws of conservation of energy and momentum.

Further, the law of conservation of angular momentum also appeared to be violated. To see this, we consider the beta-decay of the ${}^{12}_{5}$B nucleus, i.e.,

$${}^{12}_{5}B \rightarrow {}^{12}_{6}C + e^-. \tag{9.50}$$

The spin angular momenta of $^{12}_{5}B$, $^{12}_{6}C$, and e^- are 1, 0, and $\frac{1}{2}$ (in units of $\hbar$), respectively. In reaction (9.50), the total angular momentum on the left-hand side is 1, whereas it is $\frac{1}{2}, \frac{3}{2}, \ldots$ on the right-hand side, depending on the orbital angular momentum of the emitted electron. Thus, the initial angular momentum, which is an integral, does not equal the final angular momentum, which is a half-integral.

This disturbing situation, namely, the violation of the fundamental laws of conservation of energy, linear momentum, and angular momentum in β-decay, was salvaged by W. Pauli in 1931. He postulated that a particle, now called the neutrino, other than an electron or a positron is also emitted in beta-decay but is not detected. He proposed that this particle must have a charge zero, an intrinsic spin $\frac{1}{2}$, and a rest mass zero. Whenever a decay results in three particles (namely, a beta-particle, a neutrino, and a daughter nucleus), the available energy can be shared among the particles in a variety of ways while still conserving the linear momentum. The neutrino takes away some energy and thus the energy of the β-particle will, in general, be less than that it would have had were the decay a two-body one. The intrinsic spin of the neutrino allows the angular momentum to be conserved. The existence of the neutrino was experimentally confirmed in 1956.

QUESTION 9.9 Show that the law of conservation of angular momentum is not violated in β-decay if the intrinsic spin of the neutrino S is $\frac{1}{2}$.

One of the exciting observations in the beta-decay is the nonconservation of parity (see Chapter 11).

9.12 GAMMA-DECAY

A nucleus, like an atom, exists in one of a number of quantized energy states. A stable nucleus is normally in its lowest (i.e., ground) state. If a nucleus, just as an atom, is in an excited state (denoted by an asterisk, e.g., Ni*), it decays into one of its lower states by the emission of electromagnetic radiation. The electromagnetic transition of a nucleus involves an energy difference between a few tens of keV and a few MeV. The radiation so emitted is known as gamma-ray. Thus, in the gamma-decay, a nucleus reduces its excitation energy without changing the number of neutrons or protons in it. Through the gamma-decay, a nucleus can attain its ground state straightaway or a lower excited state which in turn decays to a still lower excited state, and so on, emitting, in the process, a cascade of photons (see Fig. 9.9). In this cascade, only those nuclear transitions occur that satisfy the laws of conservation of energy, momentum, angular momentum, and

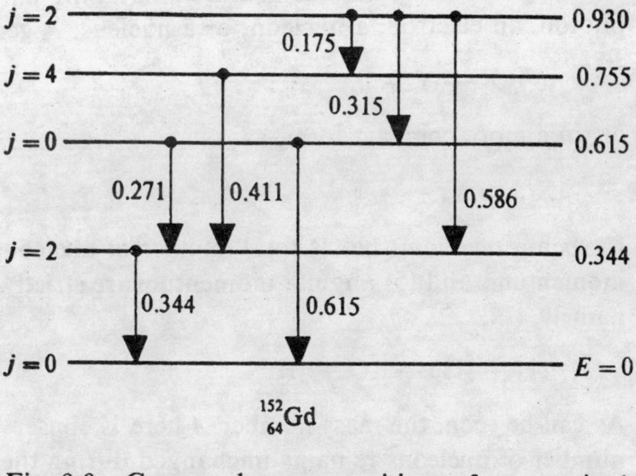

Fig. 9.9 Gamma-decay transitions from excited states of $^{152}_{64}Gd$ nucleus. Numbers by the side of arrows indicate energies in MeV of gamma-rays emitted.

parity. In other words, such a transition is governed by selection rules in the same manner as an atomic transition. An excited state of a nucleus generally has a very short half-life ($\sim 10^{-14}$ sec) if it decays by the emission of a gamma-ray. On the other hand, some excited states have a relatively long half-life. Such an excited state is called an *isomer*. Some isomers are known to have a half-life as long as a few years.

If we analyze the gamma-ray spectrum of the excited states of a nucleus, we can determine the energy levels, such as those shown in Fig. 9.9, of the nucleus.

A process that leaves the daughter nucleus in an excited state leads to the emission of gamma-rays. Most radioisotopes after the α-decay or β-decay, in general, leave the daughter nucleus in an excited state which in turn gives one or more gamma-rays. Figure 9.10 shows one such example. Here, the $^{60}_{27}$Co nucleus decays with $T_{1/2} = 5.2$ yr by electron emission into an excited state of the daughter nucleus $^{60}_{28}$Ni which in turn emits two gamma-rays of energies 1.17 MeV and 1.33 MeV. The $^{60}_{27}$Co nucleus is used as one of the standard sources of gamma-rays.

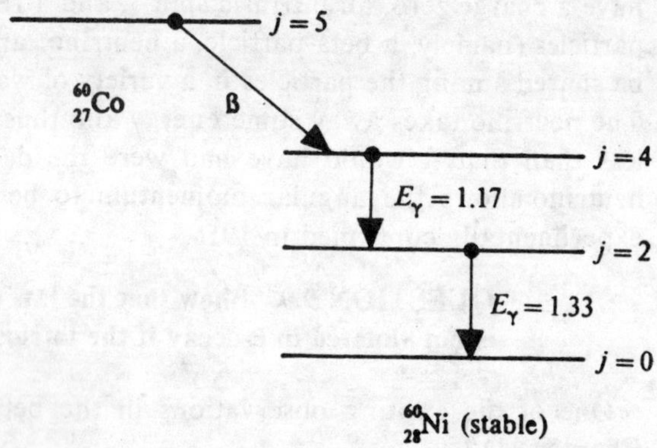

Fig. 9.10 Gamma-rays from $^{60}_{27}$Co source.

9.13 NUCLEAR REACTIONS

In a nuclear reaction, the characteristics of the nucleus can be changed by bombarding it with energetic particles. In a typical reaction, a projectile a strikes the target nucleus X and produces a residual nucleus Y and an outgoing particle b. Each of the particles a and b can be a photon, an electron, a nucleon, or a nucleus. A general nuclear reaction is represented as

$$a + X \rightarrow Y + b, \tag{9.51}$$

or, in a more compact form, as

$$X(a, b)Y. \tag{9.52}$$

In such a reaction, the (1) total number of nucleons, (2) total charge, (3) total energy, (4) linear momentum, and (5) angular momentum are strictly conserved. Let us consider a typical example, namely,

$$^{14}_{7}N + ^{4}_{2}He \rightarrow ^{17}_{8}O + ^{1}_{1}H. \tag{9.53}$$

As can be seen, the mass number A here is conserved. This is the same as saying that the total number of nucleons remains unchanged during the reaction. Also, the total charges on the two sides of the reaction are equal.

Energetics: *Q*-Value

Consider the general nuclear reaction X(a, b)Y. The total initial energy in this reaction is

$$E_i = K_a + M_a c^2 + K_X + M_X c^2, \tag{9.54}$$

where K_a and K_X are the kinetic energies of a and X, respectively. Recall that the total potential energy is zero when a and X are far apart. The energy of the final system, well after the reaction, can be written as

$$E_f = K_b + M_b c^2 + K_Y + M_Y c^2. \tag{9.55}$$

The law of conservation of total energy demands that $E_i = E_f$. The difference between the initial and final rest mass energies is called the ***reaction energy*** or the ***Q-value*** of the reaction. The Q-value is given by

$$Q = (M_a + M_X - M_b - M_Y)c^2. \tag{9.56a}$$

Then, from Eqs. (9.54) and (9.55), we get

$$Q = (K_b + K_Y) - (K_a + K_X). \tag{9.56b}$$

We write Eq. (9.56a) in the form

$$(M_a + M_X)c^2 = (M_b + M_Y)c^2 + Q. \tag{9.57}$$

When $Q > 0$, the reaction is exoergic and the energy liberated appears in the form of the kinetic energy of the reaction products. When $Q < 0$, the reaction is endoergic, that is, it can occur only if the required energy is supplied by the kinetic energy of the projectile.

EXAMPLE 9.10 Calculate the Q-value of the reaction ${}^2_1\text{H}(\text{d, n}){}^3_2\text{He}$.

The given reaction is ${}^2_1\text{H} + {}^2_1\text{H} = {}^3_2\text{He} + \text{n}$. From Eq. (9.56a), we get

$$Q/c^2 = [2M({}^2_1\text{H}) - M({}^3_2\text{He}) - m_n] = [2 \times 2.0141 - 3.01603 - 1.008665]\ \text{u}$$

or

$$Q = .0035\ \text{u} \times 931\ \text{MeV/u} = 3.26\ \text{MeV}.$$

Thus, the given reaction is exoergic and releases 3.26 MeV energy which appears as the kinetic energy of the neutron and the ${}^3_2\text{He}$ nucleus.

CROSS-SECTION

The probability that a bombarding particle will interact with the target nucleus to produce a reaction is expressed as a *cross-section.* We can visualize the probability in terms of the area presented by the nucleus to the incident particle. The incident particle impinging on this area interacts with the target nucleus, whereas a particle not impinging on this area will escape the interaction. Hence, the larger the cross-section, the greater the likelihood of a reaction taking place. This cross-section should not be confused with the geometrical area of the nucleus. The reaction cross-section depends on the nature of the interaction process which in turn depends on the energy and the type of the incident particle and the target involved. We shall not be concerned with the theoretical calculation of the cross-section, but will only elaborate the significance of the concept of cross-section.

Consider a beam of monoenergetic incident particles arriving at a rate of I_0 particles per unit area per second on a thin sheet (of thickness t) of the target material of area A. If the target contains n nuclei per unit volume, then the total number of nuclei in the target is nAt. Let us represent each nucleus, for the purpose of the reaction, by a disk of area σ normal to the direction of incident particles. If the incident particles strike a disk, the reaction occurs, otherwise it does not. We assume that the target is very thin so that the intensity of the incident beam is not appreciably attenuated as it traverses the target. Thus, the number of incident particles is essentially the same at all depths in the target.

The number of nuclei per unit area is nt and a fraction of the area within the interaction range is σnt. If the number of incident particles per unit area per second is I_0, then

$$\text{number of reactions per unit area per sec} = I_0\sigma nt. \tag{9.58}$$

The cross-section is expressed in the unit barn (1 barn $= 10^{-28}$ m^2).

In general, a beam of incident particles may produce several types of reactions with the given target nuclei. For example,

$$\begin{aligned} {}^{35}_{17}\text{Cl} + \text{n} &\rightarrow {}^{35}_{17}\text{Cl} + \text{n} \qquad \text{(elastic scattering)} \\ &\rightarrow {}^{36}_{17}\text{Cl}^* + \gamma \qquad \text{(capture reaction)} \\ &\rightarrow {}^{35}_{16}\text{S}^* + \text{p} \\ &\rightarrow {}^{32}_{15}\text{P}^* + \alpha. \end{aligned}$$

The probability of occurrence of each of these reactions can be separately measured and the cross-section for each reaction can be defined. Such a cross-section is known as the *partial cross-section* of the reaction. The sum of all the partial cross-sections equals the *total cross-section* σ_t. The total cross-section minus the scattering cross-section (σ_s) is known as the absorption cross-section (σ_a).

EXAMPLE 9.11 A 0.01-mm-thick ${}^{7}_{3}$Li target is bombarded with 10^{13} protons per second. As a result, 10^8 neutrons per second are produced. What would be the cross-section for this reaction? (The density of Li $= 500$ kg/m^3.)

The reaction to produce neutrons is ${}^{7}_{3}\text{Li} + \text{p} \rightarrow {}^{7}_{4}\text{Be} + \text{n}$. The number of neutrons produced per second is

$$\frac{\text{number of target atoms}}{\text{m}^2} \times \frac{\sigma \times \text{number of protons}}{\text{sec}}$$

$$= \frac{6 \times 10^{26} \times 10^{-5} \times 500 \times \sigma \times 10^{13}}{7} = \frac{3 \times 10^{37}}{7}\sigma = 10^8.$$

Thus, the cross-section is $\sigma = 0.23 \times 10^{-28}$ m$^2 = 0.23$ barn.

Let us now suppose that the target is so thick that the intensity of the incident particle beam gets attenuated as it traverses the target. Let the intensity of the beam impinging on the target be I (i.e., number of particles/m^2-sec). As the beam passes through, the particles in it interact with the target nuclei and are absorbed by the target or get scattered in some direction. Thus, the beam loses these particles, resulting in a decrease in its intensity. This decrease $-dI$ in the intensity as the beam of intensity I passes through a thickness dx of the target is the number of incident particles/m^2-sec interacting with the target nuclei. Thus, from Eq. (9.58), we get

$$-dI = I\sigma n\, dx. \tag{9.59}$$

Integrating, the intensity $I(t)$, after the beam traverses a distance t, we get is

$$I(t) = I_0 \exp(-n\sigma t), \tag{9.60}$$

where I_0 is the initial intensity. It should be noted that σ here is the total reaction cross-section.

EXAMPLE 9.12 Consider the reaction ${}^{197}_{79}\mathrm{Au}(\mathrm{n}, \gamma)\, {}^{198}_{79}\mathrm{Au}$. If the ${}^{197}_{79}\mathrm{Au}$ target of area 5 cm² and thickness 0.3 mm is bombarded with a beam of 2×10^{12} neutrons per m² per sec, what would be the number of ${}^{198}_{79}\mathrm{Au}$ nuclei produced per second, assuming the cross-section of this reaction to be 94 barns?

The number of ${}^{198}_{79}\mathrm{Au}$ nuclei produced per second would be equal to the number of reaction events taking place, which is $I_0 - I$. The intensity I is

$$I = I_0 \exp(-n\sigma t),$$

where

$$n = \text{number of } {}^{197}_{79}\mathrm{Au} \text{ nuclei per unit volume}$$

$$= \frac{\text{density}}{\text{atomic weight}} \times \text{Avogadro number}$$

$$= \frac{19.3 \times 10^3}{197} \times 6.02 \times 10^{26} = 5.89 \times 10^{28}/\mathrm{m}^3,$$

$$I_0 = \text{number of neutrons hitting the target per second}$$

$$= \text{number of neutrons per m}^2 \text{ per second} \times \text{target area}$$

$$= 2 \times 10^{12} \times 5 \times 10^{-4} = 10^9 \text{ neutrons per sec.}$$

Then,

$$I_0 - I = I_0[1 - \exp(-n\sigma t)].$$

Substituting for $n\sigma t$ and I_0, we get

$$I_0 - I = 10^9[1 - \exp(-.66)] = 1.5 \times 10^8.$$

This is the number of ${}^{198}_{79}\mathrm{Au}$ nuclei produced per second.

9.14 NUCLEAR MODELS

In any fundamental approach to understand the behaviour of a nucleus, the problem is to identify the nature of the forces between the nucleons that are responsible for the existence of the nucleus. In nuclear physics, this is the central problem still to be completely solved. Its solution has been and is being sought in two ways, namely, by studying

(1) simple nucleon systems, e.g., a two-nucleon system, assuming some simple nucleon-nucleon potential and comparing the results so obtained with the experimental observations and

(2) a heavy nucleus, e.g., $^{238}_{92}U$, assuming a simple nuclear potential experienced by each nucleon in the nucleus and comparing the results with the experimental observations.

The second way has led to models which account for many properties of the nucleus. We shall discuss here only this method, and describe two of these models—the liquid drop model and the shell model.

LIQUID DROP MODEL

As we know, the binding energy per nucleon is nearly constant in all nuclei except those with a very small A. Also, the density of nuclear matter is independent of the size of the nucleus. These properties, as we shall now learn, are similar to those found in a drop of a liquid. Therefore, the liquid drop model is based on these properties. In this model, the nucleus is assumed to be similar to a liquid drop.

The energy required to convert a liquid into vapour, as we know, is directly proportional to the mass of the liquid. This means that the total binding energy of all the molecules in the liquid is directly proportional to the mass of the liquid. In other words, the binding energy per unit mass is constant. This situation is similar to that in a nucleus, namely, the binding energy per nucleon is constant. Further, the density in an incompressible liquid is constant and does not depend on the size or shape of the liquid drop. This property, again, is similar to that of a nucleus, namely, the density of nuclear matter is independent of the size of the nucleus.

The liquid drop model approximates the nucleus as a sphere with uniform density which drops to zero at a distance equal to the radius $R(=R_0A^{1/3})$ of the sphere. We have already seen that the mass of a nucleus is not exactly equal to the total mass of the nucleons. The difference is due to the binding energy B of the nucleus. Thus, the mass of a nucleus having A nucleons, of which Z are protons, is

$$^{A}_{Z}M = ZM_p + (A - Z)M_n - B/c^2. \tag{9.61}$$

According to the liquid drop model, the total binding energy of a nucleus is mainly due to the following three effects:

(1) *Volume effect* The nucleons inside the nucleus interact only with the neighbours within the range of the nuclear force. Thus, the total binding energy of the nucleus will have a contribution proportional to the number of nucleons in the nucleus. If a_1 is the binding energy per nucleon, then the contribution to the total binding energy is

$$B_1/c^2 = a_1A. \tag{9.62}$$

The contribution to the total binding energy of a liquid drop is proportional to the number of molecules in the drop, and is similar to that given by Eq. (9.62).

(2) *Surface effect* Equation (9.62) assumes that all nucleons are surrounded by an equal number of neighbours. But, as we know, the nucleons on the surface of the nucleus experience the attractive forces only from one side of the surface, so their binding energy will not be the same as that of the nucleons inside the nucleus. Thus, the nucleons on the surface have a lesser binding energy. The number of such nucleons depends on the surface area of the nucleus, which is proportional to R^2. Using $R = R_0A^{1/3}$, we see that the surface area is proportional to $A^{2/3}$. Thus, we add a correction term B_2 to B_1, the latter being proportional to the surface area, and

write

$$B_2/c^2 = -a_2 A^{2/3}. \tag{9.63}$$

The negative sign here indicates that the total binding energy gets reduced.

(3) *Coulomb effect* The protons inside the nucleus, because of their charge, also interact through the Coulomb force. This force makes a contribution to the total binding energy. If the number of protons distributed uniformly in the nucleus [of radius $R(=R_0 A^{1/3})$] is Z, then the electrostatic energy is given by

$$E_c \propto \frac{Z^2 e^2}{R} \propto \frac{Z^2 e^2}{R_0 A^{1/3}}. \tag{9.64}$$

Therefore, the contribution to the binding energy is

$$B_3/c^2 = -a_3 \frac{Z^2}{A^{1/3}}. \tag{9.65}$$

Here, again, the negative sign indicates that the total binding energy gets reduced. This reduction is due to the Coulomb repulsion.

The total binding energy B of a nucleus is then the sum of contributions (1), (2), and (3), i.e.,

$$B/c^2 = a_1 A - a_2 A^{2/3} - a_3 Z^2 A^{-1/3}. \tag{9.66}$$

The mass of a nucleus, using Eq. (9.61), is

$${}^A_Z M = ZM_p + (A - Z)M_n - a_1 A + a_2 A^{2/3} + a_3 Z^2 A^{-1/3}. \tag{9.67}$$

The constants a_1, a_2, and a_3 can be determined by substituting in Eq. (9.67) the known masses of some nuclei. Once this is done, Eq. (9.67) can be used to predict the mass of any nucleus reasonably well. However, this formula does not successfully explain some of the finer details of the nuclear structure. If such details have also to be taken into account, we then need to include two more terms in the formula. The formula we so obtain can then correctly predict the mass of a nucleus, and is given as

$${}^A_Z M = ZM_p + (A - Z)M_n - a_1 A + a_2 A^{2/3} + a_3 Z^2 A^{-1/3} + a_4 \frac{(A - 2Z)^2}{A} + \delta, \tag{9.68}$$

where

$$\begin{aligned} \delta &= 0 && \text{for odd } A \\ &= \pm a_5 A^{-1/2} && + \text{ for even } A, \text{ even } Z; - \text{ for even } A, \text{ odd } Z. \end{aligned}$$

This is called the semiempirical mass formula because the values of the parameters a_1 to a_5 can be obtained by empirically fitting in it the experimentally measured masses of some nuclei. It was first proposed by C. F. Weizsacker, and thus is also known as the Weizsacker mass formula. Substituting the values of the parameters so obtained in Eq. (9.68), the mass of a nucleus (in atomic mass units) we get is

$$\begin{aligned} {}^A_Z M = {} & 1.007825Z + 1.008665(A - Z) - 0.01691A + 0.01911A^{2/3} + 0.000763Z^2 A^{-1/3} \\ & + 0.10175(Z - A/2)^2 A^{-1} + \left({}^{-1}_{+1}0\right) \times 0.012A^{-1/2}. \end{aligned} \tag{9.69}$$

In addition to very accurately determining the mass of a nucleus, this mass equation predicts the α-decay, β-decay, and other properties of a nucleus. For example, using this equation, we can predict why $^{238}_{92}U$ emits an α-particle and not an electron. Also, we can calculate the energy of the emitted particle. The liquid drop model has been successful not only because it leads to an excellent mass equation, but also because it satisfactorily explains the nuclear fission process (discussed in Chapter 10).

QUESTION 9.10 The agreement of the semiempirical mass formula with experimental observations is better for a heavy nucleus than for a light nucleus. Why?

SHELL MODEL

It has been observed that many properties of the nucleus vary periodically in a manner similar to that for the properties of elements. The systematic study of the stability of nuclei has shown that a stable nucleus, in particular, results when either the number of protons Z or the number of neutrons N in the nucleus is equal to one of the numbers

2, 8, 20, 28, 50, 82, 126.

These numbers are called the *magic numbers*. Each of these magic numbers represents a closed shell in the nucleus and the nucleus may be tied together in much the same kind of shell structure as the electrons in the atom. There exists a lot of experimental evidence that supports the shell structure hypothesis of the nucleus.

A nucleon in a nucleus can be thought of as a single particle moving in a potential well produced by the other nucleons in the nucleus. The average potential well 'seen' by each nucleon is the same. Thus, we can treat each nucleon as moving independent of the other nucleons in this potential well. We can then picture the building up of the structure of the nucleus as the gradual filling up of single-particle orbits by neutrons and protons in the same way as electrons build up the atom. In other words, each neutron and proton in a nucleus has its own energy level, which is filled in accordance with the Pauli exclusion principle.

It may seem surprising, at first sight, that, according to the shell model, a nucleon in a nucleus can maintain an independent motion when it is interacting through a strong short range force with the other nucleons in the nucleus. But the central theme underlying this model is that a nucleon in a nucleus moves in a manner such that it does not collide with the other nucleons in the nucleus but is confined in its own orbit with well defined quantum numbers (such as those associated with energy and angular momentum). This orbit remains unperturbed by many close approaches of the other nucleons. How can this be possible, particularly when the density of nucleons inside a nucleus is high and the force between the nucleons is very strong? The answer to this apparent paradox is obtained by applying Pauli's exclusion principle, as follows. When two nucleons collide or interact, they change their energy and momentum. This means they change their quantum numbers, i.e., go over to different orbits. According to the Pauli exclusion principle, such a collision is allowed only when each of the two nucleons occupies an empty level, that is, acquires an energy more than the Fermi energy. Since this process involves a relatively large energy transfer, very few collisions take place.

Let us assume that the average potential is a square well, i.e., the potential is constant up to a distance of about the range of nuclear force ($\sim 10^{-15}$ m) and becomes zero beyond this distance. Then, the nucleon will have well defined allowed energy levels. Therefore, in a nucleus, each nucleon will occupy only these levels. When the nucleus is in the ground state, each of the nucleons fills these levels in such a way that the total energy of the nucleus is minimized without the violation of Pauli's exclusion principle. Since the protons can be distinguished from the neutrons, the exclusion principle operates independently on these two types of nucleons. If we ignore the Coulomb interaction between the protons, then the average potential due to the nuclear force for the neutrons and the protons will be identical. Since the nuclear attractive force is very strong as compared to the Coulomb force, it is reasonable to neglect the latter as a first approximation. In this approximation, the two types of nucleons have identical allowed orbits. Thus, we can assume that each allowed energy state is occupied by at the most four nucleons without violating Pauli's exclusion principle—two neutrons with $m_s = \pm\frac{1}{2}$ and two protons with $m_s = \pm\frac{1}{2}$.

EXAMPLE 9.13 Assume that the nucleons inside a nucleus constitute a Fermi gas. Estimate the depth of the nuclear potential.

Let us assume that the nucleons move freely inside a spherical volume V of radius R. When a nucleus is in the ground state, all the allowed energy levels up to $E = E_F$ (Fermi energy) are occupied, the rest being unoccupied. The Fermi energy E_F (see Example 7.1) is

$$E_F = \frac{h^2}{2m}\left(\frac{3n}{8\pi}\right)^{2/3},$$

where n is the nucleon density and is given by

$$n = \frac{A}{\frac{4}{3}\pi R^3} = \frac{A}{\frac{4}{3}\pi R_0^3 A} = \frac{3}{4\pi R_0^3}$$

since $R = R_0 A^{1/3}$. Thus,

$$E_F = \frac{h^2}{2m}\left(\frac{9}{32\pi^2 R_0^3}\right)^{2/3} = \frac{h^2}{8m}\left(\frac{9}{4\pi^2}\right)^{2/3}\frac{1}{R_0^2}.$$

Substituting the value $R_0 = 1.2 \times 10^{-15}$ m, we get

$$E_F = \frac{(6.6 \times 10^{-34})^2}{8 \times 1.6 \times 10^{-27}} \times \left(\frac{9}{4 \times (3.14)^2}\right)^{2/3}$$

$$\times \frac{1}{(1.2 \times 10^{-15})^2}\ \text{J}$$

$$\sim 40\ \text{MeV}.$$

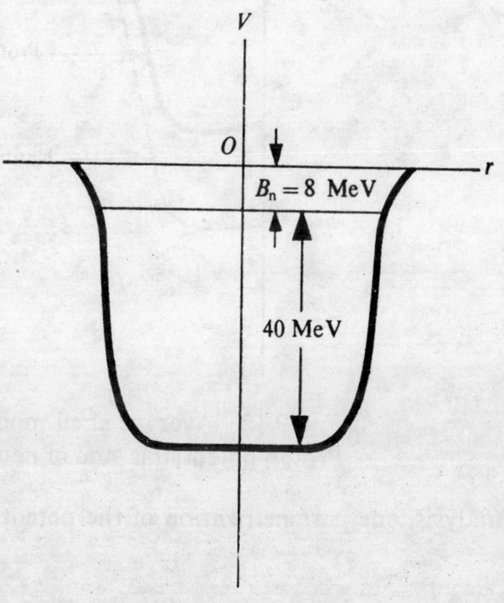

Fig. 9.11 Potential well, treating nucleons as Fermi gas.

We know that the binding energy of a nucleon in a nucleus is about 8 MeV. Thus, the depth of the potential V_0 (see Fig. 9.11) will be

$$V_0 \sim E_F + B_n \sim 40 + 8 = 48\ \text{MeV}.$$

Shell Model Potential

If we assume that a nucleon moves as if it is in a potential well, we must know the shape of this potential. We have obtained some theoretical knowledge about the shape from Example 9.13. To get the experimental information on this potential, a thin foil can be bombarded with high energy nucleons. As a result of this bombardment, the neutrons and protons are deflected at various angles. (As is clear, such an experiment is conducted in a manner similar to that for the Rutherford scattering experiment we discussed in Chapter 4.) An analysis of the deflections for neutrons then gives the potential energy of a neutron approaching a nucleus as a function of the distance of separation between the neutron and the nucleus (see Fig. 9.12). From this

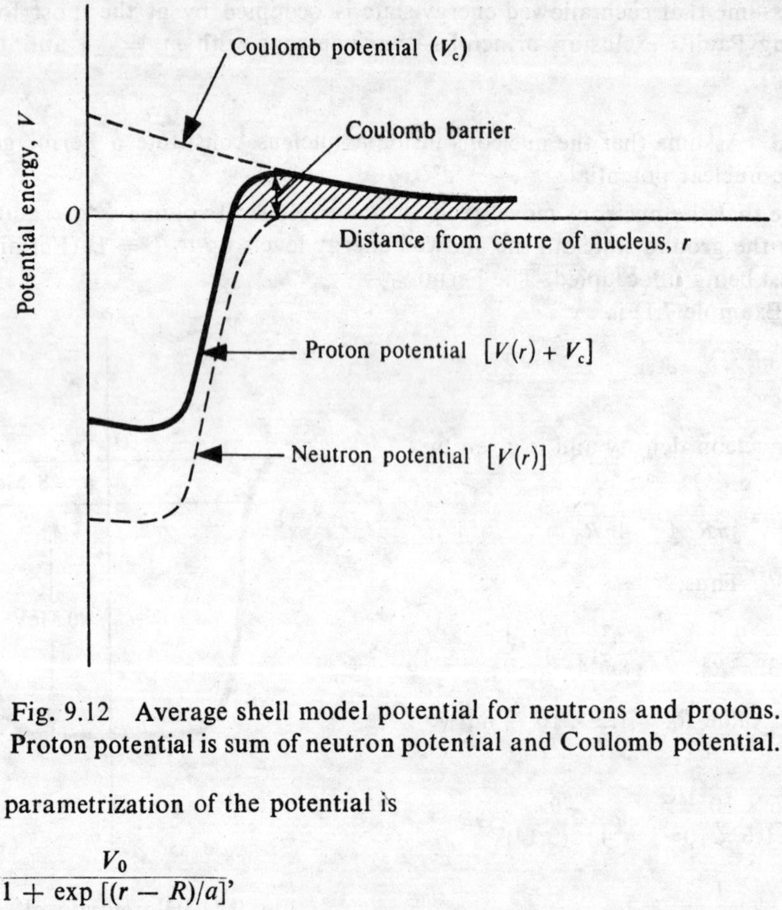

Fig. 9.12 Average shell model potential for neutrons and protons. Proton potential is sum of neutron potential and Coulomb potential.

analysis, one parametrization of the potential is

$$V = -\frac{V_0}{1 + \exp[(r - R)/a]}, \tag{9.70}$$

where the values of the constants V_0, R, and a are

$$V_0 \sim 57 \text{ MeV}, \qquad R \sim 1.25A^{1/3} \text{ F}, \qquad a \sim 0.65 \text{ F}.$$

The potential energy of a neutron at an infinite separation from the nucleus is arbitrarily chosen as zero. At a distance greater than a few fermis from the nuclear surface, the neutrons experience no force.

The potential energy of a proton approaching a nucleus increases as the distance between the proton and the nuclear surface decreases. This is because of the long range Coulomb force between the proton and the nucleus. Both the Coulomb potential and nuclear potential are separately shown in Fig. 9.12. The proton potential is the sum of these two potentials.

The nuclear shell model pictures a nucleus as a potential well in which each nucleon moves, to a large extent, independent of the others, the potential well itself somehow being the net effect of the interaction between the nucleons. We shall use this description of the model to learn all we can about the observable features of the potential well and the nucleon orbits. The nucleon orbits can be calculated quantum mechanically by solving the Schroedinger equation for the motion of a nucleon confined in a spherical potential well (see Fig. 9.12). However, the form of the potential well [i.e., Eq. (9.70)] is such that it is not possible to get an analytical solution of the Schroedinger equation, and therefore it has to be solved numerically. Such a solution yields the eigenvalues and eigenfunctions of the nucleon.

The Schroedinger equation required to be solved is

$$\nabla^2\psi + \frac{2M}{\hbar^2}[E - V(r)]\psi = 0.$$

Since the potential here is spherically symmetric, the wave function can be separated and written as a product of two parts, namely, the radial part and the angular part, the latter consisting of the spherical harmonics. Thus,

$$\psi(r, \theta, \phi) = R(r)Y_l^{m_l}(\theta, \phi).$$

The radial part is obtained by solving the radial equation

$$\frac{1}{r^2}\frac{d}{dr}\left[\frac{r^2\,dR(r)}{dr}\right] + \frac{2M}{\hbar^2}[E - V(r)]R(r) = \frac{l(l+1)R(r)}{r^2}.$$

The solution so obtained with the boundary condition $R(r) \to 0$ as $r \to \infty$ gives the allowed values of the total energy E of the nucleon. Each state with energy E is specified by the quantum numbers n, l, m_l, where

n (principal quantum number) $= 1, 2, 3, \ldots,$

l (orbital quantum number) $= 0, 1, 2, 3, \ldots, n-1$ (the numbers 0, 1, 2, 3, ... are also referred to as the orbits s, p, d, f, ..., respectively),

m_l (magnetic quantum number) $= l, l-1, \ldots, 0, -1, \ldots, -l.$

Thus, the solution of the Schroedinger equation for the nucleon in the average potential due to the other nucleons gives the stationary states (i.e., allowed orbits) of the nucleon, each characterized by the quantum numbers n, l, and m_l whose significance is the same as in the stationary states of an atomic electron. When describing the allowed orbits in nuclear physics, it is customary to use a set of quantum numbers slightly different from these quantum numbers. Thus, in place of the principal quantum number n, the radial quantum number N, defined as

$$N = n - l, \tag{9.71}$$

is used, the other two quantum numbers l and m_l remaining the same. Therefore, the allowed orbits of a nucleon are characterized by the quantum numbers N, l, and m_l. In this notation, the allowed orbits are

$N = 1,\quad l = 0, 1, 2, 3, \ldots$ or $1s, 1p, 1d, 1f, \ldots,$

$N = 2,\quad l = 0, 1, 2, 3, \ldots$ or $2s, 2p, 2d, 2f, \ldots .$

The sequence of these energy levels in the order of increasing energy has been found to be

$$1s,\ 1p,\ 1d,\ 2s,\ 1f,\ 2p,\ 1g,\ 2d,\ 3s,\ 1h,\ \ldots .$$

Each of these is known as a single-particle orbit. In a nucleus, the neutrons and protons occupy separate orbits since the protons encounter the Coulomb force in addition to the nuclear force. Thus, in general, the spacing of the energy levels for neutrons and protons is different. In our discussion, we shall ignore the Coulomb force and assume that the single-particle orbits of neutrons and protons are identical.

As can be seen, in contrast to the energy levels of an electron in a hydrogen atom, the energy levels of a nucleon in a nucleus depend on the quantum number l. In other words, the degeneracy associated with l of an energy level no longer exists. However, the energy levels with the same l but different m_l are still degenerate. Thus, each energy level in the foregoing sequence has a $(2l + 1)$-fold degeneracy associated with m_l. When we incorporate the spin quantum number, the degeneracy then is $2(2l + 1)$. Thus, the level $1p$ can accommodate up to six neutrons and six protons.

The spectroscopic notation in nuclear physics is similar to that in atomic physics (see Section 4.16). In particular, an energy level is labelled [see expression (4.119)] as $N^{(2s+1)}l_{(2j+1)}$, where N is the radial quantum number, s the total spin angular momentum quantum number, j the total angular momentum quantum number, and l the orbital angular momentum quantum number written as $s, p, d, f, \ldots$ to denote $0, 1, 2, 3, \ldots$. For example, the state $1^2s_{1/2}$ corresponds to $N = 1$, $s = 1/2$, $l = 0$, and $j = 1/2$. In Fig. 9.13, this is represented by $1s_{1/2}$ because the multiplicity factor $(2s + 1)$ for a single nucleon is always 2 and is therefore omitted in the notation.

In Fig. 9.13, the left-hand side shows the energy levels, each with the degeneracy associated with it. Once we know the allowed energy levels, we can fill them with nucleons to form a nucleus. As we can notice from this figure, there are gaps between the energy levels. The occurrence of each magic number shown in the energy level scheme can be explained in terms of the closed shell it represents. Thus, the first three magic numbers, namely, 2, 8, and 20, are accounted for. According to this scheme, the next higher magic numbers should be 40, 70, 112. But these numbers are different from those shown in the figure, viz., 28, 50, and 82. In 1949, J. R. Mayer and J. H. D. Jensen explained this discrepancy by introducing the concept of nuclear spin-orbit interaction. They proposed that each nucleon experiences a force proportional to the dot product of its spin angular momentum and orbital angular momentum vectors. Due to this interaction, the intrinsic angular momentum $\mathbf{s}$ combines with the orbital angular moment $\mathbf{l}$ to give the total angular momentum $\mathbf{j}$ as

$$\mathbf{j} = \mathbf{l} + \mathbf{s} = \mathbf{l} + \frac{1}{2}.$$

The quantum number j can then have one of two possible values, namely,

$$j = l + \tfrac{1}{2}, \tag{9.72a}$$

$$j = l - \tfrac{1}{2}, \tag{9.72b}$$

except when $l = 0$. The nature of the spin-orbit interaction is such that the level with $j = l + 1/2$ has an energy less than that of the level with $j = l - 1/2$. The magnitude of the difference in energy is proportional to l. That is,

$$E_{j=l+1/2} - E_{j=l-1/2} \propto l. \tag{9.73}$$

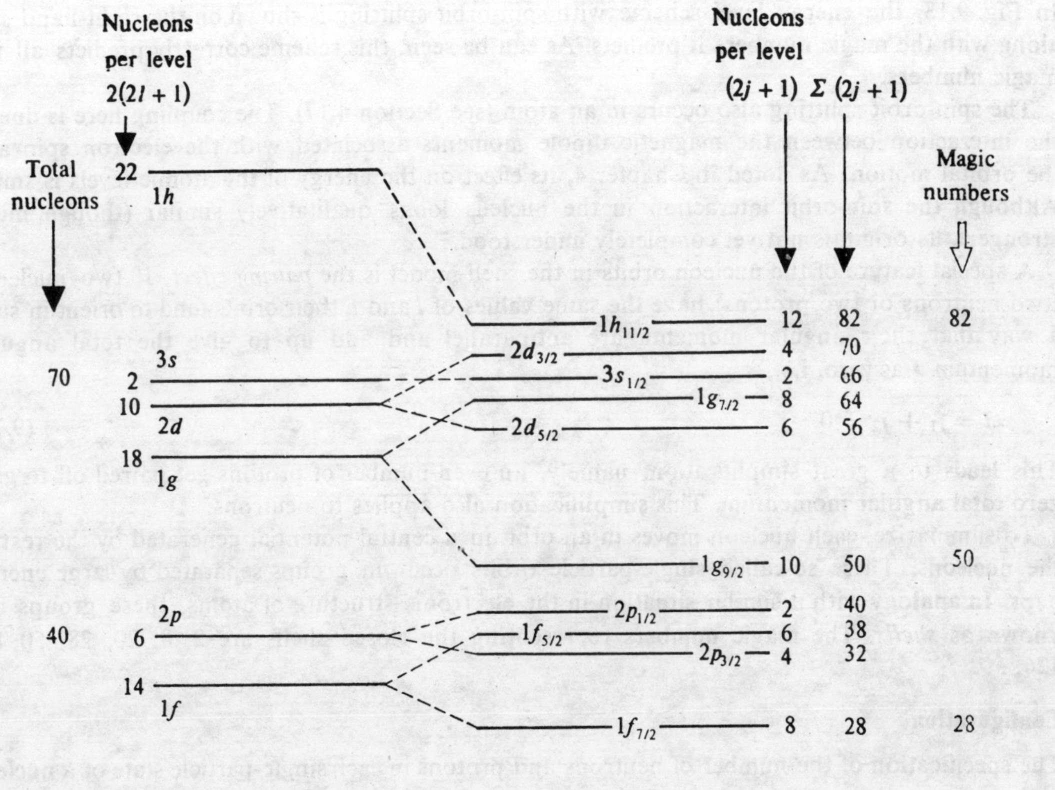

Fig. 9.13 Single-particle levels of nuclear shell model with and without spin-orbit coupling.

In Fig. 9.13, the energy level scheme with spin-orbit splitting is shown on the right-hand side along with the magic numbers it predicts. As can be seen, this scheme correctly predicts all the magic numbers.

The spin-orbit splitting also occurs in an atom (see Section 4.17). The coupling here is due to the interaction between the magnetic dipole moments associated with the electron spin and the orbital motion. As noted in Chapter 4, its effect on the energy of the atomic levels is small. Although the spin-orbit interaction in the nucleus looks qualitatively similar (though much stronger), its origin is not yet completely understood.

A special feature of the nucleon orbits in the shell model is the *pairing effect*. If two nucleons (two neutrons or two protons) have the same values of l and j, their orbits tend to orient in such a way that their angular momenta are antiparallel and add up to give the total angular momentum $\boldsymbol{J}$ as zero, i.e.,

$$\boldsymbol{J} = \boldsymbol{j}_1 + \boldsymbol{j}_2 = 0. \tag{9.74}$$

This leads to a great simplification, namely, an even number of protons get paired off to give zero total angular momentum. This simplification also applies to neutrons.

To summarize, each nucleon moves in an orbit in a central potential generated by the rest of the nucleons. These so-called single-particle orbits occur in groups separated by large energy gaps. In analogy with a similar situation in the electronic structure of atoms, these groups are known as *shells*. The magic numbers representing the closed shells are 2, 8, 20, 28, 50, 82, 126, ...

Configuration

The specification of the number of neutrons and protons in each single-particle state of a nucleus is known as the configuration of the nucleus. For example, the configuration of $^5_2$He is $(S^2_{1/2})_p(S^2_{1/2}p_{3/2})_n$. This indicates that there are two protons in the orbit $S_{1/2}$, and of the three neutrons, two are in the orbit $S_{1/2}$ and one is in the orbit $p_{3/2}$. In such a configuration, the completely filled energy levels are normally not indicated. For example, the configuration of $^{18}_8$O is written as $(1d^2_{5/2})_n$. In this configuration, it is implied that all the single-particle levels below the level $1d_{5/2}$ are completely filled and the only partly filled level is $d_{5/2}$ with 2 neutrons. The energy of a nucleus in the shell model is then just the sum of the energies of all the nucleons in the nucleus.

The shell model can predict the spin of a nucleus in its ground state by taking into account the following facts:

(1) In each completely filled shell, the spins and the orbital angular momenta of nucleons add together to give the resultant zero.

(2) The protons in an orbit form pairs and each such pair gives a resultant total angular momentum zero. The same is true of the neutrons.

The spin of the nucleus is then given by the unpaired particles. These results can be derived quantum mechanically.

EXAMPLE 9.14 Find the spin of each of the nuclei $^5_2$He, $^{17}_8$O, $^{33}_{16}$S, $^{41}_{19}$K, $^{43}_{20}$Ca, and $^{129}_{53}$I in their ground state.

The $^5_2$He nucleus has 2 protons and 3 neutrons. The last odd nucleon here is the third neutron which is in the ($p_{3/2}$)-state. The spin of this nucleus in the ground state is the angular momentum j of the last odd nucleon which, in this case, is 3/2. Thus, the spin of ${}^5_2$He is 3/2.

The spins of the other nuclei can be determined in the same way as that for the ${}^5_2$He nucleus, and are listed in Table 9.5.

TABLE 9.5 Spins of given nuclei in their ground state

Nucleus	Unpaired nucleon	Level of unpaired nucleon	Spin (J) of nucleus
${}^{17}_{8}$O	9th neutron	$d_{5/2}$	5/2
${}^{33}_{16}$S	17th neutron	$d_{3/2}$	3/2
${}^{41}_{19}$K	19th proton	$d_{3/2}$	3/2
${}^{45}_{20}$Ca	25th neutron	$f_{7/2}$	7/2
${}^{129}_{53}$I	53rd proton	$g_{7/2}$	7/2

It should be noted that the spin of a nucleus is independent of the number of nucleons in a given shell. It is the last unpaired nucleon that determines the spin. For certain nuclei, this method of predicting the spin does not agree with the experimental observations, but by and large the method gives fairly accurate results. The application of the method to a nucleus of the odd-odd type, i.e., a nucleus with both an odd proton and an odd neutron, is complicated.

EXCITED STATES

So far, we have restricted our discussion to the ground state of the nucleus. However, a nucleus, like any other quantum mechanical system, can also exist in a series of discrete excited states. The spacing between such states provides valuable information about the system. We recall that the principal success of the quantum theory of the atom is in its ability to predict that an excited atom emits a line spectrum. If a nucleus has a series of bound levels as predicted by the quantum theory, then an excited nucleus should emit radiation of discrete energy similar to an optical line spectrum. Experimentally, it has been found that an excited nucleus does indeed emit electromagnetic radiation of discrete energy, the so-called gamma-rays. Figure 9.14 shows some of the discrete levels of ${}^{17}_{8}$O and ${}^{17}_{9}$F observed experimentally.

According to the shell model, an excited state arises when a nucleon is promoted to an unfilled single-particle level with an energy higher than the initial energy of the nucleon. This description of the single-particle mode of excitation gives a reasonably correct account of the low lying excited states of a nucleus having one nucleon outside one of its closed shells. Thus, the spin of an excited state can be predicted. For example, the spin of each of the levels 5/2, 1/2, and 3/2 observed in ${}^{17}_{8}$O and ${}^{17}_{9}$F can be predicted if we assume that it is the odd nucleon that gets excited.

An excited state can also occur when two nucleons move over to levels with energies higher than the initial energies of the nucleons. The method of arriving at the properties of such a state is complicated, and therefore we shall not discuss it.

Some recent calculations on the shell model predict that some very heavy nuclei, not hitherto observed, are stable due to the closed shells with $Z = 110$ and $N = 182$. Such a nucleus is known as a superheavy nucleus. A considerable effort is being made to look for these nuclei.

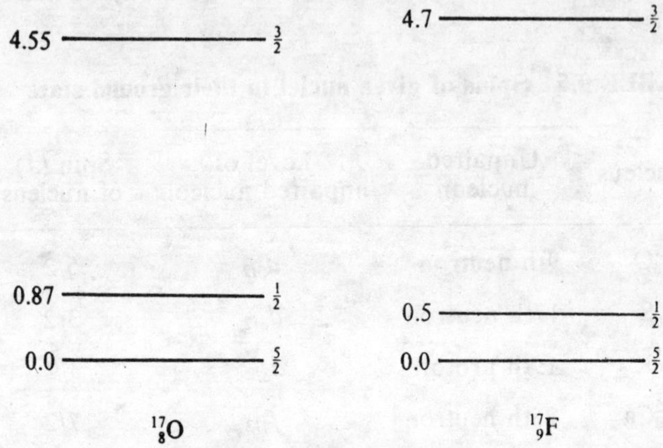

Fig. 9.14 Experimentally observed states of $^{17}_{8}O$ and $^{17}_{9}F$ nuclei. Spin of each such state and its energy (in MeV) are also shown.

PROBLEMS

9.1 In a mass spectrometer, an ion beam moves in a circle of radius 5 cm due to the presence of a magnetic field of 0.38 weber/m^2. How much voltage must be applied to the ion source (see Fig. 9.1) to accelerate the singly-charged $^{6}_{3}Li$ ion so that ions are recorded on the photographic plate?

9.2 Calculate the relative abundances in nature of $^{10}_{5}B$ and $^{11}_{5}B$ on the basis of the chemical atomic weight (=10.881) of boron. Assume that these are the only isotopes of boron.

9.3 Use the uncertainty principle to show that an electron cannot be a constituent of a nucleus.

9.4 Calculate the mass defect and the binding energy per nucleon of the $^{108}_{47}Ag$ nucleus (see Table 9.1 for the masses of isotopes).

9.5 Consider two separate sets of 8 protons and 8 neutrons at rest. How much energy will be released if we form the $^{16}_{8}O$ nucleus? (see Table 9.1 for the masses of isotopes).

9.6 Calculate the energy required to remove a neutron from the $^{13}_{6}C$ nucleus.

9.7 Calculate the mass in grams of one curie of $^{210}_{84}Po$, an alpha-particle emitter with a half-life of 138 days.

9.8 The half-life of a certain radioactive material is 15 hours. How long will it take for 93.75 per cent of this material to disintegrate?

9.9 A radioactive material initially contains 3 mg of $^{234}_{92}U$. How much of $^{234}_{92}U$ will remain unchanged after 62,000 yr? Also, what will be its activity at the end of this period of time? (The half-life $T_{1/2}$ of $^{234}_{92}U = 2.47 \times 10^5$ yr.)

9.10 Consider 1 g of pure radium emitting alpha-particles. Calculate the electric current represented by this emission.

9.11 The natural uranium found on the earth has 99.3% of $^{238}_{92}U$ and 0.7% of $^{235}_{92}U$. The half-life of these two isotopes of uranium is 6.5×10^9 yr and 1.02×10^9 yr, respectively. Assuming that at the time of formation of the universe the two isotopes were equally abundant, estimate the age of the universe.

9.12 A uranium mineral contains 0.093 g of lead ($^{206}_{82}Pb$) to 1 g of $^{238}_{92}U$. Calculate the age of the mineral [$T_{1/2}(^{238}_{92}U) = 4.5 \times 10^9$ yr], assuming that the entire $^{206}_{82}Pb$ in the mineral is due to the decay of uranium.

9.13 Consider a chain disintegration which starts with a pure radioactive source containing N_0 parent nuclei at time $t = 0$. If N_1 and N_2 are the number of parent and daughter nuclei at a time t, show that

$$N_2 = \frac{\lambda_1}{\lambda_1 - \lambda_2} N_0[\exp(-\lambda_1 t) - \exp(-\lambda_2 t)],$$

where λ_1 and λ_2 are the decay constants of the parent and daughter nuclei, respectively. Further, if $\lambda_1 \ll \lambda_2$ and the half-life of the parent nucleus is very long as compared to t, then show that the daughter nucleus approaches a constant value given by $N_2 = (\lambda_1/\lambda_2)N_0$.

9.14 Calculate the kinetic energy of the $^{208}_{81}Tl$ recoil atoms formed by the alpha-particle decay of $^{212}_{83}Bi$. The kinetic energy of an alpha-particle is 6.802 MeV.

9.15 When a $^{13}_{7}N$ atom decays into a $^{13}_{6}C$ atom by the emission of a β^+-particle, the maximum kinetic energy of a positron is 1.202 MeV. Calculate the mass of the $^{13}_{7}N$ atom. (The mass of a $^{13}_{6}C$ atom = 13.003354 u.)

9.16 If I is the difference in the electronic binding energies of the parent and daughter nuclei, then show that the Q-values of the two types of beta-particle decay are

$$Q_{\beta^-} = (^A_ZM - {}^A_{Z+1}M)c^2 - I, \qquad Q_{\beta^+} = (^A_ZM - {}^A_{Z-1}M)c^2 - 2m_ec^2 + I.$$

9.17 Show that the electron capture from the K-shell of a parent atom can occur only if the difference in the atomic masses of the parent and daughter atoms is greater than B_e, where B_e is the binding energy of an electron in the K-shell of the daughter atom. Neglect the binding energy of the electron in the last occupied shell of the daughter atom.

9.18 A Na target (of thickness 3 mg-cm^{-2}) is bombarded with a 5-μA beam of deuterons of energy 10 MeV from an accelerator. After 10 minutes of bombardment, the activity of $^{24}_{11}Na$ produced by the $^{23}_{11}Na\,(d, p)\,^{24}_{11}Na$ reaction is found to be 10^5 disintegrations/sec. What is the cross-section of this reaction (half-life of $^{24}_{11}Na$ = 15 hr)?

9.19 Use the semiempirical mass formula to calculate the binding energy per nucleon of $^{40}_{20}Ca$.

9.20 Calculate the binding energy of the last neutron in the nuclei $^{208}_{82}Pb$ and $^{209}_{82}Pb$, using the semiempirical mass formula.

9.21 List the ground state angular momenta predicted by the shell model for $^{17}_{8}O$, $^{29}_{14}Si$, and $^{60}_{28}Ni$.

9.22 The angular momentum of the ground state of the $^{17}_{8}O$ nucleus is 5/2. The first excited state of this nucleus is observed at an excitation energy of 0.871 MeV with respect to the ground state and has a spin $\frac{1}{2}$. The angular momentum of the second excited state is found to be 3/2 at an excitation energy of 5.1 MeV. What information on the single-particle orbits can you obtain from these results? Also, check if these results agree with those the shell model predicts.

10
Nuclear Energy

10.1 INTRODUCTION

We know that it is not possible to create energy and that we can only convert it from one form to another. Until recently, we have met our energy requirements to a large extent by converting chemical energy from fossil fuels such as coal and oil. It has been realized, however, that the supply of such fuels is limited and that we need to look for the other sources of energy. We know from Einstein's theory of relativity that mass and energy are equivalent and each can be converted into the other. Such a conversion occurs in a nuclear reaction. In an exoergic reaction (i.e., $Q > 0$), the mass difference between the nuclei is converted into the kinetic energy of the end products. This energy can fulfill our requirements. In this chapter, we shall discuss two types of nuclear reactions, namely, fission and fusion, each of which has a large Q-value, and thus can be exploited for producing power. First, we shall investigate the *fission* reactions, where a large amount of energy is released when the heavy nuclei split, and their use in a nuclear reactor. Second, we shall discuss the *fusion* reactions in which light nuclei combine to release a large amount of energy.

10.2 NUCLEAR FISSION

The nuclear reactions we considered in Chapter 9 involve the absorption or emission of light particles such as protons, neutrons, and α-particles. In such a reaction, generally, the Q-value is of the order of 10 MeV. In nuclear fission, the reaction is completely different. It can be induced by bombarding with neutrons a certain nucleus, e.g., ${}^{235}_{92}U$ and ${}^{239}_{94}Pu$. If ${}^{235}_{92}U$ nucleus is hit by a slow neutron, the latter gets absorbed, forming a compound nucleus, namely, ${}^{236}_{92}U$, which splits (fissions) into two nuclei of medium weight. An example of the fission reaction is

$$ {}^{235}_{92}U + n \rightarrow {}^{236}_{92}U^* \rightarrow {}^{144}_{56}Ba + {}^{89}_{36}Kr + 3n + Q. \qquad (10.1)$$

The Q-value in this reaction is about 200 MeV, i.e., twenty times the disintegration energy encountered in a normal nuclear reaction. The fission of ${}^{235}_{92}U$ does not always produce barium and krypton. A variety of other pairs of nuclei may also be produced, but in each fission reaction the fragments are nuclei with a medium A ($50 < A < 150$). For example,

$$ {}^{235}_{92}U + n \rightarrow {}^{236}_{92}U^* \rightarrow {}^{133}_{51}Sb + {}^{99}_{41}Nb + 4n + Q. \qquad (10.2)$$

This type of fission which is induced by the absorption of a particle [neutron in reactions (10.1) and (10.2)] is known as *induced fission.* A heavy nucleus (with $A > 220$) fissions by itself. This type of fission is known as *spontaneous fission.* For example,

$$ {}^{235}_{92}U \rightarrow {}^{144}_{56}Ba + {}^{89}_{36}Kr + 2n + Q. \qquad (10.3)$$

However, the probability of spontaneous fission in a nucleus with $A < 240$ is generally low.

FISSION PROCESS

The fission process takes place in three distinct stages as illustrated in Fig. 10.1. In the first stage, the fissioning compound nucleus (or the original nucleus in spontaneous fission) splits

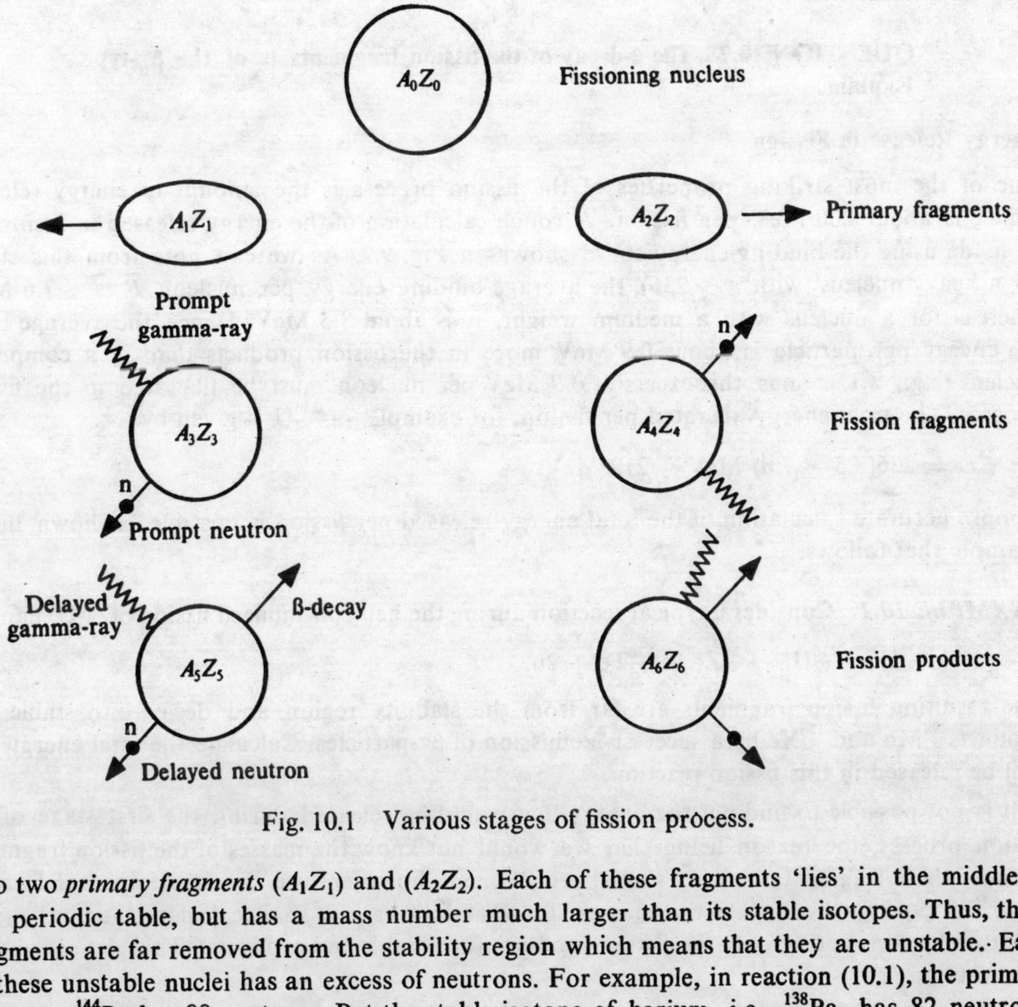

Fig. 10.1 Various stages of fission process.

into two *primary fragments* (A_1Z_1) and (A_2Z_2). Each of these fragments 'lies' in the middle of the periodic table, but has a mass number much larger than its stable isotopes. Thus, these fragments are far removed from the stability region which means that they are unstable. Each of these unstable nuclei has an excess of neutrons. For example, in reaction (10.1), the primary fragment ${}^{144}_{56}Ba$ has 88 neutrons. But the stable isotope of barium, i.e., ${}^{138}_{56}Ba$, has 82 neutrons. Thus, the primary fragment ${}^{144}_{56}Ba$ has six neutrons in excess.

QUESTION 10.1 Explain the neutron excess in the fission fragments using the stability curve in Fig. 9.4.

In the second stage, the excess neutrons in the primary fragments, and gamma-rays are emitted. These neutrons and gamma-rays are known as *prompt neutrons* and *prompt gamma-rays*, respectively. The end product of this stage then is the fragments (A_3Z_3) and (A_4Z_4). The product nuclei (A_3Z_3) and (A_4Z_4) are called the *fission fragments*. These fragments are far from the stability region.

In the third stage, the fission fragments decay by a successive emission of β-particles, finally forming stable *fission products*. In this process, several gamma-rays, called the *delayed gamma-rays*, are also emitted. However, we may have a situation where the β-decay of the fission fragments is forbidden or inhibited due to some selection rules. In such a situation, each fragment may emit a neutron and become a stable nucleus. The neutrons so emitted are known as the *delayed neutrons*.

QUESTION 10.2 The β-decay of the fission fragments is of the β^--type. Explain.

Energy Release in Fission

One of the most striking properties of the fission process is the amount of energy released which is about 200 MeV per fission. A rough calculation of the energy released in fission can be made using the binding energy curve shown in Fig. 9.2. As we can note from this curve, for a heavy nucleus (with $A \sim 236$), the average binding energy per nucleon $\bar{B}$ is $\sim$7.6 MeV, whereas for a nucleus with a medium weight, it is about 8.5 MeV. Hence, the average binding energy per particle is about 0.9 MeV more in the fission products than in a compound nucleus (e.g., ${}^{236}_{92}\text{U}$). Thus, this excess of 0.9 MeV per nucleon must be liberated in the fission process. The total energy liberated per fission, for example, of ${}^{236}_{92}\text{U}$ is given by

$$E_f = 236(8.5 - 7.6) \text{ MeV} \sim 212 \text{ MeV}.$$

A more accurate calculation of the total energy released per fission is possible as shown in the example that follows.

EXAMPLE 10.1 Consider a typical reaction during the neutron-induced fission of ${}^{235}_{92}\text{U}$, namely,

$$n + {}^{235}_{92}\text{U} \rightarrow {}^{236}_{92}\text{U}^* \rightarrow {}^{98}_{40}\text{Zr} + {}^{136}_{52}\text{Te} + 2n.$$

The resulting fission fragments are far from the stability region and decay into stable end products ${}^{98}_{42}\text{Mo}$ and ${}^{136}_{54}\text{Xe}$ by a successive emission of β^--particles. Calculate the total energy that will be released in this fission reaction.

It is not possible to find out the energy that would be released during the first stage of the fission process, the reason being that we would not know the masses of the fission fragments because these fragments are very short-lived. We can, however, use the semiempirical formula [i.e., Eq. (9.68)] to calculate the masses. In what follows, we shall calculate the total energy released by considering only the fission products (i.e., stable end products). Thus, let us consider the reaction

$$n + {}^{235}_{92}\text{U} \rightarrow {}^{98}_{42}\text{Mo} + {}^{136}_{54}\text{Xe} + 2n.$$

In this reaction, the total mass of the nuclei before fission is

$$235.0439 + 1.0087 = 236.0526 \text{ u},$$

and that after fission is

$$97.9054 + 135.9170 + 2.0174 = 235.8398 \text{ u}.$$

The mass difference is 0.2128 u, and therefore the energy release is approximately

$$0.2128 \times 931 \text{ MeV} = 207 \text{ MeV}.$$

The total amount of energy released per fission is the sum of the kinetic energies of the fission fragments, neutrons emitted, and beta-particles, and the energy of prompt gamma-rays. The values of these energies obtained experimentally by fissioning $^{236}_{92}U$ are as follows:

average energy of fission fragments	=	168 MeV
average energy of neutrons emitted	=	5 MeV
average energy of β-particles	=	16 MeV
average energy of prompt gamma-rays	=	5 MeV
average energy of delayed gamma-rays	=	6 MeV
total fission energy		200 MeV

The total fission energy obtained experimentally for the other nuclei, e.g., $^{239}_{94}Pu$ and $^{233}_{92}U$, is similar to that for $^{235}_{92}U$.

EXAMPLE 10.2 Calculate the amount of energy that will be released in completely fissioning 1 kg of $^{235}_{92}U$. Also, how much power will be generated if the same quantity of $^{235}_{92}U$ is fissioned in one year?

The number of nuclei (N) in 1 kg of $^{235}_{92}U$ is

$$N = \tfrac{1000}{235} \times 6.03 \times 10^{23} = 2.56 \times 10^{24}.$$

The energy released in fissioning one nucleus is 200 MeV. Therefore, the total energy that will be released in fissioning N nuclei is

$$200 \times 2.56 \times 10^{24} \text{ MeV} = 5.12 \times 10^{26} \times 1.6 \times 10^{-13} \text{ J} = 8.2 \times 10^{13} \text{ J}.$$

The power output will be

$$\frac{8.2 \times 10^{13}}{3.15 \times 10^{7}} \text{ W} \sim 2.6 \text{ MW}.$$

10.3 SPONTANEOUS FISSION AND POTENTIAL BARRIER

Consider a nucleus in the ground state, with mass $M(A, Z)$, fissioning into two identical fragments. The energy released E_f is given by

$$E_f = [M(A, Z) - 2M(\frac{A}{2}, \frac{Z}{2})]c^2. \tag{10.4}$$

If E_f is positive, the reaction is exoergic, and thus represents the possibility of spontaneous fission. It has been found that E_f is positive for a nucleus with $Z > 36$ and $A > 85$ and increases in magnitude as Z and A increase. At this stage, the question that arises is: Why is a nucleus with $A > 85$ stable at all? The answer to this question is that a potential barrier prevents the immediate splitting of such a nucleus. It is similar to the barrier encountered when the α-particles are emitted (see Chapter 9). The only difference between the two is that the former is wider and higher than the latter. This is so because the fission fragments are much bigger than an α-particle and carry a charge greater than the charge of the α-particle. In the light of the difference between the two barriers, the lifetime of a nucleus undergoing spontaneous fission is very long.

Let us consider two symmetrical (i.e., identical) fission fragments separated by a large distance. Assume that the potential energy is zero when the two fragments are infinitely far apart. When the two fragments are brought closer, this energy will increase according to Coulomb's law as $(Ze/2)^2/(4\pi\epsilon_0 r)$, where r is the separation between the fragments, each having a charge $Ze/2$. This expression is valid only until before the nuclear force starts operating between the two fragments. When r is continued to be decreased, the fragments will merge and form a nucleus $M(A, Z)$. The curve showing the potential energy as a function of r is given in Fig. 10.2. The

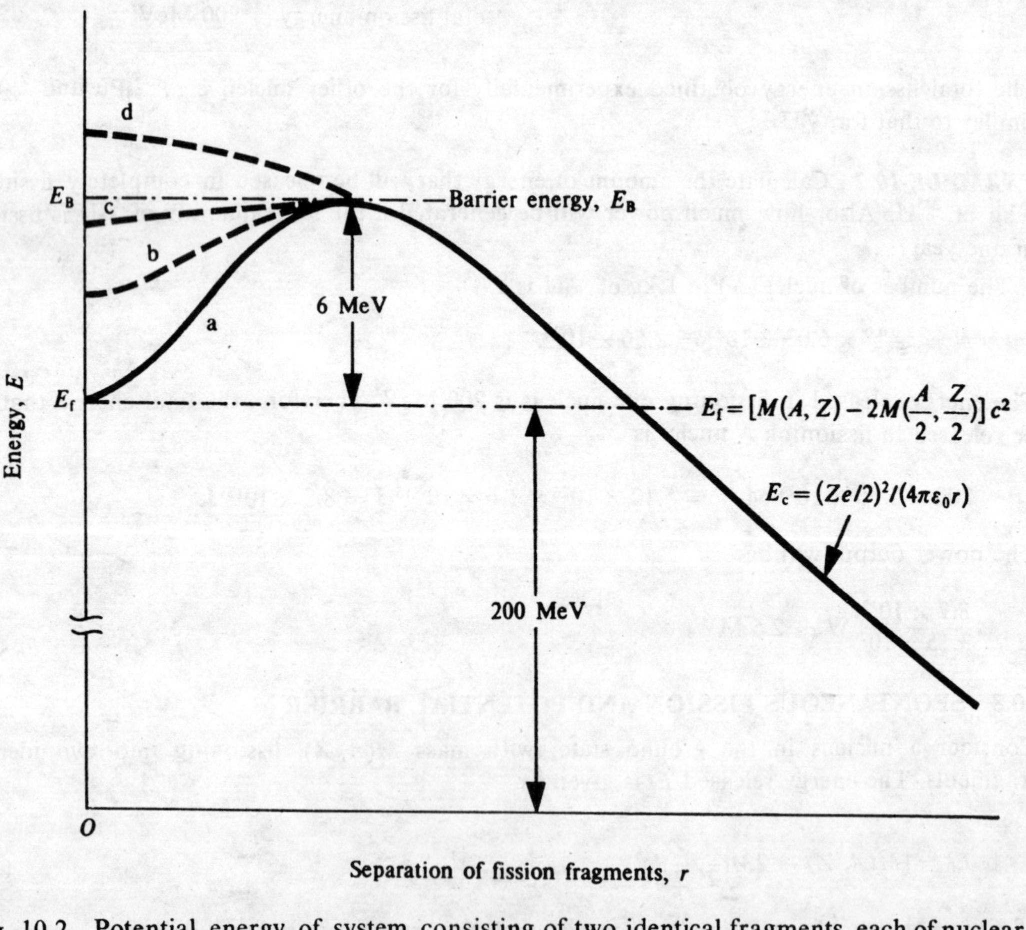

Fig. 10.2 Potential energy of system consisting of two identical fragments, each of nuclear charge $Ze/2$ and radius R, as function of their separation.

height of the barrier E_B for two symmetric fission fragments corresponds to the Coulomb potential between the fragments when they are just in contact with each other (see Example 9.10). The smallest separation between the fragments is $2R$, where R is the radius of each fragment.

Thus,

$$E_B = \frac{(Z/2)^2 e^2}{4\pi\epsilon_0(2R)} = \frac{Z^2 e^2}{32\pi R_0 A^{1/3}\epsilon_0} \tag{10.5}$$

since $R = R_0 A^{1/3}$.

In a fission process, the fragments encounter a barrier height given by $E_B - E_f$. Therefore, according to the classical concepts, this much energy has to be supplied to a nucleus for it to fission spontaneously. For a heavy nucleus, the height of the potential barrier the fragments encounter decreases as shown by the barriers b, c, d in Fig. 10.2. Let us consider the following three different situations:

(1) $E_f > E_B$ This applies to a nucleus with $A > 250$. Here, the fission is spontaneous since no energy is needed to separate the fragments. For this reason, such a heavy nucleus does not occur in nature.

(2) $E_f < E_B$ This situation occurs, for example, in $^{235}_{92}U$ where $E_B - E_f$ is about 6 MeV, and is brought about as follows. A low energy neutron absorbed by such a nucleus gives an excitation energy equal to 6 MeV, and thus induces fission in the nucleus. Such a nucleus at times also exhibits spontaneous fission which occurs due to the tunneling of the fission fragments through the barrier. But the probability of tunneling is small. For example, the spontaneous fission per hour in a gram of $^{235}_{92}U$ is about one and corresponds to a mean life approximately equal to 3×10^{21} yr.

(3) $E_f \ll E_B$ This situation corresponds to a nucleus with a medium weight, and therefore a large amount of energy has to be supplied to fission the nucleus. The possibility of spontaneous fission in such a nucleus is extremely low because the barrier has a large height and width. In a nucleus with $A < 85$, $E_f < 0$. Therefore, no fission can take place in such a nucleus.

EXAMPLE 10.3 Calculate the excitation energy of the $^{236}_{92}U$ nucleus when a thermal neutron ($E_n \sim .025$ eV) is absorbed by a $^{235}_{92}U$ nucleus.

The compound nucleus $^{236}_{92}U$ is formed as $^{235}_{92}U + n \rightarrow {}^{236}_{92}U^*$. To obtain the excitation energy, we need to know the masses of the initial nuclei and the final nucleus. Adding the mass of $^{235}_{92}U$(=235.0439 u) and that of the neutron (=1.0087 u), we get 236.0526 u. Since the mass of $^{236}_{92}U$ = 236.0457 u, the mass difference is

$$.0069 \text{ u} = .0069 \times 931 \text{ MeV} = 6.42 \text{ MeV}.$$

Therefore, the excitation energy is $(6.42 \times 10^6) + .025 \text{ eV} \sim 6.42$ MeV.

How Fission May Be Achieved

Earlier in this section, we discussed the effect of the barrier the fragments encounter when a nucleus splits into two fragments. We shall now use the liquid drop model of the nucleus to understand how the two fragments are formed. In this model, the nucleus is regarded as analogous to a uniformly charged drop of liquid. As such the volume energy E_v, surface energy E_s, and Coulomb energy E_c of the nucleus constitute the binding energy E of the nucleus, i.e.,

$$E = E_v + E_s + E_c = a_1 A - a_2 A^{2/3} - a_3 \frac{Z(Z-1)}{A^{1/3}}. \tag{10.6}$$

The surface energy of a nucleus is similar to the ordinary surface tension of a liquid drop and tends to minimize the surface area of the nucleus, that is, it makes the nucleus acquire the spherical shape. In a very compact spherical shape, the Coulomb energy is the largest. Thus, the Coulomb force tries to make a nucleus (or liquid drop) assume a nonspherical shape. In a light nucleus, the Coulomb repulsion is small, and hence such a nucleus retains a nearly spherical shape. In a very heavy nucleus, however, this repulsion is very large, and a slight gain in its internal energy by the nucleus can make it oscillate and change its shape.

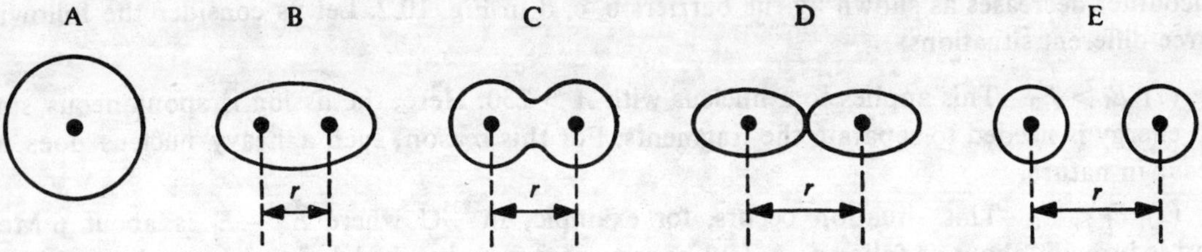

Fig. 10.3 Possible changes in shape during fission process.

The uranium nucleus can be considered to be similar to a spherical liquid drop (see A in Fig. 10.3). If some energy is supplied to the drop to excite it, it starts oscillating and changes its shape to assume the form of an ellipsoid (see B in the figure). Let us assume that in this process the volume of the drop remains constant, i.e., the term a_1A representing the volume in Eq. (10.6) does not change. When the surface area increases, then, according to the liquid drop model, the surface energy also increases. However, with the deformation of the nucleus, the average separation between the protons in it increases, thus decreasing the Coulomb energy. Therefore, the competing influences at work are the nuclear surface tension, which tends to keep the nucleus spherical, and the Coulomb repulsion, which tends to deform it. If the excitation is sufficient, the Coulomb force succeeds in making the nucleus acquire a dumbbell shape (see C in Fig. 10.3). With such a large distortion, the surface tension fails to restore the nucleus to sphericity, and the Coulomb force increases the separation between the two ends of the nucleus (see D in the figure), until the ends split into two fragments (E in the figure).

In a nucleus with $Z > 90$, the barrier height $E_B - E_f$ (see Fig. 10.2) calculated according to the liquid drop model is about 6 MeV. Such a nucleus therefore is normally stable. The fissioning in it can take place by tunneling, but since the probability of such a penetration is very small, the nucleus fissions spontaneously with a large half-life (10^{16} yr). Therefore, the rate of spontaneous fission is much too small to be of any practical use.

The breakthrough in understanding the fission process was achieved when, in 1939, O. Hahn, L. Meitner, and F. Strassmann observed the neutron-induced fission. In this process, a nucleus, after it captures a neutron, receives enough energy to take it over the top of the fission barrier (see Example 10.3). In other words, the nucleus gets sufficiently elongated to fission. For each such nuclei as $^{233}_{92}U$, $^{235}_{92}U$, and $^{239}_{94}Pu$, the capture of a thermal neutron brings in an excitation energy of about 6.5 MeV which is enough to take it over the top of the barrier. But in $^{238}_{92}U$, the capture of a thermal neutron results in an excitation energy of only about 5 MeV and is insufficient to induce the fission process. A fast neutron, i.e., of energy 1 MeV or more, when captured can induce fission in $^{238}_{92}U$.

QUESTION 10.3 Why do $^{235}_{92}U$ and $^{238}_{92}U$ require different energies for neutron-induced fission? Give a physical reason.

10.4 EMISSION OF NEUTRONS IN FISSION

As already mentioned, each primary fragment contains an excess of neutrons and these neutrons are emitted as soon as the fragments are formed. One other way to understand the process of emission of these (prompt) neutrons is as follows. Immediately after the fission, each fragment is considerably deformed. As a result, the energy associated with deformation is released and gets added to the excitation energy of the fragment as the fragment moves towards a more stable configuration. This excitation energy is, to a large extent, responsible for the emission of the neutrons. Each fission fragment emits approximately two or three neutrons and its subsequent decay by a successive emission of β-particles brings its neutron/proton ratio to the stable values.

Of the total number of neutrons released by the fission fragments, about 0.6% are emitted long (of the order of seconds) after the fission. Such neutrons, as already noted, are called delayed neutrons. Normally, the neutrons are not emitted because of the large energy binding them to the resulting stable nuclei.

QUESTION 10.4 Why is neutron emission 'preferred' by primary fragments?

The average number of prompt neutrons emitted per fission $\bar{\nu}$ is a parameter of utmost importance in designing a nuclear reactor. The values of $\bar{\nu}$ for some nuclei are listed in Table 10.1.

TABLE 10.1 Average number of prompt neutrons per fission for some nuclei

Target nucleus	$\bar{\nu}$
$^{233}_{92}U$	$2.497 \pm .008$
$^{235}_{92}U$	$2.426 \pm .006$
$^{239}_{94}Pu$	$2.892 \pm .011$

Fissionable Materials

The only naturally-occurring nuclei that can be fissioned with thermal neutrons are $^{235}_{92}U$ which constitute 0.7% of the uranium found in nature. The other nuclei that undergo fission with thermal neutrons are $^{233}_{92}U$ and $^{239}_{94}Pu$. These nuclei do not occur in nature but can be artificially produced by allowing the neutrons to interact with $^{232}_{90}Th$ and $^{238}_{92}U$, respectively. Because of their ability to convert themselves into the nuclear fuels $^{233}_{92}U$ and $^{239}_{94}Pu$, the two substances $^{232}_{90}Th$ and $^{238}_{92}U$ are called *fertile materials*.

The fissile nuclei $^{233}_{92}U$, $^{235}_{92}U$, and $^{239}_{94}Pu$ can also be fissioned with fast neutrons. But in such a fission, the possibility of a high energy neutron getting absorbed is less than when any of these

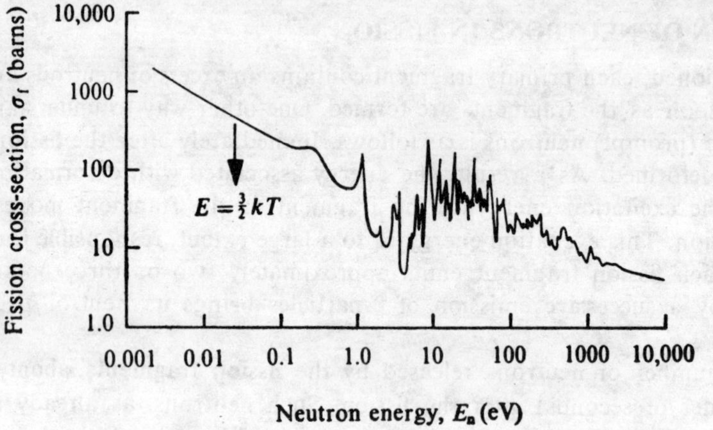

Fig. 10.4 Fission cross-section of $^{235}_{92}U$ as function of neutron energy.

nuclei is fissioned with a thermal neutron. In other words, the probability of fission (i.e., fission cross-section σ_f) decreases with increase in neutron energy, as shown in Fig. 10.4. (For a thermal neutron, $\sigma_f \sim 500$ b, and for a neutron with an energy ~ 1 MeV, $\sigma_f \sim 1$ b.) However, there are a number of heavy nuclei which can be fissioned only with fast neutrons. Of these, $^{238}_{92}U$, as already noted, has a fission threshold ~ 1 MeV. The variation in the fission cross-section of $^{238}_{92}U$ with neutron energy is depicted in Fig. 10.5.

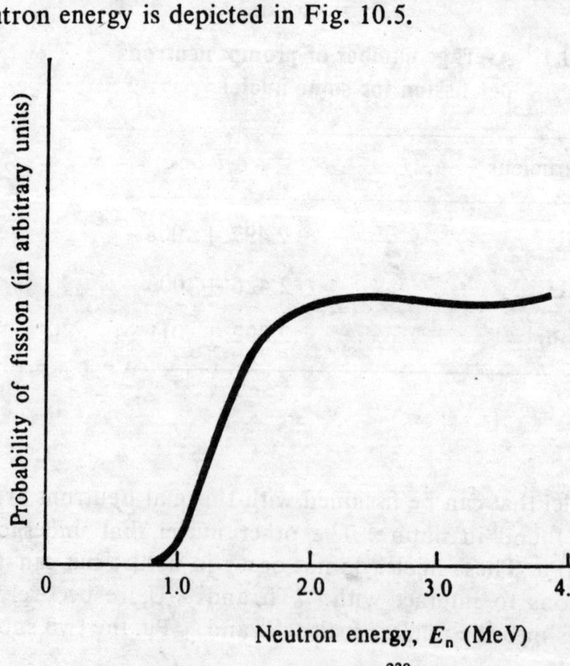

Fig. 10.5 Probability of fission of $^{238}_{92}U$ as function of neutron energy.

When a neutron is absorbed by the $^{235}_{92}U$ nucleus, the $^{236}_{92}U$ nucleus produced is in an excited state and is known as the compound nucleus. The compound nucleus can either undergo fission or just reduce its excitation energy by emitting γ-rays. In other words, the reaction can go through one of two possible channels, namely, fission and capture, besides the elastic scattering. Thus, the total absorption cross-section σ_a of the reaction can be written as

$$\sigma_a = \sigma_f + \sigma_r,$$

where σ_f is the fission cross-section and σ_r is the radiative capture cross-section. The fraction of the neutrons absorbed (ξ) leading to fission is

$$\xi = \frac{\sigma_f}{\sigma_f + \sigma_r}. \tag{10.7}$$

For a thermal neutron-induced fission of $^{235}_{92}U$, $\sigma_f \sim 550$ barns and $\sigma_r \sim 100$ barns. Thus, the fraction of absorbed neutrons which gives rise to fission is

$$\xi = \frac{550}{550 + 100} \sim .85 = 85\%.$$

Since $\bar{\nu}$ is the average number of neutrons produced per fission, the average number of neutrons emitted per neutron absorbed in a fissionable material η is

$$\eta = \xi\bar{\nu} = \frac{\bar{\nu}\sigma_f}{\sigma_f + \sigma_r} = \bar{\nu}\frac{\sigma_f}{\sigma_a}. \tag{10.8}$$

The parameter η is also important in designing a nuclear reactor. The cross-sections for fissile materials and the values of η for thermal neutrons are given in Table 10.2.

TABLE 10.2 Reactor parameters

Nucleus	Thermal neutrons			Fast neutrons ($E_n \sim 2$ MeV)	
	σ_f (barns)	σ_r (barns)	η	σ_f (barns)	σ_r (barns)
$^{235}_{92}U$	549	101	2.078	1.3	<.01
$^{233}_{92}U$	525	49	2.292	2	<.01
$^{238}_{92}U$	.0005	2.8	0	0.5	2
$^{239}_{94}Pu$	740	361	2.116	1.9	<.01
Uranium*	4.1	3.4	1.3	0.6	~.01

*Natural uranium contains $^{238}_{92}U$ and $^{235}_{92}U$ in the ratio 140 : 1.

EXAMPLE 10.4 Calculate the number of fission neutrons produced per thermal neutron absorbed for natural uranium (i.e., 99.3% $^{238}_{92}U$ and 0.7% $^{235}_{92}U$).

A thermal neutron can be absorbed either by the nucleus $^{235}_{92}U$ or $^{238}_{92}U$. In natural uranium, if there are N atoms of ${}^{235}_{92}U$, there will then be $(99.3/0.7)\times N = 142N$ atoms of ${}^{238}_{92}U$. Let the absorption cross-sections for ${}^{235}_{92}U$ and ${}^{238}_{92}U$ be σ_a^{235} and σ_a^{238}, respectively. Then, the probability P that the neutron will be absorbed by the ${}^{235}_{92}U$ nucleus is

$$P = \frac{N\sigma_a^{235}}{N\sigma_a^{235} + 142N\sigma_a^{238}} = \frac{650}{650 + 142 \times 2.8} = 0.62.$$

Further, in the case of ${}^{238}_{92}U$ nucleus, the only reaction possible is radiative capture. Thus, neutrons are produced only when a neutron is absorbed by a ${}^{235}_{92}U$ nucleus. Each fission of a ${}^{235}_{92}U$ nucleus produces ~2.43 neutrons on the average (see Table 10.1). The number of neutrons produced for each neutron captured, η, is (using Table 10.2)

$$\eta = \frac{2.43\sigma_f}{\sigma_f + \sigma_r} = \frac{2.43 \times 549}{650} \sim 2.05.$$

Thus, the number of fission neutrons produced per thermal neutron captured in natural uranium is $2.05 \times P = 2.05 \times .62 \sim 1.3$.

10.5 SELF-SUSTAINING CHAIN REACTION: NUCLEAR REACTOR

As we have already observed, a number of prompt neutrons are produced in a fission. The neutrons so generated can be made to induce fission in nuclei other than the nucleus participating in the initial fission, i.e., each fission can be made to produce one or more fissions, leading to a chain reaction. Not all the neutrons emitted in a fission can be used in bringing about a further fission since some are inevitably lost in nonfission nuclear processes and some others leak out. The condition for maintaining a chain reaction is that, on an average, at least one neutron created in a fission process should cause another fission. This condition is expressed in terms of a *multiplication factor*, K, defined as the ratio of the number of fissions produced by a given generation of neutrons (see Fig. 10.6) to the number of fissions leading to the given generation of neutrons. In other words, K gives the net *effective* fission neutrons produced per generation. If $K > 1$, the neutron population increases with each generation, resulting in a divergent chain reaction. If $K = 1$, the chain reaction is maintained at a steady state. If $K < 1$, the neutron population decreases and the chain reaction fails to develop. In a nuclear reactor, K is constantly maintained at 1 so that energy can be released in a controlled manner.

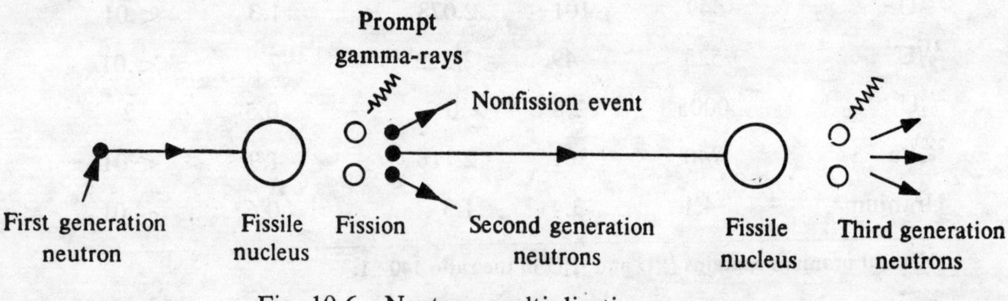

Fig. 10.6 Neutron multiplication process.

Since one neutron per fission is required to maintain the chain reaction, the number of neutrons increases by the factor $K - 1$ in each generation. Thus, if the number of neutrons initially is N, the rate of increase of neutrons will be $N(K-1)$ per generation. If τ is the average time between successive neutron generations, then

$$\frac{dN}{dt} = \frac{N(K-1)}{\tau}. \tag{10.9}$$

Integrating this equation, we get

$$N = N_0 \exp\left[\frac{(K-1)t}{\tau}\right], \tag{10.10}$$

where N is the number of neutrons present at a time t and N_0 is the number of neutrons present at the time $t = 0$. Therefore, as we can see, if the multiplication factor is greater than unity, the number of neutrons will increase exponentially with time.

Suppose $K = 1.005$, and τ is of the order of 10^{-8} sec. Then, if N fissions take place during one time interval τ, then NK fissions will take place during the next τ, NK^2 during the next τ, and so on. Thus, after a lapse of 1 sec, the number of neutrons will increase by a factor $\exp(.005/10^{-8}) \sim \exp(5 \times 10^5)$.

The possibility of a chain reaction depends on a balance among the following four competing processes:

(1) Fission with the emission of more than one neutron.

(2) Nonfission capture of neutrons by fissile materials.

(3) Nonfission capture of neutrons by materials other than fissile materials or by the structure holding the fissile materials.

(4) Complete loss or escape of neutrons (generally referred to as leakage).

In each of processes (2)–(4), neutrons are lost, but in process (1), that is, in the fission reaction, neutrons are generated. If the loss of neutrons in processes (2)–(4) is less than or equal to the surplus neutrons produced in process (1), the chain reaction occurs, otherwise it does not.

In general, the loss of neutrons in processes (2) and (3) can be reduced by appropriately choosing the materials for building the reactor, but the loss in process (4) can be diminished only by increasing the reactor size. In a reactor, if there is a regular distribution of the fissile materials, then the neutron production will depend on the volume, and the neutron escape probability on the area, of the reactor. If the reactor is a sphere (of radius r), the neutron leakage is proportional to r^2 whereas the neutron generation is proportional to the number of fissionable nuclei which are in turn proportional to the volume or to r^3. So, we see that the neutron economy can be improved by increasing the sphere radius. Thus, there is a certain *critical size* for a reactor in which the neutrons produced by fission are just equal to the neutrons lost by nonfission capture and leakage, making a chain reaction possible. A reactor below the critical size is called *subcritical*, and the one above the critical size is known as *overcritical*.

CLASSIFICATION OF REACTORS

A nuclear reactor is a system from which energy can be obtained in a controlled way by sustaining the chain reaction. This is achieved by maintaining the multiplication factor K very close to unity. The design of a reactor depends on the type of fissile materials used. There are basically two kinds of reactors, namely, (1) fast and (2) thermal.

As we know, the neutrons produced in fission have an average energy approximately equal to 2 MeV. The fission cross-section at such an energy is low. Thus, to achieve the neutron balance for sustaining the chain reaction, a large quantity of the fissionable material is required. A reactor using fast neutrons for sustaining the chain reaction is known as a *fast reactor*. In a reactor, the amount of fissionable material required can be reduced by slowing down the speed of the neutrons. This can be achieved by making the neutrons go through elastic collisions with light nuclei. A light nucleus on collision with a neutron takes away some of its energy, thus reducing the neutron speed and increasing the fission cross-section. Such a nucleus is called a *moderator*. The moderators commonly used are water, heavy water (D_2O), and graphite. If a very high fission cross-section at a thermal energy has to be made use of, a large quantity of a moderator would then be required to slow down the neutrons to thermal energies. A reactor requiring thermal neutrons for fission is called a *thermal reactor*. Such a reactor is most common, but tends to be large in size. Thus, its use is limited to a large electric power station or, possibly, ship or submarine propulsion.

QUESTION 10.5 In a reactor, why is a heavy nucleus unsuitable as a moderator?

A chain reaction can also be brought about with natural uranium ($^{238}_{92}U$, 93.3%; $^{235}_{92}U$, 0.7%). As already noted, $^{238}_{92}U$ can be fissioned only with a fast neutron (because of its threshold requirement), but the fission cross-section here is low. Therefore, to sustain the chain reaction in natural uranium, a moderator is needed. Thus, once the neutrons are thermalized by using a moderator, the large fission cross-section of $^{235}_{92}U$ sustains the chain reaction.

10.6 NEUTRON BALANCE IN REACTOR

We shall now study the behaviour of neutrons in a reactor assembly consisting of enriched uranium and a moderator. To do this, we start with the fission of a $^{235}_{92}U$ nucleus by a thermal neutron. In this fission, the number of fast neutrons produced is $\bar{\nu}$. Of these, some will cause a fission in $^{238}_{92}U$, and therefore create additional fast neutrons. We take account of this small increase in the neutrons by introducing a factor, namely, ϵ, called the *fast fission factor*. Thus, instead of $\bar{\nu}$ fast neutrons, we will have $\bar{\nu}\epsilon$ fast neutrons, where $\epsilon > 1$. Of these $\bar{\nu}\epsilon$ fast neutrons, some escape the reactor assembly. If the number of such neutrons, i.e., the escaping fraction, is l_f, then we are left with $\bar{\nu}\epsilon(1 - l_f)$ neutrons. These remaining fast neutrons have a mean energy of about 2 MeV. Since the fission cross-section at such an energy is ~1 b, it is desirable that the speed of these neutrons be reduced so that a large fission cross-section (~549 b) of $^{235}_{92}U$ is obtained. This, as already noted, can be done by allowing the neutrons to have elastic collisions with the atoms of a moderator (i.e., a light nucleus), e.g., graphite and heavy water. During the time the fast neutrons are slowing down, their energy steadily reduces, making them pass through intermediate energy stages before they acquire thermal energies. Some of the intermediate energy neutrons are captured (without producing a fission), particularly when their energies correspond to the resonances in $^{238}_{92}U$ or in other nuclei present. This is known as resonance absorption. Therefore, we define *resonance escape probability*, *p*, as the fraction of neutrons escaping resonance absorption and any other such capture. Hence, the number of neutrons surviving thus far and reaching thermal energies is

$$\nu\epsilon(1 - l_f)p. \tag{10.11}$$

Of these thermal neutrons, some can again escape from the assembly. We denote the fraction of such neutrons by l_t. Then the number of thermal neutrons we are left with is

$$\bar{\nu}\epsilon(1 - l_f)p(1 - l_t). \tag{10.12}$$

The fraction of thermal neutrons absorbed by the fuel, i.e., enriched uranium, as compared to all thermal neutron absorptions in the assembly is called the *thermal utilization factor* and is denoted by f. Thus, the number of thermal neutrons actually absorbed by the fuel is

$$\bar{\nu}\epsilon(1 - l_f)p(1 - l_t)f. \tag{10.13}$$

When these neutrons are absorbed by $^{235}_{92}U$ in the fuel, some cause fission, whereas some produce other kinds of reactions such as capture. The fraction of neutrons causing fission is given by the ratio of fission cross-section (σ_f) to the total absorption cross-section (σ_a). Therefore, the total number of fissions produced is given by

$$\bar{\nu}\epsilon(1 - l_f)p(1 - l_t)f\frac{\sigma_f}{\sigma_a}. \tag{10.14}$$

Using the definition of η [i.e., Eq. (10.8)], we get

$$\eta = \bar{\nu}\frac{\sigma_f}{\sigma_a}.$$

The ratio K of the number of fissions produced to a single neutron is

$$K = \eta\epsilon pf(1 - l_f)(1 - l_t). \tag{10.15}$$

This is the *multiplication factor*. For simplicity, if we assume that the reactor assembly is of an infinite size so that there is no leakage of neutrons (i.e., $l_f = l_t = 0$), then the multiplication factor, K_∞, becomes

$$K_\infty = \eta\epsilon pf. \tag{10.16}$$

This is known as the *four-factor formula*. In designing a nuclear reactor, the fundamental problem is to maintain K at 1 in order to achieve a self-sustaining chain reaction. Of the four factors, η and ϵ are fixed for a given fuel. The other two factors p and f are made as large as possible by properly designing the reactor geometry, arranging the fuel, and choosing the moderator.

In a reactor where the fuel is pure $^{235}_{92}U$, both the fast fission factor ϵ and the resonance escape probability p differ very slightly from unity. In such a reactor, the amount of fuel needed is much smaller than that of the moderator.

Reactor Control

When the multiplication factor K is unity, the chain reaction operates at a steady state. In real life, it is difficult to fix K at unity, and if it is slightly greater than unity, the reaction rate and the reactor power increase exponentially. If K is less than unity, the reaction rate decreases exponentially. The reaction rate, and thus the neutron generation, can be controlled by introducing a neutron-absorbing material. Therefore, if K tends to become greater than one, the neutron-absorbing material will bring down its value. Thus, K can be controlled by withdrawing or reinserting *control rods* of the neutron-absorbing material, in the process keeping the reactor power level fixed. The element cadmium has a large neutron absorption cross-section, and is therefore used for control rods. Besides, a reactor has *safety rods*. These rods can, when required, be rapidly inserted to reduce K well below unity. In an emergency, these rods make it possible to very quickly shut down the reactor, a process known as *scramming*.

Breeding

One important problem of current interest is that of developing a reasonable method for generating more nuclear fuel in a reactor than is consumed in its operation. The reason this is important is that the nuclear fuels are available only in a small amount. Once such a method is evolved, the nuclear fuels can be generated in a suitably designed reactor from the fertile materials (i.e., ${}^{238}_{92}U$ and ${}^{232}_{90}Th$) which are obtainable in a large quantity. Such a reactor is called a *breeder reactor*. The breeding reactions for the two fertile materials are

$$ {}^{238}_{92}U + n \rightarrow {}^{239}_{92}U^* \xrightarrow[23\text{ min}]{\beta^-} {}^{239}_{93}Np \xrightarrow[2.3\text{ d}]{\beta^-} {}^{239}_{94}Pu\ (T_{1/2} = 24{,}000\text{ yr}), \tag{10.17} $$

$$ {}^{232}_{90}Th + n \rightarrow {}^{233}_{90}Th^* \xrightarrow[23\text{ min}]{\beta^-} {}^{233}_{91}Pa \xrightarrow[27\text{ d}]{\beta^-} {}^{233}_{92}U\ (T_{1/2} = 1.6 \times 10^5\text{ yr}). \tag{10.18} $$

Both ${}^{238}_{92}U$ and ${}^{232}_{90}Th$ cannot be fissioned with thermal neutrons. But when they capture neutrons, the reactions lead to ${}^{239}_{94}Pu$ and ${}^{233}_{92}U$ which are fissile materials.

In a breeder reactor, there are two fuels, one fissionable (such as ${}^{239}_{94}Pu$), and the other fertile (e.g., ${}^{238}_{92}U$). In the fission of ${}^{239}_{94}Pu$, three neutrons on an average are released. Of these, one neutron must sustain the chain reaction by producing further fissions. Of the remaining two neutrons, at least one must be captured by ${}^{238}_{92}U$ to produce the same amount of ${}^{239}_{94}Pu$ as before its fission. If on an average both the remaining two neutrons can be made to be captured by ${}^{238}_{92}U$, the reactor can then breed ${}^{239}_{94}Pu$, that is, produce ${}^{239}_{94}Pu$ in a quantity more than the quantity in which it is consumed. Clearly, therefore, neutron economy has to be very carefully managed when designing a breeder reactor.

10.7 UNCONTROLLED CHAIN REACTION—BOMB

A bomb is simply a device which releases a large amount of energy in a very short time, leading to a violent explosion. As already noted, if the multiplication factor $K > 1$, the chain reaction makes the reactor overcritical. In such a reactor, the rate of build up of neutrons, and consequently the rate of energy release, is such that a powerful explosion is produced. In order to achieve the maximum value of K, a fissile material with maximum fission cross-section is chosen. After this has been done, the reactor is so designed that the number of neutrons escaping it without interacting with the nuclei of the fissile material is minimized.

Critical Mass

As we know, for a spherical shape, the ratio of area to volume (A/V) is minimum. Therefore, such a shape provides the most efficient geometry for maximizing K. The ratio A/V for a sphere of radius R is $3/R$. Thus, the fraction of neutrons that escape without interacting with the fissile material decreases as R increases. The value of R for which a neutron captured in a sample of a fissile material makes at least one more neutron available for further fissions gives the minimum size of the sphere required to generate a self-sustaining chain reaction. The mass of the sample for this minimum value of R is known as the critical mass. For a mass less than this, the chain reaction will not be sustained and die out. On the other hand, for a sphere of a mass greater than the critical mass, the chain reaction will not only be sustained but will also build up, leading to an explosion.

EXAMPLE 10.5 Estimate the critical mass of a sphere of $^{235}_{92}U$, assuming that the fission and radiative capture cross-sections are equal. The absorption cross-section for fission neutrons is 5 barns.

Of the approximately two neutrons produced in each fission, one is lost in radiative capture and the other, unless it escapes, leads to more fissions. If the radius of the $^{235}_{92}U$ sphere is more than the mean free path of a neutron (i.e., the distance travelled by the neutron between two fissions), then it is possible to sustain the chain reaction. The mean free path can be estimated as follows. When a neutron travels a distance λ, it interacts with the medium once (on an average). If the neutron sweeps out a volume $\lambda\sigma$, where σ is the fast neutron absorption cross-section, then it interacts once with the medium. Thus,

$$\lambda\sigma = \frac{1}{n} \quad \text{or} \quad \lambda = \frac{1}{n\sigma},$$

where n is the number of uranium nuclei per m^3. Therefore,

$$\lambda = \frac{235}{6 \times 10^{26} \times 18 \times 10^3} \times \frac{1}{5 \times 10^{-28}} = 0.04 \text{ m},$$

18×10^3 kg/m^3 being the density of uranium.

To sustain the chain reaction, the radius of the $^{235}_{92}U$ sphere, as already noted, should be greater than the mean free path. To get the order of magnitude of the critical mass, let $R = \lambda$. Then, the critical mass M is

$$M = \frac{4\pi R^3 \times \text{density}}{3} \sim 5 \text{ kg}.$$

From Table 10.2, we note that $^{235}_{92}U$ has a very large fission cross-section σ_f (=549 b), but its radiative capture cross-section σ_r is also large (=101 b). From these two values, the fraction of the neutrons absorbed, leading to fission, is $\xi = .85$. However, since, for an explosion, the energy release should be rapid, it is not feasible to wait for the neutrons to slow down. The neutrons produced in a fission process have an energy 2 MeV, and it takes 10^{-3} sec to slow them down to a thermal energy value. Thus, it is essential that the fast neutrons generated in fission be used to induce further fissions. The value of ξ for the fast neutrons produced in fissioning $^{235}_{92}U$ is

$$\xi = \frac{\sigma_f}{\sigma_f + \sigma_r} = \frac{1.3}{1.3 + .01} \sim 1.$$

Thus, $^{235}_{92}U$ is ideally suited for obtaining an uncontrolled chain reaction.

Let us consider the common isotope of uranium, i.e., $^{238}_{92}U$. Then, from Table 10.2,

$$\xi = \frac{\sigma_f}{\sigma_f + \sigma_r} = \frac{0.5}{2.5} = 0.2.$$

If only two neutrons are emitted in each fission, this value then implies that, irrespective of the size of the $^{238}_{92}U$ aggregate, the fission reaction will not be self-sustaining.

QUESTION 10.6 What should be the value of ν, the prompt neutrons emitted per fission, if $^{238}_{92}U$ is used for making a bomb?

EXAMPLE 10.6 Assume that 100 neutron generations are needed to fission all the nuclei in 5 kg of $^{235}_{92}U$. Estimate the total time required for and the energy emitted in this process.

The velocity of the fast neutrons (of energy ~2 MeV) produced in fission is

$$v = (\frac{2E}{m})^{1/2} = [\frac{2 \times 2 \times 1.6 \times 10^{-13}}{1.6 \times 10^{-27}}]^{1/2} \sim 2 \times 10^7 \text{ m/sec.}$$

Since the mean free path is ~4 cm (see Example 10.5), the time between fissions in each neutron generation will be $t = 2 \times 10^{-9}$ sec. Thus, the total time for 100 generations will be 2×10^{-7} sec. The total energy released will be

$$E = \frac{5 \times 6 \times 10^{26}}{235} \times 200 \text{ MeV} \sim 2.5 \times 10^{27} \text{ MeV}$$

$$\sim 2.5 \times 10^{27} \text{ MeV} \times 3.83 \times 10^{-14} \text{ cal/MeV} \sim 10^{14} \text{ cal.}$$

In order to appreciate and get an idea of the amount of energy release we have calculated, let us compare this energy with the energy that will be released in exploding 5 kg of TNT. One kilogram of TNT produces roughly 10^6 cal energy. Thus, the total energy produced by fissioning 5 kg of $^{235}_{92}U$ will be equivalent to exploding 10^8 kg (100 kilotons) of TNT.

As noted (see Example 10.5), the critical mass of $^{235}_{92}U$ is not large. That is, a 4-cm-diameter sphere of $^{235}_{92}U$ can be made to explode spontaneously. Starting the reaction is not a problem since it requires only one neutron, and there are a few neutrons always available, either as a result of the spontaneous fission of $^{235}_{92}U$ or as components of the ever-present cosmic rays.

A bomb can possibly consist of two or more pieces of $^{235}_{92}U$, each less than the critical mass, arranged in such a way that they can be assembled into an overcritical mass. Such an assembly has to be as rapid as possible, otherwise the fission rate that builds up during the assembly would blow off the pieces arriving late even before such pieces have had the time to become part of the assembly. A rapid assembly can be accomplished by dividing the critical mass into two hemispheres and shooting the two pieces together with an ordinary explosive. Once this has been done, the chain reacting mass must be kept overcritical, long enough for the reaction to proceed before the temperature generated by fission and the accompanying pressure blow off $^{235}_{92}U$ into subcritical masses.

A bomb is enclosed in an envelope, called *temper*, which serves two purposes. One, it reflects back the escaping neutrons into the bomb, and two, it holds together the reacting masses long enough for most of the fissile material to undergo fission. This time is extremely short (~microsec), but since the energy released is so large, the fissile material can get vaporized. Hence, if fission is not contained, the bomb will expand and the chain reaction will stop. The inertia of the envelope is sufficient to retard such an expansion for a microsecond or so, the time required for the explosion to occur. When the explosion does occur, an enormous temperature (comparable to that of the interior of the sun) is produced, leading to a devastating blast.

10.8 NUCLEAR FUSION

In Chapter 9, we learnt that when two light nuclei (each with $A < 20$) combine (i.e., fuse) with each other, energy is released. This is because the total binding energy of the resulting nucleus increases. A reaction in which two light nuclei combine with the release of energy is called a *fusion reaction.* Since the nuclei are charged and repel each other, the fusion can be brought about only by making one nucleus collide with another very energetic nucleus, keeping the kinetic energy of each nucleus so high that the repulsion effect equivalent to the potential barrier between the two nuclei is overcome. If a fusion reaction has to be produced to generate a large amount of power, then a large number of light nuclei must be brought together to make the whole reaction self-sustaining (just as in fission). For this, we must first know the fusion reactions that lead to energy production. It is clear that the primary kinetic energy necessary to overcome the Coulomb repulsion increases rapidly with the atomic number. Thus, the lighter the nuclei, the better the chance for the nuclei to approach closely. In this context, the hydrogen isotopes appear to be the most promising. The other factor governing the selection of a fusion reaction is the value of Q, that is, the energy release in the reaction.

It has been experimentally established that the collision of two protons does not lead to a fusion reaction with a measurable cross-section. However, in the experiments conducted by bombarding the particles ${}^2_1\text{H}$, ${}^3_1\text{H}$, and ${}^3_2\text{He}$ with the protons or deuterons, the following reactions have been found to occur:

$$ {}^1_1\text{H} + {}^2_1\text{H} \rightarrow {}^3_2\text{He} + \gamma, $$

$$ {}^2_1\text{H} + {}^2_1\text{H} \rightarrow {}^3_2\text{He} + \text{n} + Q \quad (=3.25 \text{ MeV}), \tag{10.19} $$

$$ {}^2_1\text{H} + {}^2_1\text{H} \rightarrow {}^3_1\text{H} + {}^1_1\text{H} + Q \quad (=4.0 \text{ MeV}), \tag{10.20} $$

$$ {}^2_1\text{H} + {}^3_1\text{H} \rightarrow {}^4_2\text{He} + \text{n} + Q \quad (=17.6 \text{ MeV}), \tag{10.21} $$

$$ {}^2_1\text{H} + {}^3_2\text{He} \rightarrow {}^4_2\text{He} + {}^1_1\text{H} + Q(=18.3 \text{ MeV}). \tag{10.22} $$

Each of the deuteron-deuteron reactions (10.19) and (10.20) has a small cross-section. However, the deuteron-triton reaction (10.21) is much more probable and also liberates a large amount of energy. As can be noted, the simplest fusion reaction

$$ {}^2_1\text{H} + {}^2_1\text{H} \rightarrow {}^4_2\text{He}, $$

in which the energy released $Q = 24$ MeV, is not among the reactions we have listed. The reason is that this reaction is forbidden because the resulting helium nucleus cannot have the right velocity to simultaneously satisfy the laws of conservation of momentum and energy.

EXAMPLE 10.7 Consider the fusion reaction

$$ 6\,{}^2_1\text{H} \rightarrow 2\,{}^4_2\text{He} + 2\text{p} + 2\text{n} + Q\,(=43 \text{ MeV}) $$

which is the net effect of reactions (10.19) to (10.22). Determine the amount of energy that will be generated by fusing all the nuclei in 1 kg of deuterium in such a reaction. Compare the energy so obtained with the energy that can be generated by fissioning 1 kg of ${}^{235}_{92}\text{U}$.

The number of deuterons (N) in 1 kg of deuterium is

$$ N = \frac{6.02 \times 10^{26}}{2} \sim 3 \times 10^{26}. $$

Since 6 deuterons fuse together to give an energy 43 MeV, the total energy that will be emitted in fusing 1 kg of deuterium is

$$E = \frac{43 \times 3 \times 10^{26}}{6} = 2.15 \times 10^{27}\ \text{MeV} = 3.44 \times 10^{14}\ \text{J} = \frac{3.44 \times 10^{14}}{3.6 \times 10^{6}} \sim 10^{8}\ \text{kWh}.$$

In $^{235}_{92}U$, the energy emitted per fission is 200 MeV. Therefore, the total energy that will be emitted from 1 kg of $^{235}_{92}U$ through fission will be

$$E = 200 \times \frac{6.02 \times 10^{26}}{235} \sim 5.1 \times 10^{26}\ \text{MeV} \sim 2.3 \times 10^{7}\ \text{kWh}.$$

We should like to note here that although the energy emitted per fission is much more than that emitted in fusion, the total energy generated from $^{235}_{92}U$ is less because the number of nuclei per gram in it is considerably less. One kilogram of deuterium, the basic fuel in a fusion reaction, yields 10^8 kWh of energy, which is about 10 million times the energy released by burning 1 kg of a fossil fuel.

In Fig. 10.7, the fusion cross-section of some reactions measured experimentally as a function of energy has been shown. As can be noticed, each of these cross-sections rapidly decreases

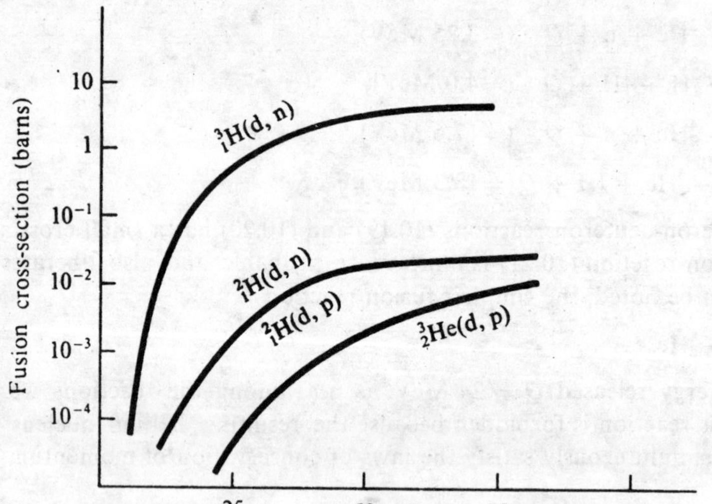

Fig. 10.7 Fusion cross-section for some reactions.

at a low energy although each reaction shown is exoergic and no threshold is involved. The reason this is so, as we know, is the presence of the Coulomb potential barrier. According to classical physics, when the energy of a bombarding particle is less than the energy of the barrier (i.e., barrier height), the particle cannot penetrate the nucleus, and therefore fails to induce a reaction. However, according to quantum physics, the probability that the particle can penetrate the barrier is finite. This is the reason each curve (in Fig. 10.7) shows that there is a finite probability of occurrence of a fusion reaction even at a very low energy.

THERMONUCLEAR PROCESSES

We shall now consider the problem of fusing a large number of deuterium nuclei. In such a fusion, some energy has to be supplied to overcome the Coulomb repulsion which is quite small (~100 keV) as compared to the energy released (~20 MeV). This situation is similar to the one encountered in combining H_2 and O_2 which react chemically only when a little energy is supplied to them (say, in the form of heat from a lighted match stick), releasing a large amount of energy in the form of an explosion. The energy required to induce a fusion reaction can also be supplied in the form of heat. A nuclear reaction that occurs as a result of an increase in the thermal motion of the particles interacting at a high temperature is called a *thermonuclear reaction*. Up to now, the only way in which a temperature high enough to cause a thermonuclear reaction has been achieved is through a fission bomb, e.g., a uranium or plutonium bomb. The thermonuclear reaction forms the basis of the hydrogen bomb. In such a bomb, perhaps some deuterium compound, e.g., lithium deuteride, is made to surround a uranium or plutonium bomb. The energy output of a hydrogen bomb, in principle, depends on the amount of deuterium used and has no upper limit.

EXAMPLE 10.8 Consider the reaction ${}^3_1\text{H}(\text{d, n}){}^4_2\text{He}$, $Q = 17.6$ MeV. Estimate the height of the Coulomb barrier of the ${}^3_1$H nucleus (triton) the deuteron has to penetrate. What must be the temperature of a gas of deuterons if the average kinetic energy of the deuterons has to be greater than the estimated barrier height?

The height of the potential barrier is

$$B = \frac{Z_d Z_t e^2}{(R_d + R_t)4\pi\epsilon_0},$$

where R_d and R_t are the radii of the deuteron and triton, respectively. The charge of the deuteron and triton is unity. Thus,

$$B = \frac{e^2}{4\pi\epsilon_0(R_d + R_t)}.$$

The value of $(R_d + R_t)$ is approximately 4 F, and therefore $B \sim 350$ keV.

As we know, when a gas has a temperature T, its molecules have velocities given by the Maxwell-Boltzmann distribution law. The average kinetic energy of such molecules is given by

$$\tfrac{1}{2}M\bar{v}^2 = \tfrac{3}{2}kT$$

or

$$T = \tfrac{2}{3}\,\frac{\text{energy}}{\text{Boltzmann constant}} = \tfrac{2}{3}\,\frac{350\ \text{keV} \times 1.6 \times 10^{-16}\ \text{J/keV}}{1.38 \times 10^{-23}\ \text{J/K}}.$$

Thus, $T \sim 10^9$ K.

10.9 CONTROLLED FUSION

As already mentioned, the probability of occurrence of a reaction involving the lightest and the most abundant nucleus, namely, ^1_1H, is very small. In a reaction involving the nucleus ^3_1H, such a probability is comparatively large, but this nucleus does not exist in nature and has to be produced in a nuclear reactor. This is an expensive proposition. In the circumstances, the best alternative to bring about fusion is to use the deuteron-deuteron reactions $^2_1H(d, n)^3_2He$ and $^2_1H(d, p)^3_1H$. The probability of occurrence of these two reactions is about equal. In a fusion reaction, the major problem is to achieve a high temperature ($\sim 10^8$ K). It has not yet been possible to maintain such a temperature long enough to generate energy on a large scale although this temperature has been realized for a small fraction of a second. The reason this is so is that it is difficult to find a container suitable for holding deuterium since all known materials would vaporize at this high temperature.

At the temperature required for a fusion, the atoms are fully ionized and these ions (i.e., nuclei) and the electrons move more or less independently but in the form of interpenetrating gases. This form, obviously, remains electrically neutral, and is called the plasma state.

It has been possible to experimentally produce a high temperature. To do this, a very intense electric current (of the order of a million amperes) is made to pass through a low pressure deuterium gas contained in a tube. The high temperature so generated sets up the plasma, and also concentrates it in a narrow beam, known as the *pinch effect*. This effect is analogous to that between two parallel wires carrying a current in the same direction, where the parallel streams of current forming the discharge attract each other, pinching the discharge into a narrow beam. A magnetic field is used to keep the discharge in the centre of the tube. This is called the *magnetic bottle*. It has so far not been possible to maintain such a pinched discharge for more than a fraction of a second.

10.10 FUSION IN STARS

As we have seen, a fusion reaction is potentially an extremely powerful source of energy. Also, to generate this energy, the reaction has to be maintained at a very high temperature. Such a temperature exists in a star and fusion is the prime source which maintains its temperature.

H. A. Bethe and his coworkers have suggested the fusion reactions possible in a star. From the known Q-values of these reactions, it is possible to compute the total energy evolved per unit time from these reactions in the interior of a star. It has been found that this rate of energy release is about the same as that of the energy radiated from the surface of the star, and that the temperature necessary for further reactions to occur is maintained.

One possible set of fusion reactions is the p-p chain *(proton-proton chain)*

$$^1_1H + {}^1_1H \rightarrow {}^2_1H + e^+ + \nu_e$$

or

$$^1_1H + {}^1_1H + e^- \rightarrow {}^2_1H + \nu_e$$

and

$$^2_1H + {}^1_1H \rightarrow {}^3_2He + \gamma,$$

$$^3_2He + {}^3_2He \rightarrow {}^4_2He + 2\,{}^1_1H$$

or

$$ {}^{3}_{2}He + {}^{4}_{2}He \rightarrow {}^{7}_{4}Be + \gamma $$

and

$$ {}^{7}_{4}Be + e^- \rightarrow {}^{7}_{3}Li + \nu_e, $$

$$ {}^{7}_{3}Li + {}^{1}_{1}H \rightarrow 2\,{}^{4}_{2}He $$

or

$$ {}^{7}_{4}Be + {}^{1}_{1}H \rightarrow {}^{8}_{5}B + \gamma, $$

$$ {}^{8}_{5}B \rightarrow {}^{8}_{4}Be^* + e^+ + \nu_e, $$

$$ {}^{8}_{4}Be^* \rightarrow 2\,{}^{4}_{2}He. $$

Another possible set of fusion reactions, suggested by Bethe, is the *carbon-nitrogen cycle*

$$ {}^{12}_{6}C + {}^{1}_{1}H \rightarrow {}^{13}_{7}N + \gamma, $$

$$ {}^{13}_{7}N \rightarrow {}^{13}_{6}C + e^+ + \nu_e, $$

$$ {}^{13}_{6}C + {}^{1}_{1}H \rightarrow {}^{14}_{7}N + \gamma, $$

$$ {}^{14}_{7}N + {}^{1}_{1}H \rightarrow {}^{15}_{8}O + \gamma, $$

$$ {}^{15}_{8}O \rightarrow {}^{15}_{7}N + e^+ + \nu_e, $$

$$ {}^{15}_{7}N + {}^{1}_{1}H \rightarrow {}^{12}_{6}C + {}^{4}_{2}He. $$

In the carbon-nitrogen cycle, the carbon nuclei act as catalysts in converting hydrogen into helium. The net effect of each of the two sets of reactions is

$$ 4\,{}^{1}_{1}H \rightarrow {}^{4}_{2}He + 2c^+ + 2\nu_e + 27 \text{ MeV}. $$

PROBLEMS

10.1 Suppose that ${}^{233}_{92}U$ when fissioned with thermal neutrons splits into ${}^{139}_{56}Ba + {}^{86}_{36}Kr + 8$ neutrons. Calculate the energy liberated. The atomic masses in u are ${}^{233}_{92}U = 233.0395$, ${}^{139}_{56}Ba = 138.909$, ${}^{86}_{36}Kr = 85.911$, n = 1.0087.

10.2 (a) Calculate the potential energy in MeV of a system consisting of a spherical ${}^{86}_{36}Kr$ nucleus touching a spherical ${}^{139}_{56}Ba$ nucleus.

(b) Compare the energy obtained in Problem 10.2a with that calculated in Problem 10.1.

10.3 How much coal must be burned to obtain the same amount of energy as would be released in fissioning 1 kg of ${}^{235}_{92}U$? Assume that the energy liberated in each fission of ${}^{235}_{92}U$ is 200 MeV and that the heat of combustion of coal is 8 kcal/g.

10.4 Find the energy that would be released (a) in fissioning 1 kg of ${}^{235}_{92}U$ and (b) in fusing a 1-kg mixture of ${}^{2}_{1}H$ and ${}^{3}_{1}H$.

10.5 What fraction of the kinetic energy of a fission neutron with energy 2 MeV would get transferred in its head-on elastic collision with a moderator when the moderator is (a) deuterium? (b) graphite?

10.6 A submarine requires a 100-MW engine to run for one day. Calculate the mass of $^{235}_{92}U$ required to generate this much energy, assuming there are no losses.

10.7 In any fission process, there are a few delayed neutrons. Why are there no (a) delayed protons? (b) positrons?

10.8 What is the power density (W/kg) in a $^{235}_{92}U$ reactor with a slow neutron flux = 10^{18} neutrons/m^2-sec?

10.9 Estimate the critical mass of a $^{239}_{94}Pu$ sphere, using the data given in Tables 10.1 and 10.2 and making assumptions similar to those in Example 10.5. Given the density of Pu = 17×10^3 kg/m^3.

10.10 In the explosion of the first atom bomb, the energy released was equivalent to the energy generated in exploding 20 kilotons of TNT. Estimate the amount of $^{235}_{92}U$ that would have been consumed during this explosion. Given that 1 kg of TNT produces an energy of 10^6 kcal and the density of U = 18.5×10^3 kg/m^3.

11 Elementary Particle Physics

11.1 INTRODUCTION

We have until now been studying the various physical phenomena involving atoms, molecules, and nuclei in terms of the fundamental constituents of matter, namely, electrons, photons, protons, and neutrons. These constituents are called the *elementary particles* because, to understand the properties of matter, these particles are assumed to be the basic building blocks of matter in much the same way as atoms and molecules are considered to be the building blocks of elements and compounds. In the last thirty years, many new particles have been found. Of these, most have been discovered using a cyclotron or some other accelerator. Thus, about one hundred particles are currently known, and all of them have been dubbed elementary. It is believed that once the nature of such particles is better understood, many of their properties will be explainable in terms of only a few of them. Therefore, these particles are very actively being studied presently. This effort is leading to a deeper understanding of nature. In this chapter, we shall discuss some of the developments such work is bringing about.

11.2 PROPERTIES OF ELEMENTARY PARTICLES

The basic properties of some of the elementary particles are listed in Table 11.1. We shall begin our study using this table. In it, the first particle is the familiar photon. It is a stable neutral particle with rest mass zero and spin quantum number $s = 1$. The reason the photon has zero rest mass can be explained using the theory of relativity. The remaining particles in the table are broadly classified as leptons and hadrons. The hadrons are further categorized as mesons and baryons.

LEPTONS

The neutrino and electron are leptons. As we have already noted, the neutrino was first discovered in the β-particle emission from the nucleus. Subsequently, it has been observed in many nuclear reactions. In fact, there are different kinds of neutrinos, namely, electron-neutrino (ν_e) and muon-neutrino (ν_μ), each of which is stable.

The first elementary particle to be discovered is of course the electron. Therefore, its physical properties have been extensively investigated. It is negatively charged and has a mass 9.11×10^{-31} kg. The mass (m) of an electron is normally expressed in MeV/c^2 obtained from the relation $m = E/c^2$, where E is the energy. Thus,

$$m = 9.11 \times 10^{-31} \times (3 \times 10^8)^2\ \text{J}/c^2 = \frac{9.11 \times 10^{-31} \times 9 \times 10^{16}}{1.6 \times 10^{-13}}\ \text{MeV}/c^2$$

$$= 0.511\ \text{MeV}/c^2. \qquad (11.1)$$

The electron is also a stable particle and does not decay. Its spin, as already noted, is 1/2.

The muon is quite similar to the electron except that it is about two hundred times heavier, is unstable, and decays very quickly (mean life = 2.2×10^{-6} sec) into one electron and two neutrinos. It was first observed in cosmic rays and its detection marked the beginning of the discovery of several new particles. It is negatively charged and its spin is the same as that of the electron.

The four particles, namely, electron-neutrino, muon-neutrino, electron, and muon, are called leptons. Here, we have not discussed the more recently discovered lepton, namely, τ, and the neutrino associated with it.

Antiparticles

When studying the properties of the electron using the special theory of relativity and quantum mechanics, P.A.M. Dirac predicted that the particle electron has an *antiparticle*. The anti-

TABLE 11.1 Some elementary particles and their properties

Particle (symbol)	Mass (MeV/c^2)	Mean life (sec)	Charge (in units of proton charge)	Spin ($\hbar$)
Photon (γ)	0	Stable	0	1
LEPTONS				
Electron-neutrino (ν_e)	0	Stable	0	$\frac{1}{2}$
Muon-neutrino (ν_μ)	0	Stable	0	$\frac{1}{2}$
Electron (e^-)	0.51	Stable	−1	$\frac{1}{2}$
Muon (μ^-)	105.66	2.2×10^{-6}	−1	$\frac{1}{2}$
HADRONS				
Mesons				
π^+	139.57	2.60×10^{-8}	+1	0
π^-	139.57	2.60×10^{-8}	−1	0
π^0	134.96	0.83×10^{-16}	0	0
K^+	493.67	1.24×10^{-8}	+1	0
K^-	493.67	1.24×10^{-8}	−1	0
K^0	497.67		0	0
$\overline{K^0}$	497.67		0	0
η^0	548.8	$< 10^{-18}$	0	0

TABLE 11.1 Some elementary particles and their properties (cont.)

Particle (symbol)	Mass (MeV/c^2)	Mean life (sec)	Charge (in units of proton charge)	Spin ($\hbar$)
Baryons				
Proton (p)	938.28	Stable	1	$\frac{1}{2}$
Neutron (n)	939.57	920	0	$\frac{1}{2}$
Lambda (Λ^0)	1115.60	2.63×10^{-10}	0	$\frac{1}{2}$
Sigma (Σ^+)	1189.36	0.80×10^{-10}	+1	$\frac{1}{2}$
Sigma (Σ^0)	1192.46	5.8×10^{-20}	0	$\frac{1}{2}$
Sigma (Σ^-)	1197.34	1.48×10^{-10}	−1	$\frac{1}{2}$
Xi (Ξ^0)	1314.9	2.90×10^{-10}	0	$\frac{1}{2}$
Xi (Ξ^-)	1321.32	1.64×10^{-10}	−1	$\frac{1}{2}$

particle of the electron is known as the positron. The mass and spin of the antiparticle are the same as those of the particle but its charge is opposite to that of the particle. Thus, the positron is positively charged and has a mass 0.511 MeV/c^2 and a spin 1/2. It was found in the β-particle emission from the nucleus and its discovery was a great step forward in confirming a theoretical prediction, namely, each particle in nature has a corresponding antiparticle. The antiparticle corresponding to the proton is the antiproton with a charge ($-e$) and a mass the same as that of the proton (=938.28 MeV/c^2). Consider now a particle with no charge, e.g., a neutrino and a neutron. Then, the question is: What distinguishes a particle from the antiparticle corresponding to it since both have the same mass, spin, and charge? The answer to this question will be clear as we go along. The photon is its own antiparticle. This is also true of a few other particles.

HADRONS

Mesons

As we know, the force between two electrons or between an electron and a proton is caused by the electromagnetic field. In such a situation, the potential energy V is

$$V(r) \propto 1/r, \tag{11.2}$$

where r is the separation between the two particles. The particle picture of the electromagnetic force is that the photons are exchanged between the particles. In his theory of nuclear forces, H. Yukawa assumed that the potential energy between any two protons in a nucleus is

$$V(r) \propto \exp(-\mu r)/r. \qquad (11.3)$$

The reason he chose the factor $\exp(-\mu r)$ was to ensure that the range of the nuclear force is short (see Fig. 11.1). He named the field causing this potential the mesonic field and suggested that this field should also exhibit a particle-like behaviour in the same way as the electromagnetic field. Further, he showed that the particles, called π-mesons, constituting the field are not massless as photons, but have a mass given by (see Problem 11.7)

$$m = \mu\hbar/c. \qquad (11.4)$$

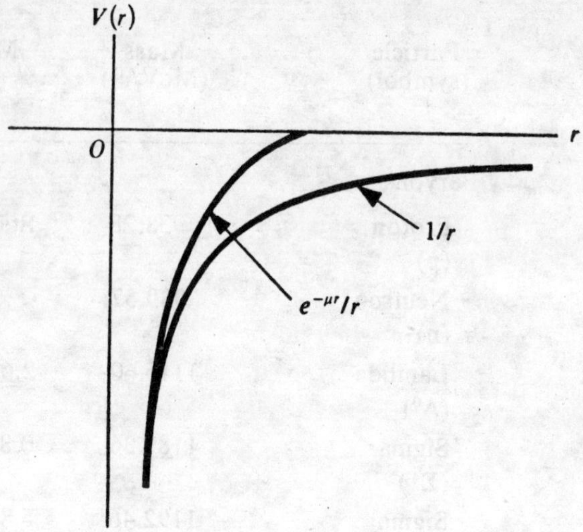

Fig. 11.1 Comparison of Yukawa and Coulomb potentials.

In order to obtain the numerical estimate of the mass of one such particle, he assumed $1/\mu = 1$ F, which approximately is the range of the nuclear force. Substituting this in Eq. (11.4), he estimated the mass of the particle to be 100 MeV/c^2. The subsequent investigations showed that there are three kinds of π-mesons, namely, π^+, π^-, and π^0. Each of these particles decays rapidly, and therefore has a very short life. After the three types of π-mesons were found, an extensive search for more such particles was undertaken both in the cosmic rays and in the collisions occurring in high energy accelerators. Some of the mesons so detected, along with their properties, have also been listed in Table 11.1.

Baryons

The first two baryons (see Table 11.1), i.e., protons and neutrons, are now well known. As we have already learnt, a free neutron is unstable. It has a mean life of 920 sec, and decays into a proton, an electron, and an electron-antineutrino. The other baryons, namely, lambda, sigma's, and chi's, were detected in the cosmic rays and in the experiments using an accelerator.

11.3 EXACT CONSERVATION LAWS

In a collision or a reaction, the total momentum, as by now we know so well, is always conserved. This is a fundamental law. Similarly, the total energy in a reaction is also conserved. The reactions involving the elementary particles obey several such conservation laws. These are the laws of conservation of

(1) energy,

(2) linear momentum,

(3) angular momentum,

(4) charge,

(5) baryon number,

(6) lepton numbers, and

(7) *CPT*.

CONSERVATION OF ENERGY AND LINEAR MOMENTUM

As already stated, in a reaction, the total energy and the linear momentum are conserved. We shall illustrate these conservation laws through an example. Consider a reaction where a π^--meson decays into a μ^--meson and an antineutrino $\bar{\nu}_\mu$, i.e.,

$$\pi^- \rightarrow \mu^- + \bar{\nu}_\mu. \tag{11.5}$$

It should be noted that we have represented the antiparticle here by placing a bar on the symbol of the particle corresponding to it. The energy of the π^--meson is equal to the sum of the energies of the particles μ^- and $\bar{\nu}_\mu$. It is necessary that the rest energy of the π^--meson due to its mass be taken into account in accordance with the special theory of relativity. Thus, if the π^--meson were initially at rest, its total energy would have been

$$E_\pi = m_\pi c^2, \tag{11.6}$$

where m_π is its rest mass. This energy is shared by the two final particles μ^- and $\bar{\nu}_\mu$. Thus,

$$m_\pi c^2 = E_\mu + E_\nu, \tag{11.7}$$

where the total energy of μ^- is given by Eq. (8.77), i.e.,

$$E_\mu = (|\boldsymbol{p}_\mu|^2 c^2 + m_\mu^2 c^4)^{1/2} \tag{11.8}$$

and that of $\bar{\nu}_\mu$ [using $m_{\bar{\nu}} = 0$ and Eq. (8.77)] is

$$E_\nu = |\boldsymbol{p}_\nu| c. \tag{11.9}$$

Here, $\boldsymbol{p}_\mu$, $\boldsymbol{p}_\nu$ are the momenta of μ^-, $\bar{\nu}_\mu$, respectively. The initial momentum of the π-meson is zero because we have assumed it to be at rest. Then, applying the law of conservation of momentum to reaction (11.5), we get

$$0 = \boldsymbol{p}_\mu + \boldsymbol{p}_\nu \tag{11.10}$$

or

$$\boldsymbol{p}_\mu = -\boldsymbol{p}_\nu. \tag{11.11}$$

From Eqs. (11.7) to (11.10), we get

$$(|\boldsymbol{p}_\mu|^2 c^2 + m_\mu^2 c^4)^{1/2} + |\boldsymbol{p}_\mu| c = m_\pi c^2 \tag{11.12}$$

or

$$(|\boldsymbol{p}_\mu|^2 c^2 + m_\mu^2 c^4)^{1/2} = m_\pi c^2 - |\boldsymbol{p}_\mu| c. \tag{11.13}$$

Writing $|\boldsymbol{p}_\mu|$ as p, we have

$$p^2 c^2 + m_\mu^2 c^4 = (m_\pi c^2 - pc)^2 = m_\pi^2 c^4 + p^2 c^2 - 2 m_\pi p c^3$$

or

$$p = \frac{(m_\pi^2 - m_\mu^2)c}{2m_\pi} \quad (11.14)$$

Then, the energy of the neutrino is

$$E_\nu = cp = \frac{(m_\pi^2 - m_\mu^2)c^2}{2m_\pi} \quad (11.15)$$

and that of the μ^--meson is

$$E_\mu = m_\pi c^2 - E_\nu = \frac{(m_\pi^2 + m_\mu^2)c^2}{2m_\pi}. \quad (11.16)$$

In the foregoing example, the energy of the particles μ^- and $\bar{\nu}_\mu$ has been determined in terms of the masses m_π and m_μ. This is characteristic of a reaction finally leading to two particles (two-body decay). The energies of such particles are, in a general n-body decay ($n > 2$), not fixed, but are only constrained by the two conservation laws. In β-decay, the energy of the electron and the antineutrino, as we have already learnt, can take continuous values. This is because it is a three-body decay.

EXAMPLE 11.1 A proton beam hits a target of protons kept at rest and produces a π^0-meson through the reaction $p + p \rightarrow p + p + \pi^0$. Show that the minimum kinetic energy of the incident protons must be greater than $m_\pi c^2$ for this reaction to occur.

From the law of conservation of energy, it may appear that the kinetic energy of an incident proton should be $m_\pi c^2$ in order to produce a π-meson at rest. However, if the incident proton has a total energy $m_p c^2 + m_\pi c^2$, then the total initial energy will be $2m_p c^2 + m_\pi c^2$, showing that the final products (i.e., two protons and a π-meson) of the reaction have to be at rest. This means that the final momentum is zero. The initial momentum, on the other hand, is not zero as the incident beam has a kinetic energy. Therefore, the linear momentum is not conserved. Thus, the kinetic energy of the incident protons has to be larger than $m_\pi c^2$.

QUESTION 11.1 Consider a reaction $A \rightarrow B + C$. If the rest mass of A is less than the sum of the rest masses of B and C, show that the reaction cannot occur, irrespective of the kinetic energy of A.

CONSERVATION OF ANGULAR MOMENTUM

If an isolated system is in a definite state of angular momentum, then, according to the law of conservation of angular momentum, the initial angular momentum is the same as the final angular momentum. The angular momentum of a particle is the sum of its orbital angular momentum and its spin. For example, consider the decay of a K^+-meson at rest, i.e.,

$$K^+ \rightarrow \pi^+ + \pi^0.$$

Initially, the total angular momentum of K^+ is zero since its orbital angular momentum is zero (particle at rest) and the spin of K^+ is zero (see Table 11.1). Thus, the total angular momentum of the system ($\pi^+\pi^0$) must also be zero. Since a π-meson has no intrinsic spin, its orbital angular momentum must be zero.

CONSERVATION OF CHARGE

In a reaction, the total charge is also conserved. For example, an electron cannot decay into neutral particles such as photons and neutrinos. If such a reaction occurs, then it would violate the law of conservation of charge. The electron is the lightest charged particle and since a particle can decay only into particles lighter than itself (i.e., its energy must be conserved), the conservation of charge implies the electron must be a stable particle. In other words, an isolated electron cannot decay.

An isolated π^--meson can go through the following types of decays:

$$\pi^- \to \mu^- + \bar{\nu}_\mu,$$
$$\pi^- \to e^- + \bar{\nu}_e, \tag{11.17}$$
$$\pi^- \to \pi^0 + e^- + \bar{\nu}_e.$$

In each of these reactions, the charge is conserved. Similarly, a free neutron decays through the reaction

$$n \to p + e^- + \bar{\nu}_e. \tag{11.18}$$

QUESTION 11.2 Is the positron a stable particle? Explain.

CONSERVATION OF BARYON NUMBER

As we have seen, the electron, which is the lightest charged particle, is stable because it conserves the charge. The proton is another known stable particle.* The conservation laws we have so far studied do not explain the proton stability since these laws allow, for example, the reactions

$$p \to e^+ + \pi^0,$$
$$p \to e^+ + \gamma. \tag{11.19}$$

But these reactions do not occur in nature. The stability of proton is explained by a new conservation law, known as the law of conservation of baryon number. Each particle is assigned a baryon number B. Thus,

$B = 0$ for the particles photon, leptons, mesons,

$B = 1$ for the particles p, n, Λ^0, Σ^+, Σ^0, Σ^-, Ξ^0, Ξ^-,

$B = -1$ for the antiparticles $\bar{p}$, $\bar{n}$, $\overline{\Lambda^0}$, $\overline{\Sigma^+}$, $\overline{\Sigma^0}$, $\overline{\Sigma^-}$, $\overline{\Xi^0}$, $\overline{\Xi^-}$.

The baryon number is an additive quantum number. That is, the baryon number of a set of particles is the sum of the baryon number of each particle in the set. This is similar to the charge of a system of particles, which is equal to the sum of the charges of the particles in the system. Thus, the baryon number of a proton-antiproton system is $1 + (-1)$, i.e., 0. According to the law of conservation of baryon number, in a reaction, the initial baryon number is the same as the final baryon number. Thus, the reactions

*A recent hypothesis is that the proton may not be stable. However, this is yet to be experimentally verified.

$$\begin{aligned} &p + n \to p + n + \pi^0, \\ &p + n \to p + n + \gamma, \\ &p + n \to p + p + \pi^- \end{aligned} \tag{11.20}$$

are allowed, but the reactions

$$\begin{aligned} &p + n \to e^+ + \gamma, \\ &p + n \to \pi^+ + \pi^0 \end{aligned} \tag{11.21}$$

are forbidden because the final baryon number is zero, whereas the initial baryon number is two. On the other hand, a system $(p\bar{n})$ has the baryon number zero, and hence the reaction

$$p + \bar{n} \to \pi^+ + \pi^0 \tag{11.22}$$

is allowed. This shows that the neutron and the antineutron are very different particles even though both have the same mass and the same charge (charge = 0). As we have already noted, the system (pn) cannot form two π-mesons, but the system $(p\bar{n})$ can, indicating that the two particles n and $\bar{n}$ interact with the proton very differently.

The proton is the lightest baryon, and hence is stable. A decay such as

$$p \to e^+ + \gamma$$

is not allowed because here the initial baryon number is one and the final baryon number is zero.

QUESTION 11.3 What is the baryon number of $p + \overline{\Lambda^0} + \pi^+$**?**

CONSERVATION OF LEPTON NUMBER

It has been found that the interaction of leptons is very different from that of baryons. Just as we have a law of conservation of a baryon, so too we have a law of conservation of a lepton. In fact, for each kind of lepton, there is an associated lepton number and a conservation law corresponding to the lepton number. Accordingly, we have the conservation laws for the electron lepton number (L_e) and the muon lepton number (L_μ). These are

$L_e = 1$ for electron (e^-) and electron-neutrino (ν_e),

$L_e = -1$ for positron (e^+) and electron-antineutrino ($\bar{\nu}_e$),

$L_e = 0$ for all other particles,

$L_\mu = 1$ for muon (μ^-) and muon-neutrino (ν_μ),

$L_\mu = -1$ for antimuon (μ^+) and muon-antineutrino ($\bar{\nu}_\mu$),

$L_\mu = 0$ for all other particles.

The lepton number like the baryon number is also additive. Thus, if we have a two-electron system, then the electron lepton number is 2. According to the laws of conservation of electron lepton and muon lepton numbers, in a reaction, the initial and the final electron lepton and muon lepton numbers must be the same. Thus, the reactions

$$\mu^- \rightarrow e^- + \nu_\mu + \bar{\nu}_e, \tag{11.23}$$

$$\pi^+ \rightarrow \mu^+ + \nu_\mu,$$
$$\pi^+ \rightarrow e^+ + \nu_e \tag{11.24}$$

are allowed, whereas the reactions

$$\mu^- \rightarrow e^- + \gamma,$$

$$\mu^- \rightarrow e^+ + e^- + e^-$$

are forbidden because the final electron lepton number is one, and the initial electron lepton number is zero.

11.4 ORIGIN OF EXACT CONSERVATION LAWS

The conservation laws are intimately connected with some symmetries in nature. In particular, a conservation law applicable to a system results from some symmetry the system possesses. Consider, for example, a particle in a spherically symmetric potential. The angular momentum of such a particle, as we have already noted, is conserved because the potential has a rotational symmetry. In general, it can be shown that the total angular momentum of an isolated system is conserved if the system possesses a rotational symmetry, that is, the physical properties of the system remain the same when it is rotated.

The law of conservation of momentum results when a system has a translational symmetry, that is, the physical properties of the system remain unchanged when it is translated by an arbitrary displacement.

We shall not consider the symmetries associated with the other conservation laws because such a treatment is beyond the scope of this text.

11.5 FUNDAMENTAL INTERACTIONS

The atomic and molecular structures and the properties of a solid, as we have already seen, can be understood in terms of the electromagnetic force between the charged constituents, namely, electrons and protons, of the solid. However, the electromagnetic force does not explain the attractive force between a neutron and a proton or between two neutrons. In Chapter 9, we studied some of the properties of this force and discovered that its nature is different from that of the electromagnetic force.

In Chapter 1, we discussed the four basic kinds of forces, namely, gravitational, electromagnetic, weak, and strong (nuclear). In Table 11.2, we have again summarized the basic properties of these forces. Since the range of these interactions varies widely, a comparison of their absolute strengths is not physically meaningful. For example, between the earth and the moon, the force is predominantly gravitational, the weak and strong interactions, being of extremely short range, do not play any role, and the electromagnetic force is zero because the two bodies are neutral. On the other hand, in a hydrogen atom, the force between a proton and an electron is predominantly electromagnetic. For a hydrogen atom, however, we can compare the strength of the gravitational interaction with that of the Coulomb interaction. Here, the gravitational binding energy is 10^{-40} eV (see Problem 4.4), and the binding energy due to the Coulomb force is $\sim$10 eV. The ratio of the two strengths thus is $\sim 10^{-41}$. This ratio differs from one system to another. We can similarly compare the different interactions by considering some suitable system. Thus, the interaction strengths in Table 11.2 do not have a precise meaning, but are only meant to indicate the wide differences among them.

As is clear from Table 11.2, all particles experience the gravitational interaction. However, in the microscopic world, its strength is negligible as compared to that of the other interactions. Thus, we shall not consider it in our subsequent discussion.

TABLE 11.2 Fundamental interactions

Interaction	Strength	Range	Particles which have interaction
Gravitational	$\sim 10^{-40}$	∞	All particles
Weak	$\sim 10^{-12}$	10^{-18} m	Leptons and hadrons
Electromagnetic	$\sim 10^{-2}$	∞	All particles except neutrinos
Strong	~ 10	10^{-14} m	Hadrons

It has been found that the leptons, antileptons, and photons interact only through the weak force and/or the electromagnetic force. They do not have the strong interaction. If such particles interact through the electromagnetic force, the effect of weak force is normally negligible. All the other particles (hadrons) react through the strong, weak, and electromagnetic interactions. In such a reaction, the most predominant is the strong interaction and the effects of electromagnetic and weak interactions are normally negligible. However, the characteristics of the various interactions are very different and quite often a reaction or decay through the strong interaction is forbidden. In such a situation, the decay occurs through the electromagnetic interaction. Further, if the decay through this interaction is also forbidden, then it occurs via the weak interaction. We shall explain these points in terms of the approximate conservation laws.

11.6 APPROXIMATE CONSERVATION LAWS

There are certain conservation laws which are not obeyed by all the interactions. These laws, known as the approximate conservation laws, are useful in situations where the contribution of one or the other interaction is negligible. Some such laws are the laws of conservation of

(1) isospin (strong),

(2) strangeness (strong and electromagnetic),

(3) parity (strong and electromagnetic),

(4) charge conjugation (strong and electromagnetic),

(5) time-reversal (strong and electromagnetic).

In this listing, the interaction for which a law is valid is stated in parentheses. For example, the law of conservation of isospin is valid only for the strong interaction. In what follows, we shall not consider isospin because of the mathematical complexity involved. Therefore, we shall begin by discussing strangeness.

STRANGENESS

It has been discovered that a large number of reactions involving the hadrons can be understood systematically only when a new quantum number is assigned to each such particle. This is known as the strangeness quantum number (S). The strangeness number assigned to each hadron is as follows:

Hadrons	*Strangeness S*
$K^+, K^0, \overline{\Lambda^0}, \overline{\Sigma^+}, \overline{\Sigma^0}, \overline{\Sigma^-}$	1
$\overline{\Xi^0}, \overline{\Xi^-}$	2
$K^-, \overline{K^0}, \Lambda^0, \Sigma^+, \Sigma^0, \Sigma^-$	−1
Ξ^0, Ξ^-	−2
Other hadrons in Table 11.1	0

The strangeness number is an additive quantum number. In a reaction, if the strangeness is conserved, the reaction has a large cross-section. Consider the reactions

$$\pi^- + p \rightarrow \Sigma^- + K^+, \tag{11.25}$$

$$\pi^- + p \rightarrow \Sigma^+ + K^-. \tag{11.26}$$

Here, the total strangeness of each of (π^-p) and (Σ^+K^-) is zero, whereas that of ($\Sigma^+ + K^-$) is 2. Thus, according to the law of conservation of strangeness, reaction (11.25) is allowed but reaction (11.26) is forbidden. Experimentally, it has been observed that reaction (11.25) has a large cross-section. Reaction (11.26) can, however, occur through weak interaction which does not obey the law of conservation of strangeness. Since the strength of the weak interaction as compared to that of the strong interaction is lower by a factor 10^{13}, the probability of occurrence of this forbidden reaction is negligible.

Let us consider, as another example, the decay of Λ^0 into two particles. It is the lightest baryon with a nonzero strangeness quantum number. The laws of conservation of baryon number and energy demand that the final products must contain a baryon lighter than Λ^0. This means that a proton or a neutron must be produced in this decay. This, as we shall presently see, implies that, in the decay of Λ^0, strangeness cannot be conserved. Consider the strangeness conserving reactions

$$\Lambda^0 \rightarrow p + K^-$$
$$\rightarrow n + \overline{K^0}.$$

In each of these reactions, the energy is not conserved since the sum of the rest masses of the products is larger than the rest mass of Λ^0. Thus, the only way the decay can occur is through the violation of the law of conservation of strangeness. In other words, the decay of Λ^0 into two particles can take place only through the weak interaction. The possible reactions then are

$$\Lambda^0 \rightarrow p + \pi^-$$
$$\rightarrow n + \pi^0.$$

PARITY

To understand the law of conservation of parity, we shall first examine the following question:

Do the laws governing a physical phenomenon hold even when the phenomenon is observed in a mirror? If the answer is yes, then the laws are said to be invariant under mirror reflection. We shall illustrate this through an example.

Consider a particle of mass m, starting from rest, under the action of gravity. By studying its motion, we can determine the law governing its motion as

$$m\frac{d^2z}{dt^2} = -mg, \tag{11.27}$$

where z is the vertical coordinate of the particle measured from the surface of the earth. The mirror image of the motion of the particle is shown in Fig. 11.2. If we study this motion, we would find that the downward acceleration of the image is identical with the downward acceleration of the particle. Thus,

$$m\frac{d^2z'}{dt^2} = -mg, \tag{11.28}$$

where z' is the z-coordinate of the image. Comparing Eqs. (11.27) and (11.28), we find the law of motion is invariant under mirror reflection. The question that now arises is: Is this true of all the physical laws of nature we know? As we shall subsequently see, the law of weak interaction is not, but the laws of strong, electromagnetic, and gravitational interactions are, invariant under mirror reflection.

In our discussion, we shall, for convenience, discuss the invariance of a law in terms of the parity transformation instead of mirror reflection. The former is closely related to the latter. In a parity transformation, the coordinates (x, y, z) of a particle become $(-x, -y, -z)$. In particular, if $\boldsymbol{r}$ refers to a point P in a coordinate frame, then, after the parity transformation, it becomes $-\boldsymbol{r}$ (see Fig. 11.3). Such a transformation is related to the transformation brought

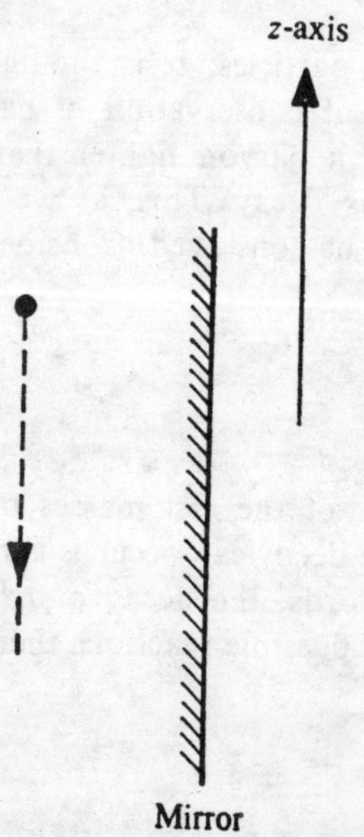

Fig. 11.2 Mirror reflection of falling body.

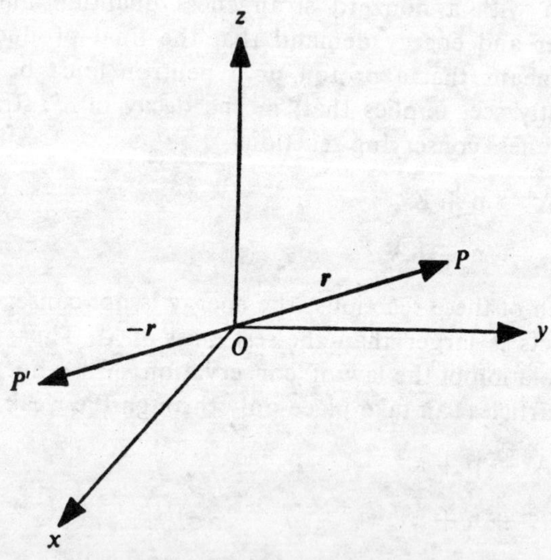

Fig. 11.3 Parity transformation.

about by a reflection in a mirror, as we shall now see. Suppose a mirror is kept on the *x*-*y* plane of a coordinate system. The mirror image of $\boldsymbol{r}$ then is $\boldsymbol{r}'$, where

$$\boldsymbol{r} = x\hat{i} + y\hat{j} + z\hat{k}, \qquad \boldsymbol{r}' = x\hat{i} + y\hat{j} - z\hat{k}.$$

If we now rotate $\boldsymbol{r}'$ about the *z*-axis through 180°, then it will be transformed into $-\boldsymbol{r}$ because the *x*- and *y*-component will change sign, and the *z*-component will remain unchanged. This is shown in Fig. 11.4. Thus, we can think of a parity operation as a product of a reflection about the *x*-*y* plane followed by a rotation about the *z*-axis.

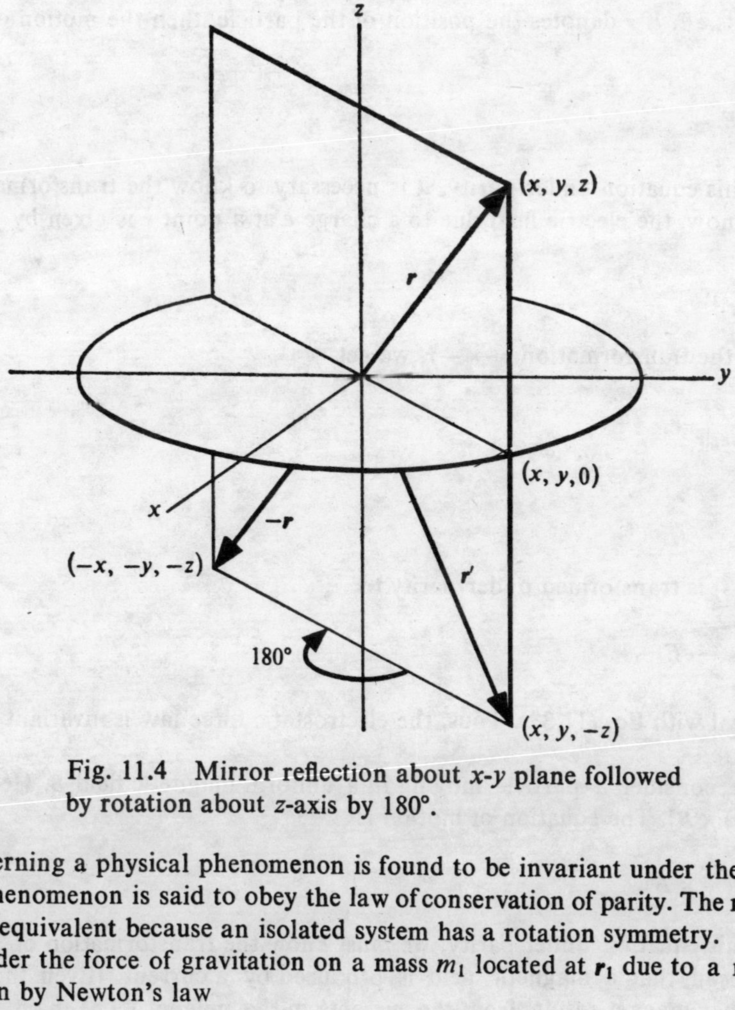

Fig. 11.4 Mirror reflection about *x*-*y* plane followed by rotation about *z*-axis by 180°.

If a law governing a physical phenomenon is found to be invariant under the transformation $\boldsymbol{r} \to -\boldsymbol{r}$, the phenomenon is said to obey the law of conservation of parity. The mirror symmetry and parity are equivalent because an isolated system has a rotation symmetry.

Let us consider the force of gravitation on a mass m_1 located at $\boldsymbol{r}_1$ due to a mass m_2 located at $\boldsymbol{r}_2$. It is given by Newton's law

$$\boldsymbol{F} = -\frac{Gm_1m_2}{|\boldsymbol{r}_1 - \boldsymbol{r}_2|^3}(\boldsymbol{r}_1 - \boldsymbol{r}_2). \tag{11.29}$$

The motion of m_1 is given by

$$m_1\frac{d^2\boldsymbol{r}_1}{dt^2} = -\frac{Gm_1m_2}{|\boldsymbol{r}_1 - \boldsymbol{r}_2|^3}(\boldsymbol{r}_1 - \boldsymbol{r}_2). \tag{11.30}$$

Under the transformation of parity, we have

$$\boldsymbol{r}_1 \to -\boldsymbol{r}_1, \qquad \boldsymbol{r}_2 \to -\boldsymbol{r}_2, \tag{11.31}$$

transforming Eq. (11.30) into

$$-m_1 \frac{d^2\boldsymbol{r}_1}{dt^2} = -\frac{Gm_1m_2}{|\boldsymbol{r}_1 - \boldsymbol{r}_2|^3}(-\boldsymbol{r}_1 + \boldsymbol{r}_2) \tag{11.32}$$

which is the same as Eq. (11.30). Thus, the law of gravitation is invariant under parity.

Now, let us consider a particle with charge e moving in a uniform electric field $\boldsymbol{E}$. The force on the particle is $e\boldsymbol{E}$. If $\boldsymbol{r}$ denotes the position of the particle, then the motion of the particle is given by

$$m\frac{d^2\boldsymbol{r}}{dt^2} = e\boldsymbol{E}. \tag{11.33}$$

To transform this equation under parity, it is necessary to know the transformation of $\boldsymbol{E}$ under parity. As we know, the electric field due to a charge e at a point $\boldsymbol{r}$ is given by

$$\boldsymbol{E} = \frac{e}{4\pi\epsilon_0|\boldsymbol{r}|^3}\boldsymbol{r}. \tag{11.34}$$

Now, if we use the transformation $\boldsymbol{r} \to -\boldsymbol{r}$, we get

$$\boldsymbol{E} \to -\frac{e}{4\pi\epsilon_0|\boldsymbol{r}|^3}\boldsymbol{r}, \tag{11.35}$$

and hence

$$\boldsymbol{E} \to -\boldsymbol{E}. \tag{11.36}$$

Thus, Eq. (11.33) is transformed under parity to

$$-m\frac{d^2\boldsymbol{r}}{dt^2} = -e\boldsymbol{E} \tag{11.37}$$

which is identical with Eq. (11.33). Thus, the electrostatic force law is invariant under the parity transformation.

We shall now consider a particle moving in a uniform magnetic field $\boldsymbol{B}$. Here, the magnetic force is $e[(d\boldsymbol{r}/dt) \times \boldsymbol{B}]$. The equation of motion is

$$m\frac{d^2\boldsymbol{r}}{dt^2} = e\left(\frac{d\boldsymbol{r}}{dt} \times \boldsymbol{B}\right). \tag{11.38}$$

To transform this equation under parity, we must know the transformation of $\boldsymbol{B}$ under parity. For this, we recall that a magnetic field is produced by a current. (Even the magnetic field produced by a bar magnet results from the currents in the magnet.) The magnetic field $d\boldsymbol{B}$ at a point $\boldsymbol{r}$ due to a current element $d\boldsymbol{l}$ at the origin is given by the Biot-Savart law

$$d\boldsymbol{B} = \frac{\mu_0}{4\pi}\frac{d\boldsymbol{l} \times \boldsymbol{r}}{|\boldsymbol{r}|^3}. \tag{11.39}$$

Under the parity transformation $\boldsymbol{r} \to -\boldsymbol{r}$, the current element $d\boldsymbol{l} \to -d\boldsymbol{l}$. This is because (1) a current element is specified by the difference between two coordinate vectors, one for each end point of the element, and (2) the signs of these two coordinate vectors are reversed. Thus,

$$d\boldsymbol{B} \to d\boldsymbol{B}$$

or

$$\boldsymbol{B} \to \boldsymbol{B}. \tag{11.40}$$

We have here obtained an important result, namely, a magnetic field produced by a current does not change under the parity transformation. This is opposite to [see Eq. (11.36)] that obtained for an electric field. Using this result, i.e., transformation (11.40), we find that Eq. (11.38) transforms as

$$-m\frac{d^2\boldsymbol{r}}{dt^2} = e\left(-\frac{d\boldsymbol{r}}{dt} \times \boldsymbol{B}\right) \tag{11.41}$$

which is identical with Eq. (11.38). Thus, again, we see that the magnetic force law is invariant under the parity transformation.

Would the magnetic force law be invariant under the parity transformation if there exists a free magnetic charge (q_m), usually referred to as a magnetic monopole? In this situation, the magnetic field $\boldsymbol{B}$ is given by

$$\boldsymbol{B} \propto \frac{q_m}{|\boldsymbol{r}|^3}\boldsymbol{r}. \tag{11.42}$$

Under the parity transformation, $\boldsymbol{B}$ would transform (in the same manner as an electric field $\boldsymbol{E}$) as

$$\boldsymbol{B} \to -\boldsymbol{B}. \tag{11.43}$$

Equation (11.38) would then transform as

$$-m\frac{d^2\boldsymbol{r}}{dt^2} = e\left[\left(-\frac{d\boldsymbol{r}}{dt}\right) \times (-\boldsymbol{B})\right].$$

This is not the same as Eq. (11.38). Thus, in the presence of a free magnetic monopole, the force law is not invariant under the transformation of parity. There have been various attempts to find a magnetic monopole. However, none of these attempts has been successful. It is clear that if a magnetic monopole is detected, it would be a very important discovery and would have far-reaching effects.

The electromagnetic force has been experimentally found to be invariant under the parity transformation. But this is not true of the weak interaction. In particular, the law of weak interaction governing the β-decay is not invariant under parity. This result was arrived at after conducting an experiment involving the β-decay of cobalt nuclei. Before we describe this experiment, we should like to point out that a cobalt nucleus has a spin angular momentum and undergoes the decay

$${}^{60}_{27}\mathrm{Co} \to {}^{60}_{28}\mathrm{Ni} + e^- + \nu_e. \tag{11.44}$$

In the experiment, the spins of cobalt nuclei were aligned along a specific direction. When these nuclei underwent the β-decay, the electrons emitted in the direction opposite to that of the spins of these nuclei were more in number than those emitted in the direction of the spins. If we choose the spin direction of the cobalt nuclei to be the z-axis, then we can find the number of electrons $N(\theta)\, d\theta$ emitted along an angle lying between θ and $(\theta + d\theta)$, where θ is the angle between the momentum p of each of these electrons and the spin direction J. It has been found that $N(\theta)$ is larger for $\theta > \pi/2$ than for $\theta < \pi/2$. In particular, the data for $N(\theta)$ fits the law

$$N(\theta) = A + B\frac{\boldsymbol{J}\cdot\boldsymbol{p}}{|\boldsymbol{J}|\,|\boldsymbol{p}|} = A + B\cos\theta, \tag{11.45}$$

where A and B are constants. Let us now find out the transformation of law (11.45) under parity. Since $\boldsymbol{J} = \boldsymbol{r}\times\boldsymbol{p}$, and $\boldsymbol{r}\to -\boldsymbol{r}$ and $\boldsymbol{p}\to -\boldsymbol{p}$, we have

$$\boldsymbol{J}\to\boldsymbol{J}, \tag{11.46}$$

$$\boldsymbol{J}\cdot\boldsymbol{p}\to -\boldsymbol{J}\cdot\boldsymbol{p}. \tag{11.47}$$

Thus, the law [i.e., Eq. (11.45)] under the parity transformation becomes

$$N(\theta) = A - B\cos\theta. \tag{11.48}$$

The two equations, namely, (11.45) and (11.48), would have been identical were $B = 0$. However, $B \neq 0$.

The experiment we have described shows that the law of weak interaction governing the β-decay is very different from the laws of gravitation and electromagnetic interaction. Several other experiments have been performed, and it has been found that, in general, a reaction involving the weak interaction violates parity invariance.

QUESTION 11.4 How does an electric dipole moment $\boldsymbol{p}$ transform under parity?

QUESTION 11.5 How does a magnetic moment $\boldsymbol{m}$ transform under parity?

CHARGE CONJUGATION

As we have already seen, each particle has an antiparticle corresponding to it with the same mass but opposite charge, baryon number, strangeness, and lepton number. If a physical law for a set of particles also holds for the antiparticles corresponding to the particles, then the law is said to be invariant under charge conjugation. For example, if the cross-section of the reaction

$$\mathrm{p} + \pi^- \to \Sigma^- + \mathrm{K}^+ \tag{11.49}$$

at a given energy is σ, then the law of charge conjugation invariance implies that the cross-section of the reaction

$$\overline{\mathrm{p}} + \pi^+ \to \overline{\Sigma^-} + \mathrm{K}^-, \tag{11.50}$$

which involves the antiparticles corresponding to the particles in reaction (11.49), at the given energy would also be σ.

We shall not discuss this law in detail. All we should like to mention is that it has been found to be true for the electromagnetic and strong interactions, but not for the weak interaction.

TIME-REVERSAL

A time-reversal transformation requires replacing t by $-t$. A physical law which is invariant under such a transformation is said to be invariant under time-reversal. One physical way of realizing such a transformation is to make a movie of an experiment illustrating a given law and then running the movie backwards. If the latter operation yields the given law, then the given law is invariant under time-reversal. For example, consider a ball, initially at rest, falling through a height h under the action of gravity (see Fig. 11.5). This is described by the equation

$$d^2z/dt^2 = -g, \tag{11.51}$$

where the gravitational force is along the negative z-axis. If we make a movie of this operation and then run the movie backwards, the ball would appear to slow down as it moves up till it comes to rest. In other words, the ball decelerates as it moves up and this process is described by the same law [see Eq. (11.51)], i.e., $d^2z/dt^2 = -g$. Another way of obtaining this result is to replace t by $-t$ in Eq. (11.51). Then,

$$\frac{d}{dt}\left(\frac{d}{dt}\right) = \frac{d}{d(-t)}\,\frac{d}{d(-t)}$$

which is the same as Eq. (11.51).

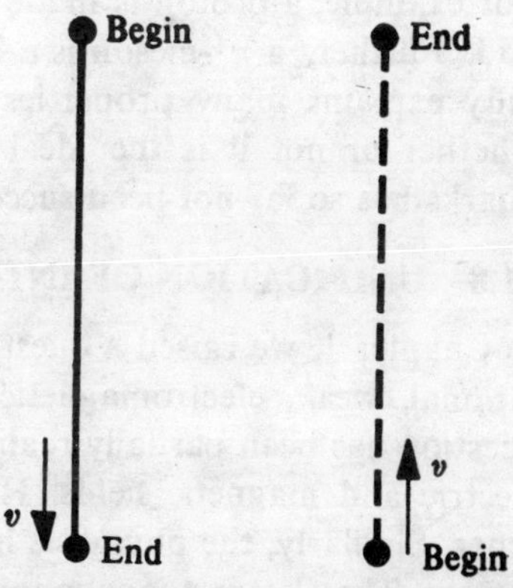

Fig. 11.5 Time-reversed situation of freely falling body.

It has been found that the electromagnetic and strong interactions are invariant under time-reversal. Further, it is believed that the weak interaction is not invariant under time-reversal.

Many laws of macroscopic physics do not conform to the law of time-reversal invariance. For example, according to the second law of thermodynamics, heat flows from a hot body to a cold body, but the time-reversed situation would imply that the heat can flow from the cold body to the hot body without a simultaneous change in the surroundings, thus contradicting the law of increase of entropy. A detailed treatment of time-reversal associated with macroscopic phenomena is outside the scope of this text.

QUESTION 11.6 How do the electric fields and magnetic fields transform under time-reversal?

CPT

The combined operation of charge conjugation (C), parity (P), and time-reversal (T) is referred to as the CPT transformation. It is believed that all interactions in nature are invariant under the CPT transformation. This is an exact conservation law.

11.7 QUARK MODEL

As we noted earlier in this chapter, there are now a very large number of elementary particles. In fact, more than one hundred such particles and antiparticles have been found. This situation is similar to the one that prevailed during the initial stages of the study of chemistry when many new elements were being discovered. The subsequent efforts to systematically arrange the elements resulted in the periodic table which in turn led to the formulation of the atomic structure of matter. Currently, similar attempts are being made to see if the elementary particles can be understood in terms of more simple objects. An outcome of such attempts is the quark model. According to this model, each hadronic particle is made up of basic units called quarks. Thus, to understand the particles in Table 11.1 in terms of this model, we need only three types of quarks, namely, u, d, and s.* The charges of these are $\frac{2}{3}e$, $-\frac{1}{3}e$, and $-\frac{1}{3}e$, respectively. For example, a proton is made up of three quarks, namely, uud, and a neutron is made up of udd. Further, a π^+-meson is a bound state of u and $\bar{d}$ (the antiquark of d). This model successfully explains many properties of the elementary particles, and the investigations to verify whether or not it is the ideal model are continuing. The search for experimentally finding the quarks has so far not been successful.

11.8 UNIFICATION OF INTERACTIONS

In Chapter 1, we raised a question about whether or not the four interactions, namely, gravitational, weak, electromagnetic, and strong, can be related to one another. The answer to this question has been partially realized. Maxwell, in writing down his four equations, unified the electric and magnetic fields. He did so by observing that a variation of one field produces the other. Similarly, the physicists have succeeded in unifying the weak and electromagnetic interactions. The electromagnetic interaction, say, between two electrons, is, as already noted, due to the exchange of photons. Likewise, the weak interaction, for example, between an electron and a neutrino and between a proton and a neutrino, occurs through the exchange of one of a set of particles, called W^+, W^-, and Z^0 particles. For instance, in the β-decay of a neutron, a virtual W^- is emitted, that is,

$$n \rightarrow p + W^- \text{ (virtual)} \rightarrow p + e^- + \bar{\nu}_e.$$

Since the weak interaction has a short range, each of the particles W^+, W^-, and Z^0 is massive. The masses of W^+ and W^- are about 80,000 MeV/c^2 and the mass of Z^0 is about 90,000 MeV/c^2. The existence of these particles was predicted by A. Salam and S. Weinberg and was subsequently confirmed during the proton-antiproton collisions conducted at CERN, Geneva, in 1983.

Theoretically, the strong interaction has been unified with the weak and electromagnetic interactions. This has been done by ensuring that the strong interaction is mediated by a set of particles, called gluons, which, like photons, are massless. However, the presence of gluons has not yet been experimentally confirmed though there is a strong evidence that such particles exist.

PROBLEMS

11.1 Calculate the energy of the π^+-meson emitted in the reaction $\Sigma^+ \rightarrow n + \pi^+$, assuming that the particle Σ^+ is initially at rest.

11.2 A positron of kinetic energy 1 MeV collides with an electron having the same kinetic energy but travelling in the direction opposite to that of the positron. The two particles annihilate to produce two photons. Calculate the energies of these two photons.

*The other types of quarks proposed recently are c (charm), b (beauty), and t (top).

11.3 A ρ^+-meson decays into a π^+- and a π^0-meson. The spin of the ρ^+-meson is known to be 1. What would be the orbital angular momentum of the two pions, assuming the ρ^+-meson to be initially at rest?

11.4 A Σ^+-baryon when at rest decays into a neutron and a π^+-meson. What are the possible values for the orbital angular momentum of the particles n and π^+?

11.5 The reactions

$$n \rightarrow e^+ + e^-, \tag{a}$$

$$n \rightarrow p + e^- + \gamma, \tag{b}$$

$$\Sigma^+ \rightarrow K^- + p + p \tag{c}$$

do not take place. Explain in terms of the conservation laws why this is so.

11.6 Consider the relativistic equation $E^2 = c^2p^2 + m^2c^4$ for a particle. Use the quantum mechanical operators for E and p to show that the wave equation for the relativistic particle is

$$\frac{1}{c^2}\frac{\partial^2 \Phi}{\partial t^2} - \left(\frac{\partial^2}{\partial x^2} + \frac{\partial^2}{\partial y^2} + \frac{\partial^2}{\partial z^2}\right)\Phi + \frac{m^2c^2}{\hbar^2}\Phi = 0,$$

where $\Phi(\mathbf{r}, t)$ is the wave function describing the particle. This equation is known as the Klein-Gordon equation.

11.7 If the wave function $\Phi(\mathbf{r}, t)$ in Problem 11.6 is independent of time, then show that $\Phi(\mathbf{r}, t) = \exp(-\mu r)/r$ is a solution of the Klein-Gordon equation with $\mu = mc/\hbar$. Evaluate the mass of the particle, assuming $\mu = 10^{13}\ \text{cm}^{-1}$.

11.8 Of the reactions

$$\bar{p} + p \rightarrow K^+ + K^0 + \overline{\Lambda^0} + \Sigma^-,$$

$$\bar{p} + p \rightarrow K^+ + K^- + \overline{\Lambda^0} + \Lambda^0,$$

$$\bar{p} + p \rightarrow K^+ + K^- + \Lambda^0 + \overline{\Lambda^0},$$

$$\bar{p} + p \rightarrow K^+ + K^- + K^+ + K^-,$$

which ones will go through the strong interaction?

11.9 Give an example of a reaction to show that Σ^- is not the antiparticle of Σ^+.

11.10 From the three momentum vectors $\boldsymbol{p}_1$, $\boldsymbol{p}_2$, and $\boldsymbol{p}_3$, construct a scalar quantity which changes its sign under the parity transformation.

11.11 If the force on a charged particle moving with a velocity $\boldsymbol{v}$ in an electric field $\boldsymbol{E}$ is proportional to $\boldsymbol{v} \times \boldsymbol{E}$, would the force be invariant under (a) the parity transformation? (b) time-reversal?

11.12 What would be the structure of K^+ and Σ^+ in terms of the quark model? The strangeness quantum number of each of the quarks u and d is zero and that of the quark s is -1.

12

Standard Model of the Universe

12.1 INTRODUCTION

The question of the origin of the universe has been a topic of speculation throughout human history. Since the development of modern science with the discovery that the earth is not the centre of the universe, the physicists and astronomers have been time and again re-examining this question. As a result, a new theory on the origin of the universe, referred to as the *big bang model* or *standard model*, has emerged. This model is an outstanding example of the application of the different concepts of physics we learnt in the preceding chapters.

12.2 GEOGRAPHY OF THE UNIVERSE

The astronomical observations, starting from the sixteenth century, have led the astronomers to a definite picture of the universe. According to this account, the universe contains $\sim 10^{10}$ galaxies, each galaxy having $\sim 10^{10}$–10^{12} stars and extending over $\sim 10^5$ light years (ly). The average separation between the galaxies is $\sim 10^6$–10^7 ly and these are distributed in all directions in space. The average distance between the stars within a galaxy is ~ 10 ly. The present observational universe extends to about 10^{10} ly. We shall discuss in some depth how the picture we have described has emerged.

OUR GALAXY: THE MILKY WAY

If we look at the night sky, we can see a band of light stretching from horizon to horizon. This is known as the *milky way*. Today, we know that this band of light comprises many stars. Our solar system is a member of the milky way (also known as *our galaxy*).

The distance of the sun, the star nearest to us, from the earth was first measured by studying the motion of the planet mars. (For a brief description of the method applied in doing this, see Appendix 17.) The distance from the earth of any other star near it is determined from the angular shift of the star with respect to the stars distant from it, as it revolves around the sun. This method, known as trigonometric parallax, can be used to measure a distance up to 100 ly.

EXAMPLE 12.1 Calculate the maximum distance of a star from the earth that can be measured if the minimum measurable angular shift is 0.03′ and the radius of earth's orbit around the sun is 150×10^9 m.

If r is the distance to the star, then, from Fig. A17.2, we have

$$150 \times 10^9/r = \theta/2,$$

where

$$\theta = \frac{.03}{60 \times 60 \times 57.29} = 1.45 \times 10^{-7} \text{ rad}$$

or

$$r = \frac{300 \times 10^9}{\theta} = 2.07 \times 10^{18} \text{ m} = 218 \text{ ly}.$$

In Table 12.1, we have given the various methods used for measuring the distance, indicating the range of distance and nature of object(s) to which each method is applicable. For the principles underlying the methods, see Appendix 17.

TABLE 12.1 Methods of measuring distance of stars and star clusters

Method	Range (ly)	Object(s)
Trigonometric parallax	100	Nearest stars
Moving-cluster	150	Hyades open cluster
Main sequence-fitting-to-Hyades-cluster	3×10^5	Open clusters
Statistical-calibration-of-luminosity	10^6	RR Lyrae stars
Period-luminosity	10^7	Classical cepheids

For more than a century, the distribution of the number of stars as a function of distance and direction was painstakingly studied. This led to the existing picture of our galaxy (see Fig. 12.1 for its schematic representation). It consists of three major components, namely,

(1) a flat disc of stars, with a radius of about 40,000 ly and a thickness of about 2000 ly,

(2) a central bulge, and

(3) a halo of old stars (i.e., globular clusters), uniformly distributed, with its centre identified as the centre of our galaxy.

The sun is at a distance of 30,000 ly from the centre and lies on the flat disc. This disc has considerable interstellar dust which prevents an optical probe of the centre. However, the astronomers have been studying the central bulge using the X- and γ-rays.

THE KNOWN UNIVERSE

The astronomers of the eighteenth century noticed that the sky contains, besides stars and planets, fuzzy and extended objects, originally called nebulae. Some of these objects were thought to be similar to our galaxy, lying far away from it. The first concrete evidence supporting the presence of galaxies other than the milky way was provided by E. Hubble. Using the 100-inch telescope at Mt. Wilson, he detected a star (in fact, a cepheid variable) in the Andromeda nebula and measured the distance of this nebula to be about one million light years. This distance is over ten times the distance of the farthest known object in our galaxy. More recent measurements suggest that the distance of the Andromeda nebula is close to two million light years.

Currently, this nebula is known as the Andromeda galaxy, the word nebula being reserved for gas and dust cloud. Since Hubble's pioneering work, several nebulae have been studied and many of these have been found to be galaxies like our own galaxy.

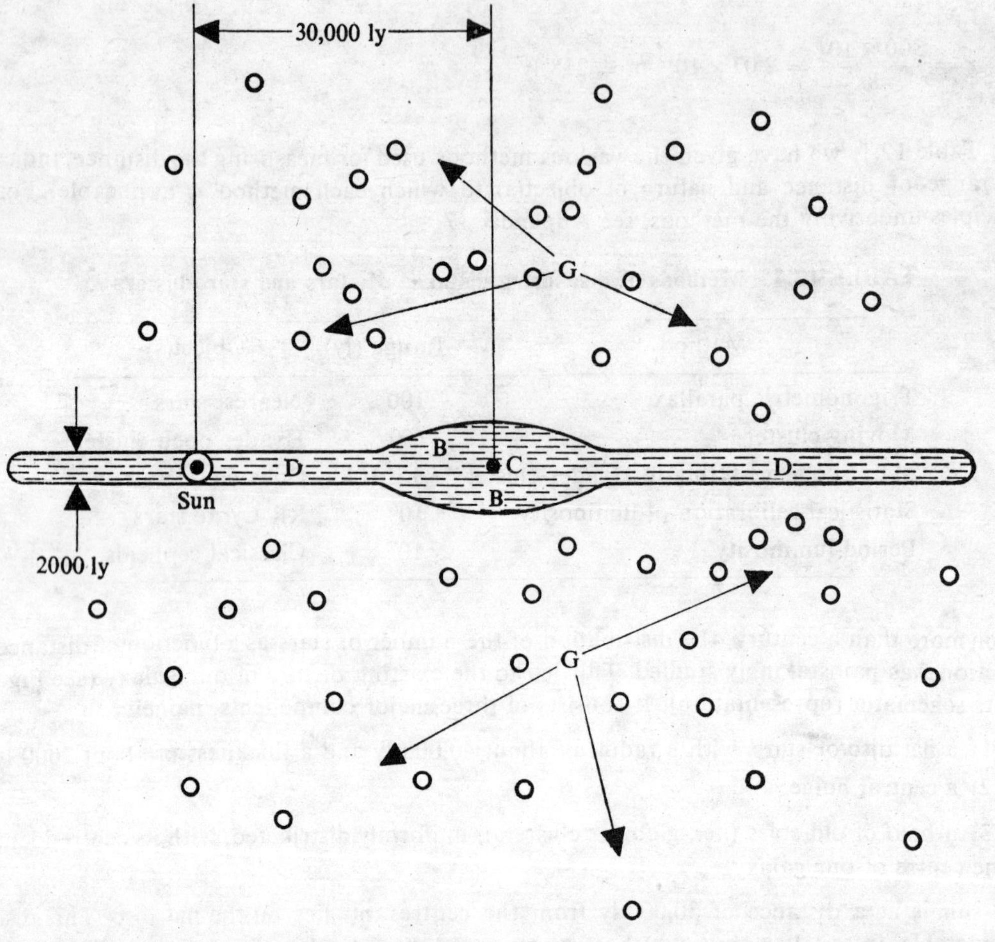

Fig. 12.1 Our galaxy—the milky way. B, central bulge; C, centre of galaxy; D, disc stars; G, globular clusters.

The problem of measuring the distance of the various galaxies has occupied the astronomers since their discovery. As already noted (see Table 12.1), the observation of a classical cepheid helps us measure a distance up to about 10 million light years. Thus, it has been possible to study such nearby galaxies as Magellanic clouds, Ursa Major, M31, and M33, all included in and known as the local group of galaxies. Beyond this distance, it is difficult to identify a cepheid, and therefore other methods, namely, the standard candle method and the standard ruler method (see Table 12.2), have to be applied. The former involves locating an astronomical object of known luminosity (see Appendix 17 for its definition). Such an object could be a nova

TABLE 12.2 Methods of measuring distance of extragalactic objects

Method	Range (ly)	Object
Standard candle	10^7–10^8	Brightest star or nova
Standard ruler	10^7–10^8	Diameter of H II region
Standard candle	10^8–10^{10}	Brightest galaxy in cluster of galaxies
Standard ruler	10^8–10^{10}	Diameter of brightest galaxy in cluster of galaxies

or globular cluster in a galaxy. Once this has been done, its distance is calculated by measuring the energy flux from it.

QUESTION 12.1 Is the distance given by formula (A17.7) an overestimate or underestimate if we take into account the absorption and scattering of light by the interstellar dust?

Applying the standard candle method, the distance of the Virgo cluster of galaxies, which contains the galaxy M87, has been measured to be $\sim 5.5 \times 10^7$ ly. (A cluster of galaxies contains 10^2–10^4 galaxies.) This cluster has 2500 galaxies. Using the brightest galaxy as a standard candle, that is, assuming that the brightest galaxy in a cluster of galaxies is similar to the brightest galaxy in any other cluster, we can measure a distance up to 10^{10} ly.

The standard ruler method requires locating an object (in a galaxy or cluster of galaxies) of known linear dimension. One such object is the large interstellar clouds of ionized hydrogen, known as the H II regions, each having the same size. The true size of these regions has been calibrated by studying them in the galaxies M101 and M51. Once the size of the object is known, its distance can be determined by measuring its angular size. The H II regions have been used to measure a distance up to 2×10^8 ly.

12.3 EXPANSION OF THE UNIVERSE: HUBBLE'S LAW

In the early part of this century, V. M. Slipher of the Lowell Laboratory discovered that the spectral lines of many elements in galaxies are shifted either towards the high frequency (blue shift) or towards the low frequency (red shift) in relation to those observed in the laboratory. This shift is due to the Doppler effect, indicating that the galaxies are either moving towards (blue shift) or away from (red shift) the earth. For example, the Andromeda galaxy is moving towards the earth with a speed of 300 km/sec, whereas the members of the Virgo cluster of galaxies are moving away at a speed of 1000 km/sec. (A mathematical discussion on the Doppler shift for light is given in Appendix 18.) It has been found that, besides the nearby galaxies, e.g., Andromeda galaxy, most galaxies show a red shift. This indicates that the galaxies are moving away from our galaxy. According to Hubble, the velocity $\boldsymbol{v}$ of recession of a galaxy is directly proportional to the displacement $\boldsymbol{r}$ from our galaxy, i.e.,

$$\boldsymbol{v} = H_0 \boldsymbol{r}, \qquad (12.1)$$

where H_0 is referred to as Hubble's constant. This is known as Hubble's law. The numerical value of this constant obtained from the observed data is beset with uncertainties. However, we shall adopt the value $H_0 = 15$ km/sec/million ly.

The simplest explanation of Hubble's law is that the universe is undergoing an expansion. This expansion could be the result of an explosion that may have occurred at what we shall call the beginning of the universe. Prior to this, all the galaxies were together. After the explosion, popularly referred to as the big bang, each galaxy has been following a cosmic path. If we ignore the gravitational pull for the time being, a galaxy, say, A, having a velocity $\boldsymbol{v}_A$ would in a time t be displaced by $\boldsymbol{d}_A$, i.e.,

$$\boldsymbol{d}_A = \boldsymbol{v}_A t. \tag{12.2}$$

Similarly, for a galaxy B, we would have

$$\boldsymbol{d}_B = \boldsymbol{v}_B t. \tag{12.3}$$

On the other hand, if we were travelling on the galaxy A, the galaxy B would be displaced with respect to us by

$$\boldsymbol{d}_B - \boldsymbol{d}_A = (\boldsymbol{v}_B - \boldsymbol{v}_A)t = \boldsymbol{v}_{B/A}t, \tag{12.4}$$

where $\boldsymbol{v}_{B/A}$ is the velocity of B with respect to A. This is the same as Hubble's law. The foregoing derivation shows that this law is valid when the universe is viewed from any galaxy in it. That is, if we were studying the universe from a galaxy other than our own, we would still find that most of the galaxies are receding away. This only shows that our galaxy does not enjoy any special status in the universe. It has been further observed that there are no special directions in the universe, i.e., the galaxies seen in different directions all obey Hubble's law. In other words, the universe appears the same in all directions (isotropic) from any fixed position (homogeneity). Expressed simply, the universe is both isotropic and homogeneous, a statement usually referred to as the cosmological principle. However, to apply this principle, we must really use the cosmological distances, that is, distances of the order of 10^9 ly.

12.4 HISTORY OF THE UNIVERSE

NEWTONIAN COSMOLOGY

What does the present data on the expansion of the universe tell us about the past? If the galaxies are all receding from one another, then, considering Hubble's law and assuming the velocity to be constant, we expect all the galaxies to have been together at a time $t = r/v = H_0^{-1}$ which is 20 billion years. Actually, the velocity of recession earlier must have been larger than that at present since the galaxies have been decelerating because of the gravitational attraction between them. Accordingly, the age of the universe should be less than 20 billion years.

To study the dynamics of the system of galaxies, we require the general theory of relativity, a theory of gravity developed by Einstein in 1916. However, in 1923, G. D. Birkhoff realized that the motion of a galaxy can be obtained without the application of this theory. To do this for a galaxy located at a distance r from us, a sphere of radius r with us as centre is drawn and, assuming that the mass of the rest of the universe consists only of matter inside the sphere, Newton's equation for the galaxy is written. The motion thus calculated is similar to the one arrived at using Einstein's equation of the general theory of relativity, even though there are deep conceptual differences. Thus, if $\rho_m(t)$ is the average mass density in the universe at a time t, then we have

$$m_g \frac{d^2\mathbf{r}(t)}{dt^2} = -m_g \frac{GM}{r^2}\hat{r}, \tag{12.5}$$

where G is the gravitational constant, $M = (4\pi/3)\rho_m(t)r^3(t)$, and m_g is the mass of the galaxy. If no mass is created, M remains constant as $\mathbf{r}$ changes. Therefore, the total energy E of the galaxy is the sum of its kinetic and potential energies, i.e.,

$$E = \tfrac{1}{2}m_g v^2 - m_g MG/r, \tag{12.6}$$

which is a constant of motion. Using Hubble's law, we can write Eq. (12.6) for the present time $t = t_0$, with $v(t_0) = H_0 r(t_0)$ and $\rho_m(t_0) = \rho_{m_0}$, as

$$E = m_g[\tfrac{1}{2}H_0^2 r(t_0)^2 - (4\pi/3)G\rho_{m_0} r^2(t_0)]. \tag{12.7}$$

If $E > 0$, the galaxy will move away to infinity. If $E < 0$, it will come to rest and fall back to the origin, in much the same way as an object with a velocity less than the escape velocity falls back on the earth. The critical situation is $E = 0$ in which case the critical value of the mass density is

$$\rho_{m_0}^{\text{cr}} = 3H_0^2/(8\pi G) \sim 5\times 10^{-27}\ \text{kg/m}^3. \tag{12.8}$$

If the density is greater than the critical value $\rho_{m_0}^{\text{cr}}$, the galaxy is bound in the universe and will eventually fall back. On the other hand, if the density is less than the critical value, the galaxy will move away, and thus keep the universe expanding. Based on the astronomical observations, the present mass density has been estimated to be $\sim 5\times 10^{-28}$ kg/m^3, about ten times less than the critical value. This seems to indicate that the universe will go on expanding for ever. However, it is likely that we have not been able to detect some of the mass present in the universe and the mass density is larger than the current estimate. If this is so, we shall then need to consider the other possibilities.

QUESTION 12.2 Is the distance to an astronomical object measured using the standard candle method the distance to the current location of the object or the distance at the time the light left the object? Remember, the universe is expanding.

EXAMPLE 12.2 Calculate the age of the universe, assuming that the present mass density is given by the critical value $3H_0^2/(8\pi G)$.

From Eq. (12.6), we have $v^2 = 2MG/r$ or

$$dr/dt = (2MG)^{1/2}/r^{1/2}.$$

Integrating this equation from $t = 0$ to the present time t_0, we get

$$\tfrac{2}{3}[r^{3/2}(t_0) - r^{3/2}(0)] = (2MG)^{1/2}t_0.$$

Using $r(0) = 0$ and $M = (4\pi/3)\rho_m(t_0)r^3(t_0)$, we obtain

$$t_0 = \tfrac{2}{3}\left[\frac{3}{8\pi G\rho_m(t_0)}\right]^{1/2}.$$

For $\rho_m(t_0) = 3H_0^2/(8\pi G)$, we have

$$t_0 = 2/(3H_0). \tag{12.9}$$

SPACE-TIME STRUCTURE

The big bang theory seems to imply that the maximum distance any galaxy can be currently away from us would be at the most t_0 light years if t_0 is the present time of the universe since the big bang. If there are no galaxies or stellar objects other than those that have been observed, is there a material edge to the universe then? If yes, is not the cosmological principle violated? If there are galaxies beyond t_0 light years and if all the galaxies started together, how did they get there without travelling faster than light? These questions can be answered if we consider the space-time curvature, a concept that is essential for understanding Einstein's general theory of relativity. According to this concept, space is not euclidean but curved and its curvature depends on time. Most cosmologists believe that there indeed are galaxies beyond t_0 light years from our galaxy, although we cannot observe them.

To understand the concept that the space is curved, we shall use an analogy. Imagine a two-dimensional world consisting of a surface identical with that of a sphere and assume that this surface has matter, say, in the form of two-dimensional galaxies. We also assume that the galaxies can freely move on the surface. Further, imagine that this sphere has an exceedingly small radius to start with (i.e., at $t = 0$) and expands as the time passes. In other words, its radius of curvature r is a function of time. Thus, with the passage of time, any two galaxies, which were together at the time $t = 0$, would, as seen in three dimensions, seem to move in different directions and remain on the surface. When viewed from the surface, the galaxies would appear to move away from each other. If θ is the angle between the two directions, then the distance $d(t)$ between the two galaxies, measured along the surface, would be

$$d(t) = r(t)\theta. \tag{12.10}$$

The relative velocity $v(t)$ is given by

$$v(t) = \theta\frac{dr(t)}{dt} = \frac{1}{r(t)}\frac{dr(t)}{dt}d(t) = H(t)d(t), \tag{12.11}$$

leading to Hubble's law with $H(t_0) = H_0$. It should be noted that the galaxies will be uniformly distributed over the surface. Since there is no material edge on the surface, the sphere will expand in the third dimension, enlarging the surface.

The foregoing ideas can be extended to three-dimensional space, except that here we would need to use tensor calculus. The result so obtained is similar to that arrived at in the analogy we have given, i.e., the universe is expanding and the cosmological principle is satisfied. As already noted, the universe looks the same in any direction from any position. However, the universe is curved and its radius of curvature is a function of time. In the expanding universe, this radius increases with time.

The equations representing the time-dependence of curvature are based on the distribution of mass and were formulated by Einstein. Their solution for the expanding universe, obtained by A. Friedmann, is identical with the one we have already derived, namely, (12.5), except that $r(t)$ should not be treated as the distance to a galaxy, but related to the radius of curvature of the space. It can be shown that, for $E \geqslant 0$, $r(t)$ increases, whereas, for $E < 0$, it first increases to a maximum and then decreases to zero. Further, for $E \geqslant 0$, the universe is expanding for ever and has an infinite volume, but, for $E < 0$, it is closed and has a finite volume. As the universe expands, its volume increases homogeneously and isotropically. Recall that a two-dimensional spherical surface (see the analogy given earlier in this section) moves to a new position when seen in three dimensions. Similarly, a three-dimensional volume can be imagined to be moving to a new location when seen in the unobservable fourth dimension.

In terms of the foregoing language, the red shift occurs when the universe expands. Such an expansion implies the distance between any two points increases. If the distance increases by a factor f, then the wavelength λ of a wave becomes λf.

EXAMPLE 12.3 Calculate the red shift resulting from a cosmological expansion and relate $\dot{r}(t)$ to Hubble's constant at a time t.

Let an electromagnetic wave emitted by a galaxy at a time t_1 reach another galaxy at a time t_2. If the wavelength at t_1 is λ_1, then, by the time the wave reaches the other galaxy, the wavelength would have changed to λ_2 given by

$$\lambda_2 = \lambda_1 r(t_2)/r(t_1).$$

Expanding $r(t_2)$ in terms of $r(t_1)$ and substituting the expansion in this equation, we get, neglecting the higher order terms in the expansion,

$$\lambda_2 = \lambda_1\left[1 + (t_2 - t_1)\frac{\dot{r}(t_1)}{r(t_1)}\right]. \tag{12.12}$$

If we attribute this shift in wavelength to the Doppler effect, we obtain, using Eq. (A18.4) for $v \ll c$,

$$\frac{v}{c} = (t_2 - t_1)\frac{\dot{r}(t_1)}{r(t_1)}$$

or

$$v = d\frac{\dot{r}(t_1)}{r(t_1)}. \tag{12.13}$$

Comparing Eqs. (12.13) and (12.1), we find $H_0 = \dot{r}(t_0)/r(t_0)$.

COSMIC MICROWAVE BACKGROUND RADIATION

In Section 6.7, we mentioned that A. A. Penzias and R. W. Wilson had measured the background radiation from space. The wavelength of this radiation, as already noted, lies in the microwave region. The experiments that have thereafter been carried out, using the wavelength ranging from 73.5 cm to 0.33 cm, to measure the temperature of the background radiation have shown that this temperature is 3 K. Further, this radiation, being isotropic, must have an origin far larger than the galactic volume. In what follows, we shall see how the presence of this radiation gives us a clue to the nature of the beginning of the universe.

To start with, we shall study the effect of the expansion of the universe on the microwave radiation. The number of photons in a volume V at a temperature T in the wavelength interval λ to $(\lambda + d\lambda)$ is

$$dN = \frac{8\pi V}{\lambda^4}\frac{1}{\exp\,[hc/(kT\lambda)] - 1}\,d\lambda. \tag{12.14}$$

If the length scale of the universe changes by f due to an expansion, then the new volume $V' = f^3V$ and the new wavelength $\lambda' = f\lambda$. The corresponding number of photons dN' is

$$dN' = \frac{8\pi V'}{\lambda'^4} \frac{d\lambda'}{\exp[hc/(kT'\lambda')] - 1}. \tag{12.15}$$

Since during an expansion the photons are not absorbed or created, $dN' = dN$. Comparing Eqs. (12.14) and (12.15), we see $T'\lambda' = T\lambda$ or

$$T' = T/f. \tag{12.16}$$

This shows that, as the universe expands, Planck's distribution law does not change, but the temperature varies by T/f (i.e., radiation temperature decreases). This means that the radiation temperature must have been higher during the early part of the formation of the universe.

The energy density (see Section 6.7) is given by

$$U = aT^4, \tag{12.17}$$

where $a = 8\pi^5k^4/(15h^3c^3) = 7.56 \times 10^{-16}$ J/m³. The mass density is then defined as

$$\rho_{rad} = U/c^2 = 8.4 \times 10^{-33}T^4. \tag{12.18}$$

If in this relation we substitute $T = 3$ K, we get the existing radiation density ($= 6.8 \times 10^{-31}$ kg/m³), showing that the radiation density is smaller by a factor 1.36×10^{-3} than the matter density, i.e., the density is dominated by matter. However, this must have not been so at the beginning of the universe. To see this, we note, from Eqs. (12.17) and (12.18),

$$\rho_{rad}(t) \propto T^4 \propto 1/f^4. \tag{12.19}$$

On the other hand,

$$\rho_m(t) \propto 1/V \propto 1/f^3. \tag{12.20}$$

Thus, the ratio $x(t) = \rho_{rad}(t)/\rho_m(t) \propto 1/f$. Taking the present time as standard, we have

$$x(t)/x(t_0) = 1/f \tag{12.21}$$

with $x(t_0) = 1.36 \times 10^{-3}$. At the beginning of the universe (i.e., $t \ll t_0$), $f \ll 1$, and therefore $x(t) = x(t_0)/f > 1$. Thus, the radiation density dominated the matter density.

EXAMPLE 12.4 The mass density at t is proportional to $r(t)^{-n}$, where $n = 3$ for the matter-dominated era and $n = 4$ for the radiation-dominated era [see Eqs. (12.19) and (12.20)]. Show that the time-dependence of the mass density is given, for the beginning of the universe, by $\rho(t) = 12/(8\pi n^2Gt^2)$.

From Eq. (12.6) and $v(t) = H(t)r(t)$, we have

$$E = mr^2(t)\left[\frac{H^2(t)}{2} - \frac{4\pi}{3}\rho(t)G\right], \tag{12.22}$$

where $H(t)$ is Hubble's constant at the time t. Since E is a constant and

$$\rho(t)r^2(t) \propto r(t)^{2-n} \to \infty \quad \text{as} \quad r \to 0 \qquad \text{for } n = 3, 4,$$

we must have, for $t \ll t_0$,

$$H^2(t) \to (8\pi/3)\rho(t)G. \tag{12.23}$$

Thus, for $t \ll t_0$,

$$dr/dt = v(t) = H(t)r(t) \propto [\rho(t)r^2(t)]^{1/2}$$

or

$$dr/dt = kr(t)^{1-n/2}, \tag{12.24}$$

where k is a proportionality constant. Integrating Eq. (12.24) between the limits t_1 and t_2 ($t_1 < t_2$), we obtain

$$t_2 - t_1 = \frac{2}{n}\left[\frac{r(t_2)}{v(t_2)} - \frac{r(t_1)}{v(t_1)}\right] = \frac{2}{n}\left(\frac{3}{8\pi G}\right)^{1/2}\left[\frac{1}{\{\rho(t_2)\}^{1/2}} - \frac{1}{\{\rho(t_1)\}^{1/2}}\right]. \tag{12.25}$$

Choosing $t_1 = 0$ and $t_2 = t$, we get the expression connecting the age of the universe and the mass density, namely,

$$\rho(t) = 12/(8\pi n^2 G t^2). \tag{12.26}$$

Consider now the radiation-dominated universe. Here, using Eqs. (12.17), (12.18), and (12.26) and $\rho_{rad}(t) = \rho(t)$, we get

$$T = \left(\frac{45h^3c^5}{256\pi^6k^4G}\right)^{1/4}\frac{1}{\sqrt{t}} \tag{12.27}$$

$$= \frac{1.52 \times 10^{10}\ \text{K}}{\sqrt{t}} \tag{12.28a}$$

if t is in seconds. Thus, if $t = 1$ sec, the temperature of the universe would be 1.52×10^{10} K, and so on. In the derivation of Eq. (12.28a), we assumed that only the photons (i.e., radiation) are present. But if we assume that other massless particles, e.g., neutrinos, are also present, then expression (12.28a) would be close to

$$T = 1.04 \times 10^{10}\ \text{K}/\sqrt{t}. \tag{12.28b}$$

12.5 THE FIRST TWO HUNDRED SECONDS AFTER THE BIG BANG

We are now in a position to describe the development of the universe after the big bang. For this, we shall start with the picture that exists at one-hundredth of a second after the big bang. Also, we shall restrict our attention to the particles baryons, e^+, e^-, ν_e, $\bar{\nu}_e$, and photons since assuming the presence of other species does not significantly change the picture.

In the time interval 10^{-1} sec $> t \geq 10^{-2}$ sec, the temperature of the universe is in the range 10^{11} K and the average kinetic energy per particle is $\sim$10 MeV. Since the baryons have a rest mass $\sim 10^3$ MeV, these have nonrelativistic velocities. On the other hand, e^+ and e^- have a rest mass ~0.5 MeV, and hence travel almost at the speed of light. Further, ν_e, $\bar{\nu}_e$, and γ, being massless, travel with the speed of light. The kinetic energy is sufficient for the reaction

$$\gamma + \gamma \rightarrow e^+ + e^- \tag{12.29}$$

to occur. Since the matter and radiation densities are very large and different particles collide with one another, thermal equilibrium is established among all species of matter. Even ν_e and $\bar{\nu}_e$, which interact only through the weak force, are in thermal equilibrium, leading to the reactions

$$p + e^- \rightarrow n + \nu_e, \tag{12.30}$$

$$p + \bar{\nu}_e \rightarrow n + e^+. \tag{12.31}$$

At $t = 1$ sec, the temperature drops to 10^{10} K and the average kinetic energy of a particle is ~1 MeV. At this temperature, the particles ν_e and $\bar{\nu}_e$ behave like free particles and are no longer in thermal equilibrium with the rest of the matter. At $t = 10$ sec, the temperature drops to 5×10^9 K and the average kinetic energy per particle is below the threshold energy for reaction (12.29) to take place. Thus, most of e^-e^+ annihilate, leaving behind a 'small' amount of electrons (which forms part of the universe today). At $t = 200$ sec, the universe cools down (the temperature being 10^9 K) to such an extent that the first nucleosynthesis takes place. In particular, the reaction

$$p + n \rightarrow d + \gamma \tag{12.32}$$

occurs. At a higher temperature, the energetic photons present in the radiation had disintegrated the deuterons through the inverse reaction $d + \gamma \rightarrow p + n$. At this temperature (i.e., 10^9 K), not enough energetic photons (their energy being 2 MeV) exist for this inverse reaction to be significant. At this time, the ratio (as has been shown by detailed calculations) of the number of neutrons to the number of protons is 13 : 87. That is, there are more protons than neutrons, because (1) the neutrons, being heavier, have a higher energy, and hence the Boltzmann factor $e^{-E/(kT)}$ reduces their number, and (2) some neutrons are converted into protons through β-decay. Very soon, further nucleosynthesis takes place, converting most of deuterium into helium.

Now, there are no free neutrons, the reason being that all the neutrons are in a bound state and exist as helium nuclei. Since there are two protons and two neutrons in a He nucleus, the fraction by weight of He to hydrogen is 26 : 74.

The universe expands further and continues to cool for 7×10^5 years when the atoms are formed, the temperature being about 3000 K (see Problem 12.9). At this stage, the photons are decoupled from the matter, because their scattering or absorption becomes negligible. The photon gas expands and its temperature is not the same as the rest of the matter. The matter, consisting mostly of hydrogen and helium, forms galaxies. Much later, the other stellar objects are formed.

The helium, produced within minutes after the universe began, forms part of the matter in galaxies and stars. The measurement of the ratio of He to hydrogen confirms that it indeed is 26 : 74, indicating that the standard model of the universe gives the correct picture.

PROBLEMS

12.1 A star of luminosity L at a distance r_0 will deliver an energy flux $e_0 = L/(4\pi r_0^2)$. Another star of the same luminosity at a distance $r < r_0$ will deliver an energy flux $e > e_0$. Suppose the number density of stars of luminosity L is $n(L)$, $n(L)$ being independent of the spatial coordinates. Prove that the number of stars of luminosity L with $e > e_0$ is

$$N_L(e > e_0) = \frac{n(L)L^{3/2}}{3(4\pi)^{1/2}e_0^{3/2}}.$$

12.2 Writing $D = r/r_0$ and $\tau = H_0 t$ when $r_0 = r(t_0)$, $H_0 = H(t_0)$, t_0 being the present time, show that Eq. (12.6) can be expressed as

$$\left(\frac{dD}{d\tau}\right)^2 - \frac{\Omega_0}{D} = 1 - \Omega_0,$$

where $\Omega_0 = 8\pi G\rho_m(t_0)/H_0^2$.

12.3 Writing

$$\eta = \frac{|1 - \Omega_0|}{\Omega_0}D, \qquad \theta = \frac{|1 - \Omega_0|^{3/2}}{\Omega_0}\tau,$$

show that the equation in Problem 12.2 becomes

$$(d\eta/d\theta)^2 - 1/\eta = \pm 1,$$

where the right-hand side is $+1$ for $\Omega_0 < 1$ (bound universe) and -1 for $\Omega_0 > 1$ (unbound universe). This equation can be solved by substituting in it $\eta = \sin^2 \alpha/2$ for $\Omega_0 < 1$ and $\eta = \sinh^2 (\alpha/2)$ for $\Omega_0 > 1$. Show that the solutions so obtained are

$$\eta = \tfrac{1}{2}(1 - \cos \alpha), \quad \tau = \tfrac{1}{2}(\alpha - \sin \alpha) \qquad (\Omega_0 < 1),$$

$$\eta = \tfrac{1}{2}(\cosh \alpha - 1), \quad \tau = \tfrac{1}{2}(\sinh \alpha - \alpha) \qquad (\Omega_0 > 1).$$

12.4 Use the result of Problem 12.3 to show that the present age of the universe t_0 is

$$t_0 = \frac{\Omega_0}{2(\Omega_0 - 1)^{3/2}}(\alpha_0 - \sin \alpha_0),$$

where

$$\cos \alpha_0 = \frac{2 - \Omega_0}{\Omega_0} \quad \text{for } \Omega_0 > 1,$$

and

$$t_0 = \frac{\Omega_0}{2(1 - \Omega_0)^{3/2}}(\sinh \alpha_0 - \alpha_0),$$

where

$$\cosh \alpha_0 = \frac{2 - \Omega_0}{\Omega_0} \quad \text{for } \Omega_0 < 1.$$

12.5 Assuming that the energy distribution for the ν_e's at a temperature T is given by the Fermi-Dirac distribution, show that, as the universe expands, the distribution remains the same with the temperature changing from T to $T' = T/f$, where f is the scale factor associated with the expansion.

12.6 For the universe dominated by matter ($n = 3$) or radiation ($n = 4$), show that $r(t) \propto t^{2/n}$.

12.7 Use Eq. (12.21) and Problem 12.6 to show that, at $t_1 \sim 5 \times 10^{-5} t_0$, the ratio $x(t_1) = 1$, assuming the universe to be matter-dominated during the period t_1 to t_0.

12.8 Calculate the neutron-to-proton ratio, assuming it arises solely due to the Boltzmann factor $N_n/N_p = \exp[-(m_n - m_p)c^2/(kT)]$, considering that the baryons are at rest. Take $T = 10^{10}$ K and $(m_n - m_p)c^2 = 1.3$ MeV.

12.9 Calculate the time taken by the universe to cool to 3000 K, assuming the universe to be radiation-dominated.

Appendix 1

Electromagnetic Waves

MAXWELL'S EQUATIONS

Light, X-rays, and gamma-rays are all examples of the electromagnetic waves. An electromagnetic wave represents a solution of Maxwell's equations. These equations give us the relations between the space and time derivatives of the electric and magnetic fields. In this appendix, we shall consider only the plane electromagnetic waves.

Maxwell's equations in free space are

$$\nabla \cdot \boldsymbol{E} = \text{div } \boldsymbol{E} = \frac{\partial E_x}{\partial x} + \frac{\partial E_y}{\partial y} + \frac{\partial E_z}{\partial z} = 0, \tag{A1.1}$$

$$\nabla \cdot \boldsymbol{B} = \text{div } \boldsymbol{B} = \frac{\partial B_x}{\partial x} + \frac{\partial B_y}{\partial y} + \frac{\partial B_z}{\partial z} - 0, \tag{A1.2}$$

$$\nabla \times \boldsymbol{E} = \text{curl } \boldsymbol{E} = -\partial \boldsymbol{B}/\partial t$$

or

$$(\frac{\partial E_z}{\partial y} - \frac{\partial E_y}{\partial z})\hat{i} + (\frac{\partial E_x}{\partial z} - \frac{\partial E_z}{\partial x})\hat{j} + (\frac{\partial E_y}{\partial x} - \frac{\partial E_x}{\partial y})\hat{k} = -\frac{\partial}{\partial t}(B_x\hat{i} + B_y\hat{j} + B_z\hat{k}) \tag{A1.3}$$

and

$$\nabla \times \boldsymbol{B} = \text{curl } \boldsymbol{B} = \mu_0\epsilon_0(\partial \boldsymbol{E}/\partial t)$$

or

$$(\frac{\partial B_z}{\partial y} - \frac{\partial B_y}{\partial z})\hat{i} + (\frac{\partial B_x}{\partial z} - \frac{\partial B_z}{\partial x})\hat{j} + (\frac{\partial B_y}{\partial x} - \frac{\partial B_x}{\partial y})\hat{k} = \mu_0\epsilon_0\frac{\partial}{\partial t}(E_x\hat{i} + E_y\hat{j} + E_z\hat{k}). \tag{A1.4}$$

Here, $\hat{i}, \hat{j}, \hat{k}$ are the unit vectors along the x-, y-, z-axis, and μ_0 and ϵ_0 are the permeability and permittivity of the vacuum, which satisfy the relation $\mu_0\epsilon_0 = 1/c^2$, where c is the velocity of light. As can be noticed from Eqs. (A1.3) and (A1.4), a time-varying magnetic field produces an electric field and a time-varying electric field generates a magnetic field. It is possible to so combine two time-varying magnetic and electric fields that one field produces the other which not only varies with time but also sustains the first. This resulting field is an electromagnetic wave. We shall not prove these statements mathematically, but rather assume that an electromagnetic wave can be formed and show that this wave is consistent with Maxwell's equations.

Suppose an electromagnetic wave of wavelength λ is travelling along the z-axis. Then, each of the electric and magnetic fields constituting the wave has a sinusoidal form and travels along the z-axis with a velocity c. Accordingly, we can write

$$\boldsymbol{E} = \boldsymbol{E}_0 \sin (kz - \omega t), \tag{A1.5}$$

$$\boldsymbol{B} = \boldsymbol{B}_0 \sin (kz - \omega t + \phi), \tag{A1.6}$$

where $k = 2\pi/\lambda$ and $\omega = 2\pi\nu = 2\pi c/\lambda = kc$. As can be noticed, we have introduced a phase difference ϕ in $\boldsymbol{B}$ but not in $\boldsymbol{E}$. This can be done because it is possible to so choose the origin of the z-axis that the electric field at $z = 0$, $t = 0$ is zero. Using Eq. (A1.1) and realizing that $\boldsymbol{E}$ depends only on z and t, we get, from Eq. (A1.5), $\partial E_z/\partial z = 0$ or

$$E_z = \text{constant}. \tag{A1.7}$$

If we assume there are no constant electric fields in the entire space, then $E_z = 0$. Similarly, we can prove that $B_z = 0$. Thus, we obtain our first important result, namely, the electric and magnetic fields lie in the x-y plane when the wave is moving along the z-axis. In other words, the wave is transverse. From Eqs. (A1.3), (A1.5), and (A1.6), we get

$$(-E_{0y}k\hat{i} + E_{0x}k\hat{j}) \cos (kx - \omega t) = (B_{0x}\hat{i} + B_{0y}\hat{j}) \cos (kx - \omega t + \phi), \tag{A1.8}$$

where E_{0x} and B_{0x} are the x-components of $\boldsymbol{E}_0$ and $\boldsymbol{B}_0$, and E_{0y} and B_{0y} are the y-components of $\boldsymbol{E}_0$ and $\boldsymbol{B}_0$, respectively. Equating the coefficients of $\hat{i}$ in the two sides of Eq. (A1.8), we have

$$E_{0y}k \cos (kx - \omega t) = -B_{0x} \cos (kx - \omega t + \phi). \tag{A1.9}$$

This equation can be satisfied for all values of x and t only if

$$\phi = 0. \tag{A1.10}$$

Thus,

$$E_{0y} = -(\omega/k)B_{0x} = -cB_{0x}. \tag{A1.11}$$

Similarly, equating the coefficients of $\hat{j}$ in the two sides of Eq. (A1.8), we obtain

$$E_{0x} = cB_{0y}. \tag{A1.12}$$

Calculating the scalar product of $\boldsymbol{E}_0$ and $\boldsymbol{B}_0$, we find

$$\boldsymbol{E}_0 \cdot \boldsymbol{B}_0 = E_{0x}B_{0x} + E_{0y}B_{0y} + E_{0z}B_{0z} = c(B_{0y}B_{0x} - B_{0x}B_{0y} + 0) = 0. \tag{A1.13}$$

Thus, we have proved that $\boldsymbol{E}$ and $\boldsymbol{B}$ are mutually perpendicular and that they oscillate in phase (since $\phi = 0$). Suppose the electric field $\boldsymbol{E}$ points along the x-axis, that is, $\boldsymbol{E}_0$ has only the x-component. Then (since $E_{0x} = E_0$),

$$\boldsymbol{E} = E_0\hat{i} \sin (kx - \omega t). \tag{A1.14}$$

According to Eq. (A1.11), B_{0x}, and hence B_x, is zero. Further, Eq. (A1.12) implies

$$B_{0y} = E_{0x}/c = E_0/c,$$

$$\boldsymbol{B} = B_0\hat{j} \sin (kx - \omega t) \tag{A1.15}$$

with

$$B_0 = E_0/c. \tag{A1.16}$$

ENERGY DENSITY AND INTENSITY

The energy density U associated with an electromagnetic field is

$$U = \epsilon_0 E^2/2 + B^2/(2\mu_0). \tag{A1.17}$$

For the electromagnetic wave we are considering, using Eqs. (A1.14) and (A1.15), we find it is

$$U = \frac{\epsilon_0}{2}(E_0^2 + c^2 B_0^2) \sin^2 (kx - \omega t). \tag{A1.18}$$

Substituting for B_0 from Eq. (A1.16), we obtain

$$U = \epsilon_0 E_0^2 \sin^2 (kx - \omega t). \tag{A1.19}$$

The average energy density U at any point can be obtained by averaging Eq. (A1.19) over time. Thus,

$$\langle U \rangle = \epsilon_0 E_0^2 \frac{1}{T} \int_0^T \sin^2 (kx - \omega t)\, dt,$$

where T is the time period $(=2\pi/\omega)$. After evaluating the integral, we get

$$\langle U \rangle = \epsilon_0 E_0^2/2. \tag{A1.20}$$

The intensity I of a wave is defined as the energy passing per unit area per unit time. Therefore,

$$I = Uc = \epsilon_0 c E_0^2/2, \tag{A1.21}$$

indicating that the intensity is proportional to the square of the amplitude.

We shall now show that an electromagnetic wave has a momentum associated with it. In particular, for the plane electromagnetic wave travelling along the z-axis if W is the energy and p the momentum, we shall show that

$$p = W/c. \tag{A1.22}$$

Let the plane electromagnetic wave fall on a thin sheet of a material kept normal to the beam. Then, an electron inside the sheet will experience a force along the x-axis due to the electric field. This will give rise to a drift velocity $\boldsymbol{v}_d$ to the electron. This drift velocity is proportional to the electric field, and we can write

$$\boldsymbol{v}_d = e\eta \boldsymbol{E} = e\eta \boldsymbol{E}_0 \sin \omega t, \tag{A1.23}$$

where η is the proportionality constant and e is the magnitude of the electron charge. As the electron drifts along the x-axis, it will be acted upon by the magnetic field with a force $e\boldsymbol{v}_d \times \boldsymbol{B}$. Since the magnetic field is along the y-axis, the electron will experience a force along the z-axis. This force is given by

$$F_z = dp_z/dt = ev_d B = e^2 \eta E B, \tag{A1.24}$$

where p_z is the momentum of the electron along the z-axis and $B = B_0 \sin \omega t$. The electromagnetic field hence imparts a momentum equal to $e^2\eta BE\Delta t$ to the electron along its direction of propagation in a time Δt. The electromagnetic field also imparts energy to the electron. This is brought about by the electric field. The magnetic field, on the other hand, does not do any work on the electron since the force due to it on the electron is always perpendicular to the electron motion. The work done ΔW in the time Δt by the electric field is

$$\Delta W = \boldsymbol{F} \cdot \boldsymbol{v}_d \Delta t = e\boldsymbol{E} \cdot \boldsymbol{v}_d \Delta T.$$

Substituting for $\boldsymbol{v}_d$ from Eq. (A1.23), we get

$$dW/dt = e^2\eta E^2. \tag{A1.25}$$

From Eqs. (A1.16), (A1.24), and (A1.25),

$$\frac{1}{c}\frac{dW}{dt} = \frac{dp_z}{dt}. \tag{A1.26}$$

Integrating this equation, we obtain

$$p_z = W/c, \tag{A1.27}$$

where W and p_z are the energy and the momentum transferred to the electron by the electromagnetic field. Thus, when an electromagnetic field loses an energy W, it also gives away a momentum $p\,(=W/c)$ or, in other words, relation (A1.22) is true for an electromagnetic wave.

We have derived Eq. (A1.22) in the context of a specific example illustrating the essential physical concepts. The formal derivation of this equation requires the manipulation of the Lorentz force expression using the Maxwell equations and can be found in a standard text on electromagnetic theory.

Appendix 2

Group Velocity of Wave Packet

WAVES IN ONE DIMENSION

A wave is some disturbance travelling in a medium. For example, in a sound wave, the disturbance is pressure variation in a medium. Further, in the transmission of light in vacuum, the disturbance is the variation in the electric field and magnetic field strengths. In this appendix, we shall discuss some of the important concepts associated with a wave.

Mathematically, let the amplitude of a disturbance (travelling in one dimension) be characterized by y, a function of the coordinate x and time t. Further, let the wave travel with a velocity v along the positive x-axis. The wave can then be represented as

$$y(x, t) = f(x - vt). \tag{A2.1}$$

Here, the right-hand side is a function of a single variable, namely, $(x - vt)$. Consider the disturbance at the origin $x = 0$ at the time $t = 0$. The amplitude of the disturbance then is

$$y(0, 0) = f(0). \tag{A2.2}$$

The amplitude of the disturbance at the point x_0 at a time $t_0 = x_0/v$ is

$$y(x_0, x_0/v) = f[x_0 - v(x_0/v)] = f(0) = y(0, 0). \tag{A2.3}$$

This shows that the disturbance at $(x_0, t_0 = x_0/v)$ is the same as that at $(0, t = 0)$. That is, the disturbance has moved from $x = 0$ to $x = x_0$ in a time x_0/v, which is the same as saying that the disturbance has travelled with a velocity v.

Of the waves we deal with in physics, most are described in terms of sinusoidal waves. Let us therefore consider a sinusoidal disturbance. Mathematically, its amplitude is given by

$$y(x, t) = A \sin (kx - \omega t). \tag{A2.4}$$

This sinusoidal wave has a definite wavelength $\lambda (=2\pi/k)$ and travels with a velocity ω/k, called the phase velocity. It extends over the entire region $-\infty < x < +\infty$. In real life, a wave is confined in a finite region, that is, the amplitude of the wave is nonzero only in a finite region. Such a wave is known as a wave packet. A wave packet can be formed by a superposition of several sinusoidal waves, each of a different wavelength. The motion of the wave packet can then be represented in terms of the sinusoidal waves. The mathematical analysis of a wave packet requires the Fourier series and Fourier transforms. However, we shall not carry out this, but present a simplified treatment.

Consider a wave packet. As already noted, a wave packet is the sum of several sinusoidal waves, each of form (A2.4) and different wavelength. If all these sinusoidal waves travel with a speed v, then the wave packet will also travel with the speed v. A medium in which sinusoidal waves of different wavelengths travel with the same speed is called a *nondispersive medium*. In such a medium, a wave packet travels without changing its shape. A nondispersive medium is more of an exception than a rule. Most media are *dispersive*. In such a medium, the velocity of a wave depends on the wavelength of the wave.

A wave packet travelling in a dispersive medium changes its shape as it moves. To find the velocity of such a wave packet, we shall study how an identifiable part of the wave packet, e.g., the crest of the wave packet, moves with time.

Let us consider a single sinusoidal wave. The phase velocity associated with it is given by

$$v = \omega/k = \nu\lambda. \tag{A2.5}$$

Let us now consider a waveform consisting of two sinusoidal waves 1 and 2 of wavelengths λ and $(\lambda + d\lambda)$. Let the two waves travel with phase velocities v and $(v + dv)$, respectively. The two sinusoidal waves can be represented as

$$y_1(x, t) = A \sin (kx - \omega t), \tag{A2.6}$$

$$y_2(x, t) = A \sin [(k + dk)x - (\omega + d\omega)t], \tag{A2.7}$$

where $(\lambda + d\lambda) = 2\pi/(k + dk)$ and $(v + dv) = (\omega + d\omega)/(k + dk)$. The sum of the two waves is

$$\begin{aligned} y &= y_1(x, t) + y_2(x, t) \\ &= A \sin (kx - \omega t) + A \sin [(k + dk)x - (\omega + d\omega)t] \\ &= 2A \cos \left(\frac{x\, dk - t\, d\omega}{2}\right) \sin \left[\frac{(2k + dk)x - (2\omega + d\omega)t}{2}\right]. \end{aligned}$$

Using the approximations $2k + dk \sim 2k$ and $2\omega + d\omega \sim 2\omega$, we get

$$y = 2A \cos \frac{x\, dk - t\, d\omega}{2} \sin (kx - \omega t). \tag{A2.8}$$

As can be seen, the second term on the right-hand side here is the same as the term in Eq. (A2.6). The total amplitude of the superposed waves is shown in Fig. A2.1. In Eq. (A2.8), the first term $\cos [(x\, dk - t\, d\omega)/2]$ represents the envelope (shown by the dashed curves in this figure) which modulates the amplitude of $\sin (kx - \omega t)$. The envelope is the wave packet formed by the two sinusoidal waves. The group velocity v_g of the envelope is obtained by examining the first term of Eq. (A2.8), namely,

$$\cos [\tfrac{1}{2}(x\, dk - t\, d\omega)] = \cos \left[\frac{dk}{2}\left(x - \frac{d\omega}{dk}t\right)\right] = \cos \left[\frac{dk}{2}(x - v_g t)\right].$$

This leads to

$$v_g = d\omega/dk. \tag{A2.9}$$

The velocity of each of the two waves is obtained from the second term of Eq. (A2.8), namely,

$$v = \omega/k. \tag{A2.10}$$

When the number of waves forming a wave packet is increased, the width of the envelope reduces and the height increases (see Fig. A2.2). In other words, the wave packet becomes narrower, leading to a better localization. The group velocity of this wave packet would still

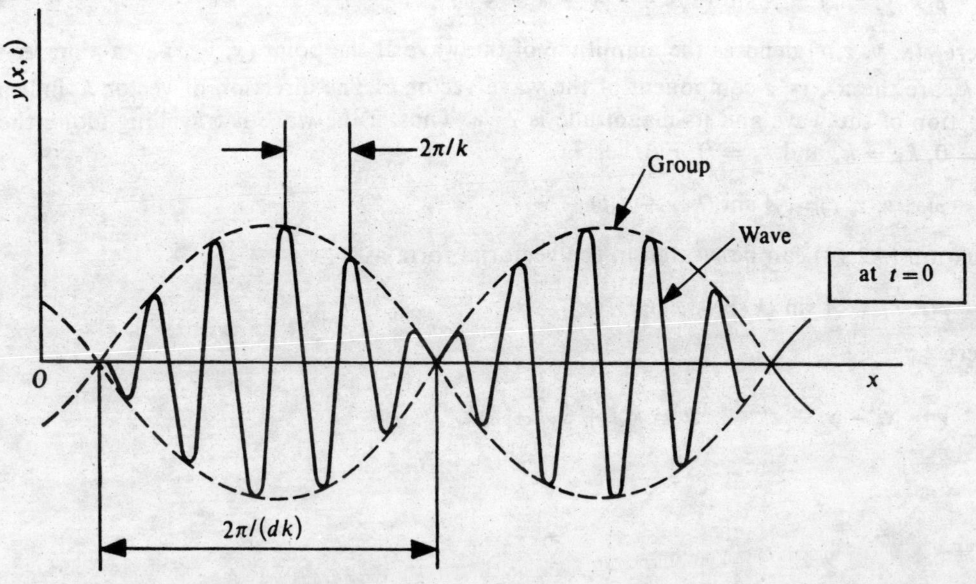

Fig. A2.1 Superposition of two sinusoidal waves of wavelengths λ and $(\lambda + d\lambda)$.

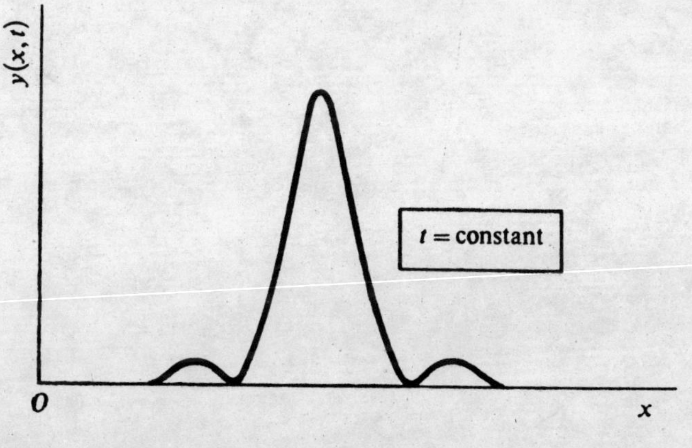

Fig. A2.2 Superposition of several sinusoidal waves giving rise to wave packet.

be the same as that given by Eq. (A2.9). Its energy, which is proportional to the square of its amplitude, is clearly localized in a small region. In quantum mechanics, such a wave packet is associated with each particle. In particular, the velocity of a particle is the same as the group velocity of the wave packet associated with it.

WAVES IN THREE DIMENSIONS

The expression for a sinusoidal wave travelling in three dimensions is a natural extension of

Eq. (A2.4). It is given by

$$\rho(x, y, z, t) = A \sin (k_x x + k_y y + k_z z - \omega t), \tag{A2.11}$$

where $\rho(x, y, z, t)$ denotes the amplitude of the wave at the point (x, y, z) at a time t, and k_x, k_y, k_z are the x-, y-, z-component of the wave vector $\boldsymbol{k}$. The direction of vector $\boldsymbol{k}$ indicates the direction of the wave and its magnitude is $2\pi/\lambda$. Thus, if the wave is travelling along the y-axis, $k_x = 0$, $k_y = k$, and $k_z = 0$. Further,

$$\rho(x, y, z, t) = A \sin (k_y y - \omega t).$$

Equation (A2.11) can be written in the vectorial form as

$$\rho(\boldsymbol{r}, t) = A \sin (\boldsymbol{k} \cdot \boldsymbol{r} - \omega t), \tag{A2.12}$$

where

$$\boldsymbol{r} = x\hat{i} + y\hat{j} + z\hat{k}, \qquad \boldsymbol{k} = k_x\hat{i} + k_y\hat{j} + k_z\hat{k}.$$

Appendix 3

Complex Numbers

A complex number is defined as

$$z = a + ib, \tag{A3.1}$$

where a and b are real numbers and

$$i^2 = -1. \tag{A3.2}$$

In Eq. (A3.1), a is referred to as the real part of z (written as Re z), i.e.,

$$\text{Re } z = a, \tag{A3.3}$$

and b is referred to as the imaginary part of z (written as Im z), i.e.,

$$\text{Im } z = b. \tag{A3.4}$$

A complex number can be represented by a point in a plane just as a real number can be represented by a point on a line. The x-coordinate of the point represents the real part of a complex number, whereas the y-coordinate represents the imaginary part. Two complex numbers $z_1 = a_1 + ib_1$ and $z_2 = a_2 + ib_2$ can be added to get a third complex number $z_1 + z_2$ given by

$$z_1 + z_2 = (a_1 + a_2) + i(b_1 + b_2). \tag{A3.5}$$

Similarly, the subtraction of two complex numbers can be expressed as

$$z_1 - z_2 = (a_1 - a_2) + i(b_1 - b_2). \tag{A3.6}$$

The operation of multiplication of two complex numbers is defined as

$$z_1 z_2 = (a_1 + ib_1)(a_2 + ib_2) = (a_1 a_2 - b_1 b_2) + i(a_1 b_2 + a_2 b_1). \tag{A3.7}$$

The modulus of a complex number $z = a + ib$ is defined as $(a^2 + b^2)^{1/2}$, i.e., as

$$\text{mod } z = |z| = +(a^2 + b^2)^{1/2}. \tag{A3.8}$$

The complex conjugate of $z = a + ib$ is defined as $a - ib$ and this is written as

$$z = a - ib. \tag{A3.9}$$

It should be noted that

$$|z|^2 = zz^* = (a + ib)(a - ib) = a^2 + b^2. \tag{A3.10}$$

Here, zz^* is a real positive number (its imaginary part is zero). We can also define the division of two complex numbers as

$$\frac{z_1}{z_2} = \frac{(a_1 + ib_1)}{(a_2 + ib_2)} = \frac{(a_1 + ib_1)(a_2 - ib_2)}{(a_2 + ib_2)(a_2 - ib_2)}$$

$$= \frac{(a_1a_2 + b_1b_2) + i(b_1a_2 - a_1b_2)}{a_2^2 + b_2^2}. \tag{A3.11}$$

In Chapter 2, we encountered the complex number $\exp(-iEt/\hbar)$. This can be written as

$$\exp\left(-\frac{iEt}{\hbar}\right) = \cos\frac{Et}{\hbar} - i\sin\frac{Et}{\hbar}. \tag{A3.12}$$

The complex conjugate of $\exp(-iEt/\hbar)$ is $\exp(iEt/\hbar)$ and is written as

$$\exp\left(\frac{iEt}{\hbar}\right) = \cos\frac{Et}{\hbar} + i\sin\frac{Et}{\hbar}. \tag{A3.13}$$

We finally note that

$$\left|\exp\left(\frac{iEt}{\hbar}\right)\right| = \left(\cos^2\frac{Et}{\hbar} + \sin^2\frac{Et}{\hbar}\right)^{1/2} = 1. \tag{A3.14}$$

Appendix 4

Operators

DEFINITIONS

We are all familiar with the operations in mathematics, e.g., addition, multiplication, differentiation, and so on. In quantum mechanics, an operator is specified by defining its operation on a wave function. Consider an operator A. When it operates on the wave function $\psi(x)$, let the result be $3\psi(x)$. This operation can be written as

$$A\psi(x) = 3\psi(x). \tag{A4.1}$$

In this simple situation, the operator A is 3. Another example of an operator is differentiation. If an operator D operates on $\psi(x)$ and gives $d\psi(x)/dx$, we write

$$D\psi(x) = d\psi(x)/dx. \tag{A4.2}$$

Here, the operator D is identified as

$$D \equiv d/dx \tag{A4.3}$$

to indicate the nature of operation. An operator O is said to be linear if it satisfies the relation

$$O[\psi(x) + \phi(x)] = O\psi(x) + O\phi(x), \tag{A4.4}$$

where $\psi(x)$ and $\phi(x)$ are two arbitrary wave functions. It can be easily verified that each of the operators 3 and $D\,(\equiv d/dx)$ is linear.

If S is an operator such that it squares the wave function, i.e.,

$$S\psi(x) = [\psi(x)]^2, \tag{A4.5}$$

then it is not linear since

$$S\psi(x) + S\phi(x) = [\psi(x)]^2 + [\phi(x)]^2,$$

$$S[\psi(x) + \phi(x)] = [\psi(x) + \phi(x)]^2,$$

i.e.,

$$S[\psi(x) + \phi(x)] \neq S\psi(x) + S\phi(x).$$

We shall deal only with linear operators since in quantum mechanics the operators are generally linear.

EIGENVALUES OF LINEAR OPERATORS

An operator O acting on a function ψ gives a function ϕ. This is written as

$$O\psi = \phi. \tag{A4.6}$$

In general, the functional form of ϕ is different from that of ψ. For example, the operator d/dx acting on sin x gives cos x. In some special situations, we may have $\phi = \alpha\psi$, where α is a number, real or complex. This means that ϕ and ψ have the same functional form. For instance, when the operator d/dx acts on exp $(2ix)$, we get

$$\frac{d[\exp(2ix)]}{dx} = 2i[\exp(2ix)]. \tag{A4.7}$$

Here, $\alpha = 2i$. In general, if

$$O\psi = \alpha\psi, \tag{A4.8}$$

then ψ is called the eigenfunction, and α the eigenvalue, of the operator O. Equation (A4.8) is known as the eigenvalue equation.

Each linear operator has a set of eigenvalues $\lambda_1, \lambda_2, \ldots$ and the eigenfunctions corresponding to these eigenvalues are $\psi_1, \psi_2, \ldots$.

EIGENVALUES OF ENERGY OPERATOR

Consider the problem of a particle in a one-dimensional box discussed in Chapter 2. The wave function of the particle is given by Eq. (2.71), i.e.,

$$\psi_n(x) = A \sin\frac{n\pi x}{L} \qquad (n = 1, 2, \ldots). \tag{A4.9}$$

Let now the energy operator

$$E = \frac{p^2}{2m} = -\frac{\hbar^2}{2m}\frac{d^2}{dx^2}$$

act on this wave function. Then,

$$\frac{p^2}{2m}\psi_n(x) = -\frac{\hbar^2}{2m}\frac{d^2\psi_n(x)}{dx^2},$$

$$\frac{p^2}{2m}A\sin\frac{n\pi x}{L} = -\frac{\hbar^2}{2m}\frac{d^2}{dx^2}\left(A\sin\frac{n\pi x}{L}\right) = \frac{\hbar^2\pi^2n^2}{2mL^2}A\sin\frac{n\pi x}{L}. \tag{A4.10}$$

Thus, we see that the eigenvalues of the energy operator are

$$E_n = \hbar^2\pi^2n^2/(2mL^2).$$

It should be noted that these are the allowed energy values [see Eq. (2.69)]. For a system, the energy operator, in general, is $[p^2/(2m) + V]$ and, if ψ_n is an eigenfunction with the eigenvalue E_n, the eigenvalue equation is

$$\left(-\frac{\hbar^2}{2m}\frac{d^2}{dx^2} + V\right)\psi_n(x) = E_n\psi_n(x) \qquad \text{or} \qquad \frac{d^2\psi_n(x)}{dx^2} + \frac{2m}{\hbar^2}(E_n - V)\psi_n(x) = 0,$$

the latter being the same as the time-independent Schroedinger equation (2.61). In other words, the Schroedinger equation is the energy eigenvalue equation. Thus, if we need to find the allowed energy value of a system, we have to solve the Schroedinger equation.

MOMENTUM EIGENVALUES

In quantum mechanics, each measurable quantity is associated with an operator. In particular, the momentum operator is given by $p = -i\hbar\partial/\partial x$ [i.e., Eq. (2.30)]. The eigenvalue equation corresponding to it is

$$p\psi = -i\hbar(\partial\psi/\partial x) = \alpha\psi. \tag{A4.11}$$

This differential equation has the solution

$$\psi = A \exp(i\alpha x/\hbar) \tag{A4.12}$$

which is the eigenfunction of the momentum operator with eigenvalue α.

ANGULAR MOMENTUM EIGENVALUES

In Eqs. (3.67), (3.49), and (3.50), we noted that

$$L_z(Y_l^{m_l}) = \hbar m_l(Y_l^{m_l}), \tag{A4.13}$$

$$L^2(Y_l^{m_l}) = \hbar^2 l(l+1)(Y_l^{m_l}). \tag{A4.14}$$

Clearly, $Y_l^{m_l}$ are the eigenfunctions of the operators L_z and L^2, the eigenvalues corresponding to the latter being $\hbar m_l$ and $\hbar^2 l(l+1)$, respectively.

When a physical system is described by a wave function $\psi_\alpha(x, t)$, which is an eigenfunction of the operator O with eigenvalue α, then a measurement of O gives the value α. Thus, a measurement of the z-component of angular momentum of a particle described by a wave function whose angular part is $Y_2^1(\theta, \phi)$ would give $\hbar$, whereas the measurement of L^2 would give $2\hbar^2$. Therefore, the system has a definite angular momentum and z-component of angular momentum. Similarly, when a system is in a definite state of energy, the wave function is an eigenfunction of the energy operator.

Appendix 5
Spherical Polar Coordinates

DEFINITIONS

The relation between the rectangular coordinates x, y, z and the spherical polar coordinates r, θ, ϕ is shown in Fig. A5.1. In this figure, PP' is parallel to the z-axis, P' is on the x-y plane,

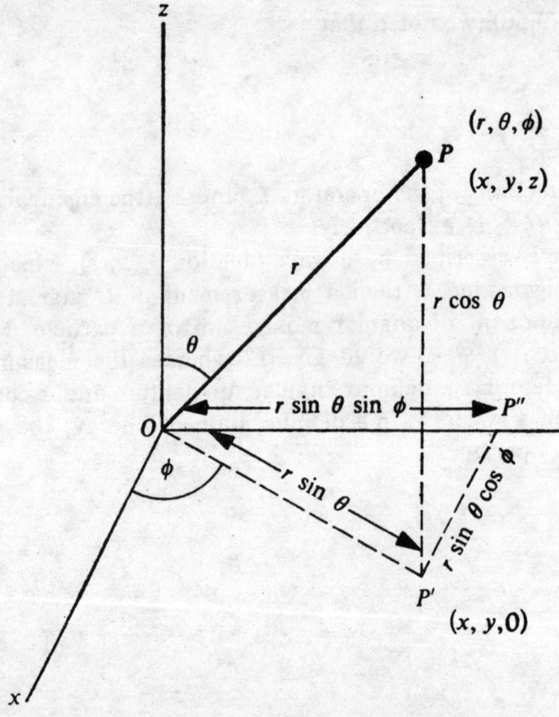

Fig. A5.1 Spherical polar coordinate representation.

and $P'P''$ is parallel to the x-axis. The two triangles OPP' and $OP'P''$ are right-angled triangles, $OP = r$, $\angle zOP = \theta$, and $\angle xOP' = \phi$. The rectangular coordinates are given by $P'P'' = x$, $OP'' = y$, and $P'P = z$. Using $OP' = r \sin\theta$, $OP'' = r \sin\theta \sin\phi$, $PP' = r\cos\theta$, and $P'P'' = r \sin\theta \cos\phi$, we get

$$x = r\sin\theta\cos\phi, \qquad y = r\sin\theta\sin\phi, \qquad z = r\cos\theta \tag{A5.1}$$

or

$$r = (x^2 + y^2 + z^2)^{1/2}, \qquad \theta = \tan^{-1}\frac{(x^2+y^2)^{1/2}}{z}, \qquad \phi = \tan^{-1}\frac{y}{x} \tag{A5.2}$$

with the restrictions $0 \leqslant \theta \leqslant \pi$ and $0 \leqslant \phi \leqslant 2\pi$. In the rectangular coordinates, $\nabla^2\psi$ is expressed as

$$\nabla^2\psi = \partial^2\psi/\partial x^2 + \partial^2\psi/\partial y^2 + \partial^2\psi/\partial z^2. \tag{A5.3}$$

To convert this into the spherical polar coordinates, the identity

$$\left.\frac{\partial\psi}{\partial x}\right|_{y,z} = \left.\frac{\partial\psi}{\partial r}\right|_{\theta,\phi}\left.\frac{\partial r}{\partial x}\right|_{y,z} + \left.\frac{\partial\psi}{\partial\theta}\right|_{r,\phi}\left.\frac{\partial\theta}{\partial x}\right|_{y,z} + \left.\frac{\partial\psi}{\partial\phi}\right|_{r,\theta}\left.\frac{\partial\phi}{\partial x}\right|_{y,z} \tag{A5.4}$$

is used. In this identity, we need to evaluate $(\partial r/\partial x)_{y,z}$, $(\partial\theta/\partial x)_{y,z}$, and $(\partial\phi/\partial x)_{y,z}$. To do this, we use Eqs. (A5.1) and (A5.2). Then,

$$\left.\frac{\partial r}{\partial x}\right|_{y,z} = \frac{x}{(x^2+y^2+z^2)^{1/2}} = \frac{x}{r}, \tag{A5.5}$$

$$\left.\frac{\partial\theta}{\partial x}\right|_{y,z} = \frac{1}{1+(x^2+y^2)/z^2}\,\frac{x}{(x^2+y^2)^{1/2}}\,\frac{1}{z} = \frac{z^2}{r^2}\,\frac{x}{(x^2+y^2)^{1/2}}\,\frac{1}{z}, \tag{A5.6}$$

$$\left.\frac{\partial\phi}{\partial x}\right|_{y,z} = \frac{1}{1+y^2/x^2}\left(-\frac{y}{x^2}\right) = -\frac{y}{x^2+y^2}. \tag{A5.7}$$

Now, from Eq. (A5.1),

$$x/r = \sin\theta\cos\phi, \tag{A5.8}$$

$$z/r = \cos\theta, \tag{A5.9}$$

$$\frac{x}{(x^2+y^2)^{1/2}} = \cos\phi, \tag{A5.10}$$

$$\frac{y}{x^2+y^2} = \frac{1}{r}\,\frac{\sin\phi}{\sin\theta}. \tag{A5.11}$$

Substituting Eq. (A5.8) in Eq. (A5.5), we get

$$\partial r/\partial x = \sin\theta\cos\phi. \tag{A5.12}$$

Substituting Eqs. (A5.9) and (A5.10) in Eq. (A5.6), we obtain

$$\frac{\partial\theta}{\partial x} = \frac{1}{r}\cos\theta\cos\phi. \tag{A5.13}$$

Similarly, using Eq. (A5.11) in Eq. (A5.7), we have

$$\frac{\partial\phi}{\partial x} = -\frac{1}{r}\,\frac{\sin\phi}{\sin\theta}. \tag{A5.14}$$

Equations (A5.12), (A5.13), and (A5.14), along with Eq. (A5.4), give

$$\frac{\partial\psi}{\partial x} = \frac{\partial\psi}{\partial r}\sin\theta\cos\phi + \frac{\partial\psi}{\partial\theta}\,\frac{1}{r}\cos\theta\cos\phi - \frac{\partial\psi}{\partial\phi}\,\frac{\sin\phi}{r\sin\theta}. \tag{A5.15}$$

Differentiating it, we get

$$\frac{\partial^2\psi}{\partial x^2} = [\frac{\partial^2\psi}{\partial r^2}\sin\theta\cos\phi - \frac{1}{r^2}\frac{\partial\psi}{\partial\theta}\cos\theta\cos\phi + \frac{1}{r^2}\frac{\partial\psi}{\partial\phi}\frac{\sin\phi}{\sin\theta} + \frac{\partial^2\psi}{\partial r\,\partial\theta}\frac{1}{r}\cos\theta\cos\phi$$

$$- \frac{\partial^2\psi}{\partial r\,\partial\phi}\frac{\sin\phi}{r\sin\theta}]\sin\theta\cos\phi + [\frac{\partial^2\psi}{\partial r\,\partial\theta}\sin\theta\cos\phi + \frac{\partial\psi}{\partial r}\cos\theta\cos\phi + \frac{\partial^2\psi}{\partial\theta^2}\frac{1}{r}\cos\theta\cos\phi$$

$$- \frac{\partial\psi}{\partial\theta}\frac{\sin\theta\cos\phi}{r} - \frac{\partial^2\psi}{\partial\theta\,\partial\phi}\frac{\sin\phi}{r\sin\theta} + \frac{\partial\psi}{\partial\phi}\frac{\sin\phi\cos\theta}{r\sin^2\theta}]\frac{1}{r}\cos\theta\cos\phi$$

$$- [\frac{\partial^2\psi}{\partial r\,\partial\phi}\sin\theta\cos\phi - \frac{\partial\psi}{\partial r}\sin\theta\sin\phi + \frac{\partial^2\psi}{\partial\theta\,\partial\phi}\frac{1}{r}\cos\theta\cos\phi - \frac{\partial\psi}{\partial\theta}\frac{1}{r}\cos\theta\sin\phi$$

$$- \frac{\partial^2\psi}{\partial\phi^2}\frac{\sin\phi}{r\sin\theta} - \frac{\partial\psi}{\partial\phi}\frac{\cos\phi}{r\sin\theta}]\frac{\sin\phi}{r\sin\theta}. \tag{A5.16}$$

The identities similar to (A5.4) are used to convert $\partial^2\psi/\partial y^2$ and $\partial^2\psi/\partial z^2$ into the spherical polar coordinates. Once this has been done, then, following several algebraic steps, Eq. (A5.3) can finally be written as

$$\frac{\partial^2\psi}{\partial x^2} + \frac{\partial^2\psi}{\partial y^2} + \frac{\partial^2\psi}{\partial z^2} = \frac{1}{r^2}\frac{\partial}{\partial r}(r^2\frac{\partial\psi}{\partial r}) + \frac{1}{r^2\sin\theta}\frac{\partial}{\partial\theta}(\sin\theta\frac{\partial}{\partial\theta}) + \frac{1}{r^2\sin^2\theta}\frac{\partial^2\psi}{\partial\phi^2}. \tag{A5.17}$$

This is the expression we used to arrive at Eq. (3.32). The foregoing method can also be used to prove relations (3.44) and (3.45). Let us, as an example, derive the operator relation [see Eqs. (3.44)]

$$L_x = i\hbar(\sin\phi\frac{\partial}{\partial\theta} + \cot\theta\cos\phi\frac{\partial}{\partial\phi}). \tag{A5.18}$$

This means

$$L_x\psi = i\hbar(\sin\phi\frac{\partial\psi}{\partial\theta} + \cot\theta\cos\phi\frac{\partial\psi}{\partial\phi}). \tag{A5.19}$$

Then, from the definition of L_x [see Eqs. (3.42)],

$$L_x\psi = -i\hbar(y\frac{\partial\psi}{\partial z} - z\frac{\partial\psi}{\partial y}) = -i\hbar(r\sin\theta\sin\phi\frac{\partial\psi}{\partial z} - r\cos\theta\frac{\partial\psi}{\partial y}) \tag{A5.20}$$

$$= -i\hbar r(\sin\theta\sin\phi\frac{\partial\psi}{\partial z} - \cos\theta\frac{\partial\psi}{\partial y}). \tag{A5.21}$$

Proceeding as before, we find

$$\frac{\partial\psi}{\partial z} = \frac{\partial\psi}{\partial r}\frac{\partial r}{\partial z} + \frac{\partial\psi}{\partial\theta}\frac{\partial\theta}{\partial z} + \frac{\partial\psi}{\partial\phi}\frac{\partial\phi}{\partial z}, \tag{A5.22}$$

$$\frac{\partial\psi}{\partial y} = \frac{\partial\psi}{\partial r}\frac{\partial r}{\partial y} + \frac{\partial\psi}{\partial\theta}\frac{\partial\theta}{\partial y} + \frac{\partial\psi}{\partial\phi}\frac{\partial\phi}{\partial y}. \tag{A5.23}$$

From Eq. (A5.2), we have

$$\frac{\partial r}{\partial z} = \frac{z}{r} = \cos\theta, \qquad \frac{\partial r}{\partial y} = \frac{y}{r} = \sin\theta\sin\phi, \tag{A5.24}$$

$$\frac{\partial\theta}{\partial z} = -\frac{(x^2+y^2)^{1/2}}{z^2+x^2+y^2} = -\frac{\sin\theta}{r},$$

$$\frac{\partial\theta}{\partial y} = \frac{z^2}{z^2+x^2+y^2}\frac{y}{z}\frac{1}{(x^2+y^2)^{1/2}} = \frac{\cos\theta\sin\phi}{r}, \tag{A5.25}$$

$$\frac{\partial\phi}{\partial z} = 0, \qquad \frac{\partial\phi}{\partial y} = \frac{x^2}{x^2+y^2}\left(\frac{1}{x}\right) = \frac{\cos\phi}{r\sin\theta}. \tag{A5.26}$$

Applying Eqs. (A5.24), (A5.25), and (A5.26) to Eq. (A5.21), we get

$$L_x\psi = -i\hbar r[\sin\theta\sin\phi\{(\frac{\partial\psi}{\partial r}\cos\theta + \frac{\partial\psi}{\partial\theta}(-\frac{\sin\theta}{r}) + \frac{\partial\psi}{\partial\phi}(0)\}$$

$$-\cos\theta(\frac{\partial\psi}{\partial r}\sin\theta\sin\phi + \frac{\partial\psi}{\partial\theta}(\frac{\cos\theta\sin\phi}{r}) + \frac{\partial\psi}{\partial\phi}\frac{\cos\phi}{r\sin\theta})]$$

$$= -i\hbar r[(\sin\theta\cos\theta\sin\phi - \sin\theta\cos\theta\sin\phi)\frac{\partial\psi}{\partial r}$$

$$+ (-\sin^2\theta\sin\phi - \cos^2\theta\sin\phi)\frac{1}{r}\frac{\partial\psi}{\partial\theta} - \frac{\cos\theta}{r\sin\theta}\cos\phi\frac{\partial\psi}{\partial\phi}]$$

$$= i\hbar[\sin\phi\frac{\partial\psi}{\partial\theta} + \cot\theta\cos\phi\frac{\partial\psi}{\partial\phi}]$$

which is identical with the right-hand side of Eq. (A5.19). We have thus derived Eq. (A5.18). Similarly, we can evaluate L_y, L_z, and L^2 $(=L_x^2 + L_y^2 + L_z^2)$.

AVERAGE OVER ANGULAR COORDINATES

In Chapter 2, we noted that the average value of an operator O (when generalized to three dimensions) is given by

$$\langle O\rangle = \frac{\int \psi^*(x, y, z, t)O\psi(x, y, z, t)\, dv}{\int \psi^*(x, y, z, t)\psi(x, y, z, t)\, dv}, \tag{A5.27}$$

where $dv = dx\,dy\,dz$ is the volume element in three dimensions. If we perform the integration in Eq. (A5.27) in the spherical polar coordinates, then the volume element is $r^2\,dr\sin\theta\,d\theta\,d\phi$ and we can write the average of the operator as

$$\langle O\rangle = \frac{\int_{r=0}^{\infty}\int_{\theta=0}^{\pi}\int_{\phi=0}^{2\pi} \psi^*(r, \theta, \phi, t)O\psi(r, \theta, \phi, t)r^2\sin\theta\,d\theta\,d\phi}{\int_{r=0}^{\infty}\int_{\theta=0}^{\pi}\int_{\phi=0}^{2\pi} \psi^*(r, \theta, \phi, t)\psi(r, \theta, \phi, t)r^2\sin\theta\,d\theta\,d\phi}, \tag{A5.28}$$

where the operator O depends only on r, θ, and ϕ. Consider a particle moving in a spherically symmetric potential. Then, $V(r, \theta, \phi) = V(r)$. As we know [see Eq. (3.74)],

$$\psi(r, \theta, \phi) = R(r) Y_l^{m_l}(\theta, \phi).$$

Substituting this in Eq. (A5.28), we get

$$\langle O \rangle = \frac{\int_{r=0}^{\infty}\int_{\theta=0}^{\pi}\int_{\phi=0}^{2\pi} R^*(r) Y_l^{m_l^*}(\theta, \phi) O(r, \theta, \phi) R(r) Y_l^{m_l}(\theta, \phi) r^2 \sin\theta \, d\theta \, d\phi \, dr}{\int_{r=0}^{\infty}\int_{\theta=0}^{\pi}\int_{\phi=0}^{2\pi} R^*(r) Y_l^{m_l^*}(\theta, \phi) R(r) Y_l^{m_l}(\theta, \phi) r^2 \sin\theta \, d\theta \, d\phi \, dr}.$$

Let the operator O be independent of r and depend only on θ and ϕ. Examples of such an operator are L_x, L_y, and L_z [see Eqs. (3.44)]. Then, we can rewrite the foregoing expression as

$$\langle O \rangle = \frac{\int_{r=0}^{\infty} R^*(r) R(r) r^2 \, dr \int_{\theta=0}^{\pi}\int_{\phi=0}^{2\pi} Y_l^{m_l^*}(\theta, \phi) O(\theta, \phi) Y_l^{m_l}(\theta, \phi) \sin\theta \, d\theta \, d\phi}{\int_{r=0}^{\infty} R^*(r) R(r) r^2 \, dr \int_{\theta=0}^{\pi}\int_{\phi=0}^{2\pi} Y_l^{m_l'}(\theta, \phi) Y_l^{m_l}(\theta, \phi) \sin\theta \, d\theta \, d\phi}.$$

The integral over r cancels, and we get

$$\langle O \rangle = \frac{\int_{\theta=0}^{\pi}\int_{\phi=0}^{2\pi} Y_l^{m_l}(\theta, \phi) O(\theta, \phi) Y_l^{m_l}(\theta, \phi) \sin\theta \, d\theta \, d\phi}{\int_{\theta=0}^{\pi}\int_{\phi=0}^{2\pi} Y_l^{m_l^*}(\theta, \phi) Y_l^{m_l}(\theta, \phi) \sin\theta \, d\theta \, d\phi}. \qquad \text{(A5.29)}$$

Using the normalization condition

$$\int_{\theta=0}^{\pi}\int_{\theta=0}^{2\pi} Y_l^{m_l^*}(\theta, \phi) Y_l^{m_l}(\theta, \phi) \sin\theta \, d\theta \, d\phi = 1, \qquad \text{(A5.30)}$$

we get

$$\langle O \rangle = \int_{\theta=0}^{\pi}\int_{\phi=0}^{2\pi} Y_l^{m_l^*}(\theta, \phi) O(\theta, \phi) Y_l^{m_l}(\theta, \phi) \sin\theta \, d\theta \, d\phi. \qquad \text{(A5.31)}$$

Appendix 6

Rutherford Scattering

The model of the atom Rutherford proposed was based on his observations of the scattering of α-particles by the nuclei in a target. The principle underlying scattering is discussed in Chapter 4. Here, we shall give the mathematical analysis of this phenomenon.

Figure A6.1 illustrates the scattering of an α-particle by a nucleus at the origin of a coordinate system. The α-particle has a charge $2e$ and the nucleus a charge Ze. The nucleus can be treated as fixed because of its large mass. The force the α-particle experiences is given by

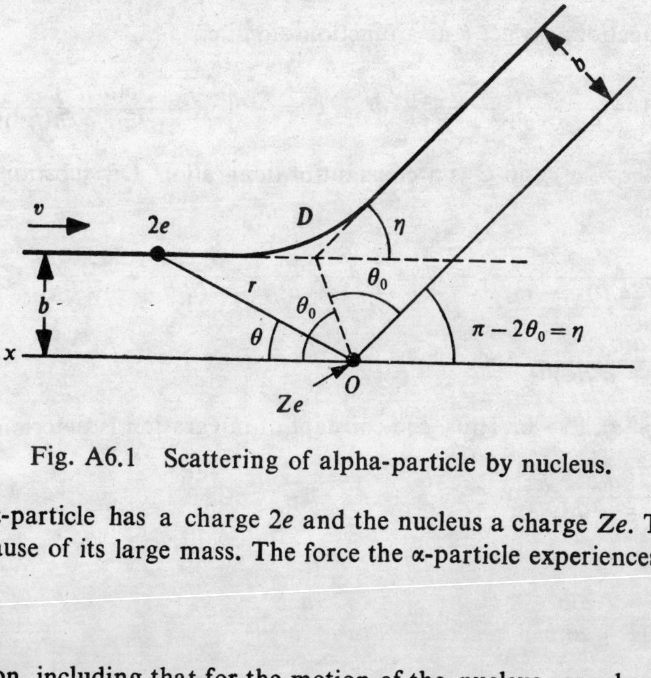

Fig. A6.1 Scattering of alpha-particle by nucleus.

$$\boldsymbol{F} = \frac{2Ze^2}{4\pi\epsilon_0 r^3}\boldsymbol{r}.$$

[The exact expression, including that for the motion of the nucleus, can be obtained by using the reduced mass of the system constituted by the nucleus and the α-particle (see Section 4.6).] Let the α-particle initially have a velocity $\boldsymbol{v}$. Consider the plane formed by $\boldsymbol{v}$ and the origin. In this plane, the force on the α-particle is along $\boldsymbol{r}$. Thus, the α-particle does not develop any acceleration in a direction normal to this plane and remains in the plane (see Fig. A6.1).

The total energy of the α-particle and its angular momentum about the origin (or the centre of mass if the nucleus is not assumed to be infinitely heavy) are conserved. The total energy of the α-particle is the sum of its kinetic and potential energies. Initially, the energy of the α-particle is $\frac{1}{2}mv^2$ and its angular momentum about O is mvb (see Fig. A6.1). Thus, we have

$$\tfrac{1}{2}m(\dot{r}^2 + r^2\dot{\theta}^2) + \frac{2Ze^2}{4\pi\epsilon_0 r} = \tfrac{1}{2}mv^2, \qquad \text{(A6.1)}$$

$$mr^2\dot{\theta} = mvb. \tag{A6.2}$$

Eliminating $\dot{\theta}$ from these equations, we get

$$\frac{dr}{dt} = -v\left(1 - \frac{4Ze^2}{4\pi\epsilon_0 mv^2 r} - \frac{b^2}{r^2}\right)^{1/2}. \tag{A6.3}$$

The reason the negative sign appears in front of v is that the α-particle is approaching the nucleus. We evaluate $d\theta/dr$ as

$$d\theta/dr = (d\theta/dt)(dt/dr).$$

In this equation, substituting for $\dot{\theta}$ from Eq. (A6.2) and $\dot{r}$ from Eq. (A6.3), we obtain

$$\frac{d\theta}{dr} = -\frac{b}{r^2}\left(1 - \frac{4Ze^2}{4\pi\epsilon_0 mv^2 r} - \frac{b^2}{r^2}\right)^{-1/2}. \tag{A6.4}$$

Integrating this equation, we get θ as a function of r, i.e.,

$$\theta = -\int \frac{b}{r^2}\left(1 - \frac{4Ze^2}{4\pi\epsilon_0 mv^2 r} - \frac{b^2}{r^2}\right)^{-1/2} dr + C = -\int \frac{b(dr/r^2)}{(1 - 2a/r - b^2/r^2)^{1/2}} + C, \tag{A6.5}$$

where $a = 2Ze^2/(4\pi\epsilon_0 m_0 v^2)$ and C is a constant of integration. On substituting $\tau = b/r$, we find the integral becomes

$$\theta = \int \frac{d\tau}{[1 - (2a/b)\tau - \tau^2]^{1/2}} + C = \int \frac{d\tau}{[1 + a^2/b^2 - (a/b + \tau)^2]^{1/2}} + C$$

$$= \sin^{-1} \frac{a/b + \tau}{(1 + a^2/b^2)^{1/2}} + C.$$

As $\tau \to 0$ (i.e., $r \to \infty$), $\theta \to 0$. Thus, the constant of integration is determined as

$$0 = \sin^{-1} \frac{a/b}{(1 + a^2/b^2)^{1/2}} + C$$

or

$$C = -\sin^{-1} \frac{a/b}{(1 + a^2/b^2)^{1/2}} = -\sin^{-1} \frac{a}{(a^2 + b^2)^{1/2}}.$$

This leads to

$$\theta = \sin^{-1} \frac{a + b\tau}{(a^2 + b^2)^{1/2}} - \sin^{-1} \frac{a}{(a^2 + b^2)^{1/2}}. \tag{A6.6}$$

At $\theta = \theta_0$, $dr/dt = 0$. From Eq. (A6.3), this implies

$$1 - \frac{2a}{r_0} - \frac{b^2}{r_0^2} = 0 \quad \text{or} \quad 1 + \frac{a^2}{b^2} - \left(\frac{b}{r_0} + \frac{a}{b_0}\right)^2 = 0,$$

where $OD = r_0$. Using $t_0 = b/r_0$, we get

$$\frac{1 + bt_0}{(a^2 + b^2)^{1/2}} = 1. \tag{A6.7}$$

Using Eq. (A6.6) for $\theta = \theta_0$, we have

$$\theta_0 = \sin^{-1} \frac{a + bt_0}{(a^2 + b^2)^{1/2}} - \sin^{-1} \frac{a}{(a^2 + b^2)^{1/2}} = \sin^{-1} 1 - \sin^{-1} \frac{a}{(a^2 + b^2)^{1/2}}$$

$$= \frac{\pi}{2} - \sin^{-1} \frac{a}{(a^2 + b^2)^{1/2}}. \tag{A6.8}$$

The angle of scattering η is

$$\eta = \pi - 2\theta_0 = 2 \sin^{-1} \frac{a}{(a^2 + b^2)^{1/2}} \quad \text{or} \quad \sin \frac{\eta}{2} = \frac{a}{(a^2 + b^2)^{1/2}}\,.$$

which can be rewritten as

$$b = a \cot (\eta/2). \tag{A6.9}$$

The scattering angle depends on b which is known as the impact parameter. If b is small, η is large.

Consider a beam of α-particles being scattered by nuclei. The α-particles will have all possible impact parameters. Those with an impact parameter between b and $(b + db)$ will have a scattering angle between η and $(\eta + d\eta)$. If the number of α-particles incident per unit area is N_0, then the number of α-particles with an impact parameter between b and $(b + db)$ is $N_0 2\pi b\, db$. Therefore, the number of α-particles scattered between the angles η and $(\eta + d\eta)$ is

$$dN = N_0(2\pi b\, db) = N_0 2\pi a \cot \frac{\eta}{2}\, d(a \cot \frac{\eta}{2}) = \pi a^2 N_0 \frac{\cos (\eta/2)}{\sin^3 (\eta/2)}\, d\eta. \tag{A6.10}$$

We can write this expression in terms of the solid angle $d\Omega\,(=2\pi \sin \eta\, d\eta)$ as

$$dN = \frac{N_0 a^2}{2} \frac{d\Omega}{\sin^4 (\eta/2)}. \tag{A6.11}$$

Substituting in it for a, we get

$$\frac{dN}{d\Omega} = \frac{N_0 Z^2 e^4}{8\pi^2 \epsilon_0^2 m^2 v^4} \frac{1}{\sin^4 (\eta/2)}. \tag{A6.12}$$

This gives the number of α-particles scattered per unit solid angle by a single nucleus in a target. For a thin target of area A and thickness t, the number of nuclei participating in scattering is nAt, where n is the number of nuclei per unit volume. The number of α-particles scattered per unit solid angle by the target will then be

$$\frac{dN}{d\Omega} = \frac{N_0 n A t Z^2 e^4}{8\pi^2 \epsilon_0^2 m^2 v^4} \frac{1}{\sin^4 (\eta/2)}. \tag{A6.13}$$

This is known as the Rutherford scattering formula. In 1913, H. Geiger and E. Marsden experimentally verified the predictions of this formula completely.

Appendix 7

Derivation of Gaussian Distribution

In this appendix, we shall show that, for a large N and n_r, the probability distribution (6.4) takes the gaussian form

$$P(n_r) = \left(\frac{2}{\pi N}\right)^{1/2} \exp\left(-\frac{2x^2}{N}\right), \tag{A7.1}$$

where $x = n_r - N/2$. To do this, we shall use the Stirling approximation for $N!$ in the limit of large N, i.e.,

$$\ln N! = N \ln N - N + \ln (2\pi N)^{1/2} + O\left(\frac{1}{N^{1/2}}\right). \tag{A7.2}$$

Here, $O(1/N^{1/2})$ means terms of the order $1/N^{1/2}$ or higher are neglected. This formula gives the value of $\ln N!$ with an error less than 1% when $N \geqslant 10$. In it, the presence of the first two terms can be understood in a nonrigorous way as follows. We can write

$$\begin{aligned} \ln N! &= \ln [N(N-1)\ldots 3\cdot 2\cdot 1] \\ &= \ln N + \ln (N-1) + \ldots + \ln 3 + \ln 2 + \ln 1 \\ &= \sum_{i=1}^{N} \ln i. \end{aligned} \tag{A7.3}$$

For a large N, we can approximate the sum here by an integral, i.e.,

$$\begin{aligned} \ln N! &\sim \int_1^N \ln x \, dx = (x \ln x - x)\Big|_1^N = N \ln N - N + 1 \\ &\sim N \ln N - N, \end{aligned} \tag{A7.4}$$

where we have neglected 1 in comparison to N. Thus, in the simplest approximation, $\ln N! = N \ln N - N$. However, if the approximation is done more carefully, we shall obtain Eq. (A7.2).

Taking the logarithm of both sides of Eq. (6.4) and using the Stirling approximation (A7.2), we get

$$\begin{aligned} \ln P(n_r) &\sim \ln N! - N \ln 2 - \ln n_r! - \ln (N - n_r)! \\ &\sim N \ln N - N + \ln (2\pi N)^{1/2} - N \ln 2 - n_r \ln n_r + n_r - \ln (2\pi n_r)^{1/2} \\ &\quad - (N - n_r) \ln (N - n_r) + (N - n_r) - \ln [2\pi(N - n_r)]^{1/2}, \end{aligned} \tag{A7.5}$$

where we have assumed $N \gg 1$ and $n_r \gg 1$. Simplifying this expression, we obtain

$$\ln P(n_r) = N \ln N - n_r \ln n_r - (N - n_r) \ln (N - n_r) + \ln (2\pi N)^{1/2} - \ln (2\pi n_r)^{1/2}$$
$$- \ln [2\pi(N - n_r)]^{1/2} - N \ln 2. \qquad \text{(A7.6)}$$

To arrive at Eq. (A7.1), it is useful to maximize $P(n_r)$. Instead of doing this, we can maximize $\ln P(n_r)$ since it is a monotonically increasing function of $P(n_r)$, i.e., if $P(n_r)$ increases, $\ln P(n_r)$ also increases, and if $P(n_r)$ decreases, $\ln P(n_r)$ also decreases. Thus, we differentiate $\ln P(n_r)$ and equate it to zero, i.e.,

$$\frac{d}{dn_r} [\ln P(n_r)] = -\ln n_r - 1 + \ln (N - n_r) + 1 - \frac{1}{2n_r} + \frac{1}{2(N - n_r)} = 0$$

or

$$\ln \frac{N - n_r}{n_r} = \frac{1}{2n_r} - \frac{1}{2(N - n_r)}. \qquad \text{(A7.7)}$$

This is satisfied when $n_r = N/2$ [i.e., both sides of Eq. (A7.7) become zero]. We can calculate the second derivative of $\ln P(n_r)$ at $n_r = N/2$ and show that the result corresponds to a maximum of $\ln P(n_r)$.

We now find the probability $P(n_r)$ near the maximum $n_r = N/2$. To do this, we introduce the variable x defined by

$$n_r = N/2 + x. \qquad \text{(A7.8)}$$

Substituting this in Eq. (A7.6), we get, after a little bit of algebra,

$$\ln P(n_r = \frac{N}{2} + x) = \ln \left(\frac{2}{\pi N}\right)^{1/2} - \frac{2x^2}{N}, \qquad \text{(A7.9)}$$

where terms of the order x^2/N^2 have been neglected. Exponentiating Eq. (A7.9), we obtain the desired formula (6.5), i.e.,

$$P(n_r) = \left(\frac{2}{\pi N}\right)^{1/2} \exp \left(-\frac{2x^2}{N}\right). \qquad \text{(A7.10)}$$

Appendix 8

Derivation of Relation $S = k \ln \Omega$

In an irreversible process, the number of microstates available to a system increases (see Section 6.3). The entropy of the system also increases in such a process. This gives us a clue that the number of microstates must be a function of entropy, i.e.,

$$S = f(\Omega). \tag{A8.1}$$

Suppose we have two macroscopic systems A and B with entropies S_A and S_B. The total entropy of the combined system then is

$$S = S_A + S_B. \tag{A8.2}$$

Let Ω_A and Ω_B be the number of microstates available to the systems A and B, respectively. Then, the total number of microstates available to the macroscopic system consisting of A and B is

$$\Omega = \Omega_A \Omega_B. \tag{A8.3}$$

To understand this, let us consider a simple situation where A has 2 states a_1, a_2, and B has 3 states b_1, b_2, b_3. The number of states for both A and B together is a_1b_1, a_1b_2, a_1b_3, a_2b_1, a_2b_2, and a_2b_3. Here, a_1b_1 means that the system A is in the state a_1 and the system B is in the state b_1, and so on. We have six states for the systems A and B considered as a whole. Therefore, for each state Ω_A of A, we have Ω_B states of B giving rise to Ω_B new states. Thus, the total number of states is $\Omega_A\Omega_B$. Combining Eqs. (A8.1) and (A8.2), we get

$$f(\Omega) = f(\Omega_A) + f(\Omega_B). \tag{A8.4}$$

From Eq. (A8.3),

$$f(\Omega_A\Omega_B) = f(\Omega) = f(\Omega_A) + f(\Omega_B). \tag{A8.5}$$

Differentiating both sides of this equation with respect to Ω_A and keeping Ω_B fixed, we obtain

$$\left.\frac{\partial f(\Omega_A\Omega_B)}{\partial \Omega_A}\right|_{\Omega_B} = \left.\frac{\partial f(\Omega_A)}{\partial \Omega_A}\right|_{\Omega_B} = \frac{df(\Omega_A)}{d\Omega_A}. \tag{A8.6}$$

Reversing the roles of Ω_A and Ω_B, we have

$$\left.\frac{\partial f(\Omega_A\Omega_B)}{\partial \Omega_B}\right|_{\Omega_A} = \left.\frac{\partial f(\Omega_B)}{\partial \Omega_B}\right|_{\Omega_A} = \frac{df(\Omega_B)}{d\Omega_B}. \tag{A8.7}$$

We also have

$$\left.\frac{\partial f(\Omega_A\Omega_B)}{\partial \Omega_B}\right|_{\Omega_A} = \Omega_A \left.\frac{\partial f(\Omega_A\Omega_B)}{\partial (\Omega_A\Omega_B)}\right|_{\Omega_A} = \Omega_A \frac{df(\Omega_A\Omega_B)}{d(\Omega_A\Omega_B)}, \tag{A8.8}$$

$$\left.\frac{\partial f(\Omega_A\Omega_B)}{\partial \Omega_A}\right|_{\Omega_B} = \Omega_B \left.\frac{\partial f(\Omega_A\Omega_B)}{\partial(\Omega_A\Omega_B)}\right|_{\Omega_B} = \Omega_B \frac{df(\Omega_A\Omega_B)}{d(\Omega_A\Omega_B)}. \tag{A8.9}$$

Substituting Eq. (A8.9) in Eq. (A8.6) and Eq. (A8.8) in Eq. (A8.7), we get

$$\frac{df(\Omega_A\Omega_B)}{d(\Omega_A\Omega_B)} = \frac{1}{\Omega_B}\frac{df(\Omega_A)}{d\Omega_A} = \frac{1}{\Omega_A}\frac{df(\Omega_B)}{d\Omega_B}. \tag{A8.10}$$

This leads to

$$\Omega_A \frac{df(\Omega_A)}{d\Omega_A} = \Omega_B \frac{df(\Omega_B)}{d\Omega_B}. \tag{A8.11}$$

In this equation, since the left-hand side depends only on Ω_A and the right-hand side only on Ω_B, the equality can hold only if each of its sides is equal to a constant. Thus, we write

$$\Omega_A \frac{df(\Omega_A)}{d\Omega_A} = \Omega_B \frac{df(\Omega_B)}{d\Omega_B} = k, \tag{A8.12}$$

where k is a constant. Therefore, we have

$$\frac{df(\Omega_A)}{d\Omega_A} = \frac{k}{\Omega_A}. \tag{A8.13}$$

Integrating both sides of this equation, we get

$$S_A = f(\Omega_A) = k \ln \Omega_A \tag{A8.14}$$

or, in general,

$$S = k \ln \Omega. \tag{A8.15}$$

Appendix 9

Derivation of the Three Statistical Distributions

INTRODUCTION

Each of the three distributions, namely, M-B, F-D, and B-E, is derived by calculating the total number of microstates available to a system under the two macroscopic conditions (6.39) and (6.40), i.e.,

$$\sum_i N_i = N, \qquad \sum_i N_i E_i = E.$$

To do this, we need to distribute N_i particles among the available g_i states. The number of states available to each particle depends on the statistics obeyed by the particle, and therefore we need to consider each statistics separately.

MAXWELL-BOLTZMANN DISTRIBUTION

In the classical statistics, the N_i particles are distinguishable and each can be put in g_i states in g_i different ways. Thus, the total number of different states for the N_i particles is

$$g_i g_i \ldots g_i = g_i^{N_i}. \tag{A9.1}$$

The number of ways in which the N particles can be divided into the groups $N_1, N_2, \ldots$ [see expression (6.44)] is

$$\frac{N!}{N_1!N_2!\ldots} = \frac{N!}{\prod_i N_i!}, \tag{A9.2}$$

and hence the total number of microstates for a system with N_1 particles in g_1 states, N_2 particles in g_2 states, ... is

$$\frac{N!}{N_1!N_2!\ldots} g_1^{N_1} g_2^{N_2} \ldots \tag{A9.3}$$

or

$$\Omega_{\text{M-B}} = N! \prod_i \frac{g_i^{N_i}}{N_i!}. \tag{A9.4}$$

The N_i's have to satisfy conditions (6.39) and (6.40).

FERMI-DIRAC STATISTICS

In the Fermi-Dirac statistics, the N identical fermions constituting a system are indistinguishable, and so the factor corresponding to expression (A9.3) is just 1. This means that each of the N_s particles can be put in the g_s states with the condition that no two particles are in the same state. This is possible only when $g_s > N_s$. Let us assume that this is so. Now, the first particle can be filled in g_s ways. The second particle cannot occupy the state already occupied by the first particle, and so has only $(g_s - 1)$ states available to it. Similarly, the third particle has $(g_s - 2)$ states available to it, and so on. Thus, the total number of possible arrangements for N_s particles in g_s states, if the particles are distinguishable, is

$$g_s(g_s - 1) \dots (g_s - N_s + 1). \tag{A9.5}$$

In this expression, we have treated the microstate in which particle 1 has been put in the first state and particle 2 in the second state as a microstate separate from the one in which particle 2 has been put in the first state and particle 1 in the second state. That is, we have counted a single state $N_s!$ times since there are $N_s!$ ways in which the N_s particles can be rearranged amongst themselves. But this is wrong because identical fermions are indistinguishable. Therefore, the total number of microstates is

$$^{g_s}C_{N_s} = \frac{g_s(g_s - 1) \dots (g_s - N_s + 1)}{N_s!}.$$

Thus, the total number of states available to the system is

$$\Omega_{\text{F-D}} = \prod_s {}^{g_s}C_{N_s}. \tag{A9.6}$$

The N_s's have to satisfy conditions (6.39) and (6.40).

BOSE-EINSTEIN STATISTICS

In the Bose-Einstein statistics also, the particles are indistinguishable. However, there is no restriction on the number of particles that can be put in a given state because the bosons do not obey the Pauli exclusion principle. As in the Fermi-Dirac statistics, so too here, the number of ways N can be divided into $N_1, \dots, N_s, \dots$ is 1 as the particles cannot be distinguished.

To calculate the number of ways the N_s particles can be distributed in the g_s states, we first consider again the simple situation of distributing two particles in three states. This gives rise to six microstates. These microstates are enumerated in Example 6.10.

We now digress a bit and consider the following different problem. A one-dimensional box has four crosses marked on it. Let us find the number of ways in which two vertical lines can be drawn on the crosses such that the box is divided into three parts (cells). The various possibilities of doing this are shown in Table A9.1. As can be noticed, the total number of such possibilities

TABLE A9.1 Different ways of drawing 2 vertical lines on 4 crosses

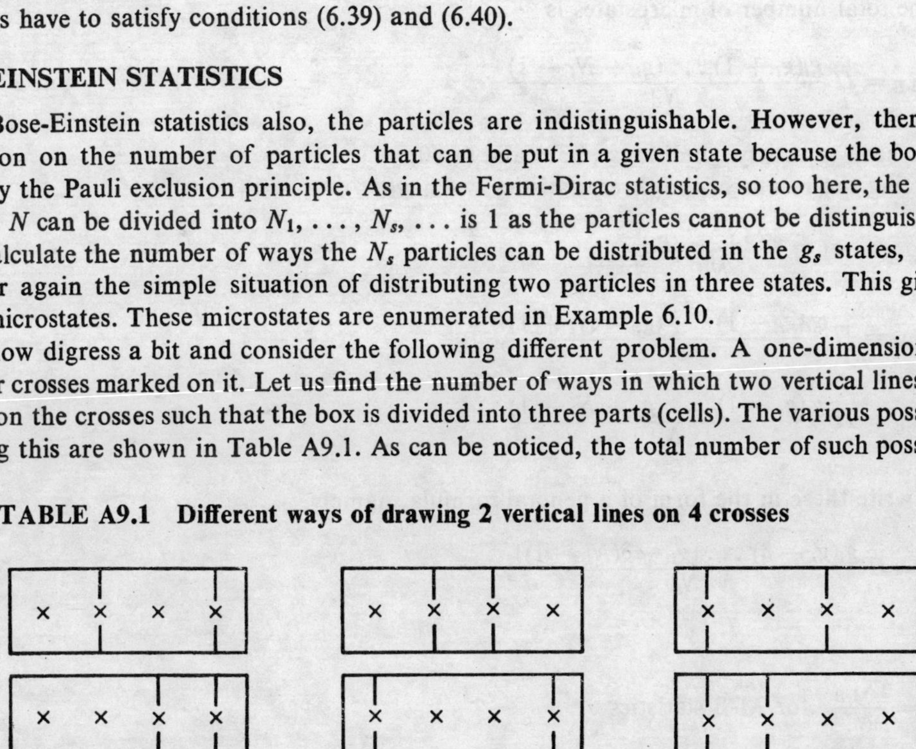

is six. Let us treat the crosses in the cells as particles (the crosses on which lines have been drawn are not to be counted as particles). In this way, the number of particles we have in the various cells are

(2, 0, 0), (0, 2, 0), (0, 0, 2), (1, 1, 0), (1, 0, 1), (0, 1, 1).

From these, we find that there is a one-to-one correspondence between the number of crosses in the various cells and the number of particles in the various states.

For the N_s particles to be distributed amongst the g_s states, we consider a box with (N_s+g_s-1) crosses and draw $(g_s - 1)$ lines. This would divide the box into g_s cells. The number of crosses that do not lie on the lines is N_s, and so we treat the crosses in cell 1 as the number of particles in the first state, and so on. As already noted, these two problems are identical, and therefore the number of ways the N_s particles can be distributed in the g_s states is equal to the number of ways the $(g_s - 1)$ lines can be drawn over the $(N_s + g_s - 1)$ crosses. Accordingly, the number of ways the N_s particles can be distributed over the g_s states is

$$^{(N_s+g_s-1)}C_{g_s-1} = \frac{(N_s + g_s - 1)!}{(g_s - 1)!N_s!} \tag{A9.7}$$

$$= \frac{g_s(g_s + 1)\ldots(g_s + N_s - 1)}{N_s!}. \tag{A9.8}$$

Thus, the total number of microstates is

$$\Omega_{\text{B-E}} = \prod_i \frac{g_i(g_i + 1)\ldots(g_i + N_i - 1)}{N_i!}. \tag{A9.9}$$

The N_i's have to satisfy conditions (6.39) and (6.40).

Collecting the results of the three statistics, we have

$$\Omega_{\text{M-B}} = N! \prod_i \frac{g_i(g_i)\ldots(g_i)}{N_i!},$$

$$\Omega_{\text{F-D}} = \prod_i \frac{g_i(g_i - 1)\ldots(g_i - N_i + 1)}{N_i!},$$

$$\Omega_{\text{B-E}} = \prod_i \frac{g_i(g_i + 1)\ldots(g_i + N_i - 1)}{N_i!}.$$

We can write these in the form of a general formula, namely,

$$\Omega = \prod_i \frac{g_i(g_i - \delta)\ldots[g_i - \delta(N_i - 1)]}{N_i!}, \tag{A9.10}$$

where

$$\begin{aligned} \Omega &= \frac{\Omega_{\text{M-B}}}{N!} \quad \text{for M-B statistics} \\ &= \Omega_{\text{F-D}} \quad \text{for F-D statistics} \\ &= \Omega_{\text{B-E}} \quad \text{for B-E statistics} \end{aligned} \tag{A9.11}$$

with

$$\begin{aligned} \delta &= 0 \quad \text{for M-B statistics} \\ &= 1 \quad \text{for F-D statistics} \\ &= -1 \quad \text{for B-E statistics.} \end{aligned} \tag{A9.12}$$

As noted in Chapter 6, we must maximize $\ln \Omega$ under conditions (6.39) and (6.40). Using the Stirling approximation, we obtain

$$\ln \Omega = \sum_i [\ln g_i + \ln (g_i - \delta) + \ldots + \ln \{g_i - (N_i - 1)\delta\}] - \sum_i (\ln N_i!). \tag{A9.13}$$

To maximize $\ln \Omega$, we cannot vary N_i in an independent way because of the two restrictions (6.39) and (6.40). To overcome this problem, we use the method of Lagrange multipliers (see the discussion on it later in this appendix) and write the function to be maximized as

$$K = \ln \Omega - \alpha \sum_i N_i - \beta \sum_i N_i E_i, \tag{A9.14}$$

where α and β are unknown constants. We can now treat all the N_s particles as independent variables and change N_1 to $N_1 + 1$, leaving $N_2, N_3, \ldots$ fixed. Then,

$$\delta K = \ln (g_1 - N_1\delta) - \ln (N_1 + 1) - \alpha - \beta E_1 = 0 \tag{A9.15}$$

is one of the conditions for Ω to be maximum. We repeat this for each particle. For the i-th particle, we make $N_i = N_i + 1$ with $N_1, N_2, \ldots, N_{i-1}, N_{i+1}, \ldots$ fixed. Then, the condition for Ω to be maximum is

$$\delta K = \ln (g_i - N_i\delta) - \ln (N_i + 1) - \alpha - \beta E_i = 0 \tag{A9.16}$$

or

$$\ln \frac{g_i - N_i\delta}{N_i + 1} = \alpha + \beta E_i$$

or

$$g_i/N_i = \exp (\alpha + \beta E_i) + \delta, \tag{A9.17}$$

neglecting 1 in comparison to N_i. This leads to

$$N_i = \frac{g_i}{\exp (\alpha + \beta E_i) + \delta} \tag{A9.18}$$

which gives the three distributions for $\delta = \pm 1$ and 0. To evaluate the constants α and β, we relate them to some physical quantity. For this, we use the relation between the entropy and the number of microstates

$$S = k \ln \Omega \tag{A9.19}$$

to fix β. Considering the Maxwell-Boltzmann distribution, we have

$$S_{\text{M-B}} = k \sum_i \ln \left(\frac{g_i^{N_i}}{N_i!}\right) = k \sum [N_i \ln g_i - N_i(\ln N_i - 1)]. \tag{A9.20}$$

Now, using Eq. (A9.18) with $\delta = 0$, we obtain

$$\ln N_i = \ln g_i - \alpha - \beta E_i$$

for the M-B distribution. This expression reduces to

$$S_{\text{M-B}} = +k \sum_i (\alpha + \beta E_i) N_i + \sum_i N_i \tag{A9.21}$$

$$= k(\alpha + 1)N + \beta k E. \tag{A9.22}$$

Using the thermodynamic relation

$$\partial S/\partial E = 1/T, \tag{A9.23}$$

where T is the temperature of the system, we get, from Eq. (A9.22), $\beta k = 1/T$ or

$$\beta = 1/(kT). \tag{A9.24}$$

In a similar fashion, it can be shown that $\beta = 1/(kT)$ also for the Bose-Einstein and Fermi-Dirac distributions. Thus, we have

$$N_i = \frac{g_i}{\exp\,[\alpha + E_i/(kT)] + \delta}. \tag{A9.25}$$

The constant α can be determined from the condition $\sum_i N_i = N$.

BLACK-BODY RADIATION

The number of microstates for a system of photons with a total energy E is given by

$$\Omega = \prod_i \frac{g_i(g_i + 1) \ldots (g_i + N_i - 1)}{N_i!}. \tag{A9.26}$$

This has to be maximized under the restriction [see Eq. (6.40)]

$$\sum_i N_i E_i = E \tag{A9.27}$$

with (see Section 6.7) no restriction on the total photon number. Thus, we maximize

$$K = \ln \Omega - \beta \sum_i N_i E_i \tag{A9.28}$$

and get

$$N_i = \frac{g_i}{\exp\,(\beta E_i) - 1}. \tag{A9.29}$$

The occupation probability $n(E)$ then is

$$n(E) = \frac{1}{\exp\,(\beta E) - 1}. \tag{A9.30}$$

By calculating the entropy S and using the thermodynamic relation $\partial S/\partial E = 1/(kT)$, we can again show that $\beta = 1/(kT)$. Thus, the occupation probability for photons we get is

$$n(E) = \frac{1}{\exp\,[E/(kT)] - 1} \tag{A9.31}$$

which is equivalent to Eq. (6.84).

THE STEFAN-BOLTZMANN LAW

In Chapter 6, we saw that the energy density of radiation at a temperature T contained in a cavity (black body) is

$$U = \frac{8\pi^5 k^4 T^4}{15c^3h^3} = U_0 T^4.$$

Let the cavity have a small hole of area ΔS to allow the radiation to leak. The amount of radiation that would be emitted through the hole per unit time can be calculated as follows. Consider the radiation incident at an angle θ and subtending a solid angle $d\Omega$ to the normal $\hat{n}$ to the surface ΔS. This radiation is from a cylinder of area of cross-section ΔS and height $c \cos\theta$ (see Fig. A9.1). Since this radiation is travelling in all directions, only the

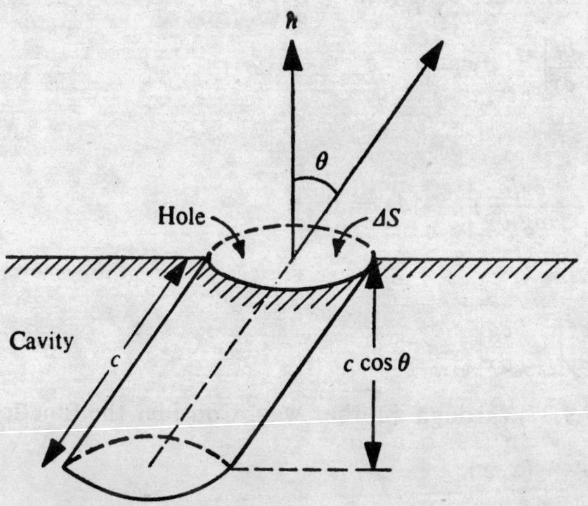

Fig. A9.1 Stefan-Boltzmann law.

fraction $d\Omega/(4\pi)$ of the radiation leaks out of ΔS. Thus, the energy radiated per unit time through the hole of area ΔS is

$$\Delta S c U \int_{\phi=0}^{2\pi} \int_{\theta=0}^{\pi/2} \cos\theta \frac{d\Omega}{4\pi} = \Delta S \frac{cU}{2} \int_0^1 \cos\theta \, d(\cos\theta) = \Delta S \frac{cU}{4}.$$

This leads to the Stefan-Boltzmann law that the power e radiated per unit area is

$$e = \frac{cU}{4} = \frac{cU_0}{4} T^4 = \sigma T^4,$$

where $\sigma = cU_0/4$.

METHOD OF LAGRANGE MULTIPLIERS

Suppose we want to find the extremum of a function $f(x, y)$ of two variables subject to the condition

$$g(x, y) = 0. \tag{A9.32}$$

This condition implies x and y are not free parameters and cannot be varied independently. In order to maintain the constraint equation (A9.32), if x changes by Δx and y by Δy, we have

$$g(x + \Delta x, y + \Delta y) = 0. \tag{A9.33}$$

Using the Taylor series expansion, we get

$$\frac{\partial g}{\partial x}\Delta x + \frac{\partial g}{\partial y}\Delta y = 0 \tag{A9.34}$$

or

$$\frac{\Delta y}{\Delta x} = -\frac{\partial g(x, y)}{\partial x}\Big/\frac{\partial g(x, y)}{\partial y}. \tag{A9.35}$$

If $f(x_0, y_0)$ is an extremum under constraint (A9.32), we have

$$\Delta f = \left.\frac{\partial f}{\partial x}\right|_{x_0, y_0} \Delta x + \left.\frac{\partial f}{\partial y}\right|_{x_0, y_0} \Delta y = 0 \tag{A9.36}$$

or

$$[\left.\frac{\partial f}{\partial x}\right|_{x_0, y_0} + \left.\frac{\partial f}{\partial y}\right|_{x_0, y_0} \left.(-\frac{\partial g/\partial}{\partial g/\partial y})\right|_{x_0, y_0}]\Delta x = 0$$

and

$$\left.\frac{\partial f}{\partial x}\right|_{x_0, y_0}\left.\frac{\partial g}{\partial y}\right|_{x_0, y_0} = \left.\frac{\partial f}{\partial y}\right|_{x_0, y_0}\left.\frac{\partial g}{\partial x}\right|_{x_0, y_0}. \tag{A9.37}$$

We can obtain result (A9.37) through another way. Consider the function

$$\psi(x, y) = f(x, y) + \lambda g(x, y), \tag{A9.38}$$

where λ is an undetermined constant. To find the extremum of this function, let us treat both x and y as free parameters, that is, the conditions for extremum are

$$\left.\frac{\partial \psi(x, y)}{\partial x}\right|_{x_0, y_0} = \left.\frac{\partial f}{\partial x}\right|_{x_0, y_0} + \lambda\left.\frac{\partial g}{\partial x}\right|_{x_0, y_0} = 0,$$

$$\left.\frac{\partial \psi(x, y)}{\partial y}\right|_{x_0, y_0} = \left.\frac{\partial f}{\partial y}\right|_{x_0, y_0} + \lambda\left.\frac{\partial g}{\partial y}\right|_{x_0, y_0} = 0$$

or

$$\left.\frac{\partial f}{\partial x}\right|_{x_0, y_0} = -\lambda\left.\frac{\partial g}{\partial x}\right|_{x_0, y_0},$$

$$\left.\frac{\partial f}{\partial y}\right|_{x_0,\, y_0} = -\lambda \left.\frac{\partial g}{\partial y}\right|_{x_0,\, y_0}.$$

Eliminating λ, we obtain condition (A9.37).

The advantage of the second method is that it can easily be generalized to an n-variable problem, namely, finding the extremum of a function $f(x_1, \ldots, x_n)$ of n variables $x_1, \ldots, x_n$ under the condition

$$g(x_1, x_2, \ldots, x_n) = 0. \tag{A9.39}$$

Consider the function

$$\psi(x_1, x_2, \ldots, x_n) = f(x_1, x_2, \ldots, x_n) + \lambda g(x_1, x_2, \ldots, x_n), \tag{A9.40}$$

where λ is an undetermined constant. On extremizing this function, we get

$$\frac{\partial f}{\partial x_i} + \lambda \frac{\partial g}{\partial x_i} = 0 \qquad (i = 1, \ldots, N). \tag{A9.41}$$

It can be shown that extremizing ψ treating the n variables $x_1, \ldots, x_n$ as independent is equivalent to extremizing f under the condition $g = 0$.

This second method can be generalized to an arbitrary number of constraints

$$\begin{aligned} &g_1(x_1, x_2, \ldots, x_n) = 0 \\ &g_2(x_1, x_2, \ldots, x_n) = 0 \\ &\vdots \\ &g_m(x_1, x_2, \ldots, x_n) = 0 \qquad (n > m) \end{aligned} \tag{A9.42}$$

For this, consider the function

$$\psi(x_1, x_2, \ldots, x_n) = f(x_1, x_2, \ldots, x_n) + \sum_{i=1}^{m} \lambda_i g_i(x_1, x_2, \ldots, x_n) \tag{A9.43}$$

and extremize ψ, treating the variables $x_1, \ldots, x_n$ as independent. Each of the constants $\lambda_1, \ldots, \lambda_m$ is called a Lagrange multiplier and such a procedure is referred to as the method of Lagrange multipliers.

Gamma-Function [$\Gamma(n)$]

The gamma-function $\Gamma(n)$ is defined as

$$\Gamma(n) = \int_0^\infty e^{-x} x^{n-1}\, dx. \tag{A9.44}$$

Integrating it by parts, we can show that it has the properties

$$\Gamma(n) = (n-1)\Gamma(n-1), \tag{A9.45}$$

$$\Gamma(n) = (n-1)! \text{ when } n \text{ is an integer.} \tag{A9.46}$$

It can also be shown that

$$\Gamma(\tfrac{1}{2}) = (\pi)^{1/2}. \tag{A9.47}$$

Thus, using the Γ-function, we can evaluate integrals of the form

$$I_n = \int_0^\infty \exp\left(-\frac{E}{kT}\right) E^n E^{1/2}\, dE. \tag{A9.48}$$

Substituting in this equation $E/(kT) = x$, we have

$$I_n = (kT)^{n+3/2} \int_0^\infty \exp(-x) x^{n+1/2}\, dx \tag{A9.49}$$

$$= (kT)^{n+3/2}\Gamma(n + 3/2). \tag{A9.50}$$

Hence,

$$I_0 = (kT)^{3/2}\Gamma(3/2) = \tfrac{1}{2}(\pi)^{1/2}(kT)^{3/2}, \tag{A9.51}$$

$$I_1 = (kT)^{5/2}\Gamma(5/2) = \tfrac{3}{2}\,\tfrac{1}{2}(\pi)^{1/2}(kT)^{5/2} = \tfrac{3}{4}(\pi)^{1/2}(kT)^{5/2}. \tag{A9.52}$$

Appendix 10

Law of Equipartition of Energy

The law of equipartition of energy as applicable to a system of particles in equilibrium at a temperature T and obeying the classical statistics is that the average kinetic energy per particle per degree of freedom is $kT/2$, where k is the Boltzmann constant. One of its simplest applications is to an ideal gas obeying the classical statistics at room temperature. Each particle of such a gas has three degrees of freedom, and so the average kinetic energy per particle is $3kT/2$.

The law of equipartition of energy can be derived by using the Maxwell-Boltzmann distribution (7.1) and calculating the average energy per particle. Thus,

$$\begin{aligned}\langle E\rangle &= \frac{1}{N}\int_0^\infty N(E)E\,dE\\ &= \frac{1}{N}\int_0^\infty \frac{2N}{(\pi)^{1/2}(kT)^{3/2}}\exp\left(-\frac{E}{kT}\right)E^{1/2}E\,dE\\ &= \frac{2}{\pi^{1/2}(kT)^{3/2}}\int_0^\infty \exp\left(-\frac{E}{kT}\right)EE^{1/2}\,dE\\ &= \frac{2}{\pi^{1/2}(kT)^{3/2}}I_1,\end{aligned}$$

where I_1 is as defined in Appendix 9. Substituting for I_1 from Eq. (A9.52), we get

$$\langle E\rangle = \frac{2}{\pi^{1/2}(kT)^{3/2}}\tfrac{3}{4}\pi^{1/2}(kT)^{5/2} = \tfrac{3}{2}kT.$$

Since the kinetic energy associated with the motion of a particle along the three directions is the same, we have

$$\langle E\rangle_{\text{per degree of freedom}} = \tfrac{1}{2}kT$$

which is the law of equipartition of energy.

We can apply the law of equipartition of energy to a system other than an ideal gas. Consider the motion of atoms about a mean position in a solid. If the amplitude of oscillation is not very large, we can treat the motion as simple harmonic. Here, again, the average kinetic energy per particle is $(3/2)kT$. We also know that for a simple harmonic oscillator the average potential energy is equal to the average kinetic energy, and hence the average total energy is $3kT$.

Appendix 11

Motion of Particle in Periodic Potential

BLOCH'S THEOREM

In this appendix, we shall obtain the energy levels of a particle in a periodic potential by solving the Schroedinger equation. The solution of this equation was simplified by F. Bloch who realized the symmetry properties of the potential in which the electron in a crystalline solid moves. The main point of Bloch's analysis is that the boundary condition satisfied by the wave function in a periodic potential is different from the boundary condition for obtaining a bound state. As we shall subsequently see, the boundary condition is also periodic.

Consider an array of N nuclei located at the lattice sites $a, 2a, \ldots, Na$, where N is a large number. The Coulomb potential due to this array is shown in Fig. A11.1a. The potential is periodic (except when we are close to the boundary atoms), that is

$$V(x + a) = V(x). \tag{A11.1}$$

Periodic Boundary Condition

Let us denote the potential at a point x due to the nucleus at x_0 by $\phi(x - x_0)$. The potential at the point x due to all the N nuclei then is

$$V(x) = \sum_{k=1}^{N} \phi(x - ka) = \phi(x - a) + \phi(x - 2a) + \ldots + \phi(x - Na). \tag{A11.2}$$

To check the periodicity of this potential, let us evaluate it at the point $(x + a)$. This can be done by substituting $(x + a)$ for x in Eq. (A11.2). Then,

$$V(x + a) = \sum_{k=1}^{N} \phi(x + a - ka) \tag{A11.3}$$

$$= \sum_{k=1}^{N} \phi[x - (k - 1)a] = \phi(x) + \phi(x - a) + \phi(x - 2a) + \ldots + \phi[x - (N - 1)a]$$

$$= \phi(x) + V(x) - \phi(x - Na). \tag{A11.4}$$

Here, so long as $\phi(x)$ and $\phi(x - Na)$ are small as compared to $V(x)$, we can neglect them, and thus get result (A11.1). The two terms $\phi(x)$ and $\phi(x - Na)$ can be small only if x and $(x - Na)$ are large. If x is any point well inside a crystal, then $\phi(x)$ and $\phi(x - Na)$ are small because these terms correspond to the potential due to the nuclei at the boundary. The most important contribution to the potential $V(x)$ at the point x is made by the nuclei close to x and not by the nuclei near the boundary. On the other hand, near the boundary, i.e., at $x = 0$, $\phi(x)$ is large, and at $x = Na$, $\phi(x - Na)$ is large and the main contribution to $V(x)$ is made by $\phi(x - Na)$. Thus, relation (A11.1) is not valid near the boundary.

As we have seen, the potential is periodic if we neglect the small contributions due to the atoms at the boundary. Mathematically, the solution of the Schroedinger equation becomes

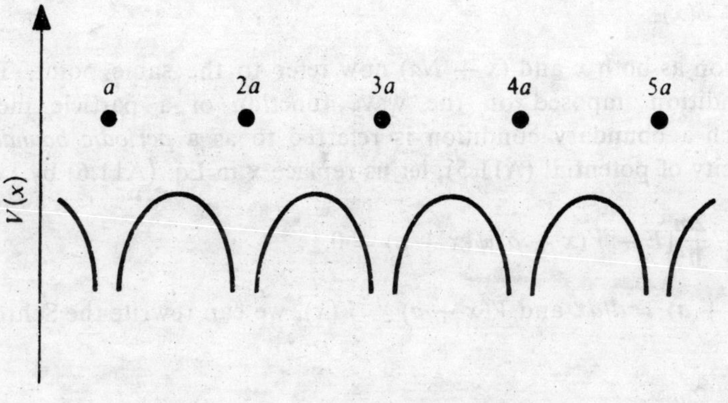

(a) Potential in crystalline solid

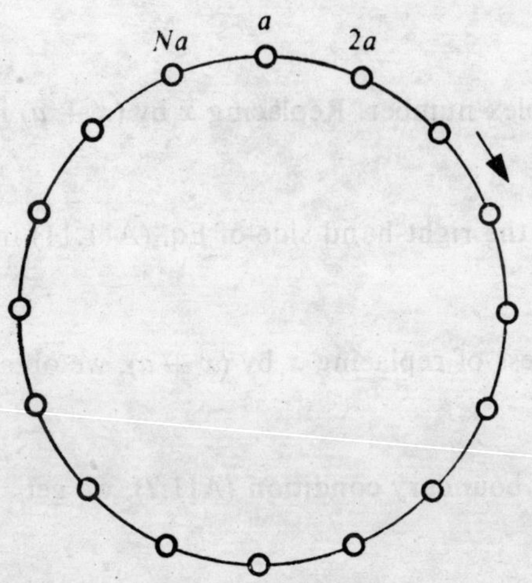

(b) Atoms in crystal arranged in the form of ring

Fig. A11.1 Periodic potential in crystal.

simpler if we make the periodicity exact. This can be done by assuming that the lattice sites in a one-dimensional crystal are arranged in the form of a ring (see Fig. A11.1b) with the $(N+1)$-th lattice site coinciding with the lattice site a. In this situation, the crystal does not have a boundary and all the lattice sites on the ring are similar. Thus,

$$V(x+a) = V(x). \tag{A11.5}$$

Consider the Schroedinger equation

$$\frac{d^2\psi(x)}{dx^2} + \frac{2m}{\hbar^2}[E - V(x)]\psi(x) = 0. \tag{A11.6}$$

Here, $\psi(x)$ represents the wave function of a particle with an energy E. We impose the boundary condition

$$\psi(x + Na) = \psi(x) \tag{A11.7}$$

on the wave function as both x and $(x + Na)$ now refer to the same point. This is similar to the boundary condition imposed on the wave function of a particle moving on a ring (Example 3.6). Such a boundary condition is referred to as a *periodic boundary condition.* To utilize the periodicity of potential (A11.5), let us replace x in Eq. (A11.6) by $(x + a)$. Thus,

$$\frac{d^2\psi(x+a)}{d(x+a)^2} + \frac{2m}{\hbar^2}[E - V(x+a)]\psi(x+a) = 0. \tag{A11.8}$$

Noting that $d/d(x + a) = d/dx$ and $V(x + a) = V(x)$, we can rewrite the Schroedinger equation as

$$\frac{d^2\psi(x+a)}{dx^2} + \frac{2m}{\hbar^2}[E - V(x)]\psi(x+a) = 0. \tag{A11.9}$$

In other words, $\psi(x + a)$ is also a solution of the Schroedinger equation with the same energy E. This suggests that $\psi(x + a)$ is related to $\psi(x)$. Let us assume

$$\psi(x + a) = A\psi(x), \tag{A11.10}$$

where A is, in general, a complex number. Replacing x by $(x + a)$ in this equation, we get

$$\psi(x + 2a) = A\psi(x + a). \tag{A11.11}$$

Using Eq. (A11.10) to express the right-hand side of Eq. (A11.11) in terms of $\psi(x)$, we have

$$\psi(x + 2a) = A^2\psi(x). \tag{A11.12}$$

If we repeat N times this process of replacing x by $(x + a)$, we obtain

$$\psi(x + Na) = A^N\psi(x). \tag{A11.13}$$

Imposing on this equation the boundary condition (A11.7), we get

$$A^N = 1 \tag{A11.14}$$

or

$$A = \exp\left(\frac{2\pi i m}{N}\right), \tag{A11.15}$$

where $m = 0, \pm1, \pm2, \ldots$. As we can see, A can take only N different values, namely,

$$\exp\left(\frac{2\pi i 0}{N}\right), \exp\left(\frac{2\pi i}{N}\right), \ldots, \exp\left[\frac{2\pi i(N-1)}{N}\right].$$

The next possible value of A with $m = N$ is the same as the first one with $m = 0$. Thus, we need to consider only the N successive values of m. In this way,

$$\psi(x + a) = \exp\left(\frac{2\pi i m}{N}\right)\psi(x), \tag{A11.16}$$

where m takes integer values. It is more convenient to write $\psi(x)$ in terms of another function, namely, $U_k(x)$, defined by the equation

$$\psi(x) = \exp(ikx)U_k(x), \tag{A11.17}$$

where

$$k = \frac{2\pi m}{Na} = \frac{2\pi m}{L} \tag{A11.18}$$

and L is the length of the crystal. If we rewrite condition (A11.16) using Eq. (A11.17), we get

$$\exp[ik(x+a)]U_k(x+a) = \exp\left(\frac{2\pi im}{N}\right)\exp(ikx)U_k(x)$$

or

$$\exp(ika)U_k(x+a) = \exp\left(\frac{2\pi im}{N}\right)U_k(x).$$

Noting that $ka = 2\pi m/N$, we get

$$U_k(x+a) = U_k(x). \tag{A11.19}$$

The wave function (A11.17) can be thought of as equivalent to a plane wave modulated by a function $U_k(x)$ with the periodicity of a lattice. The wave function (A11.17) is referred to as the *Bloch function*, and the result that $\psi(x)$ can be written in terms of e^{ikx} and a periodic function $U_k(x)$ is known as the Bloch theorem.

We have just mentioned that m takes N consecutive values, say, lying between $-N/2$ and $+N/2$. Using Eq. (A11.18), we can rewrite this condition for k as

$$-\pi/a < k \leqslant \pi/a. \tag{A11.20}$$

This is referred to as the first *Brillouin zone*. The allowed values of k in this zone are

$$k = 0, \pm\frac{2\pi}{Na}, \pm\frac{2\pi}{Na}2, \ldots, \pm\frac{2\pi}{Na}\left(\frac{N-2}{2}\right), +\frac{2\pi}{Na}\left(\frac{N}{2}\right) \tag{A11.21}$$

or

$$k = 0, \pm\frac{2\pi}{L}, \pm\frac{4\pi}{L}, \ldots, +\frac{N\pi}{L}.$$

As can be seen, the total number of states is N. It should be noted that $k = -N/2$ has not been included here since it gives the same state as $k = N/2$.

The Kronig-Penney Model

We shall now solve the Schroedinger equation for a given periodic potential to obtain the energy eigenvalues of a particle. In the Kronig-Penney model, the potential under which the particle is moving is assumed to be (see Fig. 7.11)

$$\begin{aligned} V &= 0 \qquad [0 \leqslant x \leqslant (a-b)] \\ &= V_0 \qquad [-b \leqslant x \leqslant 0], \end{aligned} \tag{A11.22}$$

$$V(x + a) = V(x). \tag{A11.23}$$

The potential has a period a. The appropriate Schroedinger equations for the two regions are

$$\frac{d^2\psi}{dx^2} + \frac{2m}{\hbar^2}E\psi = 0 \qquad [0 \leqslant x \leqslant (a - b)],$$

$$\frac{d^2\psi}{dx^2} + \frac{2m}{\hbar^2}(E - V_0)\psi = 0 \qquad [-b \leqslant x \leqslant 0]. \tag{A11.24}$$

The general solutions of these equations are

$$\psi(x) = A \exp(i\alpha x) + B \exp(-i\alpha x), \qquad 0 \leqslant x \leqslant (a - b), \tag{A11.25}$$

where $\alpha^2 = 2mE/\hbar^2$, and

$$\psi(x) = C \exp(\beta x) + D \exp(-\beta x), \qquad -b \leqslant x \leqslant 0, \tag{A11.26}$$

where $\beta^2 = 2m(V_0 - E)/\hbar^2$. Imposing the usual boundary condition at $x = 0$ that ψ and $d\psi/dx$ are continuous, we get

$$A + B = C + D, \tag{A11.27}$$

$$i\alpha(A - B) = \beta(C - D). \tag{A11.28}$$

Now, we impose Bloch's boundary condition (A11.16), i.e.,

$$\psi(x + a) = \exp\left(\frac{2\pi im}{N}\right)\psi(x) = \exp(ika)\psi(x).$$

This boundary condition is valid for all values of x. Substituting in it Eqs. (A11.25) and (A11.26), we get

$$A \exp[i\alpha(a - b)] + B \exp[-i\alpha(a - b)] = \exp(i2\pi m/N)[C \exp(-\beta b) + D \exp(\beta b)]$$
$$= e^{ika}[C \exp(-\beta b) + D \exp(\beta b)]. \tag{A11.29}$$

Since Eq. (A11.16) is valid for all values of x, differentiating both sides of Bloch's boundary condition, we get the condition

$$\frac{d\psi(x + a)}{dx} = \exp(ika)\frac{d\psi(x)}{dx}.$$

Now, substituting $x = -b$ in this equation, we obtain

$$i\alpha[A \exp\{i\alpha(a - b)\} - B \exp\{-i\alpha(a - b)\}] = \exp(ika)\beta\{-C \exp(-\beta b) + D \exp(\beta b)\}. \tag{A11.30}$$

We now have the four equations (A11.27) to (A11.30) for the four unknown constants A, B, C and D. These equations can be written as

$$A + B - C - D = 0,$$

$$+i\alpha A - i\alpha B - \beta C + \beta D = 0,$$

$$A \exp[i\alpha(a - b)] + B \exp[-i\alpha(a - b)] - C \exp(ika - \beta b) - D \exp(ika + \beta b) = 0, \tag{A11.31}$$

$$i\alpha A \exp[i\alpha(a - b)] - i\alpha B \exp[-i\alpha(a - b)] + \beta C \exp(ika - \beta b) - D \exp(ika + \beta b) = 0.$$

From the theory of equations, we know that we can solve for A, B, C and D provided the determinant

$$\begin{vmatrix} 1 & 1 & -1 & -1 \\ i\alpha & -i\alpha & -\beta & +\beta \\ \exp[i\alpha(a-b)] & \exp[-i\alpha(a-b)] & -\exp(ika-\beta b) & -\exp(ika+\beta b) \\ i\alpha\exp[i\alpha(a-b)] & \exp[-i\alpha(a-b)] & \beta\exp(ika-\beta b) & -\beta\exp(ika+\beta b) \end{vmatrix} = 0. \tag{A11.32}$$

Simplifying this determinant, we get

$$\frac{\beta^2-\alpha^2}{2\alpha\beta}\sinh(\beta b)\sin\alpha(a-b)+\cosh(\beta b)\cos\alpha(a-b)=\cos ka. \tag{A11.33}$$

This equation gives the allowed values of energy. It is a difficult equation to solve, and so we assume that $V_0 \to \infty$ and $b \to 0$ in such a way that V_0ba is a constant P, i.e.,

$$\lim_{\substack{b\to 0\\ V_0\to\infty}} [V_0ba] = P. \tag{A11.34}$$

In this limit, we have

$$\beta^2 \gg \alpha^2, \qquad \beta^2 b = 2m(V_0-E)b/\hbar^2 \sim 2mV_0b/\hbar^2 = 2mP/(\hbar^2 a).$$

Further, as $\beta b \to 0$, $\sinh \beta b \sim \beta b$ and $\cosh \beta b \sim 1$. Thus, condition (A11.33) becomes

$$\frac{mP}{\hbar^2}\frac{\sin\alpha a}{\alpha a}+\cos\alpha a=\cos ka$$

or

$$R\frac{\sin\alpha a}{\alpha a}+\cos\alpha a=\cos ka, \tag{A11.35}$$

where $R = mP/\hbar^2$. This equation gives the allowed energy values discussed in Chapter 7.

Appendix 12

Compton Scattering

While studying the photoelectric effect, we noted that a photon when it interacts with an electron gives its entire energy to the electron. An electron can also scatter an incident photon. In this situation, the photon loses only a part of its energy to the electron, and therefore the scattered photon emerges with a wavelength larger than that of the incident photon. This difference in the wavelength depends on the angle of scattering, and the effect of the change in wavelength on scattering is called the Compton effect. In what follows, we shall derive a relation between the change in wavelength and the angle of scattering.

Let us assume that an incident light beam has a frequency ν and travels along the z-axis. Thus, we have photons of energy $h\nu$ with momentum $h\nu/c$ along the z-axis. Let us also assume that the electron is initially at rest. To begin with, then, the energy is $h\nu + m_0c^2$, the sum of the energy of the photon and the rest energy of the electron. After collision, a photon of frequency ν' travels along a direction making an angle θ with the z-axis and the electron has a velocity v along a direction making an angle ϕ with the z-axis (see Fig. A12.1). Therefore, the final energy [using

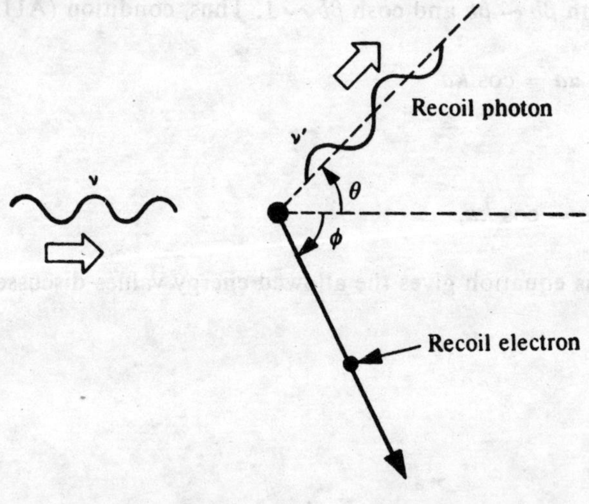

Fig. A12.1 Compton scattering.

Eq. (8.75)] is $h\nu' + m_0c^2/(1 - v^2/c^2)^{1/2}$, the sum of the energy of the photon and the energy of the electron with a speed v. It should be noted that we are using the relativistic expressions since the energy transfer, as we shall subsequently see, is large. In view of the law of energy conservation, we have

$$h\nu + m_0c^2 = h\nu' + \frac{m_0c^2}{(1 - v^2/c^2)^{1/2}}. \tag{A12.1}$$

Similarly, the law of conservation of momentum along the z-axis leads to

$$\frac{h\nu}{c} = \frac{h\nu'}{c}\cos\theta + \frac{m_0 v}{(1 - v^2/c^2)^{1/2}}\cos\phi, \quad \text{(A12.2)}$$

and along the direction perpendicular to the z-axis gives

$$0 = \frac{h\nu'}{c}\sin\theta - \frac{m_0 v}{(1 - v^2/c^2)^{1/2}}\sin\phi. \quad \text{(A12.3)}$$

From Eqs. (A12.2) and (A12.3), we get

$$\left(\frac{h\nu}{c} - \frac{h\nu'}{c}\cos\theta\right)^2 + \left(\frac{h\nu'}{c}\sin\theta\right)^2 = \frac{m_0^2 v^2}{1 - v^2/c^2} = \frac{m_0^2 c^2}{1 - v^2/c^2} - m_0 c^2. \quad \text{(A12.4)}$$

Also, from Eq. (A12.1), we have

$$(h\nu + m_0 c^2 - h\nu')^2 = \frac{m_0^2 c^4}{1 - v^2/c^2}. \quad \text{(A12.5)}$$

From Eqs. (A12.4) and (A12.5), we have

$$\left(\frac{h\nu}{c} - \frac{h\nu'}{c}\cos\theta\right)^2 + \left(\frac{h\nu'}{c}\sin\theta\right)^2 = \frac{1}{c^2}(h\nu - h\nu' + m_0 c^2)^2 - m_0^2 c^2.$$

Simplifying, we get

$$-2h^2\frac{\nu\nu'}{c^2}\cos\theta = 2hm_0(\nu - \nu') - \frac{2h^2\nu\nu'}{c^2}$$

or

$$\nu - \nu' = \frac{h\nu\nu'}{m_0 c^2}(1 - \cos\theta). \quad \text{(A12.6)}$$

Using $\nu = c/\lambda$, we can write this equation in a familiar form, namely,

$$\frac{1}{\lambda} - \frac{1}{\lambda'} = \frac{h}{m_0 c}\frac{1}{\lambda\lambda'}(1 - \cos\theta)$$

or

$$\lambda' - \lambda = \frac{h}{m_0 c}(1 - \cos\theta). \quad \text{(A12.7)}$$

Here, $h/(m_0 c)$ is the change in wavelength when the angle of scattering is 90°, and is referred to as the Compton wavelength. The numerical value of the Compton wavelength of an electron is

$$\lambda_c = \frac{h}{m_0 c} = 2.4 \times 10^{-12}\ \text{m}. \quad \text{(A12.8)}$$

We can rewrite Eq. (A12.7) as

$$\frac{\Delta\lambda}{\lambda} = \frac{\lambda_c}{\lambda}(1 - \cos\theta). \quad \text{(A12.9)}$$

It should be noted that $\Delta\lambda/\lambda \ll 1$ when $\lambda_c \ll \lambda$. Thus, for light of a wavelength much larger than 10^{-12} m, we cannot expect an appreciable shift in the wavelength. On the other hand, when $\lambda \sim \lambda_c$, the photon has an energy of the order of MeV, and hence the energy transfer to the electron is also of the order of MeV. In such a situation, as already stated, we have to use the relativistic expressions.

The Compton effect provides another evidence supporting the particle nature of electromagnetic radiation.

Appendix 13

Solution of Simple Harmonic Oscillator

The time-independent Schroedinger equation for a simple harmonic oscillator [see (3.4)] is

$$\frac{d^2\psi}{dx^2} + \frac{2m}{\hbar^2}(E - \frac{kx^2}{2})\psi = 0. \tag{A13.1}$$

To obtain its solution, we first choose a new set of dimensionless variables to represent energy and length, defined as

$$\lambda = \frac{2E}{\hbar\omega} = \frac{2E}{\epsilon_0}, \tag{A13.2}$$

$$\rho = (\frac{m\omega}{\hbar})^{1/2}x = \frac{x}{a_0}, \tag{A13.3}$$

where $\omega = (k/m)^{1/2}$, $\epsilon_0 = \hbar\omega$, and $a_0 = [\hbar/(m\omega)]^{1/2}$. Substituting these in the differential equation (A13.1), we get

$$\frac{1}{a_0^2}\frac{d^2\psi}{d\rho^2} + \frac{2m}{\hbar^2}(\frac{\epsilon_0\lambda}{2} - \frac{ka_0^2\rho^2}{2})\psi = 0 \quad \text{or} \quad \frac{1}{a_0^2}\frac{d^2\psi}{d\rho^2} + \frac{m\epsilon_0}{\hbar^2}(\lambda - \frac{ka_0^2\rho^2}{\epsilon_0})\psi = 0.$$

Using the relations

$$\frac{m\epsilon_0}{\hbar^2} = \frac{m\omega}{\hbar} = \frac{1}{a_0^2}, \qquad \frac{ka_0^2}{\epsilon_0} = \frac{m\omega^2 a_0^2}{\hbar\omega} = 1,$$

we obtain

$$\frac{1}{a_0^2}\frac{d^2\psi}{d\rho^2} + \frac{1}{a_0^2}(\lambda - \rho^2)\psi = 0$$

or

$$\frac{d^2\psi}{d\rho^2} + (\lambda - \rho^2)\psi = 0. \tag{A13.4}$$

Let us now look for the physically admissible solutions of this equation. For a bound state, this means that

$$\psi \to 0 \text{ as } \rho \to \pm\infty \; (\rho \to \pm\infty \text{ as } x \to \pm\infty).$$

The technique to solve a differential equation of the type (A13.4) which has to satisfy the asymptotic boundary condition is to first study the equation in the asymptotic limit. For Eq. (A13.4), the asymptotic solutions can be obtained by considering it for large values of ρ. In this asymptotic region, λ can be neglected in comparison to ρ^2 and the equation can then be written as

$$d^2\psi/d\rho^2 - \rho^2\psi = 0 \quad \text{for } \rho \to \pm\infty. \tag{A13.5}$$

This equation can be easily solved if we change the variable ρ to a new variable t defined by

$$t = \rho^2. \tag{A13.6}$$

Equation (A13.5) then becomes, as $t \to \infty$,

$$d^2\psi/dt^2 - \tfrac{1}{4}\psi = 0. \tag{A13.7}$$

This equation has two independent solutions and the most general solution is of the form

$$\psi = A \exp\left(-\frac{t}{2}\right) + B \exp\left(\frac{t}{2}\right) = A \exp\left(-\frac{\rho^2}{2}\right) + B \exp\left(\frac{\rho^2}{2}\right). \tag{A13.8}$$

In this solution, as $\rho \to \pm\infty$, the first term goes to zero, whereas the second term tends to ∞. If we now apply the boundary condition that the wave function goes to zero as $\rho \to \pm\infty$, then the coefficient of $\exp(\rho^2/2)$ must be zero, that is, $B = 0$. This tells us that, as $\rho \to \pm\infty$, the wave function behaves as

$$\psi \sim \exp(-\rho^2/2). \tag{A13.9}$$

With this information on the asymptotic solution, we can now attempt to solve the exact equation, i.e., (A13.4), by trying a solution of the form

$$\psi = \exp(-\rho^2/2)H(\rho), \tag{A13.10}$$

where $H(\rho)$ is a polynomial in ρ. Substituting Eq. (A13.10) in Eq. (A13.4) and using

$$\psi'' = [H'' - 2\rho H' + (\rho^2 - 1)H] \exp\left(-\frac{\rho^2}{2}\right),$$

we get

$$\frac{d^2H(\rho)}{d\rho^2} - 2\rho\frac{dH(\rho)}{d\rho} + (\lambda - 1)H(\rho) = 0. \tag{A13.11}$$

The solution of this equation can be written in the form of a series, namely,

$$H(\rho) = \sum_{i=0}^{\infty} A_i\rho^i. \tag{A13.12}$$

Substituting this series in Eq. (A13.11), we get

$$\sum_{i=0}^{\infty} \rho^i[A_{i+2}(i + 2)(i + 1) - (2i + 1 - \lambda)A_i] = 0. \tag{A13.13}$$

It should be noted that this identity must be valid for all values of ρ, and hence the coefficient of each ρ^i must be zero. Thus, we get the relation

$$A_{i+2}(i + 2)(i + 1) - (2i + 1 - \lambda)A_i = 0$$

or

$$\frac{A_{i+2}}{A_i} = \frac{2i + 1 - \lambda}{(i + 2)(i + 1)}. \tag{A13.14}$$

There are two possibilities now, namely, (1) series (A13.12) is infinite and (2) series (A13.12) is finite and terminates, say, for $i = n$; in this situation, $H(\rho)$ is an n-th degree polynomial in ρ. We shall consider these two possibilities separately:

(1) Here,

$$\lim_{i\to\infty} \frac{A_{i+2}}{A_i} \to \frac{2}{i}.$$

This series behaves as exp (ρ^2) because

$$\exp(\rho^2) = \sum_s \frac{\rho^{2s}}{s!},$$

and the ratio of the coefficients of ρ^{i+2} and ρ^i is

$$\frac{(i/2)!}{[(i/2) + 1]!} = \frac{1}{i/2 + 1} \to \frac{2}{i} \quad \text{as} \quad i \to \infty.$$

This means that the function $\psi = \exp(-\rho^2/2)H(\rho)$ behaves as exp $(\rho^2/2)$, but this is an unacceptable solution because it violates the boundary condition that $\psi \to 0$ as $\rho \to \pm\infty$. Thus, the only physically allowed possible solutions are those for which $H(\rho)$ is a finite polynomial.

(2) Here, as already stated, $H(\rho)$ is an n-th degree polynomial, i.e.,

$$H(\rho) = \sum_{i=0}^{n} A_i \rho^i$$

with $A_n \neq 0$ and $A_{k>n} = 0$. Then, condition (A13.13) reduces to

$$A_{i+2}(i + 2)(i + 1) - (2i + 1 - \lambda)A_i = 0. \tag{A13.15}$$

Since $A_{n+2} = 0$, we have, for $i = n$,

$$(2n + 1 - \lambda)A_n = 0. \tag{A13.16}$$

Since $A_n \neq 0$, we get

$$2n + 1 = \lambda, \tag{A13.17}$$

where the possible values of n are 0, 1, 2, ... Further, since $A_{n+1} = 0$, we have, for $i = n - 1$,

$$(2n - 1 - \lambda)A_{n-1} = -2A_{n-1} = 0 \tag{A13.18}$$

or $A_{n-1} = 0$. If $A_{n-1} = 0$, then

$$A_{n-3} = A_{n-5} = \ldots = 0. \tag{A13.19}$$

Thus, the alternate coefficients vanish. That is, series (A13.12) contains either even powers of ρ or odd powers of ρ. Equation (A13.17) implies that the physically acceptable solutions are possible only for some discrete values of λ. This gives the allowed values of energy as

$$E = (2n+1)\frac{\hbar\omega}{2} \qquad (n = 0, 1, 2, \ldots). \tag{A13.20}$$

This is the same as Eq. (3.6). In what follows, we shall use Eqs. (A13.15) and (A13.17) to obtain the wave functions for different values of n.

When $n = 0$, we have

$$\lambda = 1, \qquad A_0 \neq 0, \qquad A_i = 0 \quad (i > 0). \tag{A13.21}$$

Then,

$$H(\rho) = A_0, \tag{A13.22}$$

$$\psi = A_0 \exp(-\rho^2/2). \tag{A13.23}$$

This is the same as Eq. (3.7) with $A_0 = C_0$.

When $n = 1$, we have $\lambda = 3$, $A_1 \neq 0$, $A_i = 0\,(i > 1)$. Since $A_2 = 0$, $A_0 = 0$. Thus,

$$H(\rho) = A_1\rho, \tag{A13.24}$$

$$\psi = A_1\rho \exp(-\rho^2/2). \tag{A13.25}$$

This is Eq. (3.8) with $A_1 = C_1$.

When $n = 2$, we have $\lambda = 5$, $A_2 \neq 0$, $A_i = 0$ $(i > 2)$. Since $A_3 = 0$, $A_1 = 0$ and, from Eq. (A13.15), we obtain

$$2A_2 + 4A_0 = 0 \qquad \text{or} \qquad A_0 = -A_2/2.$$

Thus,

$$H(\rho) = \frac{A_2(-1+2\rho^2)}{2},$$

$$\psi = \frac{A_2(-1+2\rho^2)}{2} \exp\left(-\frac{\rho^2}{2}\right). \tag{A13.26}$$

The general solution of Eq. (A13.26), for $E = (n + 1/2)\hbar\omega$, is of the form

$$\psi_n = C_n H_n(\rho) \exp(-\rho^2/2), \tag{A13.27}$$

where $H_n(\rho)$ is a polynomial of order n. The polynomials $H_n(\rho)$ are known as the Hermite polynomials. The complete normalized wave function is

$$\psi_n = [2^n n!\,(\pi)^{1/2} a_0]^{-1/2} H_n\left(\frac{x}{a_0}\right) \exp\left(-\frac{x^2}{2a_0^2}\right) \tag{A13.28}$$

with

$$H_0(\rho) = 1, \qquad H_1(\rho) = 2\rho, \qquad H_2(\rho) = 4\rho^2 - 2,$$

$$H_3(\rho) = 8\rho^3 - 12, \qquad H_4(\rho) = 16\rho^4 - 48\rho^2 + 12.$$

Appendix 14

Spherical Harmonics

In this appendix, we shall solve the eigenvalue equation for angular momentum, i.e., (3.46), and obtain its solutions in terms of functions, called spherical harmonics. We can write this eigenvalue equation as

$$\frac{1}{\sin\theta}\frac{\partial}{\partial\theta}\left(\sin\theta\frac{\partial\Omega}{\partial\theta}\right)+\frac{1}{\sin^2\theta}\frac{\partial^2\Omega}{\partial\phi^2}+\frac{\lambda\Omega}{\hbar^2}=0. \tag{A14.1}$$

Substituting

$$\Omega(\theta, \phi)=\Theta(\theta)\Phi(\phi) \tag{A14.2}$$

in Eq. (A14.1) and separating the parts θ and ϕ [see Eq. (3.48)] of the result so obtained, we get

$$\frac{1}{\Phi(\phi)}\frac{d^2\Phi(\phi)}{d\phi^2}=\alpha, \tag{A14.3}$$

$$\frac{1}{\Theta(\theta)}\sin\theta\frac{d}{d\theta}\left(\sin\theta\frac{d\Theta}{d\theta}\right)+\frac{\lambda}{\hbar^2}\sin^2\theta=-\alpha. \tag{A14.4}$$

The solution of $\Phi(\phi)$ has been worked out in Example 3.4 and is given by Eq. (3.61), i.e.,

$$\Phi(\phi)=A\exp(im\phi), \qquad m=0, \pm1, \pm2, \ldots. \tag{A14.5}$$

The value of m is integral because $\Phi(\phi)$ must be single-valued, i.e., $\Phi(\phi+2\pi)=\Phi(\phi)$. Substituting this in Eq. (A14.3), we get

$$\alpha=-m^2, \tag{A14.6}$$

and Eq. (A14.4) then becomes

$$\sin\theta\frac{d}{d\theta}\left(\sin\theta\frac{d\Theta}{d\theta}\right)+\left(\frac{\lambda}{\hbar^2}\sin^2\theta-m^2\right)\Theta=0. \tag{A14.7}$$

Let $\cos\theta=x$. Then,

$$\frac{d}{dx}=\frac{d}{d\theta}\frac{d\theta}{dx}=-\frac{1}{\sin\theta}\frac{d}{d\theta}.$$

Equation (A14.7) can be written in terms of x as

$$\frac{d}{dx}\left[(1-x^2)\frac{d\Theta}{dx}\right]+\left(\frac{\lambda}{\hbar^2}-\frac{m^2}{1-x^2}\right)\Theta=0. \tag{A14.8}$$

For this equation, we shall consider two possibilities, namely, $m = 0$ and $m \neq 0$:

(1) $m = 0$ In this situation, Eq. (A14.8) becomes

$$\frac{d}{dx}[(1 - x^2)\frac{d\Theta}{dx}] + \frac{\lambda}{\hbar^2}\Theta = 0. \tag{A14.9}$$

This is a well known differential equation in mathematical physics and is referred to as Legendre's differential equation. As in the simple harmonic oscillator, so too here, the solution of this equation can be written in the form of a series, namely,

$$\Theta(x) = \sum_{k=0}^{\infty} a_k x^k. \tag{A14.10}$$

Substituting this in Eq. (A14.9), we get

$$\sum_{k=0}^{\infty} x^k[(k + 1)(k + 2)a_{k+2} - \{k(k + 1) - \frac{\lambda}{\hbar^2}\}a_k] = 0 \tag{A14.11}$$

or

$$a_{k+2} = \frac{k(k + 1) - \lambda/\hbar^2}{(k + 1)(k + 2)}a_k. \tag{A14.12}$$

If series (A14.10) does not terminate as $k \rightarrow \infty$, then $a_{k+2}/a_k \rightarrow 1$, and hence the series diverges for $x \rightarrow \pm 1$ [since it behaves as $\Sigma\,(1/k)x^k \rightarrow \Sigma\,(1/k) = \infty$]. However, this is not allowed because $\Theta(\cos\theta)$ must remain finite for all values of $\cos\theta$ between -1 and 1. If the series terminates for $k = l$, then

$$a_l \neq 0,$$

$$a_k = 0 \quad \text{for } k > l. \tag{A14.13}$$

Using $a_{l+2} = 0$ in Eq. (A14.12), we get (with $a_l \neq 0$)

$$l(l + 1) - \lambda/\hbar^2 = 0$$

or

$$\lambda = l(l + 1)\hbar^2. \tag{A14.14}$$

Here, l can take any integral value, i.e., $l = 0, 1, 2, \ldots$. Thus, we have derived Eq. (3.50). We can use the recurrence relation (A14.13) to obtain the solutions of Eq. (A14.11) for different values of l, as follows.

When $l = 0$, we have $\lambda = 0$, $a_0 \neq 0$, $a_i = 0$ $(i > 0)$. Then,

$$\Theta(x) = a_0.$$

When $l = 1$, we have $\lambda = 2\hbar^2$, $a_1 \neq 0$, $a_i = 0$ $(i > 1)$. Since $a_2 = 0$, we get $a_0 = 0$. Thus,

$$\Theta(x) = a_1 x = a_1 \cos\theta.$$

When $l = 2$, we have $\lambda = 6\hbar^2$, $a_2 \neq 0$, $a_i = 0$ $(i > 2)$. Since $a_3 = 0$, we get $a_1 = 0$ and, from

Eq. (A14.12), $a_2 = -3a_0$. Thus,

$$\Theta(x) = a_0(1 - 3x^2) = a_0'(3\cos^2\theta - 1).$$

This function is normalized such that $\Theta(x = 1) = 1$. Then,

$$\Theta(x) = \frac{3\cos^2\theta - 1}{2}.$$

For any l, it can be shown that

$$\Theta(x) = P_l(x) = \frac{1}{2^l l!}\frac{d^l}{dx^l}(x^2 - 1)^l, \quad \text{(A14.15)}$$

$P_l(x)$ being the Legendre polynomials.

(2) $m \neq 0$ If $\Theta(x)$ is a solution of Eq. (A14.9), then

$$(1 - x^2)^{|m|/2}\frac{d^{|m|}\Theta(x)}{dx^{|m|}}$$

is a solution of Eq. (A14.8). Thus, the solutions of Eq. (A14.8), in terms of the Legendre polynomials, can be obtained as

$$\Theta(x) = P_l^m(x) = (1 - x^2)^{|m|/2}\frac{d^{|m|}P_l(x)}{dx^{|m|}} = \frac{(1 - x^2)^{|m|/2}}{2^l l!}\frac{d^{l+|m|}(x^2 - 1)^l}{dx^{l+|m|}}. \quad \text{(A14.16)}$$

If $|m| > l$, then $P_l^m(x)$ vanishes. To see this, consider the relation

$$\frac{d^{l+|m|}}{dx^{l+|m|}}(x^2 - 1)^l = \frac{d^{l+|m|}}{dx^{l+|m|}}[x^{2l} + a_1x^{2l-1} + a_2x^{2l-2} + \ldots].$$

Clearly, if $|m| > l$, the differentiation of $P_l^m(x)$ will give 0. Thus, $|m|$ has to be restricted to values less than or equal to l, i.e.,

$$-l \leqslant m \leqslant l. \quad \text{(A14.17)}$$

The complete solution (A14.2) is

$$\begin{aligned}\Omega(\theta, \phi) &= \Theta(x)\Phi(\phi) = CP_l^m(x)e^{im\phi}\\ &= CP_l^m e^{im\phi} \qquad (m > 0)\\ &= CP_l^{|m|}e^{-i|m|\phi} \qquad (m < 0).\end{aligned} \quad \text{(A14.18)}$$

The constant C is so chosen that the function $\Omega(\theta, \phi)$ gets normalized when integrated over all the directions [see Eq. (3.57)]. The spherical harmonics are the normalized functions $\Omega(\theta, \phi)$, and are given by

$$Y_l^m(\theta, \phi) = \left[\frac{2l + 1}{4\pi}\frac{(l - m)!}{(l + m)!}\right]^{1/2}(-1)^m \exp(im\phi)P_l^m(\cos\theta), \qquad 0 \leqslant m \leqslant l, \quad \text{(A14.19)}$$

$$Y_l^m = (-1)^m Y_l^{-m*}, \qquad -l \leqslant m < 0. \quad \text{(A14.20)}$$

The spherical harmonics for $l = 0$, 1, and 2 are given by Eqs. (3.52), (3.53), and (3.54).

Appendix 15

Solution of Radial Equation of Hydrogen Atom

In this appendix, we shall discuss the solution of the radial equation of the hydrogen atom [i.e., Eq. (4.55)] and show how we can obtain the energy levels of the hydrogen atom [given by Eq. (4.56)] and the eigenfunctions corresponding to these levels. The radial equation of the hydrogen atom [i.e., Eq. (4.55)] is

$$\frac{1}{r^2}\frac{d}{dr}\left(r^2\frac{dR_n^l}{dr}\right) + \frac{2m}{\hbar^2}\left[E + \frac{e^2}{4\pi\epsilon_0 r} - \frac{\hbar^2}{2m}\frac{l(l+1)}{r^2}\right]R_n^l = 0 \tag{A15.1}$$

with the boundary condition that $R(r) \to 0$ as $r \to \infty$. We first substitute

$$R_n^l(r) = U_n^l(r)/r \tag{A15.2}$$

in Eq. (A15.1). Then, using the identity

$$\frac{d}{dr}\left(r^2\frac{dR_n^l(r)}{dr}\right) = \frac{d}{dr}\left[-\frac{dU_n^l}{dr} + r\frac{dU_n^l}{dr}\right] = r\frac{d^2U_n^l}{dr^2},$$

we obtain the equation

$$\frac{d^2}{dr^2}U_n^l(r) + \frac{2m}{\hbar^2}\left[E + \frac{e^2}{4\pi\epsilon_0 r} - \frac{\hbar^2}{2m}\frac{l(l+1)}{r^2}\right]U_n^l(r) = 0. \tag{A15.3}$$

Using the dimensionless variables

$$\rho = \frac{(8m|E|)^{1/2}}{\hbar}r, \qquad \lambda = \left(\frac{m}{2|E|}\right)^{1/2}\frac{e^2}{\hbar 4\pi\epsilon_0}, \tag{A15.4}$$

we find Eq. (A15.3) assumes the form

$$\frac{d^2U}{d\rho^2} + \left[\frac{\lambda}{\rho} - \frac{1}{4} - \frac{l(l+1)}{\rho^2}\right]U = 0. \tag{A15.5}$$

As $\rho \to \infty$ (i.e., $r \to \infty$), Eq. (A15.5) becomes $d^2U/d\rho^2 - \frac{1}{4}U = 0$. It has the solutions

$$U = \exp(\rho/2) \quad \text{or} \quad \exp(-\rho/2).$$

Of these, we choose the solution $\exp(-\rho/2)$ because the other solution does not satisfy the boundary condition $U \to 0$ as $\rho \to \infty$. Then, substituting

$$U = L\exp(-\rho/2) \tag{A15.6}$$

in Eq. (A15.4), the equation for L we get is

$$\frac{d^2L}{d\rho^2} - \frac{dL}{d\rho} + \left[\frac{\lambda}{\rho} - \frac{l(l+1)}{\rho^2}\right]L = 0. \tag{A15.7}$$

Let

$$L = \sum_{i=0}^{\infty} A_i \rho^i. \tag{A15.8}$$

Since $R_n^l = U_n^l/r$, it is necessary that $U_n^l \to 0$ as $r \to 0$. Otherwise, R_n^l will become infinite. This implies

$$L(\rho = 0) = A_0 = 0. \tag{A15.9}$$

Substituting Eq. (A15.8) in Eq. (A15.7), we get

$$l(l+1)A_1 = 0,$$

$$[i(i+1) - l(l+1)]A_{i+1} + (\lambda - i)A_i = 0 \qquad (i \geqslant 1). \tag{A15.10}$$

If series (A15.8) does not terminate, then

$$\lim_{i\to\infty} \frac{A_{i+1}}{A_i} \to \frac{1}{i+1}. \tag{A15.11}$$

This means $\Sigma A_i \rho^i \to \exp(\rho)$. Clearly, this is unacceptable as

$$U = L \exp(-\rho/2) \to \exp(\rho/2) \quad \text{as} \quad \rho \to \infty.$$

A physically acceptable solution requires that the series must terminate. Thus,

$$L = \sum_{i=1}^{n} A_i \rho^i$$

with $A_n \neq 0$. Then,

$$A_{n+1} = A_{n+2} = \ldots = A_{n+j} = \ldots = 0. \tag{A15.12}$$

If $A_{n+1} = 0$ and $A_n \neq 0$, we have, from Eq. (A15.10),

$$n = \lambda. \tag{A15.13}$$

From Eq. (A15.10), we can show that

$$\begin{aligned} A_1 &= 0 \quad \text{if } l \geqslant 1, \\ A_2 &= 0 \quad \text{if } l \geqslant 2, \\ A_3 &= 0 \quad \text{if } l \geqslant 3, \\ A_n &= 0 \quad \text{if } l \geqslant n. \end{aligned} \tag{A15.14}$$

If $l \geqslant n$, then

$$A_1 = A_2 = \ldots = A_n = 0. \tag{A15.15}$$

This condition along with Eq. (A15.12) implies the only allowed solution is the trivial solution zero. For a nontrivial solution to exist, the condition is

$$l \leqslant n - 1. \tag{A15.16}$$

The series this condition leads to is called the *associated Legendre polynomial* and can be written for different values of n and l. The energy levels, using Eqs. (A15.4) and (A15.13), are then

$$E = -\frac{mc^2\alpha^2}{2}\frac{1}{n^2} \qquad (n = 1, 2, \ldots) \tag{A15.17}$$

which are the same as those expressed by Eq. (4.56). The radial wave function for $n = 1$, $l = 0$ is

$$R_1^0(r) = \left(\frac{1}{a_0}\right)^{3/2} 2 \exp\left(-\frac{r}{a_0}\right), \tag{A15.18}$$

and the radial wave functions for $n = 2$, $l = 1, 0$ are

$$R_2^1 = \left(\frac{1}{2a_0}\right)^{3/2} \frac{1}{(3)^{1/2}} \frac{r}{a_0} \exp\left(-\frac{r}{2a_0}\right), \tag{A15.19}$$

$$R_2^0 = \left(\frac{1}{2a_0}\right)^{3/2} 2\left(1 - \frac{r}{2a_0}\right) \exp\left(-\frac{r}{2a_0}\right). \tag{A15.20}$$

These wave functions are so normalized that

$$\int_0^\infty |R_n(r)|^2 r^2\, dr = 1. \tag{A15.21}$$

The complete wave functions, including the radial and angular part, for $n = 1, 2$ are given by Eqs. (4.59) to (4.63).

Appendix 16

Some Properties of One-Dimensional Wave Functions

In this appendix, we shall prove the following two results:

(1) The wave function(s) corresponding to a definite energy can always be chosen to be real.

(2) In a one-dimensional potential, there is only one wave function for a bound state. In other words, a one-dimensional bound state is nondegenerate.

If $\psi(x)$ corresponds to a definite energy E_0, then $\psi(x)$ satisfies the Schroedinger equation

$$\frac{d^2\psi}{dx^2} + \frac{2m}{\hbar^2}(E_0 - V)\psi = 0. \qquad \text{(A16.1)}$$

Taking its complex conjugate, we get

$$\frac{d^2\psi^*}{dx^2} + \frac{2m}{\hbar^2}(E_0 - V)\psi^* = 0, \qquad \text{(A16.2)}$$

showing that ψ^* is also a solution of the Schroedinger equation (A16.1) corresponding to the same energy E_0. Further, $\phi = \psi + \psi^*$ and $\chi = (\psi - \psi^*)/i$ also satisfy the Schroedinger equation with energy E_0. Hence, ϕ and χ are equally acceptable wave functions corresponding to the energy E_0. It should be noted that

$$\phi = \phi^* = 2\,\mathrm{Re}\,\psi, \qquad \text{(A16.3)}$$

$$\chi = \chi^* = 2\,\mathrm{Im}\,\psi, \qquad \text{(A16.4)}$$

and thus ϕ and χ are real functions.

For a one-dimensional potential, it can be shown that the two functions ϕ and χ are not independent of each other if they correspond to a bound state. To do this, we start with the identity

$$\frac{d}{dx}\left(\phi\frac{d\chi}{dx} - \chi\frac{d\phi}{dx}\right) = \phi\frac{d^2\chi}{dx^2} - \chi\frac{d^2\phi}{dx^2}. \qquad \text{(A16.5)}$$

Since ϕ and χ satisfy the Schroedinger equation with energy E_0, Eq. (A16.5) can be written as

$$\frac{d}{dx}\left(\phi\frac{d\chi}{dx} - \chi\frac{d\phi}{dx}\right) = \phi\left[-\frac{2m}{\hbar^2}(E_0 - V)\chi\right] - \chi\left[-\frac{2m}{\hbar^2}(E_0 - V)\phi\right] = 0.$$

Then,

$$\phi\frac{d\chi}{dx} - \chi\frac{d\phi}{dx} = C, \qquad \text{(A16.6)}$$

where C is a constant independent of x and can be evaluated at any value of x. If ψ represents a bound state, then $\psi(x) \to 0$ as $x \to \infty$, implying $\phi = 2 \text{ Re } \psi = 0$ and $\chi = 2 \text{ Im } \psi \to 0$ as $x \to \infty$. Thus,

$$\phi\frac{d\chi}{dx} - \chi\frac{d\phi}{dx} = [\phi\frac{d\chi}{dx} - \chi\frac{d\phi}{dx}]_{x=\infty} = 0 \tag{A16.7}$$

and we get

$$\phi\frac{d\chi}{dx} = \chi\frac{d\phi}{dx} \tag{A16.8}$$

or

$$\frac{1}{\chi}\frac{d\chi}{dx} = \frac{1}{\phi}\frac{d\phi}{dx} \tag{A16.9}$$

or

$$\frac{d}{dx}(\ln \chi) = \frac{d}{dx}(\ln \phi).$$

Integrating, we obtain $\ln \chi = \ln \phi + C$ or

$$\chi = \exp (C)\phi = \text{constant} \times \phi. \tag{A16.10}$$

Thus, χ and ϕ are not independent functions.

We can extend the foregoing argument to prove, in general, that the bound state wave functions of a one-dimensional potential are always nondegenerate. In other words, if ψ_1 and ψ_2 are any two bound state wave functions corresponding to the same energy, we can use the identity

$$\frac{d}{dx}(\psi_1\frac{d\psi_2}{dx} - \psi_2\frac{d\psi_1}{dx}) = 0 \tag{A16.11}$$

and the boundary conditions that the wave functions must vanish as $x \to \pm\infty$ to show that $\psi_1 \propto \psi_2$.

Finally, it should be noted that the foregoing argument cannot be carried to a two- or three-dimensional potential because an identity similar to identity (A16.11), when written in two or a higher dimension, cannot be integrated to yield an equation similar to Eq. (A16.7). Thus, in general, a bound state in two or a higher dimension is degenerate. For instance, a particle in a two-dimensional square box (see Example 2.20) has degenerate states. Similarly, the bound state wave functions of a hydrogen atom are, in general, degenerate.

Appendix 17

Distance Measurement in Astronomy

EARTH-SUN DISTANCE

As already noted, the earth-sun distance was determined by studying the motion of mars. For this, the apparent time T_{app} between any two consecutive conjunctions of mars (it being in the same direction as the sun) is noted. Then, from this, the relative angular velocity ω_{me} of mars with respect to earth is found. If T_e and T_m are the time taken by the earth and mars, respectively, to go around the sun, the angular velocities would be

$$\omega_e = 2\pi/T_e, \qquad \omega_m = 2\pi/T_m. \tag{A17.1}$$

The relative angular velocity is given by $\omega_{me} = \omega_e - \omega_m$, i.e.,

$$2\pi/T_{app} = 2\pi/T_e - 2\pi/T_m \tag{A17.2}$$

or

$$T_m = \frac{T_e T_{app}}{T_e + T_{app}}. \tag{A17.3}$$

Now, taking $T_e = 365$ days and $T_{app} = 780$ days, the time mars takes to go around the sun is found. This works out to 686 days. This information is then used to plot the trajectory of mars with respect to that of earth. If the position of earth is E_1 and that of mars is M (see Fig. A17.1),

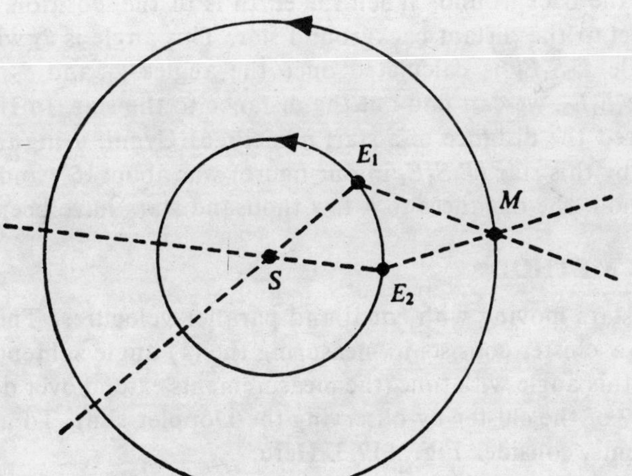

Fig. A17.1 Plotting orbit of mars.

then, after 686 days, mars would return to the same position, whereas the earth would be at E_2. Since the earth takes 365 days to travel around the sun, the angle E_1SE_2 is about 45°. Now, assuming the earth's orbit, the angles SE_1M and SE_2M are measured. This fixes M. Thereafter, following mars every day, the trajectory of mars with respect to the sun is fixed. This, in turn, fixes the ratio of the earth-sun distance and the earth-mars distance. Once the earth-mars distance is known, the earth-sun distance can be determined. More recently, the orbit of the minor planet eros was used in arriving at the earth-sun distance.

The earth-mars distance was measured using the trigonometric method, fixing a base line on the earth.

TRIGONOMETRIC PARALLAX METHOD

As the earth moves around the sun, the maximum separation between any two points on its orbit is twice the earth-sun distance ($\sim 300 \times 10^9$ m). We can use this as a base line for measuring the distance of the stars nearest to us. Figure A17.2 shows the parallax of a nearby star (S_1)

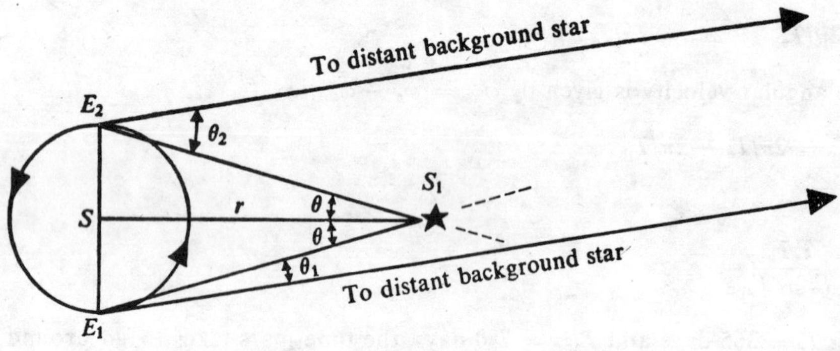

Fig. A17.2 Method of trigonometric parallax.

with a distant star in the background. When the earth is in the position E_1, the star subtends an angle θ_1 with respect to the distant background star. This angle is θ_2 when the earth is in the position E_2. The angle $E_2S_1E_1$ is calculated once the angles θ_1 and θ_2 have been measured. Knowing the distance E_1E_2, we can find out the distance to the star. In 1838, F. W. Bessel, for the first time, measured the distance of a star, namely, 61 Cygni, using trigonometric parallax. The angle subtended by this star ($E_1S_1E_2$ in our figure) was about .6″, and its distance is about 10 ly. Using this method, the distances to a few thousand stars have been determined.

MOVING-CLUSTER METHOD

There are clusters of stars moving with equal and parallel velocities. The method to calculate the distance of such a cluster consists in measuring the (1) angle subtended by the cluster and noting the change in this angle with time (the measurements extend over decades), and (2) radial component of velocity of the cluster by observing the Doppler shift. To arrive at the distance r from such measurements, consider Fig. A17.3. Here,

$$r = R \sin \theta_1, \tag{A17.4}$$

$$r + vt = R \sin \theta_2. \tag{A17.5}$$

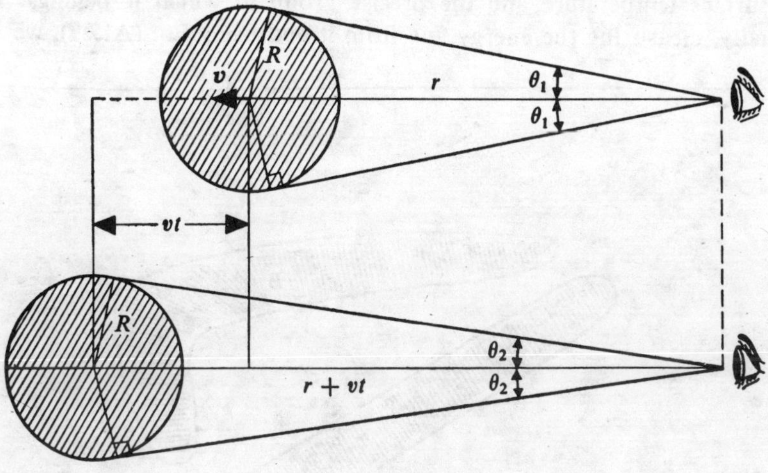

Fig. A17.3 Method of moving clusters.

Eliminating R, we get

$$r = \frac{vt \sin \theta_2}{\sin \theta_1 - \sin \theta_2} \sim \frac{vt\theta_2}{\theta_1 - \theta_2}. \tag{A17.6}$$

The Hyades cluster, whose distance was measured using this method, is at a distance of 140 ly.

MAIN SEQUENCE-FITTING METHOD

For a large number of stars of known distances (e.g., the Hyades cluster), E. N. Hertzsprung and H. N. Russell found that there is a simple relation between their luminosity (i.e., the energy emitted per second) and surface temperature. The luminosity L of a star of known distance can be determined once we measure the energy flux e to the earth. (The energy flux is defined as the energy received per unit area per unit time.) If r is the distance to the star, the luminosity is given by

$$L = 4\pi r^2 e, \tag{A17.7}$$

ignoring the interstellar absorption. To obtain the surface temperature of a star, first the ionized atoms, in the spectrum of the light from the star, are identified and their degree of ionization measured. From this measurement, the electron population of the various energy levels of the ionized atoms is computed. The surface temperature of the star is then determined by fitting this electron population to the Boltzmann distribution.

If luminosity is plotted against log T^{-1}, we get what is known as the H-R diagram (named after Hertzsprung and Russell)—see Fig. A17.4. In this diagram, the stars fall into three broad groups, namely, the main sequence (A in the figure), the red giants (B in the figure), and the white dwarfs (C in the figure). These groups actually indicate the different stages of stellar evolution and we can identify the group to which a star belongs from the spectrum of the star, a procedure referred to as the spectral classification of stars.

To find the distance to a distant star, we need to assume that it belongs to one of three groups (i.e., A, B, or C) in the H-R diagram. Then, by studying the spectrum of the star, we can determine its surface temperature and the precise group to which it belongs. This gives its luminosity. Finally, measuring the energy flux from it and using Eq. (A17.7), we can calculate

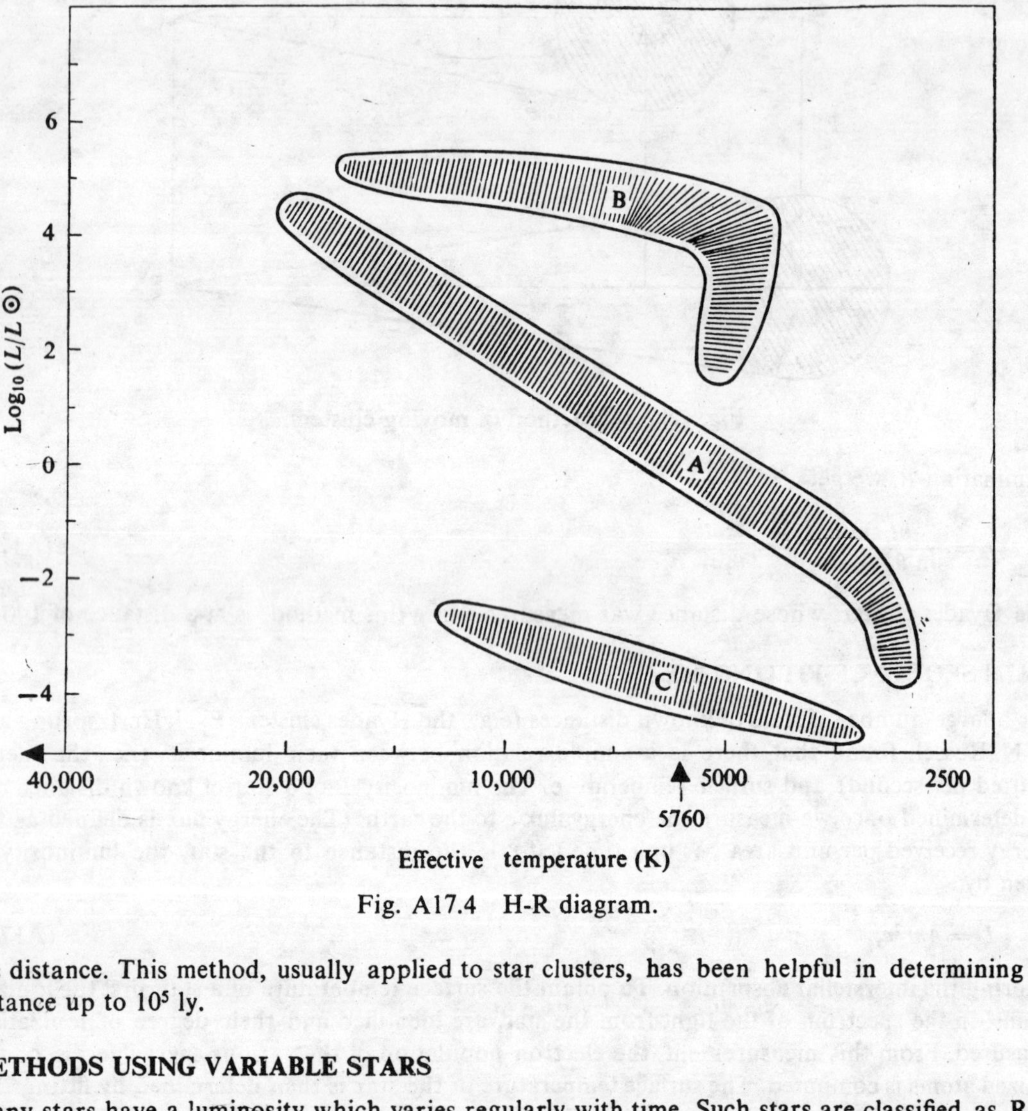

Fig. A17.4 H-R diagram.

the distance. This method, usually applied to star clusters, has been helpful in determining a distance up to 10^5 ly.

METHODS USING VARIABLE STARS

Many stars have a luminosity which varies regularly with time. Such stars are classified as RR Lyrae stars and classical cepheids. The period of luminosity of the former ranges from a few hours to a day, whereas that of the latter extends from 2 days to 40 days. It has been found that all RR Lyrae stars have a maximum luminosity 20–50 times that of the sun, showing that the luminosity of the RR Lyrae stars is more or less the same. Thus, once an RR Lyrae star is located, its distance can be determined by measuring the energy flux from it [see Eq. (A17.7)]. However, the RR Lyrae stars are not bright enough to be useful for calculating a distance beyond 10^6 ly.

The classical cepheids vary widely in luminosity. As noted by H. S. Leavitt, the luminosity of a classical cepheid is directly proportional to its luminosity period P. Since then, the astronomers have discovered that there are two kinds of cepheids, namely, type I and type II. The two types can be distinguished from each other by studying the shape of their light curves (i.e., the luminosity versus time curves). Therefore, once the period of a cepheid and the type to which it belongs are known, its luminosity can be determined and then, by calculating the energy flux from it, its distance measured. The distance to the Andromeda nebula was first estimated in 1923 by E. Hubble utilizing this method. The method has been useful in studying the local group of galaxies and is applicable for measuring a distance up to 10^7 ly.

Appendix 18

Doppler Shift for Light

The frequency of a light wave measured by an observer at rest with respect to the source emitting the wave is different from that determined by an observer moving with respect to the source. This difference in frequency is referred to as the Doppler shift. Its measurement helps determine the radial velocity of a star or galaxy. In what follows, we shall show how this frequency shift is mathematically derived.

Consider a source emitting a light wave of frequency ν_0 when measured by an observer at rest with respect to the source. This means that the wave would form two successive crests with a time interval $T_0 = 1/\nu_0$. Let an observer be stationed at a point $(-d, 0, 0)$ in a frame S with respect to which the source moves with a velocity v along the x-axis. The frame fixed to the source is S', its origin coinciding with the source. According to S', two successive crests are emitted at, say, time $t' = 0$ and $t' = T_0$. Using the Lorentz transformation, the coordinates (x, t) in S corresponding to the first crest would be

$$x = 0, \qquad t = 0, \tag{A18.1}$$

and those corresponding to the second crest would be

$$x = v\gamma T_0, \qquad t = \gamma T_0. \tag{A18.2}$$

Thus, the first crest would reach the observer at a time d/c and the second crest at a time $(d + \gamma v T_0)/c + \gamma T_0$, and hence the frequency ν measured by the observer would be

$$\nu = \frac{1}{(v/c)\gamma T_0 + \gamma T_0} = \frac{1}{(1 + v/c)\gamma}\,\frac{1}{T_0} = \left(\frac{1 - v/c}{1 + v/c}\right)^{1/2}\nu_0. \tag{A18.3}$$

As can be noted, $\nu < \nu_0$, showing red shift for the source moving away from the observer. In terms of the wavelengths, we can rewrite formula (A18.3) as

$$\lambda/\lambda_0 = [(1 + v/c)/(1 - v/c)]^{1/2}. \tag{A18.4}$$

For the Virgo cluster of galaxies, $v/c = 3 \times 10^{-3}$, making $\lambda/\lambda_0 \sim 1.003$.

Bibliography

General

Beiser, A., Perspectives of Modern Physics, McGraw-Hill, New York, 1969.

Eisberg, R. and Resnick, R., Quantum Physics, Wiley, New York, 1974.

Feynman, R. P., Leighton, R. B., and Sands, M. L., The Feynman Lectures on Physics, Vol. 3, Addison-Wesley, Reading, Massachusetts, 1965.

Sproull, R. L., Modern Physics, Wiley, New York, 1963.

Quantum Mechanics

Dicke, R. H. and Wittke, J. P., Introduction to Quantum Mechanics, Addison-Wesley, Reading, Massachusetts, 1960.

Mathews, P. M. and Venkatesan, K., A Textbook of Quantum Mechanics, Tata McGraw-Hill, New Delhi, 1976.

Atomic and Molecular Physics

Atkins, P. W., Physical Chemistry, Oxford University Press, Oxford, 1978.

Chanda, Manas, Atomic Structure and Chemical Bond, Tata McGraw-Hill, New Delhi, 1979.

Statistical Mechanics and Solid State Physics

Kittel, C., Introduction to Solid State Physics, Wiley, New York, 1976.

Reif, F., Fundamentals of Statistical and Thermal Physics, McGraw-Hill, New York, 1965.

Relativity

Resnick, R., Introduction to Special Relativity, Wiley, New York, 1968.

Wheeler, J. A. and Taylor, E. F., Space-Time Physics, W. H. Freeman, San Francisco, California, 1966.

Nuclear Physics

Cohen, B., Concepts in Nuclear Physics, McGraw-Hill, New York, 1971.

Enge, H., Introduction to Nuclear Physics, Addison-Wesley, Reading, Massachusetts, 1966.

Elementary Particles

Perkins, D. H., An Introduction to High Energy Physics, Addison-Wesley, Reading, Massachusetts, 1983.

Journals for Reviews on Recent Developments

Physics News (Bombay)
Science Today (Bombay)
Physics Today (New York, N.Y.)
Scientific American (New York, N.Y.)

Answers to Problems

CHAPTER 2

2.1 $E = 1.77$ eV **2.2** $\lambda = 12{,}400$ Å **2.4** Number $= 1.7 \times 10^{12}$ sec^{-1}

2.5 (a) $h = 6.2 \times 10^{-34}$ J-sec, (b) $\phi = 2.6$ eV **2.6** $\lambda = 1.24 \times 10^{-3}$ Å

2.7 Particle aspect **2.8** $\lambda_\gamma/\lambda_e = [E/(2m_e c^2)]^{1/2}$

2.9 (a) $v = 3.94 \times 10^4$ m/sec, (b) $V = 8.1$ V **2.10** Electron B **2.11** $E = 137$ MeV

2.12 $d = 3.96$ Å, $\theta = 0.38°$ (for electrons) $= 1.79°$ (for photons)

2.13 $\psi(x, t) = \exp[-i(kx + \omega t)]$ **2.14** $\psi(x, y, t) = \exp[i(kx + ky)/\sqrt{2} - i\omega t]$

2.15 $P = 1/8$ **2.16** $E = .23 \times 10^{-45}$ eV **2.17** $E = .14 \times 10^{-2}$ eV

2.19

$$E_n = [\hbar^2\pi^2/(2mL^2)]n^2 \qquad (n = 1, 2, 3, \ldots),$$

$$\psi_n(x) = (2/L)^{1/2} \cos(\pi x n/L) \qquad (n = 1, 3, 5, \ldots)$$

$$= (2/L)^{1/2} \sin(\pi x n/L) \qquad (n = 2, 4, 6, \ldots)$$

2.21 $\langle p \rangle = \hbar k(|A|^2 - |B|^2)/(|A|^2 + |B|^2)$ **2.22** (a) $n = 1.36 \times 10^{28}$, (b) $E = 1.47 \times 10^{-35}$ J

2.23

$$E = \frac{\hbar^2}{2mL^2}\left(\frac{n_1^2}{L_1^2} + \frac{n_2^2}{L_2^2}\right), \qquad n_1, n_2 = 1, 2, 3, \ldots,$$

$$\psi_{n_1, n_2}(x, y) = \frac{4}{L_1 L_2} \sin\frac{n_1 \pi x}{L_1} \sin\frac{n_2 \pi y}{L_2}$$

2.24

$$\langle x^2 \rangle = \frac{\langle x^2 + y^2 \rangle}{2} = \frac{L^2}{3}\left(1 - \frac{3}{2\pi^2}\right), \qquad \langle p_x^2 \rangle = \frac{\hbar^2\pi^2}{L^2}, \qquad \langle p_x^2 + p_y^2 \rangle = \frac{2\hbar^2\pi^2}{L^2}$$

2.25

$$E = \frac{\hbar^2}{2m}\left(\frac{n_1^2}{L_1^2} + \frac{n_2^2}{L_2^2} + \frac{n_3^2}{L_3^2}\right), \qquad n_1, n_2, n_3 = 1, 2, 3, \ldots,$$

$$\psi_{n_1, n_2, n_3}(x, y, z) = \left(\frac{8}{L_1 L_2 L_3}\right)^{1/2} \sin\frac{n_1 \pi x}{L} \sin\frac{n_2 \pi y}{L} \sin\frac{n_3 \pi z}{L}$$

2.26 (a)

$$E = \frac{\hbar^2\pi^2}{2mL^2}(n_1^2 + n_2^2 + n_3^2)$$

(b)

$$\psi_{n_1, n_2, n_3} = (\frac{8}{L^3})^{1/2} \sin \frac{n_1\pi x}{L} \sin \frac{n_2\pi y}{L} \sin \frac{n_3\pi z}{L}$$

(c) First excited state is triply degenerate, i.e.,

$$n_1 = n_2 = 1,\ n_3 = 2 \quad \text{or} \quad n_1 = n_3 = 1,\ n_2 = 2 \quad \text{or} \quad n_1 = 2,\ n_2 = n_3 = 1$$

Second excited state is also triply degenerate, i.e.,

$$n_1 = 1,\ n_2 = n_3 = 2 \quad \text{or} \quad n_1 = n_2 = 2,\ n_3 = 1 \quad \text{or} \quad n_1 = n_3 = 2,\ n_2 = 1$$

2.27

$$E_{n_1, n_2, n_3;\, l_1, l_2, l_3} = \frac{\hbar^2}{2}[\frac{1}{m_1}(\frac{n_1^2}{a^2} + \frac{n_2^2}{b^2} + \frac{n_3^2}{c^2}) + \frac{1}{m_2}(\frac{l_1^2}{a^2} + \frac{l_2^2}{b^2} + \frac{l_3^2}{c^2})]$$

(n_1, n_2, n_3, and l_1, l_2, l_3 are positive integers greater than zero),

$$\psi_{n_1, n_2, n_3;\, l_1, l_2, l_3}(x_1, y_1, z_1, x_2, y_2, z_2) = (\frac{8}{abc})[\sin \frac{n_1\pi x_1}{a} \sin \frac{n_2\pi y_1}{b} \sin \frac{n_3\pi z_1}{c} \sin \frac{l_1\pi x_2}{a} \times \sin \frac{l_2\pi y_2}{b} \sin \frac{l_3\pi z_2}{c}]$$

2.28 (a)

$$\psi = \frac{8}{abc} \sin \frac{\pi x_1}{a} \sin \frac{\pi y_1}{b} \sin \frac{\pi z_1}{c} \sin \frac{\pi x_2}{a} \sin \frac{\pi y_2}{b} \sin \frac{\pi z_2}{c}$$

(b)

$$\psi = \frac{8}{abc} \cdot \frac{1}{\sqrt{2}}[\sin \frac{\pi x_1}{a} \sin \frac{\pi y_1}{b} \sin \frac{\pi z_1}{c} \sin \frac{2\pi x_2}{a} \sin \frac{\pi y_2}{b} \sin \frac{\pi z_2}{c} - \sin \frac{2\pi x_1}{a} \times \sin \frac{\pi y_1}{b} \sin \frac{\pi z_1}{c} \sin \frac{\pi x_2}{a} \sin \frac{\pi y_2}{b} \sin \frac{\pi z_2}{c}]$$

CHAPTER 3

3.1 $E_g = 3\lambda^{1/3}[\hbar^2/(16m)^{2/3}]$

3.2

$$E = \hbar\omega(n + 1/2) + C \qquad (n = 0, 1, 2, \ldots),$$

$$\psi_0(x) = (\frac{\alpha}{\sqrt{\pi}})^{1/2} e^{-\alpha^2 x^2/2}, \qquad \psi_1(x) = (\frac{2\alpha^3}{\sqrt{\pi}})^{1/2} x e^{-\alpha^2 x^2/2} \qquad [\alpha = (km/\hbar^2)^{1/4}]$$

3.3 $E = \hbar\omega(n + 1/2) - a^2/(2k)$, where $\omega = (k/m)^{1/2}$ and $n = 0, 1, 2, \ldots$

3.5 $E = \hbar\omega(n + 1/2)$, where $\omega = (k/m)^{1/2}$ and $n = 1, 3, 5, \ldots$

3.7 $P(x) = 1/[\pi(a^2 - x^2)^{1/2}]$, where $a = [21\hbar/(m\omega)]^{1/2}$

3.8 $E = \hbar\omega(n_1 + n_2 + n_3 + 3/2)$, where $\omega = (k/m)^{1/2}$ and $n_1, n_2, n_3 = 0, 1, 2, \ldots$
First excited state is triply degenerate, i.e.,

$$n_1 = n_2 = 0,\ n_3 = 1 \quad \text{or} \quad n_1 = n_3 = 0,\ n_2 = 1 \quad \text{or} \quad n_1 = 1,\ n_2 = n_3 = 0$$

Second excited state is six-fold degenerate, i.e.,

$$n_1 = 0,\ n_2 = n_3 = 1 \quad \text{or} \quad n_1 = n_3 = 1,\ n_2 = 0 \quad \text{or} \quad n_1 = n_2 = 1,\ n_3 = 0 \quad \text{or}$$

$$n_1 = n_2 = 0,\ n_3 = 2 \quad \text{or} \quad n_1 = n_3 = 0,\ n_2 = 2 \quad \text{or} \quad n_1 = 2,\ n_2 = n_3 = 0$$

3.9 $E = \hbar[\omega_1(n_1 + 1/2) + \omega_2(n_2 + 1/2)]$, where $n_1, n_2 = 0, 1, 2, \ldots, \omega_1 = (k_1/m)^{1/2}$, and $\omega_2 = (k_2/m)^{1/2}$ **3.12** (a) $\langle L_z \rangle = 2\hbar$, (b) $\langle L_z \rangle = \hbar/2$

3.13 $E = \hbar^2 n^2/(ml^2)$, where $n = 0, 1, 2, \ldots$

3.14

$$E_n = \frac{\hbar^2\pi^2 n^2}{2mL^2} + \frac{\hbar^2 l^2}{2mR^2} \qquad (n = 1, 2, 3, \ldots,\ l = 0, \pm 1, \pm 2, \ldots),$$

$$\psi(z, \phi) = A \sin \frac{n\pi z}{L} e^{il\phi}$$

3.19 $\alpha^{-1} = 1.05 \times 10^{-29}\ \text{m}^{-1}$

3.20

$$R = \left[\frac{1 - (1 - V_0)^{1/2}/E}{1 + (1 - V_0)^{1/2}/E}\right]^2$$

3.22

$$R = \frac{(1 - k^2/\alpha^2)^2 \sin^2 \alpha a}{4k^2/\alpha^2 + (1 - k^2/\alpha^2)^2 \sin^2 \alpha a} \qquad \left[k = \left(\frac{2Em_e}{\hbar^2}\right)^{1/2},\ \alpha = \left\{\frac{2m_e(E + V_0)}{\hbar^2}\right\}^{1/2}\right],$$

$$T = \frac{4k^2/\alpha^2}{4k^2/\alpha^2 + (1 - k^2/\alpha^2)^2 \sin^2 \alpha a}$$

No reflection ($R = 0$) when $E_n = \hbar^2\pi^2 n^2/(2m_e a^2) - V_0$

CHAPTER 4

4.1 (b) $\nu = 2.54 \times 10^{15}$ Hz

4.2 (a) $\lambda_{min} = 9.1 \times 10^{-8}$ m, $\lambda_{max} = 1.2 \times 10^{-7}$ m, (b) $\lambda_{min} = 3.6 \times 10^{-7}$ m, $\lambda_{max} = 6.6 \times 10^{-7}$ m

4.3 (a) $E = 13.2$ eV, (b) $E = 1.51$ eV **4.4** $E = -2.6 \times 10^{-78}$ eV, error $= 1.94 \times 10^{-79}\%$

4.5 (a) $E = 186.1 E_H$, where E_H are energy levels of the hydrogen atom, (b) error $= 11.3\%$, (c) $r = 2.8 \times 10^{-13}$ m **4.6** $\Delta\lambda = .327$ Å

4.7 (a) $E = 10.2$ eV, (b) $E = 10.2(1 + 5 \times 10^{-9})$ eV

4.8 KE $= mc^2\alpha^2/(2n^2)$, $V = -mc^2\alpha^2/n^2$, magnetic moment $= [e\hbar/(2m)]n$

4.9 $\lambda = 18.7 \times 10^{-7}$ m, 6.56×10^{-7} m, 1.03×10^{-7} m, 1.21×10^{-7} m, 4.86×10^{-7} m, 0.97×10^{-7} m

4.10 $\nu = 2.93 \times 10^{15}\ \text{sec}^{-1}$, $2.47 \times 10^{15}\ \text{sec}^{-1}$, $4.6 \times 10^{15}\ \text{sec}^{-1}$

4.11 ${}^3_2He^+$, $\Delta E = 1.85 \times 10^{-3}$ eV **4.14** $\langle r^2 \rangle = 3a_0^2$

4.15 $\langle V \rangle = -e^2/(4\pi\epsilon_0 a_0)$, $\langle KE \rangle = e^2/(4\pi\epsilon_0 2a_0)$

4.16 $\langle x \rangle = L/2$, $\langle x^2 \rangle = (L^2/3)[1 - 3/(2n^2\pi^2)]$

4.17 Magnetic moment $= 2.1 \times 10^{-24}$ J/weber-m^{-2} **4.18** $J = 4\hbar$

4.19 $E = 30\hbar^2\pi^2/(m_e L^2)$ **4.20** $\nu = 3\pi\hbar/(4m_e L^2)$, $3\pi\hbar/(2m_e L^2)$

4.21 (a) Al: $1s^2 2s^2 2p^6 3s^2 3p^1$, (b) Zn: $1s^2 2s^2 2p^6 3s^2 3p^6 3d^{10} 4s^2$ **4.22** $E = 122$ eV

4.23 $\lambda = 0.15$ Å **4.24** $E = 2399$ keV

CHAPTER 5

5.1 Potential energy $= -57.6/R$ eV **5.2** $d = 8$ Å **5.3** $n = 2.6$, $C = 22.4$ eV Å$^{2.6}$

5.6 $\nu_j = 3.1 \times 10^{13} j$ sec^{-1} **5.7** $R = 1.6 \times 10^{-10}$ m **5.8** 1 : 0.51 : .30

5.9 $E = .184$ eV

5.11 (b) $\nu = \omega/(2\pi) - \hbar a^2 v/\pi$, (c) $D_e = 10$ eV, dissociation energy $= 9.82$ eV

5.12 Number of levels = 48

5.13 $\nu = (8.4 \times 10^{12} + 6.9 \times 10^9)$ sec^{-1}, $(8.4 \times 10^{12} + 13.8 \times 10^9)$ sec^{-1}, $(8.4 \times 10^{12} + 20.7 \times 10^9)$ sec^{-1}

CHAPTER 6

6.1 The different ways are

$(a_1a_2a_3;\ a_4a_5)$, $(a_1a_2a_4;\ a_3a_5)$, $(a_1a_4a_3;\ a_2a_5)$, $(a_4a_2a_3;\ a_1a_5)$, $(a_5a_1a_2;\ a_3a_4)$,

$(a_5a_1a_3;\ a_2a_4)$, $(a_5a_3a_2;\ a_1a_4)$, $(a_4a_5a_1;\ a_2a_3)$

6.2 $P = 10^{-8.6 \times 10^{18}}$ **6.3** $P = {}^N C_n (v/V)^n [(V - v)/v]^{N-n}$

6.4 (a) Larger portion: 750 molecules of Ne, 75 molecules of He, smaller portion: 250 molecules of Ne, 25 molecules of He, (b) $P = (\frac{1}{4})^{100} (\frac{3}{4})^{1000}$

6.5 $P = 10^{-0.4 \times 10^{26}}$ **6.6** Volume in phase space $= 2\pi \Delta U/\omega$

6.7 $P(x)\, dx = m\omega\, dx/[\pi(2mU - m^2\omega^2 x^2)^{1/2}]$

6.8

Energy	$-4\mu B$	$-3\mu B$	$-2\mu B$	$-\mu B$	0	μB	$2\mu B$	$3\mu B$	$4\mu B$
Microstate(s)	1	0	4	0	6	0	4	0	1

6.9 (a) $E = 2n\mu B - N\mu B$, (b) ${}^N C_n$, (c) $S = k[N \ln N - n \ln n - (N - n) \ln (N - n)]$

6.10 (a)

a	b	c	d	e
×	×	×	×	
×	×	×		×
×	×		×	×
×		×	×	×
	×	×	×	×

that is, 5 ways

(b)

All four particles in one state	5 ways
Three particles in one state, one in another state	20 ways
Two particles in one state, two in another state	10 ways
Two particles in one state, one in another state, one in still another state	30 ways
Each particle in a different state	5 ways
Total	70 ways

(c) Since the particles are distinguishable, let us label them as x_1, x_2, x_3, x_4; then,

x_1 can be in the states a, b, c, d, e,

x_2 can be in the states a, b, c, d, e,

x_3 can be in the states a, b, c, d, e,

x_4 can be in the states a, b, c, d, e;

thus, total number of ways is 5^4

6.11 Number of microstates $= dE/(\hbar\omega)$ **6.12** Number of microstates $= E\, dE/(\hbar^2\omega^2)$

6.13 Number of microstates $= 4\pi VC^{-3/n}/(nh^3) = U^{3/n-1}\Delta U$ **6.14** Ratio $= 4 \times 10^{-132}$

6.16 $\langle E\rangle = 3kT/2$ **6.17** $\langle E\rangle = \frac{3}{5}E_F$ **6.18** $\langle v\rangle = 0$, $\langle v\rangle = \frac{3}{4}(2E_F/m)^{1/2}$

6.19 (a) $E = 81\hbar^2\pi^2/(2mL^2)$, (b) $E = 30\hbar^2\pi^2/(2mL^2)$

6.21 $\lambda_{max} = (2.89 \times 10^{-3}/T)$ m, constant $= 2.89 \times 10^{-3}$ m-K

6.22 Surface temperature $= 5780$ K **6.23** Power $= 4.52 \times 10^{26}$ W

6.24 Surface temperature $= 8256$ K

6.25 (a) $\psi(x_1, x_2) = (1/\sqrt{2})[\psi_0(x_1)\psi_1(x_2) - \psi_0(x_2)\psi_1(x_1)]$, where $\psi_0(x)$ and $\psi_1(x)$ are given by Eqs. (3.7) and (3.8), E being equal to $2\hbar\omega$, (b) $\psi(x_1, x_2) = \psi_0(x_1)\psi_0(x_2)$, where $\psi_0(x)$ is given by Eq. (3.7), E being equal to $\hbar\omega$

6.26 $\langle(x_1 - x_2)^2\rangle_{\text{fermions}} = 0.175L^2$, $\langle(x_1 - x_2)^2\rangle_{\text{bosons}} = 0.06L^2$

6.27 (a)

$$\psi_{\text{fermions}}(x_1y_1;\, x_2y_2) = \frac{1}{\sqrt{2}}\left(\frac{2}{L}\right)^2\left[\sin\frac{\pi x_1}{L}\sin\frac{\pi y_1}{L}\sin\frac{\pi x_2}{L}\sin\frac{2\pi y_2}{L} - \sin\frac{\pi x_2}{L}\sin\frac{\pi y_2}{L}\right.$$
$$\left.\times \sin\frac{\pi x_1}{L}\sin\frac{2\pi y_1}{L}\right]$$

or

$$\psi_{\text{fermions}}(x_1y_1;\, x_2y_2) = \frac{1}{\sqrt{2}}\left(\frac{2}{L}\right)^2\left[\sin\frac{\pi x_1}{L}\sin\frac{\pi y_1}{L}\sin\frac{2\pi x_2}{L}\sin\frac{\pi y_2}{L} - \sin\frac{\pi x_2}{L}\sin\frac{\pi y_2}{L}\right.$$
$$\left.\times \sin\frac{2\pi x_1}{L}\sin\frac{\pi y_1}{L}\right]$$

with $E = [\hbar^2/(2mL^2)] \times 7$

(b)

$$\psi_{\text{bosons}}(x_1y_1;\, x_2y_2) = \left(\frac{2}{L}\right)^2 \sin\frac{\pi x_1}{L} \sin\frac{\pi y_1}{L} \sin\frac{\pi x_2}{L} \sin\frac{\pi y_2}{L}$$

with $E = [\hbar^2/(2mL^2)] \times 4$

CHAPTER 7

7.1 $E_F = 5.4$ eV

7.4

$(E - E_F)/(kT)$	$n(E)$	Percentage error
1	0.27	36
3	0.047	5
10	4.54×10^{-5}	4.6×10^{-3}

7.5 Electron density $= 1.6 \times 10^{28}$ per m^3

7.8 $E_F = 7.1(1 + x/150)$ eV if x is the percentage of Zn

7.9 $C_{elec} = 0.077R$, percentage contribution $= 3$

7.10 (a) Si: $\lambda = 0$–12,360 Å, (b) Ge: $\lambda = 0$–17,650 Å, (c) Ag: $\lambda = 0$–∞ (that is, totally opaque)

7.11 $\mu = 4.16 \times 10^{-3}$ m²/V-sec **7.12** Density $= 10^{25}/m^3$ **7.13** $T = 350$ K

7.14 $\sigma = 7.68$ mho/m **7.15** $\Delta E = 0.094$ eV **7.16** Number of atoms $= 35$

7.18 Number of electrons = number of holes $= 0.62 \times 10^{20}/m^3$, number of holes $= 0.17 \times 10^{20}/m^3$

7.19 Concentration of electrons $= 5.8 \times 10^{20}/m^3$, concentration of holes $= 1.72 \times 10^{19}/m^3$

7.20 Density of hole carriers $= 10^{19}/m^3$ **7.21** $E_F - E_g/2 = (kT/2) \ln (4kTB/A)$

CHAPTER 8

8.1 $x' = \gamma(x - vt),\ y' = y,\ z' = z,\ t' - \alpha = \gamma(t - vx/c^2)$

8.2 $x' = x(v_y^2 + \gamma v_x^2)/v^2 + (v_x v_y/v^2)(\gamma - 1)y - \gamma v_x t,\ y' = (\gamma - 1)(v_x v_y/v^2)x + y(\gamma v_y^2 + v_x^2)/v^2 - \gamma v_y t,$ $z' = z,\ t' = \gamma[t - (v_x x + v_y y)/c^2]$, where $\gamma = (1 - v^2/c^2)^{-1/2}$ and $v = (v_x^2 + v_y^2)^{1/2}$

8.3 $x_2' = (2 \pm 2\sqrt{7})$ m **8.4** $t' = -8.7 \times 10^{-4}$ sec **8.5** $v = 1.98 \times 10^8$ m/sec

8.6 Density $= \rho\gamma$ **8.7** Distance $= 1.23$ km

8.8 (a)

$$v_x' = (c \cos\theta - v)/[1 - (v/c)\cos\theta], \qquad v_y' = c \sin\theta (1 - v^2/c^2)^{1/2}/[1 - (v/c)\cos\theta]$$

(b)

$$\theta' = \cos^{-1}\left[\frac{\cos\theta - v/c}{1 - (v/c)\cos\theta}\right]$$

8.9 $t = 2.2 \times 10^{11}$ sec **8.11** Relative speed $= v(1 + \gamma^2)^{1/2}/\gamma$

8.12 (a) $t = 0.135 \times 10^{-8}$ sec, (b) $t = 0.108 \times 10^{-8}$ sec

8.13 $E = 11.25 \times 10^{-12}$ J ($\sim$70 MeV), $\lambda = 1.76 \times 10^{-14}$ m **8.14** $\Delta p = .015$ GeV/c

CHAPTER 9

9.1 $V = 729.3$ V **9.2** ${}^{10}_{5}B : {}^{11}_{5}B = 1 : 4$

9.4 Mass defect = 0.990391 u, binding energy/nucleon = 8.5 MeV/A

9.5 $E = 127.6$ MeV **9.6** $E = 4.95$ MeV **9.7** Mass = 0.22 mg **9.8** $t = 60$ hr

9.9 Mass of unchanged ${}^{234}_{92}U$ = 2.5 mg, activity = 5.7×10^5 disintegrations per second

9.10 $I = 0.012$ μA **9.11** Age = 8.65×10^9 yr **9.12** Age = 5.77×10^8 yr

9.14 $E = 0.13$ MeV **9.15** Mass = 13.005936 u **9.18** $\sigma = 5.3$ mb

9.19 BE/A = 8.42 MeV/A

9.20 BE (${}^{208}_{82}Pb$) = 2.88 MeV, BE (${}^{209}_{82}Pb$) = −6.23 MeV **9.21** $J = \frac{5}{2}, \frac{1}{2}, 0$

9.22 The last unpaired neutron of ${}^{17}_{8}O$ occupies $1d_{5/2}$, thus explaining the experimentally-observed ground state spin 5/2; the first two excited states give the energy of the single-particle orbits for ${}^{16}_{8}O$

CHAPTER 10

10.1 Energy liberated = 139.56 MeV **10.2** (a) Potential energy = 232.9 MeV

10.3 Mass of coal = 2.43×10^6 kg

10.4 (a) Energy released = 5.12×10^{26} MeV, (b) energy released = 21.23×10^{26} MeV

10.5 (a) Fraction = 0.89, (b) fraction = 0.28 **10.6** Mass = 105 g

10.7 After fission, the products are proton-deficient **10.8** Power density = 6.31×10^6 W/kg

10.9 Mass of ${}^{239}_{94}Pu$ = 7.25 kg **10.10** Mass of ${}^{235}_{92}U$ ≈ 1 kg

CHAPTER 11

11.1 $E = 231$ MeV **11.2** $E = 1.51$ MeV for each photon **11.3** $l = 1$

11.4 $l = 0$ or 1 **11.5** (a) Baryon number, (b) electron lepton number, (c) baryon number

11.7 $m = 3.5 \times 10^{-25}$ g (=196 MeV/c^2) **11.8** Last two reactions

11.9 $\Sigma^- + p \rightarrow K^- + n + p$ (but $\Sigma^+ + \bar{p} \not\rightarrow K^+ + \bar{n} + p$) **11.10** $(\boldsymbol{p}_1 \times \boldsymbol{p}_2) \cdot \boldsymbol{p}_3$

11.11 (a) No, (b) no **11.12** K^+ (=$\bar{s}u$), Σ^+ (=suu)

CHAPTER 12

12.8 Ratio = .22 **12.9** Time = 7×10^5 yr

Index